Approaches to Global Sustainability, Markets, and Governance

Series Editors

David Crowther, Social Responsibility Research Network, www.srrnet.org, UK

Shahla Seifi, Social Responsibility Research Network, www.srrnet.org, UK

Scopus indexed *Approaches to Global Sustainability, Markets, and Governance* book series takes a fresh and global approach to issues of corporate social responsibility, regulation, governance, and sustainability. It encompasses such issues as: environmental sustainability and managing the resources of the world; geopolitics and sustainability; global markets and their regulation; governance and the role of supranational bodies; sustainable production and resource acquisition; society and sustainability.

Although primarily a business and management series, it is interdisciplinary and includes contributions from the social sciences, technology, engineering, politics, philosophy, and other disciplines. It focuses on the issues at a meta-level, and investigates the ideas, organisation, and infrastructure required to address them.

The series is grounded in the belief that any global consideration of sustainability must include such issues as governance, regulation, geopolitics, the environment, and economic activity in combination to recognise the issues and develop solutions for the planet. At present such global meta-analysis is rare as current research assumes that the identification of local best practice will lead to solutions, and individual disciplines act in isolation rather than being combined to identify truly global issues and solutions.

This book series is part of the Social Responsibility Research Network (srrnet.org), a worldwide organisation with several thousand members. The network organises annual conferences, held at various locations around the world and many books come from these conferences. Submissions to the current conference are invited and details can be found at srrnet.org. Additionally submissions of chapters, or book proposals are always welcome at info@srrnet.org. If you are interested in contributing a monograph or curated volume to the series, please contact the Springer publishing editor.

Vinay Kandpal · Arun Kumar Tripathy ·
Nidhi S. Bisht
Editors

Developments in Corporate Governance

Emerging Market Perspective

Springer

Editors
Vinay Kandpal
Department of Management Studies
Graphic Era University
Dehradun, Uttarakhand, India

Arun Kumar Tripathy
Management Development Institute
Gurugram, Haryana, India

Nidhi S. Bisht
Management Development Institute
Gurugram, Haryana, India

ISSN 2520-8772 ISSN 2520-8780 (electronic)
Approaches to Global Sustainability, Markets, and Governance
ISBN 978-981-96-6365-1 ISBN 978-981-96-6366-8 (eBook)
https://doi.org/10.1007/978-981-96-6366-8

This Springer imprint is published by the registered company Springer Nature Singapore Pte Ltd.
The registered company address is: 152 Beach Road, #21-01/04 Gateway East, Singapore 189721, Singapore

Preface

In the past several decades, a huge amount of progress has been made in identifying corporate governance, and it has been recognized increasingly for its importance to company performance and longevity. Corporate governance entails the intricate partnership among stakeholders (shareholders, management, employees, customers, and public at large) in order to promote openness, accountability, and moral behavior in corporate decision-making. This is a book on the shifting trends in corporate governance with an eye to the emerging markets. Different market environments should be studied, as governance structures are derived from the local traditions and cultures of different countries. Further, they're rapidly integrated into world trade and finance infrastructures, providing further resonance.

Our book is based on the belief that corporate governance isn't a concept. It is critical that institutions continue to seek new modes of governance, continuously testing and adjusting practices to meet the evolving needs and demands of companies and societies. It just shows how necessary it is to continually assess and update governance strategies in order to maintain effectiveness. These readers will be aided by this book's comprehensive coverage of the topics of corporate governance in the current context. Written by practitioners, researchers and practitioners, this book aims to provide clarity on the changing nature of corporate governance, Environmental, Social and Governance (ESG) reporting and sustainable development. Companies are shifting, and they will need to implement processes to serve shareholders and more of their stakeholders—workers, communities, governments, and society. As we do so, we take governance into account when thinking about what a future for business might look like: profit and purpose. This series has four main themes on critical corporate governance and sustainability topics.

Corporate Governance and Ownership Structure: This module covers governance structures for ownership models ranging from the family business to multinational corporations. Transparency, accountability, and robustness are achievable through boards, directors, and institutional mechanisms—new economies and international corporations are explanatory.

Sustainable Governance and CSR (Corporate Social Responsibility): CSR has become no longer an optional but a fundamental business model. Chapters of this section provide an in-depth explanation of CSR strategies that drive environmental and social sustainability while fostering community and building long-term business relationships.

Corporate Governance, Compensation, and Data Protection: This part touches upon the cutting edge from executive compensation to data sovereignty and what the Digital Personal Data Protection Act 2023 means for businesses. Whether it is pay equity, technological risk, or regulatory change, these questions matter to companies looking to develop sustainable organizations moving forward.

ESG Reporting, Risk Management, and Sustainability: As ESG reporting has become a critical part of investment decisions and corporate strategy, this section will focus on the process, issues, and implications of ESG reporting on corporate attributes and financials. It also focuses on companies' strategies for climate risk, sustainability issues, and ESG indexes globally.

This book is the product of collaboration across a number of researchers and practitioners dedicated to uncovering and developing corporate governance, ESG, and sustainability strategies in developed and emerging markets. This article will be of interest to scholars, policymakers, and practitioners alike as it presents concrete guidance for building sustainability into the global corporate culture.

We hope that the information found in this book will motivate readers to see corporate governance not merely as a legal necessity, but also as a vehicle for ethical, responsible, and sustainable business change. As corporate governance evolves, so should stewardship, openness, and global collective prosperity. We would like to thank all the contributors for their experience and hard work in creating this book. Also, we would like to thank the continued assistance of our institutions that have helped us to complete this project.

Dehradun, India — Vinay Kandpal
Gurugram, India — Arun Kumar Tripathy
Gurugram, India — Nidhi S. Bisht

Acknowledgments

The book is the result of our collective commitment to advancing the literature on corporate governance by offering insights from emerging markets. We are deeply grateful to all the chapter contributors for their invaluable contributions, which have enriched this volume with diverse perspectives and broadened the understanding of corporate governance in the global context. Their willingness to go above and beyond in responding to our editorial reviews is greatly appreciated.

We extend our heartfelt appreciation to the peer reviewers whose constructive feedback enhanced the quality and depth of each chapter. Their expertise and diligence ensured that this work met the highest academic standards.

Special thanks go to Mamun Mishra, Editor for Business, Management, Economics, Finance, Law, & Statistics at Springer, Australia and New Zealand, whose support was instrumental in bringing this project to fruition. Her responsiveness and guidance throughout the editorial process were invaluable. We would also like to thank Garima Singh, Teaching and Research Assistant at Management Development Institute Gurgaon, for her assistance in compiling the book chapters and managing communications with the authors.

Finally, we are indebted to our families, friends, and colleagues for their encouragement and support in helping us stay committed to the project and meet the timelines.

Vinay Kandpal
Arun Kumar Tripathy
Nidhi S. Bisht

Contents

Contributors

Devika Agrawal Institute of Business Management, GLA University, Mathura, India

Esrafil Ali School of Business Management, SVKM's NMIMS (Deemed to Be University), Indore, India

Anchal Chanakya National Law University, Patna, India

Moumita Banerjee Department of Economics, South Calcutta Girls' College, University of Calcutta, Kolkata, West Bengal, India

Nella Bhargava Acharya Institute of Graduate Studies, Bengaluru, India

Anuj Bhatia Shri Ram College of Commerce, University of Delhi, Delhi, India

Jayanta Boruah Department of Law, Central University of Karnataka, Gulbarga, India

Deeksha Graphic Era (Deemed to Be University), Dehradun, Uttarakhand, India

G. Delina Department of Business Administration, Kalasalingam Academy of Research and Education (Deemed to Be University), Srivilliputhur, India

Kabirdoss Devi School of Management Studies, Vels Institute of Science Technology and Advanced Studies (VISTAS) Pallavaram, Chennai, India

Siddharth Dhongde Department of Finance, GITAM School of Business, GTAM (Deemed to Be University), Visakhapatnam, India

Akhil Goyal Department of Management and Commerce, NIMS University, Jaipur, India

Khushboo Gulati School of Management and Commerce, Manav Rachna University, Faridabad, India

C. P. Gupta Lal Bahadur Shastri Institute of Management, Delhi, India

Seema Gupta IIM-Sambalpur, Sambalpur, India

Rajiv Iyer Department of CSE, Amity School of Engineering and Technology, Amity University Maharashtra, Mumbai, India

Sunitha Abhay Jain School of Law, CHRIST University, Bangalore, India

Aruna Jha Shri Ram College of Commerce, University of Delhi, Delhi, India

Neelam Jhawar Department of Commerce, Faculty of Commerce and Business, Delhi School of Economics, University of Delhi, Delhi, India

Srihari Jwalapuram Bharatiya Engineering Science Technology Innovation University, Bengaluru, India

Vinay Kandpal Department of Management Studies, Graphic Era (Deemed to Be University), Dehradun, India

Siddharth Kanojia O. P. Jindal Global University, Sonipat, India

Sunaina Kanojia Department of Commerce, Faculty of Commerce and Business, Delhi School of Economics, University of Delhi, Delhi, India

Raghuveer Katragadda School of Management, V. R. Siddhartha Engineering College, Kanuru, India

Lubza Nihar Khaliq Department of Finance, GITAM School of Business, GITAM (Deemed to Be University), Visakhapatnam, India

Kirthana Singh Khurana Jindal Global Law School, O. P. Jindal Global University, Sonipat, India;
Peter A. Allard School of Law, University of British Columbia, Vancouver, BC, Canada

Hartej Singh Kochher Vinayaka Missions Law School, Vinayaka Mission's Research Foundation (DU), Chennai, India

Kishore Kumar School of Management, Bennett University, Greater Noida, India

Magesh Kumar School of Business and Management, Christ University, Bengaluru, India

Sunil Kumar Department of Entrepreneurship, GITAM School of Business, GTAM (Deemed to Be University), Visakhapatnam, India

Neetha Kurian School of Law, CHRIST University, Bangalore, India

Bikram Paul Singh Lehri University School of Business, Chandigarh University, Punjab, India

Poornima Mahesh University of Mumbai, Mumbai, India

Vijaya Kittu Manda PBMEIT, Visakhapatnam, India

Ardra Mani APJ Abdul Kalam Technological University and Saintgits College of Engineering (Autonomous), Kottayam, India

Manvee Chanakya National Law University, Patna, India

Vedprakash Maralapalle Department of Civil Engineering, Amity School of Engineering and Technology, Amity University Maharashtra, Mumbai, India

Anjali Prava Mishra School of Economics and Commerce, KIIT (Deemed to Be University), Bhubaneswar, Odisha, India

Chandra Sekhar Mishra VGSOM, IIT Kharagpur, Kharagpur, India

Subhendu Kumar Mishra Department of Entrepreneurship, GITAM School of Business, GITAM (Deemed to Be University), Visakhapatnam, India

Ranjit Kumar Mukherji Graphic Era (Deemed to Be University), Dehradun, Uttarakhand, India

Neetu Department of Management and Commerce, NIMS University, Jaipur, India

Gobinda Gopal Pahari VGSOM, IIT Kharagpur, Kharagpur, India

Debasis Pahi Model Degree College, Malkanagiri, Odisha, India

Priyanka Panday DBS Global University, Dehradun, India

Yash Pandey Galgotias University, Greater Noida, India

Sujit Kumar Patra Department of Marketing, Jaipuria Institute of Management, Lucknow, India

Srinivasan Pazhamalai CMS Business School, Jain (Deemed to Be University), Bengaluru, India

E. Prema Vellore Institute of Technology, Chennai, India

Md Safiullah Chanakya National Law University, Patna, India

Sai Aanchal Department of Sociology, SGGS College, Chandigarh, India

Rashmeet Kaur Bhupinder Singh Saluja School of Commerce, Manav Rachna International Institute of Research and Studies, Faridabad, Haryana, India

Biswajit Satpathy Department of Business Administration, Sambalpur University, Burla, Odisha, India

Nishant Sharma Jesus and Mary College, University of Delhi, Delhi, India

Preeti Sharma DIT University, Dehradun, India

Riya Sharma Graphic Era (Deemed to Be University), Dehradun, Uttarakhand, India

Shelly Department of Management, Global Group of Institutes, Amritsar, India

Abhishek K. Singh School of Liberal Studies and Media, University of Petroleum and Energy Studies, Dehradun, India

Animesh Singh University School of Business, Chandigarh University, Punjab, India

Rubee Singh Institute of Business Management, GLA University, Mathura, India; Department of Management Studies, Kumaun University, Nainital, India

Priyanka Srivastava School of Leadership and Management, Manav Rachna International Institute of Research and Studies, Faridabad, Haryana, India

K. S. Anand Sudarshan Dhanwantari Academy for Management Studies, Bengaluru, India

Jose Joy Thoppan Department of Business Administration Saintgits College of Engineering (Autonomous), Kottayam, India

V. Vardhini School of Management Studies, Vels Institute of Science Technology and Advanced Studies (VISTAS) Pallavaram, Chennai, India

Maryam Vaziri DaneshPajoohan Pishro Higher Education Institution, Isfahan, Iran

Inas Afifah Zahra Politeknik Negeri Jakarta, Depok, Indonesia

Part I
Corporate Governance and Ownership Structure

Chapter 1
Governance Practice and Efficiency Metrics in Global Microfinance Institutions

Anjali Prava Mishra and Debasis Pahi

Abstract Corporate Governance (CG) significantly impacts the performance of Microfinance Institutions (MFIs), crucially supporting their dual social and financial objectives. Despite a global dearth of research on the direct effects of CG practices on MFIs efficiencies, this study seeks to address this gap. It investigates how a nation's governance practices influence MFIs' efficiencies worldwide, analyzing Financial Efficiency (FE), Social Efficiency (SE), and Overall Efficiency (OE) using Data Envelopment Analysis (DEA) and the Malmquist index. The research utilizes panel data from 382 MFIs across 62 countries spanning 2010 to 2018 sourced from www.mixmarket.org. A Corporate Governance Index, based on dimensions like Voice and Accountability, Political Stability, Government Effectiveness, Regulatory Quality, Control of Corruption, and Rule of Law has been constructed using data from the World Governance Indicators (WGI). The study employs a fixed-effect multiple regression model to investigate the impact of governance dimensions on MFIs efficiencies along control variables at country and firm level. Results highlight the positive impact of government effectiveness on FE of MFIs, emphasizing the critical role of regulatory frameworks and governance structures. External indicators of governance—GOVEFF, POLSTA, REGQUA, CONTCOR, as well as Overall CG Index have been found having significant positive impact on SE and OE. This study provides new insights into how country-specific governance frameworks influence MFIs performance globally, enhancing understanding of their operational effectiveness across diverse regulatory landscapes.

Keywords Microfinance institutions · Malmquist index · Data Envelopment Analysis (DEA) · Financial efficiency · Social efficiency and overall efficiency

A. P. Mishra (✉)
School of Economics and Commerce, KIIT (Deemed to Be University), Bhubaneswar, Odisha, India
e-mail: anjalipravamishra@gmail.com

D. Pahi
Model Degree College, Malkanagiri, Odisha, India
e-mail: drdebasisresearch@gmail.com

V. Kandpal et al. (eds.), *Developments in Corporate Governance*,
Approaches to Global Sustainability, Markets, and Governance,
https://doi.org/10.1007/978-981-96-6366-8_1

1.1 Introduction

Microfinance encompasses various financial services like credit, savings, and insurance institutions typically aim to achieve both social outreach and financial sustainability (Hermes et al., 2011). CG plays a crucial role in instilling confidence in markets and attracting investors to both individual organizations and the broader economy. It promotes standards of transparency, accountability, and performance among corporate managers (Gaur et al., 2014; Popli et al., 2017; Singh & Delios, 2012). Good CG is fundamental to the success of MFIs, helping them meet their social and financial objectives by providing a framework for setting and achieving goals (OECD principles of CG, 2006). Non-compliance with CG practices is identified as a significant risk factor for the microfinance sector, affecting managerial decision-making quality and, consequently, future growth prospects.

The regulatory framework established by a country's CG standards directly impacts the microfinance industry, providing a stable operational environment. Effective CG practices enhance operational efficiency through robust financial reporting, internal controls, and strategic planning. Therefore, studying the impact of CG on efficiencies is crucial, utilizing a multiple regression model with control and institution-specific variables. This research contributes to existing literature by examining how CG practices influence overall firm efficiency. It aims to assist stakeholders in designing optimal organizational structures and distinguishing between firms with effective and ineffective CG policies. Effective CG not only fosters marketplace confidence but also attracts positive investors to organizations. The study has been conducted using DEA and a multiple regression model that answers the following research questions:

1. What roles do external governance indicators play in enhancing the overall efficiencies of Microfinance Institutions globally?
2. How does the Corporate Governance Index affect the efficiencies of MFIs in a global context?

Along with the stated research question, the following objectives have been developed to carry the study forward:

1. To assess the efficiencies of Microfinance Institutions in a global context.
2. To investigate the impact of a country's governance practices on the efficiencies of MFIs.

The subsequent section of the study provides a rationale for identifying the variables and undertaking the research by a structured literature review of the relevant studies conducted on CG and MFIs' efficiencies.

1.2 Literature Review

This section aims to investigate the existing literature on the complex relationship between governance practices and the efficiency of microfinance institutions (MFIs). By exploring this relationship, the study seeks to shed light on how various governance strategies impact MFIs' ability to meet their dual objectives. Furthermore, it examines how this dynamic plays out across different global settings, considering the diverse regulatory environments and operational challenges encountered by MFIs worldwide. Through critical analysis and synthesis of the literature, this review aims to contribute to the ongoing discourse on this critical issue and lay the groundwork for this research work.

1.2.1 Corporate Governance and Efficiency of MFIs

Different experts explain their ideas about how CG impacts business results. Corporate governance practices reduce ROA performance across Indian and Gulf firms when combined with a positive impact from governance effectiveness. Board structure produces poor results, but transparency and governance performance, plus leverage delivery, produce superior results (Al-ahdal et al., 2023). Economic efficiency performance in microfinance depends highly on empowerment techniques along with how an organization operates its business plan and financial tools. Research by Iqbal et al. (2018) reveals that proper management leads to better performance for MFIs and good governance helps them to succeed. Research indicates board members with insider status plus numbers and independence help financial goals, while a regular dividend payment plan does too (Rizwan et al., 2016). However, board diversity and the pay system create no such benefit.

Some studies view CG as an external factor influencing performance, while others suggest it can be shaped by prior performance, indicating its endogenous nature (Mitton, 2002; Randøy & Goel, 2003; Oxelheim et al., 2013). In the microfinance sector, several studies highlight the significant impacts of various CG indicators on the financial performance of MFIs (Aboagye & Otieku, 2010; Bassem, 2009; Galema et al., 2012; Hartarska & Mersland, 2012; Kyereboah-Coleman & Osei, 2008; Tchakoute-Tchuigoua, 2010). Additionally, research has explored the possibility of a bidirectional relationship between CG and financial performance, examining CG variables internally (Cho, 1998; Servin et al., 2012).

Wang et al. (2020), look at how certain CG attributes (legal frameworks, governance structures, and performance metrics) have an effect on firm performance. Gupta and Sharma (2014) examined CG variables' impact on firm performance in Indian and in South Korean contexts and ascertained the impact on share prices and financial performance, however with modest impacts. Danoshana and Ravivathani (2014) discussed the impact of CG on the listed financial institutions in Sri Lanka for the period, 2008–2012. Guo and Kga (2012) examine how CG measures affect performance of firms in Sri Lanka and find negative associations between board size and firm value and proportion of outside directors, and operating performance. The relationship between CG and operating performance was examined by Sami et al. (2011) as to whether operating performance would be positively correlated with a number of governance metrics in Chinese listed firms. According to Foroughi and Fooladi (2011), involvement in CG may lead to negative correlation with ROE and ROA performance measures, for example because CEO duality reduces board effectiveness. On the other hand, we found no relationship between board independence, size, ownership structure, and firm performance. 2.2 World Governance Indicators (WGI) and Firm Performance A convenient framework for comparing countries on the basis of governance quality is the set of Worldwide Governance Indicators (WGI). Nevertheless, these aggregated metrics provide valuable and comparative international insights for conducting broad international comparisons as well as time trends analysis. Since understanding how global governance can affect the microfinance sector is important, investigation of the link between WGI and MFIs is necessary. For instance, Hawaj and Buallay (2022) found differences in the relationship between sustainability reporting (ESG) and operational performance (ROA), financial performance (ROE), especially when WGI serves as a control variable. As discussed in Silva and Chalve (2015), country-level institutional and governance factors are also found to have an effect on MFIs' performance because governance in a country will influence microfinance outcome. Nawaz et al. (2018) analyze the linkage between the CG and MFIs' social performance in Asia and retort that CG does exert limited impact on some social performance indicators; however, CG stresses the social performance as a key determinant of CG mechanisms in MFIs. Building on this literature, the present study develops a Corporate Governance Index with six indicators: Voice and Accountability (VOICE), Political Stability (POLSTA), Government Effectiveness (GOVEFF), Regulatory Quality (REGQUA), Control of Corruption (CONTCOR), and Rule of Law (ROL).

1.2.2 Macro Indicators and Efficiency of MFIs

Previous research has established a connection between control variables and the efficiencies of MFIs (Batuo et al., 2017; Bibi & Balli, 2023; Gaganis, 2016; Gupta & Mirchandani, 2020; Nawaz et al., 2018). For instance, Table 1.1 presents the control variables considered by the researchers in their respective studies for studying the efficiency of MFIs.

Drawing from this literature, the current study incorporates several control variables, including macroeconomic factors such as GDP growth (annual %) (GDPG), the Human Capital Index (scale 0–1) (HCI), and inflation in consumer prices (annual %) (INFGDP). In addition, institution-specific variables are considered, including Return on Assets (ROA), Financial Revenue (FR), Portfolio at Risk over 30 days (RISK30), Capital to Total Assets (CAPTA), and Cash to Total Assets (CASHTA).

Table 1.1 List of control variables used in previous studies

Control Variables	Reference Studies
GDP growth rate (GDPG)	Mersland and Strom (2008); Strom et al. (2014)
Human capital index (scale 0–1) (HCI)	Mersland and Strom (2008); Strom et al. (2014)
Inflation in consumer prices (INFGDP)	Gaganis (2016); Batuo et al. (2017); Bibi and Balli (2023)
Return on assets (ROA)	Nawaz et al., (2018); Gaganis (2016)
Financial revenue (FR)	Nawaz et al., 2018; Gaganis (2016)
Portfolio at risk over 30 days (RISK30)	Hartarska and Mersland (2012); Bibi and Balli (2023)
Capital to total assets (CAPTA)	Gupta and Mirchandani (2020); Batuo et al. (2017)
Cash to total assets (CASHTA)	Gupta and Mirchandani (2020); Batuo et al. (2017)

Research Gap: As MFIs expand and become more regulated, the effectiveness of governance mechanisms, including board structure and ownership, becomes increasingly critical (Coleman & Osei, 2008). Despite extensive research on external governance (media coverage, external audit, and CEO duality) and internal governance factors, there remains a scarcity of studies on the impact of external corporate governance index, which includes ROL, GOVEFF, POLSTA, REGQUA, VOICE, and CONTCOR on efficiencies of MFIs in a global context.

1.3 Conceptual Framework and Hypotheses Development

Based on the literature review and gap analysis, the study has been conducted to accomplish the stated objectives along the following conceptual framework as shown in the diagrammatic presentation Fig. 1.1

The framework reveals the direction and impact relationship among the independent variables (i.e., Rule of Law (ROL), Government Effectiveness (GOVEFF), Political Stability (POLSTA), Regulatory Quality (REGQUA), Voice and Accountability (VOICE), and Control of Corruption (CONTCOR)), and dependent variables (i.e., Financial Efficiency, Social Efficiency and Overall Efficiency), and control variables as explained in Table 1.1 (i.e., GDPG, HCI, INFGDP, ROA, FR, RISK30, CAPTA, CASHTA). Therefore, a total of 21 impact relationships have been proposed to reach the inference. However, only the impact of individual CG indicators and CG index on Overall firm efficiency (OE) have been developed and presented as hypotheses:

$H_1$1: Country Governance Index has a significant positive impact on the overall efficiency of MFIs.

$H_1$1.1: Rule of Law has a significant positive impact on the overall efficiency of MFIs.

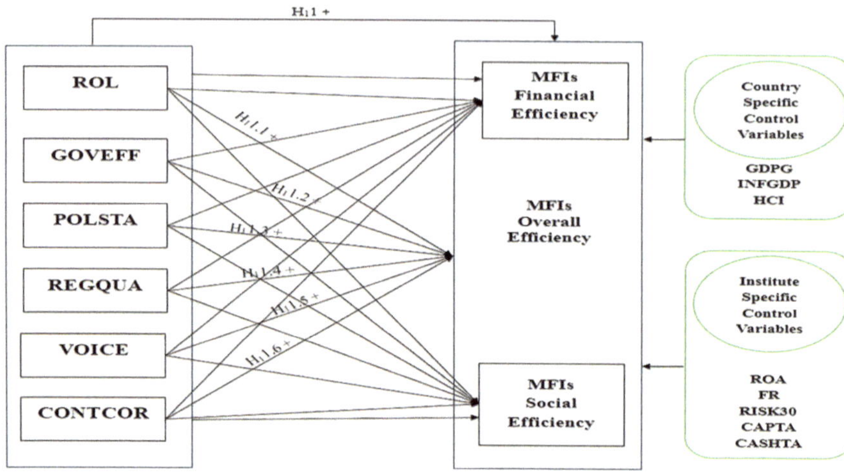

Fig. 1.1 Diagrammatic presentation of conceptual framework

$H_1$1.2: Government Effectiveness has a significant positive impact on the overall efficiency of MFIs.

$H_1$1.3: Political Stability has a significant positive impact on the overall efficiency of MFIs.

$H_1$1.4: Regulatory Quality has a significant positive impact on the overall efficiency of MFIs.

$H_1$1.5: Voice and Accountability have a significant positive impact on the overall efficiency of MFIs.

$H_1$1.6: Control of Corruption has a significant positive impact on the overall efficiency of MFIs.

These hypotheses have been tested using appropriate inferential statistical tools in compliance with the objectives stated earlier. A comprehensive methodology has been developed to undertake the study.

1.4 Research Methodology

The research analyzes the efficiency of 382 global MFIs between 2010 and 2018 within 62 different countries worldwide. The data used for this research examination extends up to 2018 because it represents the most recent release of efficiency data from firms according to Bharti et al. (2016), Wijesiri et al. (2017), Sharma and Rastogi (2021), Mia et al. (2022), and Sohn and Ju (2023). MFIs were assessed for their overall efficiency using the Data Envelopment Analysis (DEA) and measured against two primary efficiency characteristics, which included financial efficiency together with social efficiency. The Malmquist index serves as a tool for analyzing MFIs' efficiency adjustments across the investigated time periods according to findings in Wijesiri et al. (2017), Bharti et al. (2016), Sharma and Rastogi (2021), Mia et al. (2022), Sohn and Ju (2023), Bøhren and Ødegaard (2001), Méndez and Gómez-Ansón (2006), Bogan (2008), Haq et al. (2009), Strøm et al. (2014), and Abd Hamid and Nassir (2021).

The study has applied fixed-effect regressions between efficiency data from individual MFIs and their related corporate governance (CG) factors. The research utilizes ROL and GOVEFF and POLSTA and REGQUA and CONTCOR and VOICE as independent variables along with Country-specific macro variables [GDPG, HCI, INFGDP, and institution-specific variables] [ROA, FR, RISK30, CAPTA, and CASHTA] as control variables. The regression analysis was conducted using the following equation to examine the impact of CG on the efficiency of MFIs.

$$\begin{aligned} OE_{it} = \beta_0 + \beta_1 ROL_{it} + \beta_2 GOVTEFF_{it} + \beta_3 POLSTA_{it} + \beta_4 REGQUA_{it} \\ + \beta_5 VOICE_{it} + \beta_6 CONTCOR_{it} + \beta_7 LNFR_{it} + \beta_8 ROA_{it} \\ + \beta_9 CAPTA_{it} + \beta_{10} RISK30_{it} + \beta_{11} CASHTA_{it} + \beta_{12} GDPG_{it} \\ + \beta_{13} INFGDP_{it} + \beta_{14} HCI_{it} + \varepsilon_{it} \end{aligned} \tag{1.1}$$

where:

Dependent Variable:

- OE_{it}= Overall Efficiency of the MFIs *i* at the time *t* (Dependent variable)

Independent Variable:

- ROL_{it} = Rule of Law of the MFIs *i* at the time *t*
- $GOVTEFF_{it}$ = Government Effectiveness of the MFIs *i* at the time *t*
- $POLSTA_{it}$ = Political Stability and Absence of Violence/Terrorism of MFIs *i* at the time *t*
- $REGQUA_{it}$ = Regulatory Quality of the MFIs *i* at the time *t*
- $VOICE_{it}$ = Voice and Accountability of the MFIs *i* at the time *t*
- $CONTCOR_{it}$ = Control of Corruption of the MFIs *i* at the time *t*

Institution Specific Variables (MFIs Characteristics) (Control Variables):

- $LNFR_{it}$ = Logarithm of Financial Revenue of the MFIs *i* at the time *t*
- ROA_{it} = Return on Assets of the MFIs *i* at the time *t*
- $CAPTA_{it}$ = Capital to Total Assets ratio of the MFIs *i* at the time *t*
- $RISK30_{it}$ = Portfolio at Risk 30 days of the MFIs *i* at the time *t*
- $CASHTA_{it}$ = Cash to Total Assets ratio of the MFIs *i* at the time *t*

Country Specific Macro Variables:

- $GDPG_{it}$ = GDP growth (annual %) of the MFIs *i* at the time *t*
- HCI_{it} = Human capital index (HCI) (scale 0–1) of the MFIs *i* at the time *t*
- $INFGDP_{it}$ = Inflation, consumer prices (annual %) of the MFIs *i* at the time *t*

1.5 Results and Discussion

The following section presents the data analysis part of the study. Table 1.2 depicts country-wise FE, SE, OE, and CG index scores. This study has taken 6 regions, namely Africa (A), East Asia and the Pacific (EAP), Eastern Europe and Central Asia (EECA), Latin America and the Caribbean (LAC), Middle East and North Africa (MENA), and South Asia (SA). The number of MFIs in each region is written inside the brackets (). #MFI represents the number of Microfinance firms in each country, and #OBS represents the number of Firm-Year observations from each country.

Table 1.3 presents descriptive statistics for various variables across different categories. The statistics include the mean, median, standard deviation, skewness, kurtosis, minimum and maximum. SETFPC indicates a positively skewed distribution, and FETFPC and OETFPC have a left-skewed distribution.

The distribution of CGINDEX, i.e., ROL, GOVTEFF, POLSTA, REGQUA, VOICE, and CONTCOR, indicates a positively skewed distribution. The distribution of Macro variables, i.e., GDPG, is negatively skewed, whereas HCI and INFGDP have a positively skewed distribution. The distribution of all firm characteristics, i.e., ROA, FR, RISK30, CAPTA, and CASHTA, are positively skewed with a high kurtosis. Overall, the substantial differences between means and medians in some cases suggest the presence of outliers or asymmetry in the data distributions.

The results of a Fixed Effect Panel regression analysis examining the impact of various external CG indicators, institute-specific variables, and macroeconomic variables on the FE, SE, and OE of MFIs have been presented in Tables 1.4, 1.5, and 1.6. These include seven different models (M1 to M7), each introducing different sets of independent variables. In Model-1, the impact of ROL on the efficiency of MFIs has been measured, and similarly, investigation on the impact of GOVTEFF in Model-2, POLSTA in Model-3, REGQUA in Model-4, VOICE in Model-5, CONTCOR in Model-6, and CG Index (CGINDEX) in Model-7 on FE, SE & OE of MFIs.

Table 1.4 presents the results of a regression analysis investigating the impact of institute-specific variables, external country governance indicators, and macroeconomic variables on the FE of MFIs.

The coefficients for governance variables (ROL, GOVTEFF, POLSTA, REGQUA, VOICE, CONTCORR, and CGINDEX) are positive, suggesting a positive impact on FE. However, only GOVTEFF and CGINDEX are statistically significant, indicating that these variables may have a significant impact on FE. Furthermore, the table provides model fit statistics, including adjusted R^2 values ranging from 0.310 to 0.752, indicating moderate to substantial explanatory power across the models. Chi-square tests are significant for each model ($\chi 2$ range = 7581 to 7612), supporting the overall model fit. Year-fixed effects are included in all models, and multicollinearity is assessed with VIF values ranging from 1.21 to 1.48, indicating acceptable levels.

Table 1.5 presents the results of a regression analysis investigating the impact of country governance on the SE of MFIs. Coefficients for governance variables (GOVTEFF, POLSTA, REGQUA, VOICE, CONTCORR, and CGINDEX) except

Table 1.2 Distribution of samples and average of FE, SE, OE, and CG by region and country

Region	Country Name	#MFI	#OBS	FETFPC	SETFPC	OETFPC	CGINDEX
AFRICA (32)	Nigeria	4	36	0.911	0.936	0.909	1.107
	Senegal	4	36	0.950	0.894	0.930	0.244
	Benin	3	27	0.901	0.874	0.882	0.449
	Burkina Faso	3	27	1.121	0.980	1.064	0.425
	Ghana	3	27	0.919	0.899	0.905	0.183
	Cameroon	2	18	0.917	0.941	0.915	0.973
	Congo, democratic Republic of the	2	18	1.108	1.309	1.001	1.607
	Kenya	2	18	0.943	0.918	0.914	0.629
	Tanzania	2	18	0.902	0.870	0.869	0.456
	Togo	2	18	0.943	0.978	0.931	0.827
	Cote d'Ivoire (Ivory Coast)	1	9	0.885	1.029	0.894	0.798
	Madagascar	1	9	0.929	0.944	0.925	0.768
	Malawi	1	9	0.971	0.917	0.927	0.423
	Rwanda	1	9	0.846	0.971	0.875	0.385
	Uganda	1	9	0.843	1.133	0.869	0.602
EAP (32)	Cambodia	13	117	0.917	0.960	0.926	0.784
	Philippines	10	90	0.916	0.888	0.893	0.418
	Vietnam	3	27	3.259	0.907	0.902	0.519
	East Timor	2	18	0.904	0.979	0.923	0.718
	Papua New Guinea	2	18	0.931	0.994	0.937	0.606
	Fiji	1	9	1.446	0.985	1.100	0.457
	Laos	1	9	0.962	1.123	0.986	0.893
EECA (36)	Tajikistan	8	72	1.050	0.920	0.962	1.156
	Azerbaijan	5	45	0.831	0.864	0.841	0.761
	Kyrgyzstan	5	45	0.877	0.911	0.879	0.768
	Armenia	4	36	0.912	0.899	0.889	0.368
	Georgia	3	27	0.897	0.947	0.902	0.450
	Kazakhstan	3	27	0.942	0.921	0.925	0.556
	Kosovo	2	18	0.814	0.861	0.825	0.403
	Macedonia	2	18	0.878	0.903	0.892	0.226
	Bosnia and Herzegovina	1	9	0.726	0.917	1.100	0.305
	Bulgaria	1	9	0.827	0.833	0.812	0.286
	Mongolia	1	9	0.942	0.944	0.948	0.389

(continued)

Table 1.2 (continued)

Region	Country Name	#MFI	#OBS	FETFPC	SETFPC	OETFPC	CGINDEX
	Russia	1	9	0.814	1.004	0.816	0.735
LAC (172)	Ecuador	36	324	0.921	0.924	0.912	0.568
	Peru	33	297	0.898	0.927	0.897	0.402
	Mexico	28	252	0.924	1.310	1.036	0.414
	Nicaragua	18	162	0.919	0.954	0.918	0.622
	Bolivia	15	135	0.910	0.926	0.920	0.580
	Colombia	15	135	0.909	0.902	0.902	0.379
	Brazil	9	81	0.918	0.945	0.931	0.254
	Dominican Republic	4	36	0.877	1.064	0.897	0.366
	Haiti	3	27	0.981	0.992	1.010	1.160
	Honduras	3	27	0.905	0.894	0.895	0.622
	Paraguay	3	27	0.927	0.955	0.912	0.518
	Argentina	2	18	0.951	0.836	0.926	0.364
	Chile	1	9	0.926	1.355	1.230	1.068
	El Salvador	1	9	0.919	0.910	0.902	0.294
	Panama	1	9	0.899	0.888	0.884	0.328
MENA (13)	Egypt	3	27	0.905	0.907	0.904	0.790
	Jordan	3	27	0.888	0.909	0.894	0.327
	Morocco	2	18	0.884	0.925	0.898	0.322
	Iraq	1	9	0.909	0.934	0.898	1.412
	Lebanon	1	9	0.931	0.931	0.926	0.725
	Syria	1	9	2.352	1.034	1.274	1.661
	Tunisia	1	9	0.915	0.911	0.923	0.279
	Yemen	1	9	1.002	0.915	0.994	1.593
SA (97)	India	77	693	0.980	1.187	0.985	0.409
	Pakistan	8	72	0.935	0.956	0.913	1.064
	Nepal	6	54	2.884	0.974	0.937	0.760
	Afghanistan	3	27	1.118	1.026	1.017	1.590
	Bangladesh	3	27	0.952	0.889	0.917	0.842

Notes FETFPC = Financial Efficiency Total Factor Productivity Change; SETFPC = Social Efficiency Total Factor Productivity Change; OETFPC = Overall Efficiency Total Factor Productivity Change; CGINDEX = Corporate Governance Index
Source Author's compilation through DEA (Data from MIX Market)

Table 1.3 Descriptive statistic

Variables	Mean	Median	Std. Dev	Skewness	Kurtosis	Min	Max
Dependent variables							
FETFPC	0.92	0.997	0.395	−0.775	5.234	0	2.201
SETFPC	0.947	1	0.459	0.666	9.244	0	3.138
OETFPC	0.921	1.001	0.389	−0.724	5.963	0	2.322
Independent variables							
Corporate governance variables							
ROL	0.561	0.55	0.404	0.393	2.538	0.003	2.09
GOVTEFF	0.419	0.301	0.362	1.359	5.266	0	2.263
POLSTA	0.748	0.714	0.571	1.18	4.871	0	3.006
REGQUA	0.504	0.418	0.311	0.856	3.204	0	1.829
VOICE	0.435	0.387	0.391	1.44	4.62	0	1.987
CONTCOR	0.637	0.576	0.314	0.631	3.047	0	1.681
CGINDEX	0.551	0.46	0.268	1.839	7.196	0.125	2.015
Control variables							
Country-specific macro variables							
GDPG	4.792	5.118	2.815	−1.705	18.277	−27.995	19.047
HCI	0.138	0	0.235	1.169	2.494	0	0.777
INFGDP	5.371	4.066	5.389	3.26	23.569	−30.2	60.987
Institution specific variables (Firm Characteristics)							
ROA	0.302	0.026	1.281	0.85	13.191	−5.75	5.85
FR	31,400,000	6,900,000	7	4	21	83,810	450,000,000
RISK30	0.33	0.042	1.419	6.714	50.017	0	11.61
CAPTA	4.709	0.242	11.474	3.024	12.473	−0.509	62.65
CASHTA	13.951	11.25	11.526	1.569	6.033	0	60.46

Source Author's compilation through Stata (Data from MIX Market)

ROL and VOICE are generally positive and statistically significant, suggesting that higher values of these variables are associated with increased SE.

The t-values for these variables are generally moderate to high, indicating the reliability of these relationships. The models also offer fitness metrics such as adjusted R^2 values spanning from 0.385 to 0.752, indicating moderate to significant explanatory capability. Chi-square tests for model adequacy are statistically significant ($\chi2$ range = 4167 to 4509), affirming the overall adequacy of the models. Year-specific effects are integrated into all models, and multicollinearity is evaluated with VIF values ranging from 1.24 to 1.52, signaling acceptable levels.

Table 1.6 presents the results of a regression analysis investigating the impact of country governance on the OE of MFIs. The significance of governance variables implies the importance of CG in explaining variations in overall efficiency.

Table 1.4 Impact of country governance on financial efficiency of MFIs

	M1	M2	M3	M4	M5	M6	M7
$LNFR_{it}$	0.021*	0.021*	0.022*	0.021*	0.021*	0.021*	0.021*
	(7.30)	(7.55)	(7.68)	(7.46)	(7.57)	(7.50)	(7.55)
ROA_{it}	0.014*	0.014*	0.013*	0.014*	0.014*	0.014*	0.014*
	(4.34)	(4.51)	(4.12)	(4.30)	(4.23)	(4.26)	(4.36)
$CAPTA_{it}$	0.001*	0.002*	0.001*	0.001*	0.001*	0.001*	0.001*
	(4.25)	(4.66)	(3.95)	(4.21)	(4.14)	(4.15)	(4.36)
$RISK30_{it}$	−0.004	−0.004	−0.005	−0.005	−0.005	−0.005	−0.004
	(−1.57)	(−1.41)	(−1.87)	(−1.68)	(−1.77)	(−1.71)	(−1.59)
$CASHTA_{it}$	0.000	−0.000	0.000	0.000	0.000	0.000	−0.000
	(0.09)	(−0.43)	(0.06)	(0.03)	(0.14)	(0.14)	(−0.23)
$GDPG_{it}$	0.004*	0.004*	0.004*	0.004*	0.004*	0.004*	0.004*
	(2.51)	(2.43)	(2.39)	(2.56)	(2.32)	(2.45)	(2.46)
$INFGDP_{it}$	0.000	0.000	0.000	0.000	0.000	0.000	0.000
	(0.51)	(0.07)	(0.23)	(0.37)	(0.30)	(0.43)	(0.05)
HCI_{it}	−0.000	0.010	0.001	0.002	−0.005	−0.002	0.007
	(−0.00)	(0.28)	(0.02)	(0.05)	(−0.13)	(−0.07)	(0.19)
ROL_{it}	0.012						
	(1.10)						
$GOVTEFF_{it}$		**0.36***					
		(3.15)					
$POLSTA_{it}$			0.012				
			(1.60)				
$REGQUA_{it}$				0.023			
				(1.81)			
$VOICE_{it}$					0.008		
					(0.74)		
$CONTCORR_{it}$						0.008	
						(0.62)	
$CGINDEX_{it}$							**0.034***
							(2.20)
INTERCEPT	−0.328*	−0.338*	−0.339*	−0.335*	−0.327*	−0.328*	−0.342*
	(−7.18)	(−7.40)	(−7.30)	(−7.29)	(−7.15)	(−7.14)	(−7.40)
N	3438	3438	3438	3438	3438	3438	3438
$Adj.r^2$	0.320	0.473	0.386	0.362	0.385	0.310	0.752
Chi^2	7584	7612	7588	7590	7581	7581	7595
F (probability)	5.24 (0.0019)	4.83 (0.0026)	3.26 (0.0012)	3.67 (0.0029)	3.41 (0.0022)	4.25 (0.0015)	6.23 (0.0025)

(continued)

Table 1.4 (continued)

	M1	M2	M3	M4	M5	M6	M7
Durbin-Watson	2.1179	2.1182	2.1257	2.1189	2.1194	2.1191	2.1185
VIF	1.32	1.42	1.21	1.38	1.47	1.29	1.48

Notes Significance level is shown by * for $p < 0.05$
Source Author's compilation through Stata (Data from MIX Market)

Coefficients for governance variables (GOVTEFF, POLSTA, REGQUA, VOICE, CONTCORR, CGINDEX) except ROL and VOICE are generally positive and statistically significant, suggesting that higher values of these variables are associated with increased social efficiency. The t-values for these variables are generally moderate to high, indicating the reliability of these relationships.

The models also present fit indices such as adjusted R^2 values between 0.579 and 0.776, indicating they have considerable explanatory capability. Chi-square tests confirm significant model fit ($\chi 2$ range = 8219 to 8264), underscoring the overall adequacy of the models. Year-specific effects are incorporated across all models, and multicollinearity is evaluated with VIF values ranging from 1.28 to 1.41, demonstrating acceptable levels.

The results shown in Table1.6 on assessing the impact of individual CG indicators and CGI on overall firm efficiency provide a rationale for the validation of the hypotheses framed. The results of the hypotheses are shown in Table 1.7, considering the P-value at a 95% confidence level for each indicator and the CG index.

This clearly shows that all the indicators of CG have been found to have a significant positive impact on the efficiency of MFIs, other than the Rule of Law and Voice and accountability. The P-value assessment of the impact of CG Index on overall MFIs efficiency also suggests accepting the alternate hypothesis affirming that there is significant positive impact of CG Index on MFIs overall efficiency.

1.6 Findings

The result of the study affirms that among all external CG indicators, Government Effectiveness stands out as it shows a significant positive impact while the financial efficiency of MFIs is considered. It suggests that these factors have a substantial impact on enhancing the financial efficiency of MFIs when compared to other governance indicators. Governance excellence serves as a fundamental instrument to protect both corporate financial sustainability and enhanced company worth, according to (Sulimany et al., 2021). The results gain validation through institutional theory because regulatory environments contribute to barrier reduction and service delivery enhancement.

The study established a positive relationship between social efficiency of MFIs and all CG indicators except Rule of Law and Voice and accountability. Research findings

Table 1.5 Impact of country governance on social efficiency of MFIs

	M1	M2	M3	M4	M5	M6	M7
$LNFR_{it}$	−0.023*	−0.022*	−0.021*	−0.023*	−0.022*	−0.023*	−0.022*
	(−5.69)	(−5.64)	(−5.35)	(−5.72)	(−5.58)	(−5.76)	(−5.63)
ROA_{it}	−0.002	−0.002	−0.003	−0.002	−0.003	−0.002	−0.002
	(−0.46)	(−0.40)	(−0.71)	(−0.50)	(−0.58)	(−0.36)	(−0.43)
$CAPTA_{it}$	0.001	0.001	0.000	0.001	0.000	0.001	0.001
	(1.23)	(1.43)	(0.88)	(1.17)	(1.10)	(1.36)	(1.36)
$RISK30_{it}$	−0.019*	−0.019*	−0.020*	−0.019*	−0.020*	−0.019*	−0.019*
	(−4.90)	(−4.87)	(−5.23)	(−5.01)	(−5.11)	(−4.85)	(−4.90)
$CASHTA_{it}$	−0.001*	−0.001*	−0.001*	−0.001*	−0.001*	−0.001*	−0.001*
	(−2.70)	(−2.98)	(−2.84)	(−2.84)	(−2.70)	(−2.96)	(−3.07)
$GDPG_{it}$	0.005*	0.005*	0.005*	0.005*	0.004	0.005*	0.005*
	(2.08)	(2.01)	(1.98)	(2.17)	(1.91)	(2.14)	(2.05)
$INFGDP_{it}$	0.001	0.001	0.001	0.001	0.001	0.001	0.000
	(0.78)	(0.50)	(0.43)	(0.62)	(0.56)	(0.46)	(0.28)
HCI_{it}	0.061	0.069	0.067	0.067	0.057	0.072	0.075
	(1.25)	(1.41)	(1.36)	(1.37)	(1.17)	(1.46)	(1.51)
ROL_{it}	0.013						
	(0.81)						
$GOVTEFF_{it}$		**0.031***					
		(1.96)					
$POLSTA_{it}$			**0.020***				
			(2.00)				
$REGQUA_{it}$				**0.035***			
				(2.00)			
$VOICE_{it}$					0.011		
					(0.75)		
$CONTCORR_{it}$						**0.043***	
						(2.34)	
$CGINDEX_{it}$							**0.041***
							(2.39)
INTERCEPT	0.238*	0.229*	0.216*	0.226*	0.238*	0.222*	0.214*
	(3.73)	(3.59)	(3.32)	(3.52)	(3.74)	(3.46)	(3.32)
N	3438	3438	3438	3438	3438	3438	3438
r^2	0.401	0.473	0.486	0.466	0.385	0.510	0.752
Chi^2	4197	4325	4167	4205	4391	4208	4509
F (probability)	2.87 (0.0095)	5.29 (0.0048)	4.30 (0.0034)	4.69 (0.0041)	3.02 (0.0063)	5.40 (0.0029)	7.46 (0.0015)

(continued)

Table 1.5 (continued)

	M1	M2	M3	M4	M5	M6	M7
Durbin-Watson	2.0216	2.0471	2.1025	2.0798	2.0976	2.1256	2.1473
VIF	1.32	1.28	1.41	1.48	1.35	1.52	1.24

Notes t-values are shown in the Parentheses. Significance level is shown by * for $p < 0.05$, ** for $p < 0.01$, and *** for $p < 0.001$

Source Author's compilation through Stata (Data from MIX Market)

indicate that the individual components of CG and CG Index and the combined CG measurement demonstrate positive relationships that support high efficiency levels in MFIs except for Rule of Law and Voice & Accountability. The CG indicators produce dual benefits for MFI efficiency while promoting social well-being, except for the Rule of Law and Voice and accountability. Whenever government shows high effectiveness its public service delivery standards tend to improve. Tossou et al. (2023) explain how efficient infrastructure together with government support services brings benefits to MFIs.

The study confirms the World Governance theory by showing how political stability enables economic activities and sustainable development to proceed without disruption through violence or changes that could harm Microfinance Institutions. The effort to control corruption helps MFIs minimize operational risks and credit risks, resulting in a transformation of investor confidence. Political stability directly promotes investor confidence levels since it prevents disturbances that would scare MFIs from operating peacefully and avoiding sudden upheavals or violent events.

1.7 Conclusion

External CG indicators, especially Government Effectiveness, demonstrate their essential contribution in improving the financial efficiency of MFIs according to the study. A strong system of governance manages organizational success and financial solidity during the long term. The research shows most CG indicators produce beneficial effects on the social efficiency of MFIs, but Rule of Law and Voice and Accountability resist change. The efficiency outcomes of both overall performance and social improvements demonstrate substantial improvement when combining all governance indicators, including the general CG index. Widespread governance techniques that deliver excellent government services help MFIs attain better outcomes. Organizations require political stability to establish economic-friendly conditions which lower operational threats and build investor trust. Operational success and efficiency at MFIs exist because their framework keeps corruption low and maintains stable political frameworks connected to effective infrastructure. The research investigates how government actions within countries affect microfinance institution financial operations to develop this historical perspective.

Table 1.6 Impact of country governance on overall efficiency of MFIs

	M1	M2	M3	M4	M5	M6	M7
$LNFR_{it}$	0.007*	0.008*	0.009*	0.008*	0.008*	0.008*	0.008*
	(2.70)	(2.93)	(3.14)	(2.82)	(2.97)	(2.84)	(2.94)
ROA_{it}	0.007*	0.007*	0.006	0.006*	0.006	0.006*	0.007*
	(2.11)	(2.29)	(1.82)	(2.04)	(1.94)	(2.06)	(2.12)
$CAPTA_{it}$	0.001*	0.001**	0.001	0.001*	0.001	0.001*	0.001*
	(2.17)	(2.61)	(1.72)	(2.04)	(1.94)	(2.06)	(2.25)
$RISK30_{it}$	−0.006*	−0.006*	−0.007**	-0.007*	−0.007**	−0.007*	−0.006*
	(−2.32)	(−2.15)	(−2.70)	(−2.46)	(−2.59)	(−2.44)	(−2.35)
$CASHTA_{it}$	−0.000	−0.000	−0.000	−0.000	−0.000	−0.000	−0.000
	(−0.44)	(−1.05)	(−0.47)	(−0.55)	(−0.38)	(−0.47)	(−0.85)
$GDPG_{it}$	0.004*	0.004*	0.003*	0.004*	0.003*	0.004*	0.004*
	(2.34)	(2.23)	(2.20)	(2.42)	(2.10)	(2.30)	(2.27)
$INFGDP_{it}$	0.001	0.000	0.000	0.001	0.000	0.001	0.000
	(0.77)	(0.24)	(0.44)	(0.57)	(0.51)	(0.58)	(0.18)
HCI_{it}	0.019	0.030	0.019	0.022	0.014	0.019	0.028
	(0.56)	(0.90)	(0.58)	(0.66)	(0.41)	(0.57)	(0.82)
ROL_{it}	0.014						
	(1.36)						
$GOVTEFF_{it}$		**0.041***					
		(3.75)					
$POLSTA_{it}$			**0.013***				
			(2.88)				
$REGQUA_{it}$				**0.031***			
				(2.52)			
$VOICE_{it}$					0.009		
					(0.92)		
$CONTCORR_{it}$						**0.018***	
						(2.41)	
$CGINDEX_{it}$							**0.042***
							(2.84)
INTERCEPT	−0.152*	−0.164*	−0.165*	−0.162*	−0.151*	−0.156*	−0.170*
	(−3.48)	(−3.75)	(−3.70)	(−3.68)	(−3.44)	(−3.54)	(−3.84)
N	3438	3438	3438	3438	3438	3438	3438
r^2	0.579	0.624	0.618	0.603	0.591	0.650	0.776
Chi^2	8222	8264	8228	8238	8219	8223	8243
F (probability)	1.87 (0.0043)	4.23 (0.0064)	3.26 (0.0071)	3.73 (0.0083)	2.34 (0.0036)	4.25 (0.0047)	5.23 (0.0021)

(continued)

Table 1.6 (continued)

	M1	M2	M3	M4	M5	M6	M7
Durbin-Watson	2.2458	2.0394	2.0419	2.1396	2.0619	2.0802	2.0371
VIF	1.32	1.29	1.34	1.28	1.41	1.29	1.37

Notes Significance level is shown by * for p < 0.05
Source Author's compilation through Stata (Data from MIX Market)

Table 1.7 Hypotheses testing result

Hypotheses (Alternate)	Significance level is shown by * for p < 0.05	Result
H_1 1	0.042*	Accepted
H_1 1.1	**0.014**	**Rejected**
H_1 1.2	0.041*	Accepted
H_1 1.3	0.013*	Accepted
H_1 1.4	0.031*	Accepted
H_1 1.5	**0.009**	**Rejected**
H_1 1.6	0.018*	Accepted

The research shows that standards of good governance serve as fundamental support for microfinance operations in present-day governance discussions about sustainability. The world requires external governance systems for all microfinance institutions to safeguard operational integrity alongside stability and societal impact.

1.8 Implications

The way Microfinance Institutions function depends on their governance systems, which determine both financial stability and global market social outcomes. Services within the sector increase their resilience through frameworks of strong governance that also minimize financial risks while enhancing transparency. The section presents vital recommendations for policymakers and MFI leadership teams and investors to develop improved governance systems that boost operational performance together with enhanced accountability of MFIs.

The regulatory bodies directly oversee microfinance institution governance development through basic operational rules for MFIs. Policymakers need to create specific governance standards for all organizations by using accepted international criteria such as Basel Core Principles in addition to IFC Corporate Governance Guidelines. The standards must adapt to microfinance needs, so governance policies function similarly between large financial institutions and microfinance establishments. For achieving this goal, law needs to mandate the inclusion of board independence, risk management, and social impact measurement as annual report disclosures. MFIs must function under a system that modifies their regulatory responsibilities based on their risk profiles and scale and outreach dimensions to manage operational flexibility while staying compliant with regulations. Training programs should be institutionalized because they serve to enhance the decision-making authority and oversight capabilities of MFI board members as well as regulators. Governments should create intercountry peer-learning networks to allow regulators to share knowledge, which enables them to adopt best practices from multiple jurisdictions.

Policy Implications for Investors: Investors and lenders, including development finance institutions and impact investors, have the power to incentivize better governance by integrating governance performance metrics into funding decisions. Governance-linked investment criteria can serve as a tool to encourage MFIs to improve their governance structures in exchange for preferential financing terms. For instance, offering lower interest rates or priority funding to MFIs that demonstrate strong governance practices—such as having independent risk and compliance committees—can create a positive reinforcement mechanism. Moreover, transparency in financial and social impact reporting should be prioritized, with investors mandating environmental, social, and governance (ESG)—aligned disclosure standards for MFIs receiving international capital. Public–private partnerships can play a significant role in strengthening governance assessment mechanisms, such as establishing microfinance governance rating systems to evaluate and compare the governance practices of different institutions.

Policy Implications for MFI Institutions: Managers and boards of MFI entities should actively put forth necessary changes to realize governance reform. The fundamental principle establishes autonomous board structures that need professionals from outside organizations, with external non-executive members forming at least thirty percent of the total membership. The monitoring of MFI financial completeness and ethical lending practices belongs to independent committees dedicated to audit and risk management tasks that compose integral parts of governance systems. Government enterprises in the present digital transformation era should integrate financial technology solutions into their governance structures for better transparency and operational efficiency. MFIs advance their client protection mechanisms by educating clients about finances and implementing social metrics to safeguard them from unethical borrowing behaviors yet achieve financial inclusion targets.

A properly managed governance framework enables MFIs to become sustainable and efficient because regulators oversee activities, while investors support high-quality governance activities, and leadership maintains open accountability processes and innovative practices. The implementation of suitable policy measures strengthens MFIs, so they achieve better financial stability and improve stakeholder confidence alongside adherence to international governance standards for empowering economies and enhancing financial inclusion delivery.

1.9 Scope for Future Research

Future research needs a qualitative and case-study analysis of governance frameworks to build comprehensive knowledge. The data used in this study was up to 2018 because COVID-19 has substantially affected financial institutions, which include MFIs, since its onset in 2019. Research should investigate how the pandemic altered efficiency determinants in the microfinance industry to generate new knowledge about microfinance operations during periods of global crisis. Ongoing research will follow changes throughout time while adjusting to new trends because the changing environment requires continuous tracking of developments. Results from this study create important value that encourages academic and researcher investigation into corporate governance and firm performance relationships and alternative variable measurement approaches.

References

Abd Hamid, S., & Nassir, A. M. (2021). Does economic efficiency matters? Evidence from microfinance sector of Pakistan. *The Journal of Contemporary Issues in Business and Government, 27*(6), 979–988.

Aboagye, A. Q., & Otieku, J. (2010). Are Ghanaian MFIs' performance associated with corporate governance? *Corporate Governance: International Journal of Business in Society, 10*(3), 307–320. https://doi.org/10.1108/14720701011051938

Al-ahdal, W. M., Almaqtari, F. A., Tabash, M. I., Hashed, A. A., & Yahya, A. T. (2023). Corporate governance practices and firm performance in emerging markets: Empirical insights from India and Gulf countries. *Vision, 27*(4), 526–537.

Bassem, B. S. (2009). Governance and performance of microfinance institutions in Mediterranean countries. *Journal of Business Economics and Management, 10*(1), 31–43. https://doi.org/10.3846/1611-1699.2009.10.31-43

Batuo, M., Mlambo, K., & Asongu, S. (2017). Linkages between financial development, financial instability, financial liberalisation and economic growth in Africa. *Research in International Business and Finance, 45*, 168–179. https://doi.org/10.1016/j.ribaf.2017.07.148

Bharti, N. B. N., & Chitnis, A. C. A. (2016). Size and efficiency of MFIs: A data envelopment analysis of Indian MFIs. *Enterprise Development and Microfinance, 27*(4), 255–272. https://doi.org/10.3362/1755-1986.2016.017

Bibi, U., & Balli, H. O. (2023). Trade-offs in South Asian MFIs: Regulation and governance structure. *Applied Economics, 55*(53), 6310–6326.

Bogan, V. L. (2008). Microfinance institutions: Does capital structure matter? *SSRN Electronic Journal*. https://doi.org/10.2139/ssrn.1144762

Bøhren, Ø., & Ødegaard, B. A. (2001). *Corporate governance and economic performance in Norwegian listed firms*. Research Reports. http://home.bi.no/oyvind.bohren/work/mono/14-Corporate_governance_and_economic_performance_in_Norwegian_listed_firms.pdf

Cho, M. (1998). Ownership structure, investment, and the corporate value: An empirical analysis. *Journal of Financial Economics, 47*(1), 103–121. https://doi.org/10.1016/s0304-405x(97)00039-1

Danoshana, S., & Ravivathani, T. (2014). Impact of corporate governance framework on the organizational performance. A study on financial institutions in Sri Lanka. *International Journal of Technological Exploration and Learning, 16*(1), 73–78.

Foroughi, M., & Fooladi, M. (2011). Corporate ownership structure and firm performance: Evidence from listed firms in Iran. *International Conference on Humanities, Society and Culture.*

Gadedjisso-Tossou, A., Kouevi, T., & Gueyie, J. P. (2023). The influence of external governance mechanisms on the performance of microfinance institutions in Togo. *Corporate Governance: THe International Journal of Business in Society, 23*(7), 1564–1584.

Gaganis, C. (2016). Assessing the overall performance of microfinance institutions. *International Journal of Banking, Accounting and Finance, 7*(1), 52–83.

Galema, R., Lensink, R., & Mersland, R. (2012). Do powerful CEOs determine microfinance performance? *Journal of Management Studies, 49*(4), 718–742. https://doi.org/10.1111/j.1467-6486.2012.01046.x

Gaur, A. S., Kumar, V., & Singh, D. (2014). Institutions, resources, and internationalization of emerging economy firms. *Journal of World Business, 49*(1), 12–20. https://doi.org/10.1016/j.jwb.2013.04.002

Guo, Z., & Kga, U. K. (2012). Corporate governance and firm performance of listed firms in Sri Lanka. *Procedia—Social and Behavioral Sciences, 40*, 664–667. https://doi.org/10.1016/j.sbspro.2012.03.246

Gupta, N., & Mirchandani, A. (2020). Corporate governance and performance of microfinance institutions: Recent global evidence. *Journal of Management and Governance, 24*(2), 307–326.

Gupta, P., & Sharma, A. M. (2014). A study of the impact of corporate governance practices on firm performance in Indian and South Korean companies. *Procedia—Social and Behavioral Sciences, 133*, 4–11. https://doi.org/10.1016/j.sbspro.2014.04.163

Haq, M., Skully, M., & Pathan, S. (2009). Efficiency of microfinance institutions: A data envelopment analysis. *Asia-Pacific Financial Markets, 17*(1), 63–97. https://doi.org/10.1007/s10690-009-9103-7

Hartarska, V., & Mersland, R. (2012). Which governance mechanisms promote efficiency in reaching poor clients? Evidence from rated microfinance institutions. *European Financial Management, 18*(2), 218–239. https://doi.org/10.1111/j.1468-036X.2009.00524.x

Hawaj, A. Y., & Buallay, A. M. (2022). A worldwide sectorial analysis of sustainability reporting and its impact on firm performance. *Journal of Sustainable Finance & Investment, 12*(1), 62–86.

Hermes, N., Lensink, R., & Meesters, A. (2011). Outreach and efficiency of microfinance institutions. *World Development, 39*(6), 938–948.

Iqbal, S., Nawaz, A., & Ehsan, S. (2018). Financial performance and corporate governance in microfinance: Evidence from Asia. *Journal of Asian Economics, 60*, 1–13. https://doi.org/10.1016/j.asieco.2018.10.002

Kyereboah-Coleman, A., & Osei, K. A. (2008). Outreach and profitability of microfinance institutions: The role of governance. *Journal of Economic Studies, 35*(3), 236–248.

Méndez, C. F., & Gómez-Ansón, S. (2006). Does ownership structure affect firm performance? Evidence from a continental-type governance system. *Corporate Ownership and Control, 3*(2), 74–89. https://doi.org/10.22495/cocv3i2p9

Mersland, R., & Strom, R. O. (2008). Performance and trade-offs in microfinance organizations: Does ownership matter? *Journal of International Development, 20*(5), 598–612.

Mia, M. A., Rangel, G. J., Nourani, M., & Kumar, R. (2022). Institutional factors and efficiency performance in the global microfinance industry. *Benchmarking an International Journal, 30*(2), 433–459. https://doi.org/10.1108/bij-06-2021-0326

Mitton, T. (2002). A cross-firm analysis of the impact of corporate governance on the East Asian financial crisis. *Journal of Financial Economics, 64*(2), 215–241.

Nawaz, A., Iqbal, S., & Ehsan, S. (2018). Does social performance drive corporate governance mechanism in case of Asian MFIs? An issue of endogeneity. *Global Business Review, 19*(4), 988–1012.

Demise, N. (2006). OECD Principles of Corporate Governance. In *Springer eBooks* (pp. 109–117). https://doi.org/10.1007/978-4-431-30920-8_10

Oxelheim, L., Gregorič, A., Randøy, T., & Thomsen, S. (2013). On the internationalization of corporate boards: The case of Nordic firms. *Journal of International Business Studies, 44*(3), 173–194. https://doi.org/10.1057/jibs.2013.3

Popli, M., Ladkani, R. M., & Gaur, A. S. (2017). Business group affiliation and post-acquisition performance: An extended resource-based view. *Journal of Business Research, 81*, 21–30. https://doi.org/10.1016/j.jbusres.2017.08.003

Randøy, T., & Goel, S. (2003). Ownership structure, founder leadership, and performance in Norwegian SMEs: Implications for financing entrepreneurial opportunities. *Journal of Business Venturing, 18*(5), 619–637.

Rizwan, M., Asrar, H., Siddiqui, N. A., & Usmani, W. U. (2016). The Impact of corporate governance on financial performance: An Empirical investigation. *Zenodo (CERN European Organization for Nuclear Research)*. https://doi.org/10.5281/zenodo.3464888

Sami, H., Wang, J., & Zhou, H. (2011). Corporate governance and operating performance of Chinese listed firms. *Journal of International Accounting, Auditing and Taxation, 20*(2), 106–114.

Servin, R., Lensink, R., & Van den Berg, M. (2012). Ownership and technical efficiency of microfinance institutions: Empirical evidence from Latin America. *Journal of Banking & Finance, 36*(7), 2136–2144.

Sharma, A., & Rastogi, S. (2021). Impact of efficiency on voluntary disclosure of non-banking financial company—microfinance institutions in India. *Journal of Risk and Financial Management, 14*(7), 289.

Silva, A. C., & Chávez, G. A. (2015). Microfinance, country governance, and the global financial crisis. *Venture Capital, 17*(1–2), 191–213.

Singh, D. A., & Delios, A. (2012). Corporate governance, board networks and growth strategies. *Academy of Management Proceedings, 2012*(1), 10810. https://doi.org/10.5465/ambpp.2012.10810abstract

Sohn, S. Y., & Ju, Y. (2023). Mission efficiency analysis of for-profit microfinance institutions with categorical output variables. *Sustainability, 15*(3), 2732.

Strøm, R. Ø., D'Espallier, B., & Mersland, R. (2014). Female leadership, performance, and governance in microfinance institutions. *Journal of Banking & Finance, 42*, 60–75. https://doi.org/10.1016/j.jbankfin.2014.01.014

Sulimany, H. G., Ramakrishnan, S., Chaudhry, A., & Bazhair, A. H. (2021). Impact of corporate governance and financial sustainability on shareholder value. *Studies of Applied Economics, 39*(4). https://doi.org/10.25115/eea.v39i4.4318

Tchakoute-Tchuigoua, H. (2010). Is there a difference in performance by the legal status of microfinance institutions? *The Quarterly Review of Economics and Finance, 50*(4), 436–442. https://doi.org/10.1016/j.qref.2010.07.003

Wang, Y., Abbasi, K., Babajide, B., & Yekini, K. C. (2020). Corporate governance mechanisms and firm performance: Evidence from the emerging market following the revised CG code. *Corporate Governance: THe International Journal of Business in Society, 20*(1), 158–174.

Wijesiri, M., Yaron, J., & Meoli, M. (2017). Assessing the financial and outreach efficiency of microfinance institutions: Do age and size matter? *Journal of Multinational Financial Management, 40*, 63–76. https://doi.org/10.1016/j.mulfin.2017.05.004

Chapter 2
Watchdog of Watchdogs: Making of National Financial Reporting Authority

Anuj Bhatia and Aruna Jha

Abstract Auditing is a critical aspect of corporate governance. The monumental corporate frauds of Enron and Satyam, alongside other incidents have created a public distrust in auditing practices, which led to the need for a strong system to detect and prevent such fraud. The United States and India established the independent audit monitoring institutions, such as Public Company Accounting Oversight Board and National Financial Reporting Authority to supervise the auditing profession after these corporate frauds took place. The creation of National Financial Reporting Authority serves to maintain audit quality while supervising the audit profession to strengthen corporate governance frameworks. Financial oversight watchdogs are often called auditors, who now report to the watchdog of watchdogs—National Financial Reporting Authority. This study evaluates the role of an auditor and independent audit oversight authority in improving the quality of audit, financial reporting, and corporate governance. This chapter traces the developments in the field of corporate governance alongside regulatory changes in the field of auditing, which collectively led to the establishment of National Financial Reporting Authority in India. The regulatory changes in accounting and auditing after the establishment of National Financial Regulatory Authority have not received adequate academic examination through documented research studies. Further investigation is required regarding the National Financial Reporting Authority and its effects on auditing practices and corporate governance, according to this chapter.

Keywords National Financial Reporting Authority · Corporate governance · Audit · Audit quality · Independent audit oversight authority · India

A. Bhatia (✉) · A. Jha
Shri Ram College of Commerce, University of Delhi, Delhi, India
e-mail: anuj.bhatia@srcc.du.ac.in

A. Jha
e-mail: aruna.jha@srcc.du.ac.in

V. Kandpal et al. (eds.), *Developments in Corporate Governance*, Approaches to Global Sustainability, Markets, and Governance,
https://doi.org/10.1007/978-981-96-6366-8_2

2.1 Introduction

The quality of both financial and non-financial information maintains significant importance because it reduces information gaps between firms and their investors. The resulting improvement of investment efficiency occurs due to this decreased information asymmetry (Hundal et al., 2024). High-level assurance from auditing leads to improved quality of financial reporting (DeFond & Zhang, 2014). The audit process together with financial reporting activities fall under industry regulations to achieve transparent financial practices in capital markets (Hubais et al., 2023). Organizations that prioritize good financial reporting experience reduced vulnerability to economic downturns since their firm-specific risks decrease alongside enhanced capital market resource allocation (Biddle et al., 2009; Lee & Chung, 2017; Nguyen et al., 2021; Synn & Williams, 2024). Audits of high-quality deliver protection to investor interests, according to Lassoued and Bacha (2015) while simultaneously reducing frauds and accounting manipulation and creating stakeholder trust in financial reporting, which leads to improved market decision-making (Meng Chao et al., 2020). However, some major financial scandals in the recent past have harmed public perceptions of audit quality (Ishak et al., 2017).

Colossal corporate scams like Enron and WorldCom exposed the deep-rooted issues within the corporate governance mechanism such as latency of board of directors, ineffective internal controls, accounting irregularities, failure of auditors to expose the material misstatements, dominant CEOs, and exorbitant remuneration to top executives (Soltani, 2014). Such financial scams lead to significant financial losses and social costs (Montesdeoca et al., 2019; Moroke & Makatjane, 2022), emphasizing on the need for an effective fraud detection system (Albrecht et al., 2016). To reduce social costs associated with these scams, the governments should develop regulations to improve corporate governance and enhance public trust (Albrecht et al., 2016; Montesdeoca et al., 2019). As an aftermath of these scams, the Government of the United States passed the Sarbanes–Oxley Act, 2002 (SOX) and constituted the Public Company Accounting Oversight Board (PCAOB), that aims to restore investors' trust and public confidence by enhancing the reliability of corporate disclosures through an independent review of the auditing profession (Sheu, 2018).

India was not immune to financial and corporate scandals. Scams like Satyam shook the corporate landscape, and public confidence in financial markets was reduced. Strengthening of corporate governance mechanisms and regulatory oversight requires strong laws (Santhosh Kumar, 2014). The Indian Government enacted The Companies Act, 2013 and paved the way for the establishment of the National Financial Reporting Authority (NFRA) (Ramesh & Ramanathan, 2020; Santhosh Kumar, 2014). NFRA, established in 2018, has been crafted along the lines of PCAOB (Jha & Bhatia, 2024a, 2024b). In the landmark case *Re Kingston Cotton Mill Co. (1896),* the learned judge, while pronouncing the judgment, equated auditor to a "watchdog, not a bloodhound" (Baxt, 1970). However, with the advent of audit oversight authorities like NFRA, even the watchdog is watched upon, which makes such independent oversight authority the *"watchdog of the watchdogs"*.

Despite a significant shift in how accounting and auditing professions are regulated, there is a dearth of academic literature documenting and analyzing this phenomenon. This chapter attempts to document the NFRA's history and the developments in the corporate governance landscape. It will serve as an academic resource for conducting further research on NFRA and contribute as useful literature to studies comparing the functioning and impact of audit oversight authorities across different countries. Additionally, this chapter outlines the potential scope of future research with respect to integration of audit oversight and corporate governance.

This chapter discusses the role of auditors in Sect. 2 and the need for independent review of audits in Sect. 3. Building on these topics, Sect. 4 explores the history and genesis of NFRA, and Sect. 5 pertains to the discussion on the functions, powers, and jurisdiction of NFRA. Section 6 outlines the milestones achieved and some impact of NFRA. Following this, Sects. 7 and 8 review the existing academic literature on NFRA and identify areas for further research. The chapter concludes with Sect. 9, which addresses challenges and offers suggestions for improvement.

2.2 Auditors' Role in Corporate Governance

Auditors play an important role in corporate governance structure to ensure transparency, enhance financial statements' quality and reliability, and conformity with relevant laws. The auditors work together with management, audit committee, and board of directors to monitor the process of financial reporting, acting as watchdogs protecting the interests of shareholders and various stakeholders (Ashbaugh & Warfield, 2003; Cohen et al., 2010; Nasrudin et al., 2017). Within agency theory, auditors lower the information asymmetry among stakeholders and the management (Subedi, 2023). They significantly contribute to detecting fraud, upholding the integrity of the financial reporting, and mitigating the risk of material misstatements in financial statements (Ndreca & Dibra, 2023; Ţager et al., 2014). By assuring stakeholders regarding the conformity of financial reports with different accounting standards and regulations, they contribute to mitigating the agency problems (Bédard & Compernolle, 2014; Cho & Wu, 2014).

External audits are critical tools within the structure of corporate governance because of their independence (Bédard & Compernolle, 2014). The audit committee has crucial internal governance function as it oversees the company's financial reporting system and is critical for reinforcing the independence of the external auditors (Ţager et al., 2014). External auditors reduce the likelihood of earnings management, thereby enriching the investors' confidence in financial statements (Saleem et al., 2016).

Improved corporate governance is associated with the likelihood of firms appointing auditors of high-quality (Lin & Liu, 2009). Audit quality improvement enhances corporate governance quality and simultaneously minimizes monitoring costs (Han et al., 2013) and produces higher firm performance (Asghar et al., 2021; Bijalwan et al., 2019; Dash & Raithatha, 2019; Khan et al., 2021; Venkatraman & Selvam, 2014). Stock price crashes decrease when investors select high-quality auditors because these auditors effectively reduce tail risk and balance the effects of opaqueness and accounting conservatism as well as tax avoidance (Robin & Zhang, 2015). The external audit function plays a vital role in corporate governance structures for unlisted firms within developing countries because it strengthens accountability and transparency (Anaman et al., 2023). The auditors maintain an essential role in corporate governance systems because they function as independent evaluators who verify the financial state and operational management and governance of companies (Bédard & Compernolle, 2014; Subedi, 2023).

2.3 Independent Oversight of Audit and Auditing Profession

The auditing profession was primarily self-regulated until oversight authorities like PCAOB and NFRA were established. For instance, in India, auditing profession was regulated by the Institute of Chartered Accountants of India (ICAI) which is established by law but governed by the members (Peer Review Board, n.d.). Accounting scams raised questions about the sanctity of the audit and auditing profession. After these accounting scandals, independent oversight of audits became a precondition to reestablishing public trust in auditing and financial reporting (Loehlein, 2017). The independent review authorities function like judicial bodies to perform examinations which have trial-like characteristics (Westermann et al., 2019).

Independent review bodies support the adherence to financial reporting standards by raising both accountability levels and understanding of quality reporting for auditors' directors and accountants (Louw & Maroun, 2017). Inspections from independent review bodies drive auditors to use a risk-averse approach before starting or maintaining work with clients presenting high levels of danger which advances cautious auditing behavior (Elshandidy et al., 2021). The independent inspections have revealed significant audit problems leading to financial statements, according to Lennox and Pittman (2010). Through independent review processes, accounting and auditing professionals gain opportunities to learn together by both discussing evaluation results and auditing problems while giving feedback about audit quality standards (Church & Shefchik, 2011). Independent audits serve as an effective long-term evaluation tool which monitors the audit procedures at all times (Franzel, 2014). The inspections have caused significant changes to auditing procedures as well as evoked reactions from capital markets and companies (Gipper & Leuz, 2015). To sum up, independent review of audits is necessary to ensure transparency and integrity in the financial reporting and the quality of financial statements. It strengthens the credibility of published information in capital markets, thereby increasing its reliability.

While there are advantages linked to the independent review of audits, it comes with its own challenges. There is a lack of understanding about the scope of independent reviews along with several misconceptions and resistance to change (Nyambuya et al., 2021). These inspections are responsible for rise in the cost of audits, both for auditing firms as well as their clients, especially for the smaller firms (Hay et al., 2023). Inspectors may be highly critical of audit practices and impose stricter standards than peer reviews, which may decrease the overall effectiveness of the oversight process (Church & Shefchik, 2011). An independent audit review practice has concentrated audit work among several large audit firms resulting in insufficient options for clients to select auditors (Elshandidy et al., 2021). The practice of violating regulations leads to substantial fines and practice debarment as well as disciplinary penalties (Dharmasiri et al., 2021). Lack of congruent understanding regarding the review process generates misalignment between organizations' expectations and the actual implementation actions of audit firms (Ater et al., 2019).

The independent review system of audit profession proves vital to achieve better audit quality and maintain compliance with auditing standards. Even though there are challenges associated with independent review, especially regarding cost implications and perception of auditors about the inspections, efforts to balance stringent reviews with practical audit considerations will help to smoothen the process. As a retort to the mounting necessity for vigorous and independent oversight of the auditing profession, several countries have set up dedicated regulatory authorities. In India, the formation of the NFRA marked a meaningful step toward enhancing the dependability on financial reporting through independent review of audits. The following section of this chapter will discuss the genesis and evolution of NFRA in India.

2.4 History and Genesis of NFRA

Auditing is a critical element in the system of corporate governance. Auditing is a monitoring and gatekeeping mechanism in the corporate governance system (Bédard & Compernolle, 2014; Nilsson, 2018). The history of corporate governance is intertwined with the field of auditing. The auditors' role depends on the country's corporate governance systems, which encompass the legal environment and the regulations for investor protection (Nilsson, 2018). The creation of an institution like NFRA has its roots in the development of a corporate governance system where every committee report, code, and regulation have the auditors' role, and audit quality of financial statements as a critical aspect.

The need to improve corporate governance through regulatory initiatives in India began in the late 1990s post the Asian Financial Crisis (Jha & Bhatia, 2024a). The institutional initiative in India for improving corporate governance started with the Confederation of Indian Industry (CII), constituting a task force in 1996 to draft a code for corporate governance in India. The CII issued a *"Desirable Corporate Governance: A Code"*, a voluntary code for corporate governance in 1998 (Confederation of Indian Industry, 1998), which was broadly attuned to the Cadbury Committee Report in the United Kingdom (Kaur, 2021; The Committee on the Financial Aspects of Corporate Governance, 1992). The Securities and Exchange Board of India (SEBI) appointed the Kumar Mangalam Birla Committee to further strengthen the standards of corporate governance in India in 1999. In 2000, following the recommendations of the Kumar Mangalam Birla Committee, SEBI instituted clause 49 in the listing agreement of stock exchanges along with various provisions for corporate governance to enhance the transparency and integrity in the system (Kandukuri et al., 2015; Kaur, 2021). These codes and recommendations focused on auditors' independence and the audit committee's role and constitution to strengthen corporate governance.

During early 2000s, a series of scandals, including the Enron scandal, followed by other notable scandals such as WorldCom, Qwest, and Global Crossing, as well as frauds in audits that caused the collapse of Anderson, resulted in significant changes in corporate governance landscape globally (Friedrichs, 2017; Iren & Kim, 2023; Scrimgeour & Duppati, 2014; Sheu, 2018; Vagadia, 2014). The United States enacted the SOX in 2002 as an extensive transformation in corporate governance practices. SOX 2002 introduced reforms that aimed to strengthen both auditors' independence while enhancing financial reporting accuracy and dependability besides implementing corporate governance requirements (Boylan, 2015; Cohen et al., 2010; Romano, 2005; Westermann et al., 2019). Moreover, it introduced penalties for intentional misconduct by management and auditors. The PCAOB was created as an independent body to regulate and oversee the accounting profession in the United States (Westermann et al., 2019).

Department of Company Affairs (now Ministry of Corporate Affairs or MCA) of India constituted a committee by Naresh Chandra to thoroughly evaluate India's corporate governance concerns in 2002. The *"Naresh Chandra Committee Report on Corporate Audit and Corporate Governance"* recommended establishment of an independent regulatory body similar to the PCAOB to supervise the quality of audits of public companies (Ministry of Corporate Affairs, 2002). The Committee also suggested setting up an office to investigate serious corporate frauds. In 2003, upon the recommendations of Committee, the Indian Government established a Serious Fraud Investigation Office (SIFO) under MCA to inspect corporate frauds (Serious Fraud Investigation Office, n.d.). While the SFIO is engaged in post-event investigations, it is reactive and inherently limits its ability to proactively prevent corporate frauds. The Committee also suggested setting up of the Quality Review Board (QRB) under ICAI. The QRB was set up in 2007 to conduct periodic reviews of audit firms, oversee the audit quality, and provide recommendations for improvement (Quality Review Board, n.d.). Before the establishment of QRB, a Peer Review Board was set up by the ICAI in 2002 to ensure that the members of the ICAI (i.e., the auditors) adhere to all applicable technical, professional, and ethical standards and any other applicable regulatory requirements when conducting assurance service agreements (Peer Review Board, n.d.). Establishing QRB was the right step in improving the audit quality as the board members also included nominees from the government. However, QRB failed to raise the audit quality effectively because it had no disciplinary or investigative powers (Kumar, 2018).

In 2002 and 2004, the SEBI constituted two committees, both chaired by Narayana Murthy, with the mandate to enhance the provisions of Clause 49. *"The Report of the SEBI Committee on Corporate Governance"* (also known as the Narayana Murthy Committee) proposed a combination of mandatory and non-mandatory recommendations encompassing various corporate governance aspects (Securities & Exchange Board of India, 2003). These recommendations included the establishment of audit committees, stringent regulations for related party transactions, comprehensive risk management frameworks, the formulation of a code of conduct for both the Board of Directors as well as the senior management, the appointment of nominee and independent directors, compensation guidelines for non-executive directors, and the implementation of a robust whistleblower policy. Alongside these events, MCA had formed a committee under the leadership of Dr J. J. Irani in 2004 to review and strengthen the corporate regulations in India. The Irani Committee submitted the *"Report of the Expert Committee on Company Law"* (the Irani Committee Report) in May 2005. The Committee emphasized on the need to simplify the laws and make the regulations attuned to the current times (Ministry of Corporate Affairs, 2005). The Committee also made suggestions related to auditing and rotating auditors to ensure auditors' independence (Ministry of Corporate Affairs, 2005). Following the suggestions of Irani Committee, MCA proposed the Companies Bill, 2008 (Ministry of Corporate Affairs, 2008), which could not be enacted as a law at that time. Though both the Committee reports and the draft Companies Bill proposed changes to enhance the audit function, they were silent on setting up an independent audit oversight authority. In the backdrop of these ongoing efforts to improve corporate governance, India witnessed its most significant corporate scandal to date—the Satyam Scandal.

The Satyam Scandal, also known as "India's Enron", that involved massive accounting fraud, sent shockwaves through the corporate and regulatory landscape in India. At the time of Satyam Scandal, Indian companies were governed by the Companies Act, 1956. This scandal exposed the presence of significant loopholes in the regulatory framework overseen by authorities like SEBI, MCA, and SIFO. This also revealed the failure of the audit function in safeguarding the interests of stakeholders (Bhasin, 2016; Ganesh, 2015; Narayanaswamy et al., 2021). The collusion between auditors and management and the regulatory framework's inability to prevent such connivance was a significant factor in this massive accounting fraud (Santhosh Kumar, 2014). Further, this scandal also highlighted the failure of QRB to check the involvement of auditors in corporate fraud. The auditor's involvement in the Satyam scandal brought to the fore the need to improve audit quality to contain the recurrence of such scams. (Santhosh Kumar, 2014).

As an aftermath of the Satyam scandal, the CII set up a task force on corporate governance under the chairmanship of Naresh Chandra in February 2009. The task force submitted its report *"Corporate Governance—Recommendations for Voluntary Adoption"* in November 2009. MCA also proposed adopting these voluntary guidelines (Kaur, 2021). The task force gave many recommendations for enhancing corporate governance, improving audit quality, and maintaining auditors' independence. The report also emphasized on the failure of QRB to achieve the stated objectives and the need for a robust audit oversight mechanism along the lines of PCAOB in the United States (Confederation of Indian Industry, 2009).

In 2009, the *Twenty-First Report of Standing Committee on Finance* on the Companies Bill, 2009 suggested establishing an independent audit oversight authority to oversee the accounting and auditing standards adopted by the companies or their auditors (Lok Sabha Secretariat, 2010). It was proposed that this role be assigned to the National Advisory Committee for Auditing and Accounting Standards (NACAAS). The Standing Committee on Finance in its twenty-first report recommended the change in nomenclature of NACAAS to NFRA (Lok Sabha Secretariat, 2010). Subsequently, MCA introduced the Companies Bill in 2011 (Ministry of Corporate Affairs, 2011), making the first formal proposal for the creation of NFRA. This bill was later passed by the Indian Parliament and enacted as The Companies Act, 2013. This Act contained provisions for implementing and strengthening corporate governance practices and making corporate social responsibility mandatory for certain classes of companies (Aggarwal & Jha, 2019; Uzma, 2023). The Act changed how companies are governed and contained provisions for listed and unlisted public companies and consequences for non-compliance.

NFRA is India's first independent regulator of the accounting and auditing profession (National Financial Reporting Authority, 2021). Section 132 of the Act empowered the Government of India to establish NFRA as a statutory body to deal with matters relating to the accounting and auditing of companies. The Companies Act, 2013 became effective from 1st April, 2014. Though the Act contained provisions for establishing NFRA, the implementation was delayed until 2018. Some triggers nudged the Government of India to implement Sect. 132 of the Act.

In response to inquiries raised in the Rajya Sabha (the upper house of the Indian Parliament), the government attributed the delay in constituting the NFRA to the time required for drafting rules and engaging in public consultation (Ministry of Corporate Affairs, 2018b). The government also acknowledged the apprehensions raised by the ICAI, indicating that addressing these concerns through consultations contributed to the delay. Additionally, the matter was referred for consideration to the Companies Law Committee in June 2015, which submitted its report in February 2016, and to the Standing Committee on Finance, that examined the Companies (Amendment) Bill, 2016 (Ministry of Corporate Affairs, 2016a, 2016c, 2016b).

The *Report of Companies Law Committee*, in its report, acknowledged the receipt of a letter dated 18th August 2015 from the ICAI, expressing reservations regarding the establishment of the NFRA (Ministry of Corporate Affairs, 2016a). The ICAI asserted that it already possessed a robust disciplinary mechanism, overseen by a Board of Discipline and Disciplinary Committee with government nominees, to regulate its members. The profession viewed the creation of the NFRA as an encroachment upon its autonomy and argued that multiple layers of regulation could lead to delays and duplication of work. Consequently, the ICAI suggested the omission of Sect. 132, which pertains to the establishment of NFRA. However, the Committee concluded that, given the critical nature of the responsibilities involved, and the serious repercussions observed from lapses in the past, the establishment of an independent oversight body for the profession was indispensable. Citing the existence of similar regulatory bodies in major global economies, the Committee recommended the prompt establishment of the NFRA. It also suggested that consultations with the ICAI be conducted regarding the NFRA's jurisdiction and the representation of the ICAI in the new setup.

Further, the report of *Standing Committee on Finance* asserted the need for establishment of NFRA (Ministry of Corporate Affairs, 2017). The report quoted the stance of the Ministry of Corporate Affairs on the apprehensions raised by ICAI. The Ministry stated that even though the corporate laws entrust external auditors with verifying financial statements, this role faces an inherent conflict of interest as auditors are paid by the companies they audit. Following global accounting scandals, many countries have moved from self-regulation to independent regulators like the PCAOB in the United States and the Financial Reporting Council (FRC) in the United Kingdom to ensure auditor independence and improve audit quality. This shift aims to strengthen the independence of auditors, enhance the quality of audits, and boost investors' trust in the financial statements of the corporates (Ministry of Corporate Affairs, 2017).

These two reports cemented the intentions of the government to establish NFRA. However, there are two events that further nudged the pavement of NFRA. First, in the case of *S. Sukumar v. The Secretary, ICAI*, the Supreme Court of India, acknowledging the importance and critical nature of auditing, suggested the Union Government to consider drafting a regulatory framework to oversee the profession, citing the example of PCAOB under the SOX, 2002, along with Dodd-Frank Wall Street Reform and Consumer Protection Act, 2010 of the United States. The Supreme Court further directed the government to consider drafting rules for the oversight authority. Second, the case of Nirav Modi, also known as the Punjab National Bank (PNB) heist. In early 2018, the PNB reported a scam to the tune of $1.8 billion by an Indian businessman, Nirav Modi (Bloomberg, 2018). This scam further raised questions on the auditing profession and auditors' role. The massive uproar created by this scam twinned with the debate in India to regulate the auditing profession furthered the need to establish NFRA (Kumar, 2018; Ramesh & Ramanathan, 2020).

MCA, following the judgment in the case of *S. Sukumar v/s The Secretary, ICAI*, through a Gazette, notified October 1, 2018 as the date of NFRA's constitution (Ministry of Corporate Affairs, 2018a). The government also constituted a Committee of Experts (COE) that presented its findings and recommendations on *"Regulating Audit Firms and the Networks"*. The COE, in its October 2018 report, recommended that the establishment of NFRA is not inconsistent with the provisions of the Chartered Accountants Act, 1949; it will only strengthen the auditing profession by solving the concerns related to self-regulation (Ministry of Corporate Affairs, 2018c). Two weeks after the submission of this report, MCA notified the National Financial Reporting Authority Rules, 2018, specifying its powers, roles, and jurisdiction. The enactment of these rules made NFRA operational.

2.5 NFRA: Functions, Powers, and Jurisdiction

NFRA is now the apex institution for overseeing accounting and auditing practices in India. It is established to protect the interests of public, investors, creditors, and other stakeholders by upholding the highest standards in accounting and auditing practice. According to Sect. 132 of The Companies Act, 2013, the primary functions of NFRA include advising the government on formulating and implementing standards on accounting and auditing, monitoring conformity with these standards, supervising the quality of services provided by auditing professionals, and suggesting measures for improvement (Ramesh & Ramanathan, 2020; Santhosh Kumar, 2014). It also promotes awareness regarding compliance and collaborates with national and international organizations to ensure adherence to accounting and auditing standards. NFRA may also perform other necessary or incidental functions to fulfill its primary duties (Rule 9) (National Financial Reporting Authority Rules, 2018). NFRA constitutes of a chairperson, three full-time members, and nine-part time members. The chairperson must be eminent and have at least twenty-five years of experience in the fields of accounting, auditing, law, and finance. The other members are also experts in accounting, auditing, and related fields, with nine-part time members being high-rank executives in government and affiliated bodies.

NFRA has the power to initiate investigations into the activities of auditors and companies that fall under its jurisdiction (Rule 10, 11, and 12) (National Financial Reporting Authority Rules, 2018). It can examine any books of accounts, registers, and documents necessary for investigation in any matter. Moreover, NFRA can impose penalty on companies and auditors and debar any ICAI member from practice for six months to ten years.

The jurisdiction of the NFRA encompasses a broad spectrum of companies and bodies corporate, reflecting its scope to ensure compliance with standards on accounting and auditing and oversee the quality of financial reporting in India (Rule 3) (National Financial Reporting Authority Rules, 2018). Specifically, NFRA's authority extends to all companies with securities listed on stock exchanges (listed companies) and unlisted companies with a paid-up capital of ₹500 crores or above, or an annual turnover of ₹1000 crores or above, or outstanding loans, debentures, and deposits totaling ₹500 crores or above as of the preceding financial year-end. Besides, the companies that are governed by special statutes like insurance, banking, and electricity generation or supply companies, fall under the jurisdiction of NFRA. Its jurisdiction also extends to foreign subsidiaries or associate companies of Indian companies or corporate bodies, provided their income or net worth is above 20% of the parent entity's consolidated income or net worth. NFRA can investigate, on a referral from the Central Government or *suo moto*, any companies, class of companies, or auditors for any misconduct.

Since its establishment in 2018, the roles, jurisdiction, and approach of NFRA have evolved. MCA appointed a three-member panel upon the directive of the Supreme Court in response to attempts made by ICAI to scrap NFRA, terming it as "unconstitutional" and "ultra-vires to the provisions of the Chartered Accountants Act, 1949".

The expert panel highlighted the international trend, representing the change from a self-regulatory structure to an independent regulation owing to the failure of the self-regulatory model. They further stated that the formation of NFRA was a "necessary institutional reform" and an attempt to align India's auditing practices with the global landscape (Srivats, 2021).

There has been a shift in NFRA's regulatory approach from an enforcement-focused model to a supervision-oriented one. Initially NFRA gave prominence on enforcement to address the need for stricter regulatory action necessary at that time. However, NFRA has shifted its focus to improving audit quality through "firm-wide audit quality inspections" and increased engagement with audit committees and boards to better understand their perspectives and to strengthen the communication between auditors and management. This supervisory approach is intended to strengthen the auditors' role, improve the quality of audits, and enhance the corporate governance practices (Pattanayak, 2024).

Recently, the Supreme Court of India reinforced the jurisdiction of NFRA by passing an order stating that the NFRA had a retrospective jurisdiction over any alleged offenses by auditors, even if the offenses pertained to a period prior to the formation of NFRA (Prasad, 2024). These developments testify to the power and jurisdiction of NFRA, indicating its dominance in overseeing the auditing profession. This expansion in authority is expected to significantly and positively impact the auditing profession and financial reporting practices in India.

2.6 NFRA: Milestones and Impact

NFRA has been actively performing its role as a regulator of accounting and auditing profession. Irrespective of ongoing litigations, the role of NFRA in overseeing the work of auditing professionals has assumed a center stage. This is evident from the annual reports published by NFRA. With its mandate to recommend the accounting and auditing standards and policies to be adopted by the companies with the approval of Central Government, NFRA reviewed and approved a total of four (4) proposals of ICAI for amendments in Indian Accounting Standards (Ind AS) subsequent to the amendments in corresponding International Financial Reporting Standards (IFRS) during the year 2022–2023 (National Financial Reporting Authority, 2023).

NFRA is the watchdog of watchdogs. In other words, it is the auditor of auditors (Ramesh & Ramanathan, 2020). One of the core functions of NFRA is that of Audit Quality Review (AQR). The primary objective of the AQR is to examine the quality control system of the auditors and to determine how well it complies with auditing standards. NFRA initiated audit quality inspections of auditors, covering both firm-wide audit quality and individual audit engagements for five large audit firms were inspected, involving audit reports of twenty-five clients during December 2022 and January 2023. NFRA in its annual report of 2022–2023 published AQR in respect of Statutory Audit conducted by SRBC& Co LLP (the auditors) of Infrastructure Leasing & Financial Services Limited (IL&FS) pertaining to the financial year

2017–2018 (National Financial Reporting Authority, 2023). NFRA also reported the failure of the auditors in (a) complying with the Standards on Auditing (SAs) and applicable Ind AS, (b) assessing the risk of material misstatements (ROMM) in the financial statements due to fraud, (c) evaluating the fair value of investments, (d) noticing Related Party Transactions (RPT), (d) performing audit procedures to mitigate risks associated with sanctioning loans, (e) evaluating management's assertions about RPT, (f) exercising due diligence in obtaining sufficient evidence, and (g) determining the persons comprising TCWG (Those Charged With Governance) and communicating auditor's independence (National Financial Reporting Authority, 2023).

NFRA is empowered to supervise and enforce conformity with the standards on accounting and auditing. For this, it employs Financial Reporting Quality Review Reports (FRQRs), which is an important tool to assess and evaluate whether the information presented in the Financial Statements meets the key requirements for presentation and disclosure. It also guides the companies to improve quality of financial reporting to comply with the accounting standards as stipulated by law. During the year 2022–2023, NFRA published two FRQRs in respect of PSP Projects Limited and ISGEC Heavy Engineering Limited pertaining to the financial year 2019–2020. NFRA observed that both these companies failed to comply with several accounting standards and directed them to rectify the deficiencies in the financial statements as pointed out by FRQR. This is also an indication of failure of audit firms to report the non-compliance with the relevant standards.

NFRA is empowered to investigate the professional misconduct of any auditor or auditing firm or any other professional and may impose sanctions like penalties and debarment from practice. As far as disciplinary actions on auditors are concerned, as on 31st March 2023, a total of 74 disciplinary cases are in progress at various stages of the disciplinary process (National Financial Reporting Authority, 2023). Further, NFRA has issued 70 disciplinary orders for professional misconduct by auditors in various Public Interest Entities (PIEs) during the financial year 2023–2024 (National Financial Reporting Authority, 2024). In a recent case, four auditors of Dewan Housing Finance Corp. Ltd. (DHFL) branches were investigated for professional misconduct, and punishment was meted out to them by NFRA (Press Trust of India, 2023). NFRA also initiated a *suo moto* AQR for statutory auditors of DHFL. The auditors appealed to the National Company Law Appellate Tribunal (NCLAT) that only ICAI has the authority to take disciplinary action against them, which was dismissed by the tribunal. Both NCLAT and Supreme Court held that once NFRA initiates investigation in any case, no other institution has the authority to initiate parallel proceedings.

There is a visible impact of NFRA's inspection on the provision of audit services. The provision of audit and non-audit services by the same firm was a pertinent issue related to the independence of auditors (Sect. 144, The Companies Act, 2013). The inspection report of NFRA suggests that there has been an improvement in the implementation of the law after the establishment and intervention of NFRA. In the words of Ajay Bhushan Prasad Pandey, the Chairperson of NFRA, "There is now greater awareness among firms of issues related to independence. We have started seeing some improvements. They are fixing the issues which were pointed out" (Chitravanshi, 2024).

The advent of NFRA is purported to change the auditing landscape in India. NFRA has proposed the renaming of Standards on Auditing (SAs) to Indian Standards on Auditing (IndSAs) and overhauling them in line with global auditing practices. NFRA has proposed 40 auditing standards which are in tune with international auditing standards to be made effective by April 2026. Though the proposed SAs have been made in consultation with ICAI, there is a disagreement between NFRA and ICAI specifically with respect to SA 600 and SA 299. While SA 600 deals with the auditing of corporate groups, SA 299 is concerned with the audit of joint financial statements. NFRA's proposed changes make the principal auditor of a corporate group responsible for the audit of the financial statements of the entire group. This is done to "prevent the finger-pointing seen in past corporate scandals, where principal auditors and component auditors blamed each other for audit failures". ICAI is of the view that these changes are detrimental to small accounting firms and will lead to the concentration of work in the hands of big audit firms (Shah & Etcfo, 2024).

Notwithstanding the resistance from ICAI, the steps taken by NFRA in a bid to improve audit quality and transparency have found some support from the auditors and industry. Atul Gala, Partner, Bhuta Shah & Co. LLP, observed that the NFRA's recommendations for LLP auditing standards "will have a wide-ranging impact on enhancing the quality and transparency of financial reporting, ensuring greater compliance with regulatory requirements, improving the reliability of financial statements, and ultimately fostering increased investor confidence and protection of stakeholder interests across the LLP sector" (Kaushik, 2024).

Ajay Bhushan Prasad Pandey, Chairperson of NFRA, stated that it is the collective responsibility of the management, the independent directors and audit committee, statutory auditors, investors, and the government to improve corporate governance. Auditing is a critical pillar of corporate governance. Improving auditing standards and adopting global standards will help to create the trust of global investors in Indian companies. NFRA's role in improving audit quality, making auditing practices at par with global standards, and making auditors accountable for lapses has assumed a prominent position in the corporate governance landscape (Press Trust of India & Business Standard, 2024).

The establishment of NFRA has brought about noteworthy changes in the practice of accounting and auditing. Stringent oversight and disciplinary actions have forced auditors to be more accountable. Regular oversight, inspections, and quality reviews by NFRA have led to auditors being more diligent and improvements in the quality of audits. NFRA's mandates to publish the reports of inspections and enforcement actions have led to a reduction in information asymmetry and increased transparency and trust in the financial reporting.

2.7 Summary of Prior Studies on NFRA

This section provides a summary of academic research on NFRA.

Santhosh Kumar (2014) identified the nexus between the auditor and management as a key reason for the rise of corporate accounting frauds in India. He pointed out the rising number of investigations by agencies like SIFO as an indication of increasing number of scams. He suggested that forming an agency like NFRA to oversee the quality of audit and compliance by auditors would significantly address weaknesses in auditing practices. He also noted that vigilance by a quasi-judicial body would enhance the accountability of the corporates.

Roy (2015) examined the case of Satyam and contended that a lack of auditor independence reduces the quality of audits, enabling fraudulent activities by the company's management. He recommended establishing an independent oversight authority like NFRA to prevent corporate frauds, improve the quality of audits, and address corporate governance issues.

Kumar (2018) explored the auditors' and other stakeholders' perceptions regarding the establishment of an independent audit oversight mechanism (NFRA). The study revealed a difference in perceptions among stakeholders. While the auditors favored making the self-regulation mechanism more effective, other stakeholders like investors, bankers, managers, and bureaucrats supported creating an independent review mechanism.

Ramesh & Ramanathan (2020) analyzed the audit procedures in India, the UK, and the USA and compared professional qualifications and training requirements for auditors, including CA, CPA, and ACCA. They found that Indian public accountants did not accept state-run independent oversight of audits by NFRA. The Indian auditors preferred self-regulation through QRB or Peer Review Boards over an independent authority, echoing the apprehensions of ICAI about setting up NFRA. The study also noted that stakeholders perceive the current state of auditing in India as unsatisfactory compared to the UK and USA, underscoring the need for stringent regulations to improve audit quality in India.

Singh (2021) supported the view that establishing NFRA will contribute to regulating the corporate frauds and lead to improvements in governance and audit practices.

Thoppan et al., (2021) measured the impact of corporate governance, financial legislation, and global reporting on earnings management practices of 712 firm-year data consisting of 89 Indian companies across 7 sectors, over 8 years (2011–2018). The empirical study found that large listed companies have become more cautious in financial reporting. The enactment of The Companies Act, 2013, the establishment of NFRA, and the adoption of global accounting standards have led to more trust and transparency in financial reporting.

It has been more than five years since the establishment of NFRA, but studies on various aspects of NFRA are scarce.

2.8 Future Scope for Research on NFRA

NFRA is a relatively new institution; its roles and functions of NFRA are evolving, and it needs to be complemented by academic research. The long-term impact on corporate governance and audit quality is still unfolding. Future research could focus on several areas, including:

1. Evaluating the effectiveness of NFRA's regulatory and enforcement actions on audit quality and the practices of financial reporting
2. Examining the impact of NFRA's supervision on the auditing profession and corporate governance.
3. Studying the perception of auditors and other stakeholders like investors, creditors, and the public about the independent regulatory mechanism and suggesting measures to improve public confidence.
4. Analyzing the challenges faced by the NFRA in capacity building of its inspection teams, ensuring independence from ICAI, and addressing resistance from the auditing professionals.
5. Studying the potential of NFRA to adopt a more proactive and preventive approach to regulating the auditing profession.
6. Examining the impact of NFRA on the auditing procedures, compliance with the standards on accounting and auditing, and advancing the auditing practices.
7. The investigations made by NFRA should be documented as case studies to enable further research.

2.9 Conclusion

The formation of NFRA is a significant milestone in India's efforts to improve corporate governance and financial reporting quality. Exposure of accounting and auditing scams exposed the flaws in the self-regulatory model. The creation of NFRA was a step in restoring investors' confidence in the auditing profession and financial reporting. The powers vested with NFRA to investigate the misconduct by the auditors and impose severe sanctions have led to demur among the auditors. However, the presence of an independent oversight authority will also incentivize the auditors to exercise professional skepticism and fulfill their duties more diligently. Additionally, monitoring compliance with accounting and auditing standards will foster greater transparency and reliability in financial reporting and boost the confidence of both domestic and international investors.

Despite the expected gains, NFRA faces challenges like building the capacity of its team for investigations, maintaining independence from ICAI and the government, and coordinating with other regulators like SEBI and RBI. Additionally, an ambitious target of resolving disciplinary cases within 90 days requires the processes to be streamlined. Going forward, NFRA must resolve all the issues with ICAI and engage with the stakeholders, including companies, auditors, and investors, to build trust and ensure that regulations are practical and effective. NFRA should continuously improve the regulatory framework to align with the global best practices. By addressing these challenges, NFRA can effectively enhance corporate governance practices in India, restore investor confidence, and support economic growth. The recent changes in the regulations and judgments of the Supreme Court confirmed that the NFRA is the final authority in the oversight of the auditing profession and will continue to be the *'watchdog of the watchdogs'*.

References

Aggarwal, V. S., & Jha, A. (2019). Pressures of CSR in India: An institutional perspective. *Journal of Strategy and Management, 12*(2), 227–242. https://doi.org/10.1108/JSMA-10-2018-0110

Albrecht, C. C., Sanders, M. L., Holland, D. V., & Albrecht, C. (2016). The debilitating effects of fraud in organizations. In *Crime and corruption in organizations: Why it occurs and what to do about it* (pp. 163–185). https://doi.org/10.4324/9781315574776-14s

Anaman, P. D., Ahmed, I. A., Appiah-Oware, F., & Somiah-Quaw, F. (2023). External auditors' impact on corporate governance of unlisted firms: A developing country perspective. *Seisense Business Review*. https://doi.org/10.33215/sbr.v3i1.904

Asghar, M., Anwar, Z., Usman, M., & Khan, H. (2021). Better corporate governance leads to better performance: Evidence from asian countries. *Argumenta Oeconomica, 46*(1), 183–204. https://doi.org/10.15611/aoe.2021.1.09

Ashbaugh, H., & Warfield, T. D. (2003). Audits as a corporate governance mechanism: Evidence from the German market. *Journal of International Accounting Research, 2*(1). https://doi.org/10.2308/jiar.2003.2.1.1

Ater, B., Gimbar, C., Jenkins, J. G., Saucedo, G., & Wright, N. S. (2019). Audit roles and the review process: Workpaper preparers' and reviewers' differing perspectives. *Managerial Auditing Journal*. https://doi.org/10.1108/maj-05-2018-1896

Baxt, R. (1970). The modern company auditor. A nineteenth-century watchdog? *The Modern Law Review, 33*(4), 413–431. http://www.jstor.org/stable/1094298

Bédard, J., & Compernolle, T. (2014). The external auditor and the audit committee. In *The routledge companion to auditing*. https://doi.org/10.4324/9780203094921-32

Bhasin, M. L. (2016). Creative accounting scam at satyam computer limited: How the fraud story unfolded? In *Open journal of accounting* (pp. 338–347). https://doi.org/10.4236/ojacct.2016.54007

Biddle, G. C., Hilary, G., & Verdi, R. S. (2009). How does financial reporting quality relate to investment efficiency? *Journal of Accounting and Economics, 48*(2–3). https://doi.org/10.1016/j.jacceco.2009.09.001

Bijalwan, J. G., Bijalwan, A., & Amare, L. (2019). An exploratory analysis of corporate governance using supervised data mining learning. *International Journal of Recent Technology and Engineering, 8*(3), 3546–3557. https://doi.org/10.35940/ijrte.C5279.098319

Bloomberg. (2018). *Timeline to the discovery of India's biggest bank fraud*. https://economictimes.indiatimes.com/markets/stocks/news/timeline-to-the-discovery-of-indias-biggest-bank-fraud/articleshow/63012827.cms?from=mdr

Boylan, D. (2015). A review of the effects of sarbanes-oxley on stock price. *Global Business and Finance Review, 20*(1), 121–126. https://doi.org/10.17549/gbfr.2015.20.1.121

Chitravanshi, R. (2024, June 27). *NFRA second inspection of Big five firms shows signs of improvement*. https://www.business-standard.com/economy/news/nfra-second-inspection-of-big-five-firms-shows-signs-of-improvement-124062701369_1.html

Cho, C.-C., & Wu, C.-H. (2014). Role of auditor in agency conflict and corporate governance: Empirical analyses of Taiwanese firms. *Chinese Management Studies, 8*(3), 333–353. https://doi.org/10.1108/CMS-09-2012-0126

Church, B. K., & Shefchik, L. B. (2011). PCAOB inspections and large accounting firms. *Accounting Horizons, 26*(1), 43–63. https://doi.org/10.2308/acch-50077

Cohen, J., Krishnamoorthy, G., & Wright, A. (2010). Corporate governance in the post-Sarbanes-Oxley era: Auditors' experiences. *Contemporary Accounting Research, 27*(3), 751–786. https://doi.org/10.1111/j.1911-3846.2010.01026.x

Confederation of Indian Industry. (1998). *Desirable corporate governance: A code.* https://www.ecgi.global/sites/default/files/codes/documents/desirable_corporate_governance240902.pdf

Confederation of Indian Industry. (2009). *Corporate governance—Recommendations for voluntary adoption.* https://cii.in/WebCMS/Upload/CorporateGovernanceReport.pdf

Dash, S. R., & Raithatha, M. (2019). Corporate governance and firm performance relationship: Implications for risk-adjusted return behavior. *Managerial and Decision Economics, 40*(8), 923–940. https://doi.org/10.1002/mde.3080

DeFond, M., & Zhang, J. (2014). A review of archival auditing research. *Journal of Accounting and Economics, 58*(2–3). https://doi.org/10.1016/j.jacceco.2014.09.002

Dharmasiri, P., Phang, S., Prasad, A., & Webster, J. (2021). Consequences of ethical and audit violations: Evidence from the PCAOB settled disciplinary orders. *Journal of Business Ethics.* https://doi.org/10.1007/s10551-021-04786-4

Elshandidy, T., Eldaly, M. K., & Abdel-Kader, M. (2021). Independent oversight of the auditing profession: A review of the literature. *International Journal of Auditing, 25*(2), 373–407. https://doi.org/10.1111/ijau.12224

Franzel, J. M. (2014). A Decade after Sarbanes-Oxley: The need for ongoing vigilance, monitoring, and research. *Accounting Horizons, 28*(4), 917–930. https://doi.org/10.2308/acch-50868

Friedrichs, D. (2017). Enron et al.: Paradigmatic white collar crime cases for the new century. In *Transnational financial crime* (pp. 59–78). https://doi.org/10.4324/9781315084572

Ganesh, M. (2015). Corporate governance: A must in every Indian company. *International Journal of Applied Engineering Research, 10*(2), 3417–3424.

Gipper, B., & Leuz, C. (2015). Public audit oversight and reporting credibility: Evidence from the PCAOB inspection regime. In *SSRN electronic journal.* https://doi.org/10.2139/ssrn.2641211

Han, S., Kang, T., & Rees, L. (2013). The association between institutional ownership and audit properties. *Asia-Pacific Journal of Accounting and Economics, 20*(2), 199–222. https://doi.org/10.1080/16081625.2012.748449

Hay, D., Rainsbury, E., & Van Dyk, D. (2023). The effect of audit inspections on audit fees. *Pacific Accounting Review, 35*(5), 701–726. https://doi.org/10.1108/PAR-02-2023-0017

Hubais, A. S. A., Kadir, M. R. A., Bilal, Z. O., & Alam, M. N. (2023). The impact of auditor integrity to audit quality: An exploratory studies from the Middle East. *International Journal of Professional Business Review, 8*(1). https://doi.org/10.26668/businessreview/2023.v8i1.1254

Hundal, S., Dalwai, T., Mohammadi, S. S., Al Balushi, I. A. K., & Al Zawani, H. R. H. (2024). Financial reporting quality and investment efficiency: Evidence from emerging market. In *Accounting, finance, sustainability, governance and fraud: Vol. Part F2403.* https://doi.org/10.1007/978-981-99-9145-7_12

Iren, P., & Kim, M. S. (2023). How harsh should the legislation be to prevent financial crimes? Lessons after the Enron scandal. In *Concepts and cases of illicit finance.* https://doi.org/10.4018/978-1-6684-8587-3.ch003

Ishak, N. A., Nazri, S. N. F. S. M., Yusoff, N. M., & Mohamed, M. (2017). The influence of audit fee, audit tenure and growth on audit quality: Malaysian ace market. *Advanced Science Letters, 23*(8). https://doi.org/10.1166/asl.2017.9480

Jha, A., & Bhatia, A. (2024). *Taxmann's auditing.* Taxmann Publications Private Limited.

Jha, A., & Bhatia, A. (2024a). *Taxmann's corporate governance.* Taxmann Publications Private Limited.

Kandukuri, R. L., Memdani, L., & Babu, P. R. (2015). Effect of corporate governance on firm performance—A study of selected Indian listed companies. In *Research in finance* (Vol. 31). https://doi.org/10.1108/S0196-382120150000031010

Kaur, H. (2021). Developments in corporate governance in India. In *Facets of corporate governance and corporate social responsibility in India* (pp. ix–xv). Springer. https://doi.org/10.1007/978-981-33-4076-3

Kaushik, M. (2024, November 25). NFRA recommends audit standards for LLPs. *Financial Express.* https://www.financialexpress.com/business/industry/nfra-recommends-audit-standards-for-llpsnbsp/3677072/

Khan, M. T., Al-Jabri, Q. M., & Saif, N. (2021). Dynamic relationship between corporate board structure and firm performance: Evidence from Malaysia. *International Journal of Finance and Economics, 26*(1), 644–661. https://doi.org/10.1002/ijfe.1808

Kumar, A. (2018). Auditing the auditors: An Indian perspective. *International Journal of Accounting and Financial Reporting.* https://doi.org/10.5296/ijafr.v8i4.13920

Lassoued, K., & Bacha, I. (2015). The impact of quality audit on the management of earnings. *Corporate Ownership and Control, 12*(3CONT1). https://doi.org/10.22495/cocv12i3c1p3

Lee, S., & Chung, K. (2017). Financial reporting quality and amplification of financial crisis: Evidence from Korea. *International Journal of Economic Research, 14*(13), 115–125.

Lennox, C., & Pittman, J. (2010). Auditing the auditors: Evidence on the recent reforms to the external monitoring of audit firms. *Journal of Accounting and Economics, 49*(1–2), 84–103. https://doi.org/10.1016/j.jacceco.2009.04.002

Lin, Z. J., & Liu, M. (2009). The impact of corporate governance on auditor choice: Evidence from China. *Journal of International Accounting, Auditing and Taxation, 18*(1), 44–59. https://doi.org/10.1016/j.intaccaudtax.2008.12.005

Loehlein, L. (2017). Measuring the independence of audit oversight entities: A comparative empirical analysis. *Accounting Research Journal, 30*(2), 165–184. https://doi.org/10.1108/ARJ-05-2015-0071

Lok Sabha Secretariat. (2010). *Twenty-first report, standing committee on finance.* Retrived from June 15, 2024, from https://eparlib.nic.in/bitstream/123456789/64114/1/15_Finance_21.pdf

Louw, A., & Maroun, W. (2017). Independent monitoring and review functions in a financial reporting context. *Meditari Accountancy Research, 25*(2), 268–290. https://doi.org/10.1108/MEDAR-02-2017-0114

Chao, M., Abdullah, D. F., & Idris, N. (2020). Investigating the relationship between intellectual capital and audit quality based on China audit market. *Journal of Economic Info, 7*(1), 65–75. https://doi.org/10.31580/jei.v7i1.1352

Ministry of Corporate Affairs. (2002). *Naresh Chandra committee report on corporate audit and corporate governance.* Retrived from June 3, 2024, from https://dea.gov.in/sites/default/files/chandra.pdf

Ministry of Corporate Affairs. (2005). *Report of the expert committee on company law.* Retrived from June 1, 2024, from https://ibbi.gov.in/uploads/resources/May%202005,%20J.%20J.%20Irani%20Report%20of%20the%20Expert%20Committee%20on%20Company%20Law.pdf

Ministry of Corporate Affairs. (2008). *The companies bill, 2008.* Retrived from June 2, 2024, from https://www.mca.gov.in/Ministry/actsbills/pdf/Companies_Bill_2008_27oct2008.pdf

Ministry of Corporate Affairs. (2011). *The companies bill, 2011.* Retrived from June 7, 2024, from https://www.mca.gov.in/Ministry/pdf/The_Companies_Bill_2011.pdf

Ministry of Corporate Affairs. (2016a). *Report of the companies law committee, February 2016.* Retrived from June 7, 2024, from https://www.mca.gov.in/Ministry/pdf/Report_Companies_Law_Committee_01022016.pdf

Ministry of Corporate Affairs. (2016b). *The Companies (Amendment) bill, 2016.* Retrived from June 7, 2024, from https://www.mca.gov.in/Ministry/pdf/Company_AmendentBill_2016.pdf

Ministry of Corporate Affairs. (2016c). *Thirty-seventh report of standing committee on finance on companies (Amendment) Bill, 2016.* Retrived from June 9, 2024, from https://prsindia.org/files/bills_acts/bills_parliament/2016/SCR-%20Companies%20(A)%20Bill,%202016.pdf

Ministry of Corporate Affairs. (2017). *Standing committee on finance, thirty-seventh report*. Retrived from June 9, 2024, from https://prsindia.org/files/bills_acts/bills_parliament/2016/SCR-%20Companies%20(A)%20Bill,%202016.pdf

Ministry of Corporate Affairs. (2018a). *Ministry of corporate affairs notification dated 1st October, 2018*. Retrived from June 5, 2024, from https://cdnbbsr.s3waas.gov.in/s3e2ad76f2326fbc6b56a45a56c59fafdb/uploads/2023/01/2023010598.pdf

Ministry of Corporate Affairs. (2018b). *Rajya Sabha starred question No. 389*. Retrived from June 14, 2024, from https://www.mca.gov.in/Ministry/pdf/rs389_04042018.pdf

Ministry of Corporate Affairs. (2018c). *The report of committee of experts on regulating audit firms and the networks*. Retrived from June 10, 2024, from https://www.mca.gov.in/Ministry/pdf/2018_CommitteeExperts_Report_08112018.pdf

Montesdeoca, M. R., Medina, A. J. S., & Santana, F. B. (2019). Research topics in accounting fraud in the 21st century: A state of the art. *Sustainability (Switzerland), 11*(6). https://doi.org/10.3390/su11061570

Moroke, N. D., & Makatjane, K. (2022). Predictive modelling for financial fraud detection using data analytics: A gradient-boosting decision tree. In *Applications of machine learning and deep learning for privacy and cybersecurity*. https://doi.org/10.4018/978-1-7998-9430-8.ch002

Narayanaswamy, R., Raghunandan, K., & Rama, D. V. (2021). Economic theory and social pressure: Audit committee director departures in India. *Managerial Auditing Journal, 36*(8), 1053–1067. https://doi.org/10.1108/MAJ-03-2021-3081

Nasrudin, W. A. W., Mohamed, N., & Shafie, N. A. (2017). Corporate governance &Amp; auditor choice in Malaysia. In *SHS Web of Conferences*. https://doi.org/10.1051/shsconf/20173600007

National Financial Reporting Authority Rules. (2018). Retrived from June 8, 2024, from https://cdnbbsr.s3waas.gov.in/s3e2ad76f2326fbc6b56a45a56c59fafdb/uploads/2023/01/2023010520.pdf

National Financial Reporting Authority. (2021). *Enhancing engagement with stakeholders. Report of technical advisory committee (March 2021). Government of India, ministry of corporate affairs*. (Issue March). Retrived from June 8, 2024, from https://cdnbbsr.s3waas.gov.in/s3e2ad76f2326fbc6b56a45a56c59fafdb/uploads/2023/02/2023020735.pdf

National Financial Reporting Authority. (2023). *National financial reporting authority, annual report 2022–2023*. Retrived from June 9, 2024, from https://cdnbbsr.s3waas.gov.in/s3e2ad76f2326fbc6b56a45a56c59fafdb/uploads/2024/01/202401051901137882.pdf

National Financial Reporting Authority. (2024). *National financial reporting authority, annual report 2023–2024*. https://cdnbbsr.s3waas.gov.in/s3e2ad76f2326fbc6b56a45a56c59fafdb/uploads/2025/02/202502131697810144.pdf

Ndreca, P., & Dibra, R. (2023). Correct relationships between audit and corporate governance provide economic effectiveness and added value to the organization. *Quality—Access to Success, 24*(197), 35–47. https://doi.org/10.47750/QAS/24.197.05

Nguyen, L. T. M., Le, D., Tran, T., & Dang, T. (2021). Financial reporting quality, corporate governance, and idiosyncratic risk: Evidence from a frontier market. *Australasian Accounting, Business and Finance Journal, 15*(4), 28–46. https://doi.org/10.14453/aabfj.v15i4.3

Nilsson, O. (2018). The relationship between shareholder protection through regulation and the demand for external auditor services. *International Journal of Disclosure and Governance, 15*(3), 162–175. https://doi.org/10.1057/s41310-018-0044-0

Nyambuya, K., Yasseen, Y., & Soni, F. (2021). Barriers to the adoption of independent reviews by nonpublic interest entities in South Africa. *African Journal of Business and Economic Research, 16*(2), 185–206. https://doi.org/10.31920/1750-4562/2021/V16N2A9

Pattanayak, B. (2024). *NFRA changes tack, ramps up focus on supervision to boost audit ecosystem: Chairman AB Pandey*. Retrived from June 12, 2024, from https://economictimes.indiatimes.com/news/economy/policy/nfra-changes tack-ramps-up-focus-on-supervision-to-boost-audit-ecosystem-chairman-ab-pandey/articleshow/108238329.cms

Peer Review Board. (n.d.). *Peer review board*. https://icai.org/post/peer-review-board

Prasad, G. C. (2024). *NFRA gets thumbs up from SC on retrospective jurisdiction.* Retrived from June 13, 2024, from https://www.livemint.com/economy/nfra-gets-thumbs-up-from-sc-on-retrospective-jurisdiction-11716723358353.html

Press Trust of India. (2023). NFRA imposes fine, bans auditors for misconduct in audit of DHFL branches. *Business Standard.* Retrived from June 17, 2024, from https://www.business-standard.com/companies/news/nfra-imposes-fine-bans-auditors-for-misconduct-in-audit-of-dhfl-branches-123041500613_1.htm

Press Trust of India & Business Standard. (2024, October 17). *It's collective responsibility to improve corporate governance: NFRA Chief*. www.business-standard.com. https://www.business-standard.com/industry/news/it-s-collective-responsibility-to-improve-corporate-governance-nfra-chief-124101700712_1.html

Quality Review Board. (n.d.). *Quality review board.* https://www.qrbca.in/

Ramesh, V., & Ramanathan. (2020). Auditing the auditors: A comparative study about India, UK, and the USA. *Materials Today: Proceedings, 37*(Part 2). https://doi.org/10.1016/j.matpr.2020.08.304

Robin, A. J., & Zhang, H. (2015). Do industry-specialist auditors influence stock price crash risk? *Auditing, 34*(3), 47–79. https://doi.org/10.2308/ajpt-50950

Romano, R. (2005). The Sarbanes-Oxley act and the making of quack corporate governance. *Yale Law Journal, 114*(7), 1521–1611.

Roy, M. N. (2015). Statutory auditors' independence in corporate accounting scandals: A case study of Satyam Computer Services Ltd. *Prabandhan: Indian Journal of Management, 8*(2). https://doi.org/10.17010/pijom/2015/v8i2/61399

Saleem, F., Alifiah, M. N., & Tahir, M. S. (2016). The effectiveness of monitoring mechanisms for constraining earnings management: A literature survey for a conceptual framework. *International Journal of Economics and Financial Issues, 6*(3), 209–214.

Santhosh Kumar, S. (2014). Cosy auditor-management relationships and corporate frauds in india. *Prabandhan: Indian Journal of Management, 7*(3), 35–42. https://doi.org/10.17010/pijom/2014/v7i3/59288

Scrimgeour, F., & Duppati, G. (2014). Corporate governance in the public sector: Dimensions; guidelines and practice in India and New Zealand. *Corporate Ownership and Control, 11*(2 E), 364–377. https://doi.org/10.22495/cocv11i2c4p2

Securities and Exchange Board of India. (2003). *The report of Shri N R Narayana Murthy committee on corporate governance.* Retrived from June 19, 2024, from https://www.sebi.gov.in/reports/reports/mar-2003/the-report-of-shri-n-r-narayana-murthy-committee-on-corporate-governance-for-public-comments-_12986.html

Serious Fraud Investigation Office. (n.d.). *Serious fraud investigation office.* https://sfio.gov.in/en/

Shah, A., & Etcfo. (2024, November 13). *India to overhaul auditing standards by 2026, NFRA proposes key revisions amid disagreements with ICAI over. ETCFO.com.* https://cfo.economictimes.indiatimes.com/news/tax-legal-accounting/india-to-overhaul-auditing-standards-by-2026-nfra-proposes-key-revisions-amid-disagreements-with-icai-over-group-audits-and-sqm/115239527

Sheu, R. J. R. J. (2018). An investigation about origins: A brief history of the PCAOB'S regulatory model. *Research in Accounting Regulation, 30*(2), 159–165. https://doi.org/10.1016/j.racreg.2018.09.011

Singh, V. K. (2021). Corporate governance failures as a cause of increasing corporate frauds in India—An analysis. In *Accounting, finance, sustainability, governance and fraud.* https://doi.org/10.1007/978-981-33-4076-3_2

Soltani, B. (2014). The anatomy of corporate fraud: A comparative analysis of high profile American and European corporate scandals. *Journal of Business Ethics, 120*(2). https://doi.org/10.1007/s10551-013-1660-z

Srivats, K. R. (2021). *Strengthen operation of NFRA, recommends Govt-appointed panel.* Retrived from June 19, 2024, from https://www.thehindubusinessline.com/economy/strengthen-operation-of-nfra-recommends-govt-appointed-panel/article25378805.ece

Subedi, M. (2023). Independent audit matters: Mitigation of auditors' independence issues and biases. *EDPACS, 67*(2), 15–23. https://doi.org/10.1080/07366981.2022.2125538

Synn, C., & Williams, C. D. (2024). Financial reporting quality and optimal capital structure. *Journal of Business Finance and Accounting, 51*(5–6), 885–910. https://doi.org/10.1111/jbfa.12697

Ţager, L., Tušek, B., & Mališ, S. S. (2014). External auditing in the function of managing financial reporting risks. In *Risk management: Strategies for economic development and challenges in the financial system.*

The Committee on the Financial Aspects of Corporate Governance. (1992). Report of The committee on the financial aspects of corporate governance. *Gee* (a division of Professional Publishing Ltd). Retrived from June 15, 2024, from https://www.ecgi.global/sites/default/files/codes/documents/cadbury.pdf

The Companies Act. (2013). https://www.mca.gov.in/Ministry/pdf/CompaniesAct2013.pdf

Thoppan, J. J., Nathan, R. J., & Victor, V. (2021). Impact of improved corporate governance and regulations on earnings management practices—analysis of 7 industries from the Indian National Stock Exchange. *Journal of Risk and Financial Management, 14*(10). https://doi.org/10.3390/jrfm14100454

Uzma, S. H. (2023). International financial reporting standards convergence in the Indian context: Insights from practitioners. *Journal of Public Affairs, 23*(3). https://doi.org/10.1002/pa.2861

Vagadia, B. (2014). Corporate governance. In *Management for professionals* (Vol. Part F406). https://doi.org/10.1007/978-3-642-38589-6_5

Venkatraman, K., & Selvam, M. (2014). Impact of corporate governance practices and firm performance: An empirical study. *Indian Journal of Finance*, *8*(12), 34–46. https://doi.org/10.17010//2014/v8i12/71691

Westermann, K. D., Cohen, J., & Trompeter, G. (2019). PCAOB inspections: Public accounting firms on "Trial." *Contemporary Accounting Research, 36*(2), 694–731. https://doi.org/10.1111/1911-3846.12454

Chapter 3
Driving Value Creation: Mapping the Landscape of Corporate Governance in Emerging Economies

Devika Agrawal, Kishore Kumar, Animesh Singh, and Bikram Paul Singh Lehri

Abstract In the past decade, the vital role of a robust Corporate Governance (CG) mechanism in fostering value creation in emerging economies has only witnessed an upward trend. Strong corporate governance becomes pertinent for sustainable growth and building a competitive edge as these developing economies undergo major transitions as the global paradigms shift. The book chapter delves into the intricacies of corporate governance in emerging economies to unveil its profound impact on value creation for a comprehensive of stakeholders, customers, employees, and society. The data is gathered from the SCOPUS database using a relevant and logical search query. The collected data is analyzed through the R-studio package (Biblioshiny) and *VOS*viewer to conduct a bibliometric analysis during a time frame from 2014 to 2024. The term 'Corporate Social Responsibility' emerged as the dominating keyword. The growth of literature on corporate governance identified three major clusters: 'Corporate Social Responsibility and Economic Impact', 'Environmental Impact and Investments', and 'Methodological Approaches in Emerging Economies'. The study presented underlying themes and patterns, emerging areas of research, and implications for policymakers and organizations to bridge the gap from theoretical knowledge to a practice-based approach.

D. Agrawal (✉)
Institute of Business Management, GLA University, Mathura, India
e-mail: devikaagrawal.phd@gmail.com

K. Kumar
School of Management, Bennett University, Greater Noida, India
e-mail: akishore001@gmail.com

A. Singh · B. P. S. Lehri
University School of Business, Chandigarh University, Punjab, India
e-mail: animesh.researcher02@gmail.com

B. P. S. Lehri
e-mail: lehribikram@gmail.com

V. Kandpal et al. (eds.), *Developments in Corporate Governance*, Approaches to Global Sustainability, Markets, and Governance,
https://doi.org/10.1007/978-981-96-6366-8_3

Keywords Corporate governance · Corporate sustainability · ESG · Sustainable development · Value creation

3.1 Introduction

In the contemporary global landscape, the importance of corporate governance has never been more evident, particularly in emerging economies where there is a huge potential for driving growth by tapping the unutilized and underutilized resources for sustainable value creation. The topic of our study 'Driving Value Creation: Mapping the Landscape of Corporate Governance in Emerging Economies' captures the vital role played by robust corporate governance in steering the sustainable growth and long-term success of an organization as well as any economy. As these emerging economies are in their transition phase, they need to integrate the framework and practices that govern the corporate entities to combat the challenges the business environment poses in front of them (Javed et al., 2020; Prakash et al., 2023). Emerging economies such as China, India, Thailand, South Africa, Malaysia, etc., represent a very diverse market that is characterized by huge growth potential and crucial investment opportunities for investors (Driss et al., 2024). These markets are often found in the frontline to combat unique challenges such as political instability in the country, regulatory complexities, and the effects of changes in the global economy (Christensen, 2022). However, on the positive side, these emerging economies offer immense business opportunities, incorporating innovation and further value creation. Effective and robust corporate governance is not just paperwork or regulatory compliance but rather a strategic imperative that can enhance companies' financial as well as non-financial performance, boost investor confidence, and contribute to broader economic goals of society (Vishwanathan et al., 2020).

Corporate governance is simply understanding that it is a system of practices, rules, and processes through which a company is directed and controlled. Corporate governance encompasses various components such as shareholders' rights, corporate social responsibility, transparency, accountability, and board structure (Kumar et al., 2022). In emerging economies, the corporate governance landscape is crafted by the congruency of various factors that include cultural norms, market conditions, and nations' regulatory frameworks aligned with international standards (Driss et al., 2024; Kumar & Prakash, 2019). Nonetheless, understanding this landscape is crucial to identifying the competitive edge, strategic intent, and best practices that can be tailored as per the present scenario to maximize the value creation. One of the cornerstones of corporate governance regarding emerging economies is the function and the proposed structure of the board of directors. A well-designed and well-crafted board personifying diversity expert is an independent can immensely influence a company's strategic intent and risk management abilities (Cormier et al., 2024). Board diversity refers to the constitution of the board, the composition of the board, and the demographics of the board, such as gender, experience, and skills. It is becoming immensely recognized as an advocate for better decision-making. Additionally, the independence of the board members is important for setting up accountability and having a clear understanding of the objective of fostering credibility and trust in the minds of stakeholders (Fernández-Guadaño & Sarria-Pedroza, 2018).

Corporate governance is successful only when it is accompanied by two most important components: transparency and accountability. In emerging economies, asymmetry of information and lack of standardization can pose certain challenges. Therefore, internal controls and strict disclosure practices are very vital to ensure information symmetry to disseminate the information among the stakeholders on time (Cui et al., 2018). Bringing in transparency enhances the company's reputation, mitigates the risk that a business environment puts forth, and attracts investors to invest in the company. Accountability mechanisms such as strong audit committees and external audits prevent malpractices and provide checks and balances to promote ethical behavior (Christensen, 2022). Another important aspect of corporate governance is the integration of environmental, social, and governance (ESG) factors, which are emerging as a trend to hold significant frameworks to steer the value creation in emerging economies (Agrawal, 2024). ESG considerations that address issues such as social responsibility, environmental sustainability, and corporate governance are increasingly becoming important in framing corporate strategy. Companies are taking a proactive approach to incorporate ESG principles into their business processes and governance practices to achieve long-term benefits such as enhanced brand loyalty, easy access to capital, and operational as well as financial efficiency (Khandelwal et al., 2023).The congruency of global sustainability trends with the framework of emerging economies positions these companies as leaders, driving the transition to a more sustainable economy.

This chapter illustrates how companies are proactively integrating ESG principles into their business models to assure long-term financial value creation along with alignment of global sustainability norms. Engaging stakeholders is another important component and point of discussion in this chapter in emerging economies where the expectations of shareholders are ever-changing. Companies that are effectively communicating with the stakeholders and addressing their concerns are in a better position to create sustainable value for themselves (Agrawal et al., 2024a, 2024b). This chapter will further assess the various strategies for effective stakeholder engagement, which include community involvement, global partnership with local entities, and community involvement. To provide an in-depth understanding of this chapter, we present a comparative analysis of corporate governance practices in various emerging markets and highlight common challenges and innovative solutions, offering valuable insights for policymakers, business leaders, and scholars.

The landscape of corporate governance in emerging economies is multifaceted and presents both challenges and opportunities for sustainable value creation. As the framework of corporate governance in emerging economies carries some complexities, it can only be minimized by adopting best practices such as ESG integration, enhancing transparency, setting up accountability, and robust governance (Kumar et al., 2021). As emerging economies are striving to evolve and justify their presence in the global landscape, the role played by corporate governance in shaping their growth trajectories ensures long-term benefits are widely acknowledged. Furthermore, the chapter aims to provide an in-depth understanding of the mechanism of corporate governance and presents the mapping of this landscape, offering various insights for policymakers to drive value creation through effective corporate governance, specifically in the case of emerging economies. Few studies in the past have explored the relationship between corporate governance and value creation, especially in high-impact factor journals. Secondly, this topic has been studied in relation to other factors such as corporate social responsibility (CSR), sustainable development goals (SDGs), and environmental social and governance (ESG). This bibliometric study provides an in-depth overview of the present literature through the usage of various techniques such as keywords co-occurrence, publication trend, tree plot, and cluster analysis. This will unlock new doors for future research by providing a deeper understanding of the trending topics. We will be delving into the various components of the bibliometric process, such as authors with the most publications, authors with the highest citations, most impactful journals, collaboration networks, and author and index terms used.

RQ1: What are the present status and trends regarding publications, citations, and journals that probe into 'Corporate Governance' and 'Value Creation' in emerging economies?

RQ2: What is the collaboration status on the 'Corporate Governance' and 'Value Creation' in emerging economies?

RQ3: What are the most significant articles published on 'Corporate Governance' and 'Value Creation' in emerging economies?

RQ4: Who are the leading authors from the emerging economies who published on the 'Corporate Governance' and 'Value Creation'?

RQ5: What are the major themes that underline the current literature on how 'Corporate Governance' impacts 'Value Creation'?

3.2 Literature Review

The concept of corporate governance has evolved to coincide with the demands of a changing business environment, changing stakeholders' expectations, and ever-increasing regulatory compliances. This chapter aims to unfold the contemporary trends in corporate governance with its emerging frameworks and how it is integrated with environmental, social, and governance (ESG) frameworks to increase the value. There has been heightened awareness among companies and investors that it is very crucial to ensure long-term sustainability in any business process if they are aiming for long-term value creation; there has been a transition in the corporate governance framework to prioritize the long-term value over short-term gains (Agrawal, 2024; Cormier et al., 2024).

The advent of technology and the digitalization of business operations have introduced new challenges and complexities for corporate governance. The company's board is now expected to hold a deeper understanding of the technology, implications of business strategy, and risk management abilities (Chen et al., 2024). The COVID-19 pandemic has also stressed the importance of effective risk management to make the business more resilient to outside changes. As a response to this paradigm shift, companies are now putting great effort into and emphasizing the identification and management of risks related to long-term value creation, supply chain disruptions, environmental sustainability, and climate change. There is a burgeoning recognition by the companies regarding their responsibilities toward their stakeholders, ranging from shareholders, employees, suppliers, customers, and the societies in which they are operating (Kumar & Prakash, 2020). The shift in the expectations from short-term profit to long-term gains led to the revisiting of the corporate governance framework to ensure that they are considering the interests of all the stakeholders. To respond to various corporate scams and governance failures, pressure from regulatory bodies has become more stringent in bolstering corporate governance practices. Corporations are now subjected to more scrutiny regarding their disclosure requirements and governance frameworks.

Corporations cannot overlook the integration of ESG principles in corporate governance mechanisms, and they acknowledge the importance of ESG issues in sustaining their long-term performance in their decision-making process (Sharma et al., 2025). Investors have never been more aware of their decisions, and they scan transparent and strategized ESG disclosures. Companies are now subjected to publishing their annual sustainability reports, also known as non-financial reporting, along with traditional financial reporting to disclose their ESG performance and various initiatives. To facilitate this reporting, various ESG metrics and frameworks have emerged in the past decade, such as the Sustainability Accounting Standard Board (SASB), Task Force on Climate-related Financial Disclosures (TCFD), and the most prevalent Global Reporting Initiatives (GRI) (Agrawal, 2024). The contemporary trend, known as impact investing, has further incentivized corporations to improve their ESG performance as investors are seeking to invest in those companies that are committed to ensuring long-term sustainability practices for positive societal and environmental outcomes.

Board diversity, which gives us a peek into the dynamics of the board, is gaining recognition as an important element of effective corporate governance. Corporations are proactively managing to diversify their boards in terms of age, ethnicity, gender, and professional background to incorporate a broader and wider range of skills and comprehensive perspectives (Abdelkader et al., 2024). Companies are going over and beyond their efforts to engage and communicate with their stakeholders to understand their concerns and implement their views into the decision-making process. This community engagement extends beyond the owners and includes other stakeholders such as customers, suppliers, employees, and the broader community. The United Nations Sustainable Development Goals (UN-SDGs) have also become a pioneer point for companies to align their business strategies with broader societal objectives (Agrawal et al., 2024a, 2024b; Kandpal, 2020). Many companies are incorporating ESG into their corporate governance frameworks as well as their non-financial reporting practices.

The landscape of corporate governance is experiencing major transformation as corporations are rapidly adapting themselves to the changing business environment. Emerging trends such as focusing on long-term business sustainability, adoption of ESG principles, carbon emissions, and climate change are emphasizing stakeholder engagement. These factors are individually and collectively shaping the future of corporate governance. Companies that are at the forefront of embracing their strengths are in a better position to combat the challenges of the environment and can create long-term value for all of their stakeholders (Agnese et al., 2024). This interdependence and overlapping call for an extensive literature review. This book chapter will contribute to the present literature on corporate governance and unveil its impact on steering value creation.

3.3 Data and Methodology

In this chapter, we used bibliometric analysis after fetching the data from the Scopus database. The entire process is known as the bibliographic process, which starts from the collection of data from a relevant database and is then organized to facilitate the systematic analysis of the data. It consists of several steps, beginning with outlining the entire scope, recognizing the relevant keywords, and creating a valid search query. After collecting all the data from various sources, it is filtered at different levels to justify the quality, importance, and logic of the chosen topics. This multi-stage filtering helps the researcher to identify emerging trends and quantify descriptive information to structure the entire research study and uncover the underlying themes. This process further presents a base for future researchers to analyze the research gaps and future research direction to carry out potential research. As the previous studies have elaborated the role of ESG performance in value creation, it became necessary to study the role of corporate governance, a precursor to Corporate social responsibility and ESG, in driving the value creation. Despite all the merits of bibliometric studies, this technique poses a certain limitation as the database influences the results, and the thresholds and algorithms drive outcomes. A multi-database study would fetch some insightful results, and this relationship could also be tested empirically.

3.3.1 Data Collection

Our bibliographic study undertook a 5-stage review process for conducting a comprehensive bibliometric analysis of corporate governance and its relationship with value creation. Our approach is very important in providing a base for guiding the entire review process. For this purpose, we have chosen the Scopus database as it is the world's largest citation and abstract, developed and managed by Elsevier. Scopus is widely recognized for its comprehensive coverage of peer-reviewed research articles, book chapters, and journals. The rationale behind choosing the Scopus database over the other databases is its broader inclusion, particularly in business, management, and sustainability. Table 3.1 shows the five-stage Bibliographic Process and Results.

- The first stage involved deciding the logical keywords for starting the bibliography search within the Scopus database. The chosen keywords were 'corporate governance' and 'value creation'. We have chosen both 'stakeholder's value OR shareholder value' to give this bibliography process a comprehensive overview. These key terms together provided us with 13,991 documents.
- To narrow down our search results, we have further put in three filters: SUBJECT AREA (business management and accounting and economics), ARTICLES, TIMELINE (2014–2024),
- The third stage has further refined our search queries, and we have only chosen the articles that were in their 'final stage' of publication.
- In the fourth and second-to-last stage of our bibliographic process, we have limited our source type to journals to avoid conference proceedings and book chapters and limited our third queries to only the 'English' language.
- The last stage is the most crucial. In this particular stage, we have refined our search results by including only emerging economies in our search queries. This means

Table 3.1 Five-stage bibliographic process and results

Stage	Search string on 'Scopus'	Output
Stage 1	Corporate Governance AND ('Stakeholder Value' OR 'Shareholder Value')	13,911
Stage 2	Corporate Governance AND ('Stakeholder Value' OR 'Shareholder Value') AND PUBYEAR > 2013 AND PUBYEAR < 2025 AND (LIMIT-TO (SUBJAREA, 'BUSI') OR LIMIT-TO (SUBJAREA, 'ECON') AND (LIMIT-TO (DOCTYPE, 'ar')	7,017
Stage 3	Corporate Governance AND ('Stakeholder Value' OR 'Shareholder Value') AND PUBYEAR > 2013 AND PUBYEAR < 2025 AND (LIMIT-TO (SUBJAREA, 'BUSI') OR LIMIT-TO (SUBJAREA, 'ECON')) AND (LIMIT-TO (DOCTYPE, 'ar')) AND (LIMIT-TO (PUBSTAGE, 'final'))	6,619
Stage 4	Corporate Governance AND ('Stakeholder Value' OR 'Shareholder Value') AND PUBYEAR > 2013 AND PUBYEAR < 2025 AND (LIMIT-TO (SUBJAREA, 'BUSI') OR LIMIT-TO (SUBJAREA, 'ECON')) AND (LIMIT-TO (DOCTYPE, 'ar')) AND (LIMIT-TO (PUBSTAGE, 'final')) AND (LIMIT-TO (SRCTYPE, 'j')) AND (LIMIT-TO (LANGUAGE, 'English'))	6,478
Stage 5	*Search Refinement considering only Emerging Economies:*	2,324
	China, India, Malaysia, South Korea, Indonesia, Pakistan, United Arab Emirates (UAE), South Africa, South Korea, Vietnam, Saudi Arabia, Turkey, Brazil, Poland, Thailand, Egypt, Ghana, Iran, Bangladesh, Jordan, Russian Federation, Nigeria, Chile, Bahrain, Romania, Morocco, Qatar, Oman, Ukraine, Hungary, Columbia, Mexico, Philippines, Kazakhstan, Kenya, Ethiopia, Peru, Sri Lanka, Serbia, Argentina, Bulgaria, Azerbaijan, Uzbekistan, Uruguay, Ecuador, Cameroon, Bosnia and Herzegovina, Tanzania, Georgia, Zimbabwe, Trinidad and Tobago, Nepal, Kyrgyzstan, Haiti, Armenia, Algeria	

Table 3.2 Data structure

Description	Results
Main information about data	
Timespan	2014:2024
Sources (Journals, Books, etc.)	591
Documents	2324
Annual growth rate %	17.42
Document average age	3.41
Average citations per doc	19
Document contents	
Keywords plus (ID)	1403
Author's keywords (DE)	5186
AUTHORS	
Authors	4779
Authors of single-authored docs	198
Authors collaboration	
Single-authored docs	220
Co-Authors per Doc	3.06
International co-authorships %	45.09
Document types	
Article	2324

that we can only study research articles published or based on these countries, limiting our findings to developing nations.

Table 3.2 presents the breakdown of the entire search results into various important metrics, such as the period of a bibliometric study, the number of sources and documents, the annual growth rate, and the number of keywords used in the entire document.

3.3.2 *Data Analysis*

The data collected during the bibliographic process regarding various publications was further compiled into an Excel sheet to carry out the entire analysis. Bibliometric analysis served as a key tool in revealing the underlying patterns of research conducted on corporate governance and value creation. When this process was further combined with social network analysis, it provided various insights into the established networks and highlighted the major themes within the field of the study. Furthermore, bibliometric analysis facilitated the identification of the present status and potential research avenues on the chosen topic. This provided a strong motivation to conduct a comprehensive bibliometric review and for the revaluation of the published literature. Our study incorporates citations and co-citation analysis, keyword occurrence analysis, co-authorship patterns, and cluster analysis to address the research questions.

3.4 Results Analysis

3.4.1 Year Wise Publication

To respond to our RQ 1: What are the present status and trends regarding publications, citations, and journals that probe into 'Corporate Governance' and 'Value Creation' in emerging economies? Fig. 3.1 shows the current status and trends of publication on corporate governance and value creation within a time frame of 2014 to the first of July 2024 as the figure shows the upward trend of corporate governance throughout. The publications on corporate governance and value creation have never seen a drop in the entire last decade, and they are continuously growing. The concept gained significant momentum in 2020–2022 due to the outbreak of COVID-19 as firms have understood the importance of risk management. Companies are acknowledging the importance of adopting robust governance mechanisms and building sustainability practices to improve their performance. Researchers and scholars all over the emerging economies have shown a deep interesting the topics related to corporate governance and value creation. Investors are gaining knowledge and are keenly interested in knowing about how companies are managing their external risk and what strategies they are building in coping with the challenges that may arise in due course.

As in our bibliography study, we have primarily focused on emerging economies. Figure 3.2 shows country-wise publication; we have set the filter in the Scopus database to provide us with the data of emerging economies only. The figure shows the number of publications each country has on corporate governance and value creation from 2014 to 2024. As shown in the figure, China stands out among all the emerging economies, having more than 800 publications in the last decade. After China, we have Malaysia, but again, the difference between these two economies in terms of publication is over 500 publications, followed by South Korea, Indonesia, Pakistan, the United Arab Emirates, and South Africa. This figure gives us an insight into the untapped potential in emerging economies other than China to publish good quality impactful research on the topic related to corporate governance and its impact on value creation. As more and more scholars in these emerging economies publish empirical research papers, this would provide great insights to the policymakers and managers of the firms sustaining their businesses in these countries.

This segment of our studies examines the proportion of published articles across different reputed journals. We have filtered out the top 10 journals that have published

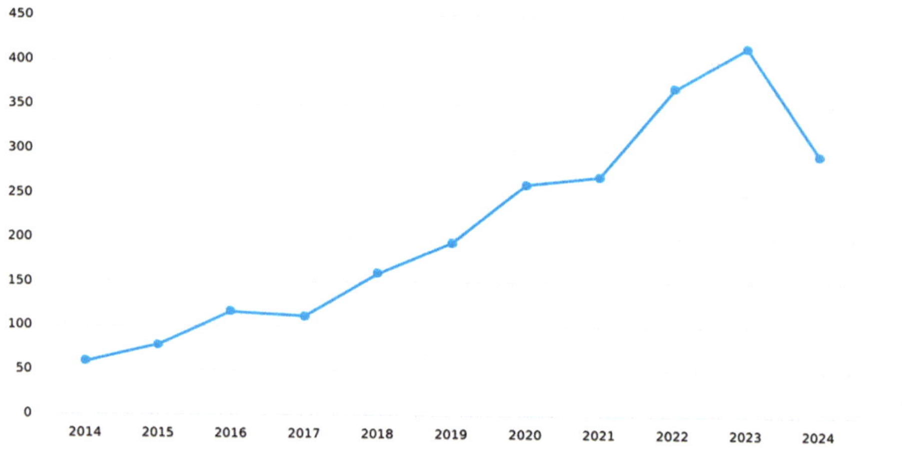

Fig. 3.1 Year-wise publication trend of corporate governance and value creation

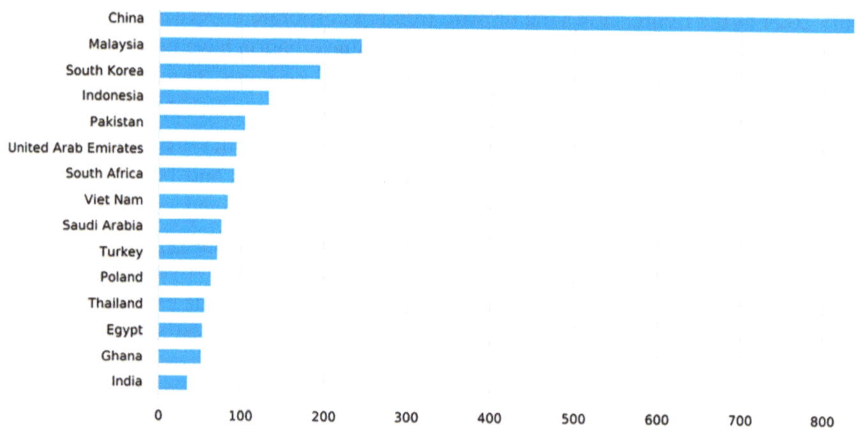

Fig. 3.2 Country-wise publication (emerging economies)

research articles on the chosen topic. Table 3.3 presents the allocation of the journals in descending order in terms of publication and the impact factor. This offers valuable insights and information to gauge which journals dominate in terms of proportion in the last 10 years. As shown above, our study takes into account 2,324 documents published within 591 sources. 'Corporate Social Responsibility and Environmental Management' is emerging as the top journal with the highest number of publications (65) carrying an impact factor of 8.3., subsequently 'Journal of Business Ethics', 'Cogent Business and Management', 'Journal of Cleaner Production', 'International Review of Financial Analysis' are within the list. These journals carry huge potential for publication in the field of corporate governance and value creation. This empowers the upcoming scholars and researchers to submit their manuscripts to these journals, as we have also mentioned the impact factor to ensure credibility.

To answer the RQ2: What is the collaboration status on 'Corporate Governance' and 'Value Creation' in emerging economies? We have put forth two figures, Figs. 3.3 and 3.4, to provide an in-depth understanding of the question. The study investigates and assesses the networking pattern among emerging economies, as shown in Fig. 3.3; we can see that China is emerging as the top emerging economy to have a strong network collaboration, followed by Malaysia, Pakistan, and Indonesia. Mexico and Chile are farther in the figure, showing that fewer collaboration efforts have been made within these emerging economies at the global level. A similar pattern can also be observed in Colombia and Ecuador. This provides opportunities for scholars and researchers in these emerging economies to come forward and establish a global network to publish in the field of corporate governance and value creation.

To examine the keyword co-occurrence and network build on author keywords in between leading authors from emerging economies has been shown in Fig. 3.4. We have used the VOSviewer software to generate a text mining-based map to examine the association between these key terms. The proximity between the keywords is directly related to the strength of the relationship which means the higher the proximity higher the strength. For example, terms such as corporate social responsibility, stakeholder theory, innovation, and ESG are closely related to each other showing great interrelatedness. The strength that any two keywords have can easily be decoded by the distance and thickness between the nodes and the linking line respectively. As also shown in the figure corporate social responsibility is taking the largest space, followed by 'Corporate Governance' showing proximity to 'Firm Value' and 'Firm Performance'.

To gauge more into the analysis of the keyword, we have used the R-studio package and Biblioshiny to delve more into the top 30 index terms that have been used in the papers related to corporate governance and value creation. As shown in Fig. 3.5,

Table 3.3 Top 10 journals publishing on corporate governance and value creation

Journal	Number of publications	Impact factor
Corporate Social Responsibility and Environmental Management	65	8.3
Journal of Business Ethics	64	5.9
Cogent Business and Management	49	3
Journal of Cleaner Production	41	9.7
International Review of Financial Analysis	40	8.23
Pacific Basin Finance Journal	40	4.8
Finance Research Letters	38	7.4
Corporate Governance (Bingley)	37	5.5
Business Strategy and The Environment	35	12.5
Journal of Business Research	33	10.5

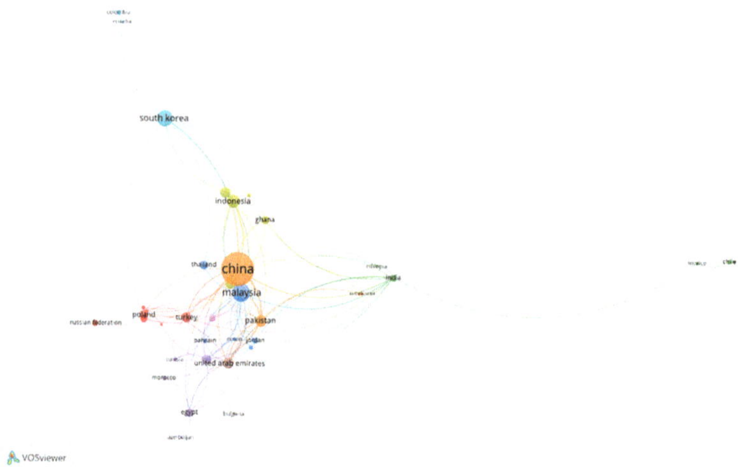

Fig. 3.3 Collaboration network of lead authors from emerging economies

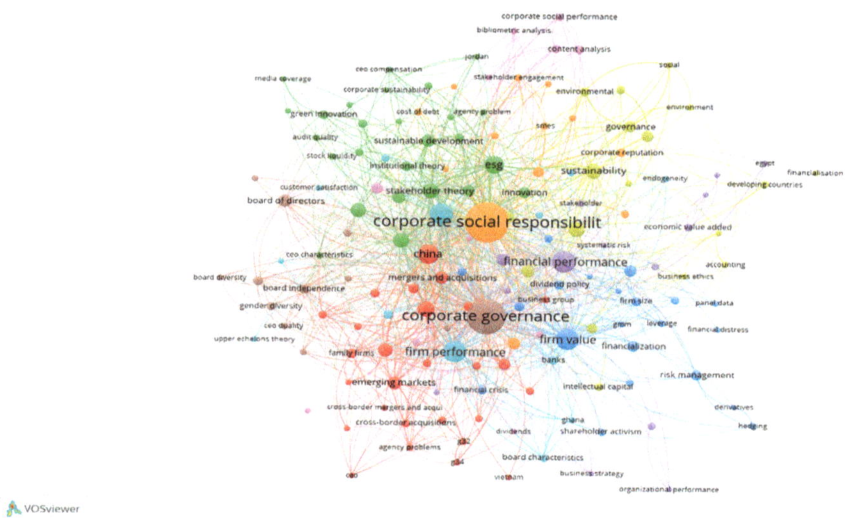

Fig. 3.4 Co-occurrence network (authors' keywords)

the keyword 'corporate social responsibility' has been used most frequently in the entire database with an occurrence of 401 times with a total 19% contribution while the term 'corporate governance' has been quoted 300 times with a 15% contribution, subsequently, firm value (113), financial performance (112), corporate social responsibility (105), and China (104) have also been used almost at the same level. By analyzing, this diagram, authors can find out the areas of focus and underline patterns to know about the research areas with the utmost attention.

Fig. 3.5 Leading 30 index terms (tree-diagram analysis)

3.4.2 Analysis of Network Citation

Research Question 3 aims to find out the most impactful articles published on corporate governance and value creation. To attend to this objective, we have examined the citation network of 2324 publications. The name of the published article is mentioned with the total citations received, as this technique is widely adopted to investigate the influence of any publication. Table 3.4 presents the top 10 most influential articles published on corporate governance and value creation across various disciplines and fields. The most frequently cited article 'A Meta-Analytic Review of Corporate Social Responsibility and Corporate Financial Performance: The Moderating Effect of Contextual Factors' generated a total of 440 citations, followed by 'Does Corporate Social Responsibility Affect Information Asymmetry?' with 397 citations authored by Cui J. Some other influential publications include 'Whose call to answer: Institutional complexity and firms' CSR reporting', 'Can political capital drive corporate green innovation? Lessons from China', Does corporate social responsibility (CSR) improve credit ratings? Evidence from geographic identification, 'Environmental, Social and Governance (ESG) Scores and Financial Performance of Multilatinas: Moderating Effects of Geographic International Diversification and Financial Slack'. This analysis will empower the readers and scholars to know about the prominence and metal of these articles within the realms of corporate governance and sustainability.

Table 3.4 Top 10 most influential articles published on corporate governance and value creation

Author	Year	Title	Source	TC	TCpY
Wang Q	2016	A meta-analytic review of corporate social responsibility and corporate financial performance: The moderating effect of contextual factors	Business and Society	440	48.89
Cui J	2018	Does corporate social responsibility affect information asymmetry?	Journal of Business Ethics	397	56.71
Luo X. R	2017	Whose call to answer: Institutional complexity and firms' CSR reporting	Academy of Management Journal	357	44.63
Lin H	2014	Can political capital drive corporate green innovation? Lessons from China	Journal of Cleaner Production	352	32.00
Jiraporn P	2014	Does corporate social responsibility (CSR) improve credit ratings? Evidence from geographic identification	Financial Management	346	31.45
Duque-Grisales E	2021	Environmental, social and governance (esg) scores and financial performance of multilatinas: Moderating effects of geographic international diversification and financial slack	Journal of Business Ethics	342	85.50
Chen T	2020	Institutional shareholders and corporate social responsibility	Journal of Financial Economics	326	65.20
Bennouri M	2018	Female board directorship and firm performance: What really matters?	Journal of Banking and Finance	281	40.14

(continued)

Table 3.4 (continued)

Author	Year	Title	Source	TC	TCpY
Husted B. W	2019	Board structure and environmental, social, and governance disclosure in Latin America	Journal of Business Research	277	46.17
Helfaya A	2017	Do board's corporate social responsibility strategy and orientation influence environmental sustainability disclosure? UK evidence	Business Strategy and the Environment	263	32.88

3.4.3 Leading Authors from Emerging Economies

To respond to our RQ4: Who are the leading authors from the emerging economies who published on 'Corporate Governance' and 'Value Creation'? Fig. 3.6 presents the leading 10 authors (in descending order) out of 4779 authors who have published on 'Corporate Governance and Value Creation in the journals of repute'. 'Jiraporn, P.' tops the list with 21 publications, and the second leading author 'Tsang, A.' published 13 articles. Additionally, Treepongkaruna, S. (13 publications), Maroun, W. and Salehi, M. (10 publications each), and Hussainey, K. (9 publications) also made their significant contribution to the literature.

A similar pattern is observed in Fig. 3.7 which has been drawn with the help of VOSviewer software, where the lead author Jiraporn, P. dominates the entire cluster, another notable author within this cluster is Treepongakaruna, S., Kyaw, K., and

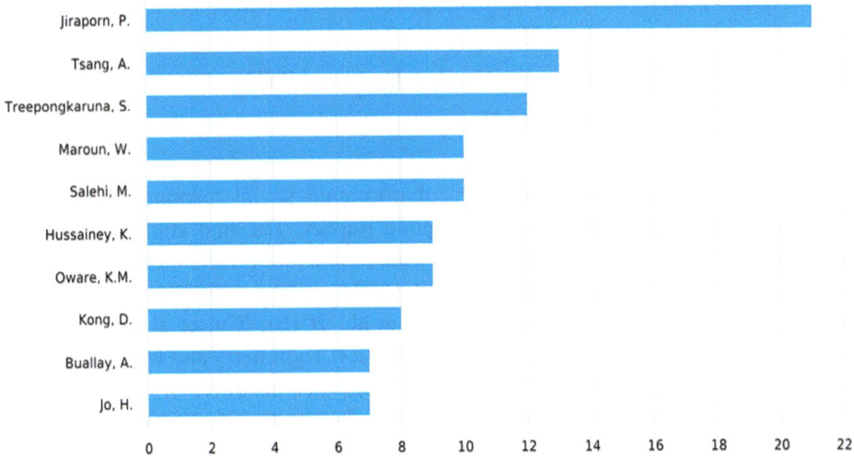

Fig. 3.6 Top 10 authors who have published on corporate governance and value creation

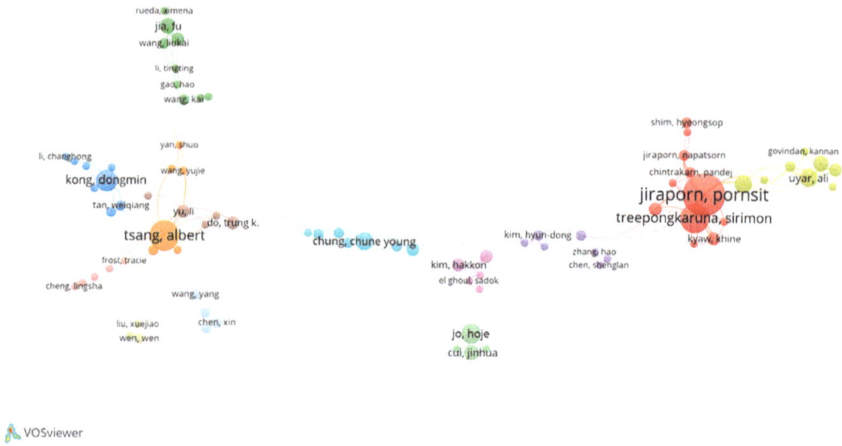

Fig. 3.7 Network sustained on top author' co-authorship

Chintrakarn, P. Another major cluster is dominated by our second lead author Tsang, A. accompanied by Wang, Y. and Yan, S. The scholars interested in this subject area can follow the work of these eminent researchers from emerging economies.

To respond to our RQ 5: What are the major themes that underline the current literature on how 'Corporate Governance' impacts 'Value Creation'?, Fig. 3.8 presents the graphical representation of the cluster formed based on the bibliographic coupling of the top 25 papers from our corpus, and Table 3.5 shows the data-related frequency, centrality, and impact of the cluster. Three major clusters have been identified from the selected papers: Corporate Social Responsibility and Economic Impact, Environmental Impact and Investments, and Methodological Approaches in Emerging Economies.

3.4.4 *Cluster 1: Corporate Social Responsibility and Economic Impact*

This cluster highlights the intricacies of corporate social responsibility and its emphasis on decision-making and economic impact. The first study reviews and quantifies the link between CSR and Corporate Financial Performance (CFP) using a meta-analytic framework and finds a positive and significant relationship advocating the instrumental stakeholder theory (Wang et al., 2016). (Cui et al., 2018) examined the relationship between CSR and information asymmetry and found an inverse association showing that CSR engagement reduces the information asymmetry and advocates the reputation-building aspect of stakeholder theory. A study undertaken to explore the impact of CSR on corporate reputation and financial performance

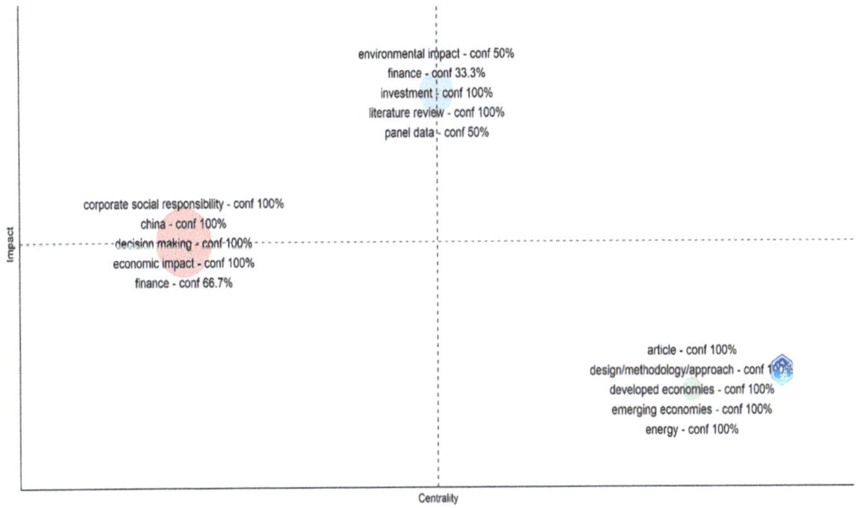

Fig. 3.8 Cluster formed to analyze centrality and impact

Table 3.5 Shows the occurrence of keywords within each cluster

Group	Cluster	% of occurrence	Freq.	Centrality	Impact
1	Corporate social responsibility and economic impact	Corporate social responsibility—conf 100% china—conf 100% decision-making—conf 100% economic impact—conf 100% finance—conf 66.7%	79	0.958	3.459
2	Environmental impact and investments	Environmental impact—conf 50% finance—conf 33.3% investment—conf 100% literature review—conf 100% panel data—conf 50%	16	0.996	4.818
3	Methodological approaches in emerging economies	Article—conf 100% design/methodology/approach—conf 100% developed economies—conf 100% emerging economies—conf 100% energy—conf 100%	5	1.679	1.522

in Pakistani firms shows a positive impact on both reputation and financial performance. However, excessive CSR can negatively affect performance (Javed et al., 2020). (Nekhili et al., 2017) reported that family firms report less CSR information but are in an advantageous position through CSR disclosure in terms of market value compared to non-family firms. Few other studies have examined the impact of CSR on financial performance, such as innovation and reputation, and on lowering the risk of financial distress (Boubaker et al., 2020; Vishwanathan et al., 2020). (Ren et al., 2024) highlighted the role of gender diversity in improving firm performance by promoting financial performance.

3.4.5 Cluster 2: Environmental Impact and Investments

This cluster focused on the integration of environmental considerations into investment decisions backed by empirical research. The first study in this cluster investigated the role played by political capital combined with stakeholder pressure in navigating the funds for facilitating green innovation in China (Lin et al., 2014). CSR policies play a pivotal role in enhancing the credit rating to boost the confidence of the shareholders to motivate them to undertake environmental initiatives. (El Ghoul et al., 2018) undertake a study by taking a multinational sample to show how the fulfillment of corporate environmental responsibility can reduce the cost of equity capital. Another study conducted in the manufacturing sector showed the inverse relationship between environmental responsibility and firm risk (Cai et al., 2016). Studies undertaken in most emerging economies in China show the linkage between ESG ratings in steering green innovation by tackling financial constraints and improving managerial environmental awareness. In another emerging economy in Latin America, (Husted & de Sousa-Filho, 2019) highlighted how board structure influences ESG disclosure concerning environmental sustainability, however, (Duque-Grisales & Aguilera-Caracuel, 2021) investigated and found a negative relationship between ESG scores and financial performance moderated by financial slack and international diversification. A study undertaken on Chinese non-financial listed firms analyzed the role of ESG disclosure in driving financial performance with a specific focus on ESG investors (Chen & Xie, 2022).

3.4.6 Cluster 3: Methodological Approaches in Emerging Economies

This cluster emphasizes the various methodological approaches that have been applied in various economic setups. The first study in this cluster analyzes CSR reporting more as an organizational response to address the institutional complexity in Chinese companies (Luo et al., 2017). The second study in this cluster has assessed how board structure and its demographics influence ESG disclosure in Latin America (Husted & de Sousa-Filho, 2019). (Jia et al., 2018) conducted a systematic literature review on sustainability supply chain management (SSCM) practices, especially in emerging economies, and identified its drivers, barriers, mechanisms, and results. The fourth study in this cluster integrates quasi-natural experiments to address the questions related to the impact of institutional shareholders on CSR-driven performance (T. Chen et al., 2020) investigated and offered a meta-analysis of family firms' internationalization that focuses on institutional context to ensure shareholder protection (Arregle et al., 2017). Another crucial study undertaken in major developed and developing economies (Eastern Europe, China, India, Poland, Russia, and Brazil) reviewed mergers and acquisition-related activities and factored in the various institutional and governance distinctions from developed economies (Lebedev et al., 2015).

3.5 Discussions and Conclusion

The growth trajectory of emerging economies is dynamic and full of challenges, presenting an opportunity to further examine the relationship between corporate governance and value creation in another timeline or from other databases. We conducted a bibliographic process on the Scopus database with 'Corporate Governance' and 'Value Creation' to gather a database of quality publications. In the data analysis of the collected corpus of the papers, we shed light on the major metrics such as most influential publications, citation scores, leading authors, major themes, and future research agenda. The chapter presented an overview of the corporate governance challenges faced by emerging economies vis-a-vis regulatory compliances, socio-political factors, and market volatility. Further, this chapter delved into the fundamental principles of corporate governance and how it enhances transparency, minimizes risks, and promotes responsible business conduct. One of the cornerstones of this chapter that we have explored is the specific corporate governance mechanism steering the value creation in emerging economies. Effective board governance is emerging to be the central theme of strategic decision-making and risk management which influences companies to innovate and foster its adaptability. Moreover, the crucial aspects of robust internal controls and audit efficiency in protecting the assets and promoting accountability have also been discussed. Integrating environmental, social and governance (ESG) factors into the CG Framework has emerged as an important tool in steering the value creation. ESG considerations are now prominent in devising corporate strategies for emerging markets where social and environmental challenges are more pertinent. Regulators can motivate corporations to enhance their corporate governance frameworks by aligning them with current trends such as ESG integration, carbon emissions, and climate change. Further, it can also be strengthened by the involvement of stakeholders and presenting a transparent and fair disclosure of companies' affairs.

References

Abdelkader, M. G., Gao, Y., & Elamer, A. A. (2024). Board gender diversity and ESG performance: The mediating role of temporal orientation in South Africa context. *Journal of Cleaner Production, 440*. Scopus. https://doi.org/10.1016/j.jclepro.2024.140728

Agnese, P., Cerciello, M., Oriani, R., & Taddeo, S. (2024). ESG controversies and profitability in the European banking sector. *Finance Research Letters, 61*. Scopus. https://doi.org/10.1016/j.frl.2024.105042

Agrawal, D. (2024). Corporate environmental, social and governance (ESG) practices. In *Reference module in social sciences*. Elsevier. https://doi.org/10.1016/B978-0-443-13701-3.00141-9

Agrawal, D., Kumar, A., & Kumar, K. (2024a). Ethical sourcing. In *Reference module in social sciences* (p. B9780443137013000165). Elsevier. https://doi.org/10.1016/B978-0-443-13701-3.00016-5

Agrawal, D., Kumar, K., & Rahman, O. (2024b). ESG investment strategy. In *Reference module in social sciences*. Elsevier. https://doi.org/10.1016/B978-0-443-13701-3.00300-5

Arregle, J., Duran, P., Hitt, M. A., & Van Essen, M. (2017). Why is family firms' internationalization unique? *A Meta-Analysis. Entrepreneurship Theory and Practice, 41*(5), 801–831. https://doi.org/10.1111/etap.12246

Boubaker, S., Cellier, A., Manita, R., & Saeed, A. (2020). Does corporate social responsibility reduce financial distress risk? *Economic Modelling, 91*, 835–851.

Cai, L., Cui, J., & Jo, H. (2016). Corporate environmental responsibility and firm risk. *Journal of Business Ethics, 139*(3), 563–594. https://doi.org/10.1007/s10551-015-2630-4

Chen, P., Chu, Z., & Zhao, M. (2024). The road to corporate sustainability: The importance of artificial intelligence. *Technology in Society, 76*. Scopus. https://doi.org/10.1016/j.techsoc.2023.102440

Chen, T., Dong, H., & Lin, C. (2020). Institutional shareholders and corporate social responsibility. *Journal of Financial Economics, 135*(2), 483–504.

Chen, Z., & Xie, G. (2022). ESG disclosure and financial performance: Moderating role of ESG investors. *International Review of Financial Analysis, 83*, Article 102291.

Christensen, H. B. (2022). Is corporate transparency the solution to political failure on our greatest problems? A discussion of Darendeli, Fiechter, Hitz, and Lehmann (2022). *Journal of Accounting and Economics, 74*(2–3), Article 101542.

Cormier, D., Gutierrez, L., & Magnan, M. (2024). The link between CSR performance and CSR disclosure quality: Does board diversity matter? *Journal of Management and Governance, 28*(1), 237–263. Scopus. https://doi.org/10.1007/s10997-022-09661-6

Cui, J., Jo, H., & Na, H. (2018). Does corporate social responsibility affect information asymmetry? *Journal of Business Ethics, 148*(3), 549–572. https://doi.org/10.1007/s10551-015-3003-8

Driss, H., Drobetz, W., El Ghoul, S., & Guedhami, O. (2024). The Sustainability committee and environmental disclosure: International evidence. *Journal of Economic Behavior and Organization, 221*, 602–625. Scopus. https://doi.org/10.1016/j.jebo.2024.02.019

Duque-Grisales, E., & Aguilera-Caracuel, J. (2021). Environmental, Social and Governance (ESG) scores and financial performance of multilatinas: Moderating effects of geographic international diversification and financial slack. *Journal of Business Ethics, 168*(2), 315–334. https://doi.org/10.1007/s10551-019-04177-w

El Ghoul, S., Guedhami, O., Kim, H., & Park, K. (2018). Coporate environmantel responsibility and the cost of capital: Interanational evidence. *Joubrnal of Business Ethics, 149*, 335–361

Fernández-Guadaño, J., & Sarria Pedroza, J. (2018). Impact of corporate social responsibility on value creation from a stakeholder perspective. *Sustainability, 10*(6), 2062. https://doi.org/10.3390/su10062062

Husted, B. W., & de Sousa-Filho, J. M. (2019). Board structure and environmental, social, and governance disclosure in Latin America. *Journal of Business Research, 102*, 220–227.

Javed, M., Rashid, M. A., Hussain, G., & Ali, H. Y. (2020). The effects of corporate social responsibility on corporate reputation and firm financial performance: Moderating role of responsible leadership. *Corporate Social Responsibility and Environmental Management, 27*(3), 1395–1409. https://doi.org/10.1002/csr.1892

Jia, F., Zuluaga-Cardona, L., Bailey, A., & Rueda, X. (2018). Sustainable supply chain management in developing countries: An analysis of the literature. *Journal of Cleaner Production, 189*, 263–278.

Kandpal, V. (2020). Reaching sustainable development goals: Bringing financial inclusion to reality in India. *Journal of Public Affairs*. https://doi.org/10.1002/pa.2277

Khandelwal, V., Tripathi, P., Chotia, V., Srivastava, M., Sharma, P., & Kalyani, S. (2023). Examining the impact of agency issues on corporate performance: A bibliometric analysis. *Journal of Risk and Financial Management, 16*(12), 497.

Kumar, K., Kumari, R., & Kumar, R. (2021). The state of corporate sustainability reporting in India: Evidence from environmentally sensitive industries. *Business and Society Review, 126*(4), 513–538. https://doi.org/10.1111/basr.12247

Kumar, K., Kumari, R., Nandy, M., Sarim, M., & Kumar, R. (2022). Do ownership structures and governance attributes matter for corporate sustainability reporting? An examination in the Indian context. *Management of Environmental Quality: An International Journal, 33*(5), 1077–1096.

Kumar, K., & Prakash, A. (2019). Developing a framework for assessing sustainable banking performance of the Indian banking sector. *Social Responsibility Journal, 15*(5), 689–709.

Kumar, K., & Prakash, A. (2020). Managing sustainability in banking: Extent of sustainable banking adaptations of banking sector in India. *Environment, Development and Sustainability, 22*(6), 5199–5217. https://doi.org/10.1007/s10668-019-00421-5

Lebedev, S., Peng, M. W., Xie, E., & Stevens, C. E. (2015). Mergers and acquisitions in and out of emerging economies. *Journal of World Business, 50*(4), 651–662.

Lin, H., Zeng, S. X., Ma, H. Y., Qi, G. Y., & Tam, V. W. (2014). Can political capital drive corporate green innovation? Lessons from China. *Journal of Cleaner Production, 64*, 63–72.

Luo, X. R., Wang, D., & Zhang, J. (2017). Whose call to answer: Institutional complexity and firms' CSR reporting. *Academy of Management Journal, 60*(1), 321–344. https://doi.org/10.5465/amj.2014.0847

Nekhili, M., Nagati, H., Chtioui, T., & Rebolledo, C. (2017). Corporate social responsibility disclosure and market value: Family versus nonfamily firms. *Journal of Business Research, 77*, 41–52.

Prakash, A., Kumar, K., Khan, W., & Siddiquei, M. I. (2023). Self-induced versus structured corporate social responsibility: The Indian context. In *Measuring sustainability and CSR: From reporting to decision-making* (pp. 167–178). Springer International Publishing.

Ren, X., Li, J., Wang, X., & Lei, X. (2024). Female directors and CSR: Does the presence of female directors affect CSR focus? *International Review of Financial Analysis, 92*. Scopus. https://doi.org/10.1016/j.irfa.2024.103101

Sharma, A., Kumar, K., & Dwesar, R. (2025). "Keenly aware of environmental issues, yet wavering in conviction": Understanding the factors influencing the consumers' willingness to pay for green energy. *Sustainable Futures, 9*, Article 100427.

Vishwanathan, P., Van Oosterhout, H. J., Heugens, P. P. M. A. R., Duran, P., & Van Essen, M. (2020). Strategic CSR: A concept building meta-analysis. *Journal of Management Studies, 57*(2), 314–350. https://doi.org/10.1111/joms.12514

Wang, Q., Dou, J., & Jia, S. (2016). A meta-analytic review of corporate social responsibility and corporate financial performance: The moderating effect of contextual factors. *Business & Society, 55*(8), 1083–1121. https://doi.org/10.1177/0007650315584317

Chapter 4
Ownership Structure and Impact on Corporate Governance

Neetu and Akhil Goyal

Abstract The ownership structure of a firm usually possesses' a significant impact on the perceptions of employees, regulations, and organizational processes. As a result, a direct association between ownership structure and corporate governance has been found by several scholars. The chapter begins with an analysis of ownership structure and the mechanics of concentrated ownership, in which one company or a few investors have the most influence on corporate decisions. Increased oversight and alignment of interests can lead to strong governance, but it risks monopolizing and repressing the minority shareholders and their relations. However, it may face weak maintenance and representation issues, leading to inefficiencies and governance issues. Furthermore, in this chapter, a critical analysis of the roles of institutional investors and state ownership has also been conducted along with emphasizing their dual potential to support conservative governance strategies and possibly pursue broader social and political objectives. The chapter also examines other developments in ownership structure, including the effects of venture capital and private equity, and how these developments affect corporate governance models. This chapter has combined theoretical perspectives related to corporate governance with empirical findings to analyze the role of ownership structure as a determinant of good corporate governance practice.

Keywords Corporate governance · Ownership structure · Concentrated ownership · Dispersed ownership

Neetu (✉) · A. Goyal
Department of Management and Commerce, NIMS University, Jaipur, India
e-mail: welcome.neetu93@gmail.com

A. Goyal
e-mail: akhil.goyal@nimsuniversity.org

V. Kandpal et al. (eds.), *Developments in Corporate Governance*,
Approaches to Global Sustainability, Markets, and Governance,
https://doi.org/10.1007/978-981-96-6366-8_4

4.1 Introduction

Corporate Governance (CG) is comprised of a set of regulations, policies, and legislative frameworks, which set the parameters regarding the running of the organization, governing, and monitoring of operations. The term corporate governance is associated with both internal as well as external components, which affect the interest of stakeholders, including stockholders, customers, vendors, government authorities, and administration. The capacity of a company to operate properly depends on its corporate governance. The reputation of a corporation is largely dependent on its ability to exhibit sound corporate governance. It acknowledges the significance of stockholders as well. In addition to providing funding for corporate operations and direct control over how the firm is run, shareholders also choose the organization's board of directors.

The analysis of the association between corporate governance and ownership structure has attracted the attention and focus of numerous scholars within the past 10 years. Scholars and professionals have endeavored to comprehend how diverse ownership arrangements influence the diversity in boards, management incentives, strategic choices, and ultimately, the overall success of corporations. This book explores the complex link between corporate governance and ownership structure by looking at different types of ownership structures, their theoretical foundations, and empirical data from other contexts.

This chapter mainly emphasizes the factors that influence the form of ownership structure to influence Corporate Governance (CG). This analysis aims to provide insights into how they can adapt a suitable type of ownership structure to improve organizational performance and shareholder value.

4.2 Literature Review

4.2.1 Ownership Structure

Different forms of ownership have been described by Demsetz and Villalonga (2001) as the attributes of equity owners and their ability to hold stocks in any organization. Forms of ownership of a corporation are contingent upon the distribution of its stock concerning voting privileges and invested capital. According to Boodhoo (2009), ownership is the authority to direct the business or select its management. "The authority to exercise control" is another definition of ownership (Kirchmaier & Grant, 2005). It is not quite so easy, though, as evolving definitions of ownership have given rise to new forms of ownership and associated problems. For instance, ownership and control are not entirely distinct; rather, the degree of separation depends on the kind of structure of ownership.

Since ownership arrangements impact managers' incentives and, in turn, the organization's efficiency, they are crucial to corporate governance. Ang et al. (2000) is a well-known source. By combining concepts from the theories of agency, intellectual property, and financing, these financial economists attempted to create a theory of the company's system of ownership.

4.2.2 *Types of Ownership Structure*

Every nation has a different type of ownership structure (Wahl, 2006). Chen et al. (2009) and Abdullah et al. (2012) identify both concentrated and scattered ownership arrangements. Because Bruton et al. (2010) divided ownership structures based on ownership size, every framework incorporates the traits, attributes, incentives, and interests of its constituent investors. According to Kirchmaier and Grant (2005), there is no ideal type since each one works well in a certain situation.

The degree of ownership rights' concentration and the owner's identification can be used to identify a system of ownership (Al-Thuneibat, 2018; Elvin & Hamid, 2016). The following section has covered the two forms of ownership structures: distributed ownership and concentrated ownership.

4.2.2.1 Concentrated Ownership

Concentrated ownership refers to a kind of corporate governance in which a majority shareholder or small group of persons has substantial control over the leadership and operations of the organization. These individuals frequently have a greater say in how the business is conducted daily (Choi, 2018).

Global ownership is dominated by major investors (Singal & Singal, 2011). This can be explained by the fact that the global legal framework safeguarding investor rights is not very strong. Major shareholders require less rights to safeguard their interests than do smaller shareholders. Big shareholders can improve performance by keeping a closer eye on and exerting more influence over management (Cho, 1998). However, there are drawbacks to concentrated ownership as well. These essentially consist of the potential for significant shareholders to appropriate other traders' and stakeholders' holdings (Morck, 2007). Large shareholders suffer an extra risk due to limited diversity, which is an additional expense of concentrated ownership (Demsetz & Lehn, 1985; Fama & Jensen, 1983). According to several recent research, supervisors may be overly scrutinized when there is a significant concentration of ownership. Consequently, there is a reduction in the managerial ambition to undertake investments relevant to the business (Manzaneque et al., 2016; Tsionas et al., 2012).

The four most prevalent forms of concentrated ownership are foreign, institutional, state, and family-owned businesses.

1. **Family-Owned Firms**

Family members possess a sizable portion of the organization's shares under this sort of concentrated ownership. As a result, close family members decide the majority of the business's choices, and other small owners agree with them. The most obvious feature of a family-owned company is that family members hold every important managerial job in these companies (FĖlix & David, 2019). Although family members often have the last say in decisions, non-family individuals may work for the organization. This is most likely the cause of the fear of losing control over the company that keeps many family firms from being proactive in implementing stringent corporate governance practices (Saridakis et al, 2018). These companies are fueled by the love, trust, and close relationships among their members, but even a small amount of discord or competition within the family can have a negative impact on the company and present a bad image of the family business to potential investors. A high percentage of family ownership is likely to broaden the organization's decision-making horizon and match the family's interests with those of the company's long-term investors. Businesses may benefit from bank loans with more enticing conditions (Weiss, 2014).

It is anticipated that family-owned businesses would follow the same corporate governance guidelines as any other company. A family-owned company cannot succeed unless it follows the same rules and guidelines as any other type of organization. A few of these initiatives involve regular board conferences, the recruitment of independent directors in addition to other executives to create an open and accountable board of executives, whistle-blower legislation, and adherence to the Accounting Standards in the development of financial reports by an organization and its external auditors (Alderson, 2015). Financial disclosure is also an indicator of transparency and accountability, offering important financial data about the organization to all of its investors and other stakeholders.

2. **Institutional Ownership**

"Institutional Ownership" reflects the situation where some financial organizations have owned majority of ownership in a company. The financial organizations include investment firms, mutual funds, and pension plans. These organizations usually acquire a substantial chunk of a business's outstanding equity, which gives them a big interest in its success. When buying mid-cap shares, institutional ownership is a crucial consideration since it can significantly affect the stock's subsequent performance (Chung & Zhang, 2011). Institutional investors have the power to influence better governance procedures, such as increased openness and disclosure, which will help management decisions better serve the interests of investors (Cornett et al., 2007). The effects of institutional ownership, however, might differ depending on the shareholder's objectives. While some, like pension plans, may prioritize long-term stability and sustainability, other scholars may prioritize short-term financial benefits.

According to scholars, institutional investors can influence Corporate Governance (CG) and fulfill their monitoring duty in two ways. Direct involvement is the first, and it has the advantage of rectifying actions taken by management. According to Boone and White (2015), there is a greater motivation for large owners, especially institutional investors, to participate in direct monitoring operations. Second, they can exercise the exit risk and transfer their stakes to show that they are unhappy with the way management is acting (Duggal & Millar, 1999; Franks, 2020). Institutions profit from being able to discipline the management as a result. As a result, institutional investors are seen to be able to uphold shareholder values in companies with poor CG quality while also improving CG standards.

3. **State Ownership**

State ownership, which can be defined as public ownership or government ownership, is the ownership of a business, asset, real estate, or sector by a public entity serving the population or the national government of a nation or state, as compared to a person or private entity (Pargendler, 2011). Public ownership is distinct from services provided by the government and public commodities that are paid for out of a state's general budget. It particularly refers to sectors that offer goods and services to customers. Public ownership may be categorized as ownership owned by the national, state, local, or municipality or as non-governmental public ownership engaged in independent public firms (Yu, 2013). Separated from private, collective/cooperative, and shared ownership, public ownership is one of the three main types of real estate ownership.

A government may hold all or a portion of a state-owned business (SOE), which is usually designated to engage in particular economic operations. Occasionally, an official organization will undergo a process known as corporatization to become an SOE (Cornett et al., 2010). This makes it possible for the agency to become a for-profit enterprise. Even though the recently established SOE is technically a commercial firm, it frequently continues to function with government objectives in mind. Occasionally, politicians in developing nations will establish a state-run enterprise in an area they want to grow or capitalize on to improve their economic position internationally. Examples of such industries include the telecommunications and oil sectors in Argentina and Brazil (Teubal, 2004).

There exist several rationales behind a government's involvement in corporate ownership. Firstly, privatization is a popular method of transferring ownership from government entities to other forms of ownership, either totally or partially (Ding et al., 2007). Second, by establishing new businesses, the government hopes to grow the economy. This can encourage companies to enter new markets and industries in which the government forms a corporation and makes a portion of its stock available to the general public (Zhang et al., 2020). Third, political factors might be at play, including the government forming alliances with certain company owners to win their support (Alok & Ayyagari, 2020). The Saudi initiative known as "Saudization" is one instance of government engagement. Its goal is to turn select significant international banks and businesses into publicly listed firms with the Saudi government holding the majority of the shares (Ramady, 2010).

It is often known that state-owned businesses typically perform worse than those with other ownership structures. Perotti (2004) has provided a comprehensive analysis of the effectiveness of state ownership. Perotti contended that the primary cause of the inefficiency of state ownership is the absence of rewards for government workers to optimize efficiency concerning both quality improvement and cost minimization.

4. **Foreign and Domestic Ownership**

Management may employ various strategies to intimidate foreign investors, such as forfeiting voting records, proclaiming some of their stock to be unlawful, and so on. According to Douma et al. (2006) and Yudaeva et al. (2003), local investors have greater options for defending their rights on behalf of, notably deeper connections to other stockholders, the judiciary, and even law enforcement.

According to agency theory, the rewards of bigger foreign business owners are better suited to carry out an efficient monitoring function since their ownership shares are greater and less dispersed than those of foreign institutional stockholders. Foreign companies that possess a portion of domestic companies also frequently invest in businesses that are associated with their main line of operations. Businesses with overseas corporate ownership have access to better managerial, monetary, and technological assets. For example, Barbosa and Louri (2005) discovered a positive correlation between the level of resource commitment to the transfer of technology and the degree of influence a foreign corporation had over a local company. Sivathaasan (2013) discovered that there is a correlation between foreign investment and the offering of particular expertise (that cannot be transmitted at arm's length) and generic information (leadership skills and quality standards).

Foreign ownership is significant due to several factors. First, exposure to new networks through foreign investment may boost sales, promotion, and distribution. Second, the benefits are obtained by local businesses due to participation by international investors, as they bring some specialization into the organization (Wang & Wang, 2015). A corporation from a developed economy, for instance, would bring its legal systems, the newest technology, and assistance in expanding into emerging markets for the local business. Third, in a market where there is competition, foreign investment will bolster the organization's position. Fourth, having a foreign owner increases the efficiency of the business as they bring with them their internal reporting frameworks, procedures, and data disclosure guidelines (Greenaway et al., 2014).

4.2.2.2 Dispersed Ownership

Dispersed ownership is a corporate governance system in which no single investor holds sufficient shares to control a firm. Rather, a board of directors and a management team oversee the operation of the company on behalf of several shareholders as well as owners (Ozili & Uadiale, 2017). This organizational structure may provide managers with a great deal of discretion, which might result in opportunistic behavior that requires oversight by an impartial board. Since it is most common in the United

States and the UK, dispersed ownership is sometimes referred to as the "Anglo-Saxon model" of corporate governance by observers. In other places, ownership is typically concentrated, with one family controlling and running the business (Coffee, 2010).

Diverse ownership frequently calls for robust corporate governance practices. To avoid any one party from gaining too much impact, multiple checks and balances are usually in place. Boards of directors and management committees are accountable for choosing options for the best benefit of all stockholders (Coffee Jr, 2001). Moreover, an additional difficulty linked to distributed ownership is the possibility of an agency issue. This arises when the company's management (agents') interests differ from those of the stockholders (principals). Dispersed ownership can make it more difficult for stockholders to collaborate and affect managerial activities, and executives may make choices that equally treat the interests of all stakeholders instead of focusing entirely on the stockholders (Burkart & Lee, 2015).

4.2.3 Corporate Governance

The significance of corporate governance for the financial development of both corporations and society at large has made it a topic of intense discussion across a broad spectrum of stakeholders, involving politicians, business management, and even academic scholars (Monks & Minow, 2011). On its definition, there is disagreement, though multiple disciplines of study have multiple definitions for it, the general public has not embraced any one term.

Corporate governance, according to Mallin (2016), is related to the composition and operation of boards of directors. According to Khan (2011), corporate governance is the framework whereby the board of members, related structures, executive promotions, and other governance and surveillance programs oversee the top tier of management inside a corporation. According to Larcker and Tayan (2020), everyone with an interest in the affairs of a business, including lenders, debt financial institutions, economists, auditing firms, and corporate regulators, needs to be present in the boardroom.

A corporation's internal structures that explicitly define the relationship between its investors and management are collectively referred to as corporate governance. According to Anand's (2007) analysis, the corporate governance framework serves as a means for the company's investors to guarantee a return on their initial investment. According to Monks and Minow (2011), corporate governance refers to the interactions among the various members of an organization that influence the organization's effectiveness and strategy. The board of directors, upper management, and beneficiaries are the corporation's main players.

As mentioned above, corporate governance can be defined narrowly. Some academics give it a far more expansive definition. According to Gregory and Simms (1999), corporate governance refers to the processes, procedures, cultures, and structures that are directly linked to a company's capacity to run successfully. According to Thomsen (2004), corporate governance aims to strike a balance between achieving individual and group objectives as well as between economic and social goals. Ensuring that resources are used as effectively as feasible, that resource management is responsible, and that the needs of people, businesses, and the community are as closely connected as is feasible are the goals of corporate governance.

Welge and Eulerich (2014) offer the concept of corporate governance that is most often recognized. It was maintained that corporate governance is a framework or system that provides rules for managing and directing company activities. All members of the corporation, particularly the board, management, and stakeholders, have clearly defined roles and duties. In addition, guidelines and protocols are delineated that must be adhered to while rendering choices about business matters. Therefore, corporate governance is a framework that includes the goals of the company, how those goals will be attained, and how to keep an eye on the company's performance.

According to two different definitions provided by the World Bank, corporate governance through the lens of a corporation is about upholding positive relationships between the founders, supervisors, and multiple stakeholders, including vendors, consumers, staff, and shareholders (Bhagat & Bolton, 2008). The board of directors is given extra consideration in particular because of their capacity to maximize the corporation's long-term worth and to fairly distribute the interests of all stakeholders. Within the framework of government policy (the second perspective), corporate governance centers on two primary concerns: (1) the survival, expansion, and advancement of a firm; and (2) responsibility concerning the utilization of control and authority over businesses (Vives, 2000).

Corporate governance encompasses a wide range of social and economic challenges, and it is becoming more and more popular in a variety of academic fields, including politics, sociological studies, economics, finance, and law. Diverse academics with varying specialties hold varying opinions about the notion of corporate governance. Nonetheless, the key query that has to be answered is related to corporate assets and influence, and the best way to manage them (Clarke, 2004).

4.2.4 Ownership Structure and Corporate Governance

One significant factor in corporate governance is ownership structure. It is an important organizational factor that affects how well a company performs (Singh & Davidson III, 2003). One key characteristic that sets financial systems apart is ownership structure (Xu & Wang, 1999). It can, thus, in conjunction with other effective and technological capabilities, have a major impact on the success of the organization since it is a main factor in shaping corporate governance and behavior (Abdallah & Ismail, 2017). Additionally, Elvin and Hamid (2016) have highlighted the significance of corporate governance and ownership structure in figuring out company strategy:

The ownership structure of an organization, debtors' and proprietors' motivation, corporate governance practices, and top managers' financial incentives all have a significant influence on the business's objectives. Publicly traded companies' objectives mirror the features of that country's capital markets (Demsetz & Lehn, 1985).

To elucidate the study of ownership structure and corporate governance outcomes, it is crucial to consider ownership structures with respect to corporate governance theories. Out of these theories, the Agency Theory, developed by Jensen and Meckling (1976), provides an insightful view of such conflicts that may arise between managers and shareholders in ownership structures in which ownership is widely dispersed. Agency risks are seen increasing as the separation of ownership from control gives managers (agents) the opportunity to pursue their own interests rather than the best interest of the shareholders (principals). This incongruence necessitates the existence of stringent governance mechanisms, such as regulatory oversight and active institutional investors' participation to mitigate managerial opportunistic behavior (Fama & Jensen, 1983). On the contrary, concentrated ownership can cut the costs of the agency by largely aligning the interests of large shareholders and managers. However, such a situation might also lead to principal-principal conflict, where large shareholders exploit their power for the detriment of minority investors (La Porta et al., 1999).

The Stewardship Theory formulated by Donaldson and Davis (1991) offers an alternative perspective by arguing that, in family-owned or small firms, managers operate as stewards of the organization. Accordingly, ownership structures-the concentrated ownership in particular-like family businesses, may favor long-term strategic planning and stability since family owners may naturally emphasize the founding of the company and its sustainability rather than immediate financial gain (Anderson & Reeb, 2003). This theory also goes further to claim that, in certain contexts, family-led firms may even outperform non-family-led firms because of some beneficial aspects of nepotism and entrenchment (Filatotchev et al., 2005). Furthermore, the family adds depth to their total commitment to ethical business practices and corporate social responsibility, primarily because their stakeholders' long-term survival is identified with the company's long-term prosperity. (Miller & Le Breton-Miller, 2005).

Moreover, Resource Dependence Theory, as stated by Pfeffer and Salancik (1978), underlines the strategic merit of institutional shareholding in corporate governance. Institutional investors, like pension funds and mutual funds, possess abundant resources and expertise to pressure firms to comply with best governance practices (Aguilera & Crespi-Cladera, 2016). These investors basically operate not only as a supply of capital but also as sources of policy interventions advocating transparency, accountability, and efficient resource distribution (Short & Keasey, 2005). Foreign ownership, too, has the capacity to upgrade governance standards by introducing global best practices and acting as conduits for knowledge transfer (Shubita & Shubita, 2019). However, Miyajima et al., (2015a, 2015b) warn that the geographical and cultural circumstances may actualize the limiting effect of foreign investors in the governance of domestic firms and, in turn, render them ineffectual. In addition, regulatory requirements, especially in terms of economic systems and institutional development standards among regions, can create a barrier to the creation of international corporate governance practices (Khanna & Palepu, 2000).

The implications of governance of ownership structures vary across institutions and regulatory environments. In developed economies, therefore, with strong legal safety nets and ownership patterns, those with dispersed responsibility work well due to external governance mechanisms such as regulation and market discipline (Clarke, 2015). Conversely, the lowest point of control is generally maintained by emerging economies because of concentrated patterns of ownership-essentially under institutional control. That said, such a structure may result in a higher strategic direction. Likewise, Bebchuk et al.'s (2017) case suggests the presence of active institutional investors to be likely to dilute such risks by imposing market discipline on the controlling shareholder. Also, having multiple large shareholders, as quoted by Baums and Scott (2005), can boost governance by bringing checks and balances among dominating sharers. Nonetheless, even in cases where the regulation is loose or enforcement is not effective, the controller might use his power to do exercises that harm minority interests, ergo increasing governance challenges (Claessens et al., 2000).

Ownership structures also have implications for firm performance. While some research highlights that dispersed ownership may foster more excellent innovation and competition in the market, others have underscored the stability and longer-term view that accompanies concentrated ownership (Gomez-Mejia et al., 2011). For instance, excessive ownership concentration can lead to entrenchment and self-dealing behaviors that reduce company value, especially when combined with insufficient governance systems (Morck et al., 1988).

Given that ownership arrangements have a fundamental influence on how firms are run and directed, the association between corporate ownership and corporate governance is an important topic of research for corporate finance and management (Connelly et al., 2010). Corporate governance is the systems and procedures that guarantee the interests of all parties involved—from executives and stockholders to consumers, suppliers, lenders, governments, and the public at large—are fairly balanced. In terms of control, supervision, and company performance, different ownership structures including concentrated, dispersed, family, institutional, and governmental ownership—have distinct effects on corporate governance.

Numerous studies have examined ownership concentration and its relationship to business performance, more specifically (Chen et al., 2006; Demsetz & Lehn, 1985; Demsetz & Villalonga, 2001; Lemmon & Lins, 2003). For instance, Akimova and Schwödiauer (2004) employ explanatory factors at the national level to clarify the ownership concentration in publicly listed companies across a broad range of industrialized and developing nations. The average ownership interest of the top three biggest stakeholders (investors) in the top ten publicly listed corporations is used to calculate the level of power distribution among shareholders in each nation. The makeup of the ownership divisions or the identities of the owners are significant factors in corporate ownership structures. Shareholders may include non-financial corporations, holding companies, banks, families, individuals, or institutional investors. Various owner types may have varying interests, which means they will have varied motivations and capacities to manage the managers in a company. As noted by Mollah et al. (2012), an owner's dedication and readiness to step in may be greatly influenced by who they are. Contrary to this, the position of rights to control may be a more relevant predictor of the level of control exercised by owners as compared to ownership concentration. Depending on the unique aspirations they carry for the company and how much they continually track it, owners might differ greatly from one another (Monks & Minow, 2011). For instance, institutional investors' significant investments are more probable than atomistic ownership to result in better oversight and management as they have sufficient motivation to bear the costs of performance monitoring (Aguilera & Crespi-Cladera, 2016; Bokpin, 2013; Dyck, 2000; Kirchmaier & Grant, 2005; Xu & Wang, 1997, 1999). Furthermore, business investors could be more intrigued in developing a long-term connection, whereas financial shareholders might focus their investment choice toward short-term profits on their stake (Reyna et al., 2012).

Every economy has a very different system of ownership. In developed economies that are prosperous and have a strong legal and regulation system, along with active supervision from reliable agents and sufficient institutional and professional infrastructure, like the United States and UK, dispersed ownership has proven to be an effective foundation for wealth accumulation and economic development (Davies, 2015). A significant portion of the literature on corporate governance is predicated on the idea of dispersed ownership. As a consequence, the literature mostly concentrates on resolving conflicts that arise from the division of ownership and control among executives (as agents) and investors (as principles).

The new emerging research bodies, nevertheless, challenge the notion of distributed ownership and contend that a concentrated structure of ownership is more common in many economies. For instance, Holderness et al. (1996) reported that management ownership increased in the United States from 13% in 1935 to 21% in 1995. According to research by Rusmin et al. (2011), the three major stakeholders of non-financial companies from 49 different nations possess 46% of the companies on average. Numerous research investigations have also confirmed that family ownership does occur.

Contrary to this, dispersed ownership involves numerous investors having a very small percentage of the total shares, with a large dispersion of shares among them. The more ownership and control are separated, the more likely it is that agency issues will arise. Since no single shareholder is actively monitoring the company, management can occasionally behave in the best intentions of the investors (He et al., 2022). As a result, corporate governance in these kinds of environments depends increasingly on outside forces, like the market for ownership, regulatory supervision, and institutional shareholders' advocacy, which might intervene to support improved governance procedures.

Prior scholars have revealed that the standard of Corporate Governance (CG) can be greatly influenced by a limited group of institutional shareholders (Correa-Mejía, 2022; del Carmen Briano-Turrent & Rodríguez-Ariza, 2016). According to Arslan et al. (2019), companies with stronger institutional ownership are more inclined to fire CEOs who perform badly and have excellent Corporate Governance (CG). They also demonstrate how institutional shareholders support positive returns on capital in their global holdings.

Nonetheless, institutional shareholders favor a different strategy, according to research by Short and Keasey (2005), which is to make investments in companies that already have strong governance procedures in place. Comparably, Hooper et al. (2009) discovered that the level of governance correlates with the total quantity of shares held by organizational institutions. Furthermore, studies conducted by Koehn and Ueng (2005) and Mohanty (2003) have demonstrated that large institutional shareholders view CG as a significant financial signal when making investment choices.

Shubita and Shubita (2019) asserted that foreign ownership improves corporate governance. Three main stages have been used by the researchers to further explain the good influence. Initially, the controlled firm will get experience and knowledge from the foreign corporation. Secondly, the organization in charge will provide effective supervision to guarantee their safety. Third, because the success of subsidiaries influences the efficiency of their parent businesses, a more concentrated ownership structure in the affiliates of multinational organizations may assist in lowering agency expenses by offering greater supervision and greater incentives to executives (Tunay & Yüksel, 2017). Contrary to this, Miyajima et al., (2015a, 2015b) contended that foreign ownership may also be detrimental to corporate governance. The experts claim that due to certain institutional and cultural conditions, the foreign corporation is not necessarily able to keep an eye on the domestic business. Furthermore, the foreign firm's capacity to monitor and obtain information may be hampered by the physical distance between it and the local corporation. As a consequence, if the country's market has substantial amounts of expertise, manufacturing technology, or administration, it could be challenging for the foreign firm to stand out in the local marketplace.

In several emerging market economies, the business climate differs significantly from that of the majority of industrialized economies. For instance, the majority of businesses in Asian nations are part of a business association that is usually run by families. Many public and private organizations might make up the group (Claessens et al., 2000). The family efficiently manages the enterprises within the group using the implementation of cross-shareholdings and equity hierarchies, that may have extremely complicated structural features. Furthermore, Claessens et al. (2000) have noted that a family's power to vote is often higher than the financial liberties of the company. Their research in nine East Asian countries revealed that more than two-thirds (67.1%) of Indonesia's publicly traded businesses are owned by family members, with just 0.6% being widely held. Furthermore, according to Nuzula and Lokuwaduge (2017), Indonesia is home to the greatest number of single-family-controlled businesses in all of East Asia.

Numerous single-economy investigations have also proven the issue of concentrated family ownership, including those conducted in South Korea by Kim and Cho (2021), Taiwan by Filatotchev et al. (2005), and Thailand by Bertrand et al.

(2008). According to these researches, business arrangements with concentrated family ownership exacerbate the issues of unequal knowledge, inadequate oversight, and opportunity-seeking conduct. They also increase the complexity of corporate governance change.

The aforementioned highlights two clear implications of ownership structure variations for corporate governance. A concentrated ownership structure may make it possible for insiders to exert strong influence over the company, but it also creates avenues for the expropriation of outside shareholders' money. Connelly et al. (2010) proposed that the expenses of distributed equity ownership and strategies for promoting greater ownership concentration have dominated discussions over corporate governance in the United States. According to Iskandar (2014), the costs related to concentrated ownership solely escalate when there is a high concentration of ownership paired with lax laws protecting outside the stockholders, a competitive financial marketplace, and opportunities for fraud and misconduct by large, powerful (insider) investors.

Research on the connection between corporate governance and ownership arrangements offers complex findings. Research suggests that in emerging economies, where safeguards for investors are less robust, ownership concentration frequently corresponds with stronger business performance. Family-owned businesses may operate better than non-family businesses because of improved interest congruence and an emphasis on the future, even though governance issues still exist. While state ownership has a broad range of effects depending on the particular governance frameworks used, institutional ownership is often associated with more effective governance.

Previous research, such as those conducted by Boateng and Huang (2017), Liu and Tian (2012), and Gao and Kling (2008), examined the impact of big and controlling owners on tunneling activities and financial choices. According to these findings, companies with concentrated ownership provide investors with excessive control powers. These stockholders then utilize their considerable leverage to obtain advantages from assets and make tunneling easier. However, Zhang et al. (2017) have looked at how controlling owners' tunneling activities are impacted by the trading activity of overseas institutional investors. They discover a reverse U-shaped correlation between tunneling and international institutional shareholders. The U-shaped relationship is caused by institutional investors' large trading, which limits tunneling by providing market control to supervisors. However, Bian et al. (2023) discovered that large shareholders could lessen the effects of tunneling in concentrated ownership.

However, Baums and Scott (2005) have looked at the impact of having many significant shareholders. They discovered that greater shareholder protection is the outcome of having several major stockholders. Additionally, controlling shareholders improves the financial health of banks, according to Mülbert (2013) study on bank corporate governance.

In summary, there is a complicated and context-dependent relationship between corporate ownership arrangements and corporate governance. Ownership arrangements shape governance procedures and results by influencing the motivations and capacities of individuals in charge of and supervising business operations. Thus, legislators, shareholders, and management must have an adequate understanding of these dynamics to improve corporate governance norms and overall business profitability. Further studies should continue to examine these links in many institutional and regulatory contexts to offer more thorough and nuanced perspectives.

4.2.5 Conclusion

Based on an in-depth assessment of the topic and some earlier research, ownership structure has a major influence on the organization's governance policy. Two other factors of ownership structure that have been looked at in this study are the percentage of shares held by management and the percentage of shares owned by the largest shareholder interests. This supports the idea that ownership arrangements that are typically observed in this chapter are the ones that best fit the operating environments of businesses.

Costs and advantages are related to concentrated ownership. Through closer oversight of senior management, concentrated ownership may lower agency expenses. On the other hand, dominant proprietors may also get personal advantages from concentrated ownership. On the other hand, a dispersed ownership system, where market forces for corporate influence serve as the fundamental disciplinary mechanism, is marked by robust securities markets, strict disclosure requirements, and a high level of market transparency. The findings add credence to the idea that regardless of how distributed or concentrated the businesses they service may be, the market is effective in creating forms of ownership that are generally appropriate for them. These structures differ among the businesses because of the differences in the circumstances that each one encounters. These situations include economies of scale, legislation, and the stability of the environments in which the businesses operate. If perfect markets determined ownership structure and corporate governance, then such arrangements would eradicate any systematic link between the two.

Based on empirical research, it appears that there isn't a single, effective method for enhancing corporate governance by means of ownership structure. The specific conditions of any governance structure, such as the market factors, legislative environment, and cultural shifts, dictate how successful the system is. Organizations need to carefully consider their ownership structure while taking into consideration contextual elements as part of corporate governance in order to build a system of governance that maintains a balance between control and responsibility.

4.2.5.1 Implications of Ownership Structure on Corporate Governance

1. Corporate Strategy: Corporate strategy is shaped by the kind of ownership structure. For instance, "family-owned businesses" may place more importance on sustainability and legacy than on short-term profit, but "institutional ownership"

frequently demands short-term gains, which affects governance practices. This influences the company's choices on capital expenditure, creativity, and taking risks.

2. Board Dynamics and Composition: The ownership structure affects the number of independent directors and the distribution of power among shareholders and management on the board. For example, fewer independent boards are typically associated with concentrated ownership, which may result in less supervision and a higher risk of conflicts of interest.
3. Legal and Regulatory Frameworks: Governments and regulators must consider ownership types when creating corporate governance regulations; therefore, the outcomes have regulatory ramifications. Policies should encourage openness in state-owned enterprises (SOEs), as they may need different governance regulations than privately held companies.

4.2.5.2 Scope for Future Research

1. Emerging Markets: Future studies should examine how ownership structure affects corporate governance in "emerging markets," paying special attention to the functions played by "state-owned enterprises (SOEs)" and "family businesses." Examining the effects of regional economic policies and governance laws on these organizations can shed light on the particular difficulties these businesses encounter.
2. Ownership and Governance in Startups: Research might examine how governance changes when ownership shifts with each financing round in the setting of "startups," where ownership is frequently dynamic owing to "venture capital" or "angel investors." An original perspective is provided by the impact of ownership on both the viability and achievement of startups.
3. Impact of Regulatory Changes: Future studies should examine the effects of modifications to corporate governance laws, such as those imposed by the US Securities and Exchange Commission (SEC) or the Securities and Exchange Board of India (SEBI), on the governance of companies with various ownership arrangements. Enhancing reforms to corporate governance initiatives can be facilitated by an understanding of the "regulatory interplay" among ownership and governance.

References

Abdallah, A. A. N., & Ismail, A. K. (2017). Corporate governance practices, ownership structure, and corporate performance in the GCC countries. *Journal of International Financial Markets, Institutions and Money, 46*, 98–115.

Abdullah, N. M. H., Ahmad, Z., & Roslan, S. (2012). The influence of ownership structure on the firm's dividend policy based Lintner model. *International Review of Business Research Papers, 8*(6), 71–88.

Aguilera, R. V., & Crespi-Cladera, R. (2016). Global corporate governance: On the relevance of firms' ownership structure. *Journal of World Business, 51*(1), 50–57.
Akimova, I., & Schwödiauer, G. (2004). Ownership structure, corporate governance, and enterprise performance: Empirical results for Ukraine. *International Advances in Economic Research, 10*(1), 28–42.
Alderson, K. (2015). Conflict management and resolution in family-owned businesses: A practitioner-focused review. *Journal of Family Business Management, 5*(2), 140–156.
Alok, S., & Ayyagari, M. (2020). Politics, state ownership, and corporate investments. *The Review of Financial Studies, 33*(7), 3031–3087.
Al-Thuneibat, A. (2018). The relationship between the ownership structure, capital structure and performance. *Journal of Accounting, Business and Management (JABM), 25*(1), 1–20.
Anand, S. (2007). *Essentials of corporate governance.* John Wiley & Sons.
Anderson, R. C., & Reeb, D. M. (2003). Founding-family ownership and firm performance: Evidence from the S&P 500. *The Journal of Finance, 58*(3), 1301–1328.
Ang, J. S., Cole, R. A., & Lin, J. W. (2000). Agency costs and ownership structure. *The Journal of Finance, 55*(1), 81–106.
Arslan, M., Abidin, S., Alqatan, D. A., &Roudaki, J. (2019). Corporate governance in the extreme institutional environment: Evidence from an emerging economy. *Corporate Ownership & Control, 17*(1).
Barbosa, N., & Louri, H. (2005). Corporate performance: Does ownership matter? A comparison of foreign and domestic-owned firms in Greece and Portugal. *Review of Industrial Organization, 27*, 73–102.
Baums, T., & Scott, K. E. (2005). Taking shareholder protection seriously-corporate governance in the United States and Germany. *American Journal of Comparative Law, 53*, 31.
Bebchuk, L. A., Cohen, A., & Hirst, S. (2017). The agency problems of institutional investors. *Journal of Economic Perspectives, 31*(3), 89–112.
Bertrand, M., Johnson, S., Samphantharak, K., & Schoar, A. (2008). Mixing family with business: A study of Thai business groups and the families behind them. *Journal of Financial Economics, 88*(3), 466–498.
Bhagat, S., & Bolton, B. (2008). Corporate governance and firm performance. *Journal of Corporate Finance, 14*(3), 257–273.
Bian, H., Kuo, J. M., Pan, H., & Zhang, Z. (2023). The role of managerial ownership in dividend tunnelling: Evidence from China. *Corporate Governance: An International Review, 31*(2), 307–333.
Boateng, A., & Huang, W. (2017). Multiple large shareholders, excess leverage and tunneling: Evidence from an emerging market. *Corporate Governance: An International Review, 25*(1), 58–74.
Bokpin, G. A. (2013). Ownership structure, corporate governance and bank efficiency: An empirical analysis of panel data from the banking industry in Ghana. *Corporate Governance: The International Journal of Business in Society, 13*(3), 274–287.
Boodhoo, R. (2009). Capital structure and ownership structure: A review of literature. *Journal of Online Education, January.*
Boone, A. L., & White, J. T. (2015). The effect of institutional ownership on firm transparency and information production. *Journal of Financial Economics, 117*(3), 508–533.
Bruton, G. D., Filatotchev, I., Chahine, S., & Wright, M. (2010). Governance, ownership structure, and performance of IPO firms: The impact of different types of private equity investors and institutional environments. *Strategic Management Journal, 31*(5), 491–509.
Burkart, M., & Lee, S. (2015). Signalling to dispersed shareholders and corporate control. *The Review of Economic Studies, 82*(3), 922–962.
Chen, G., Firth, M., & Xu, L. (2009). Does the type of ownership control matter? Evidence from China's listed companies. *Journal of Banking & Finance, 33*(1), 171–181.
Chen, G., Firth, M., Gao, D. N., & Rui, O. M. (2006). Ownership structure, corporate governance, and fraud: Evidence from China. *Journal of Corporate Finance, 12*(3), 424–448.

Cho, M. H. (1998). Ownership structure, investment, and the corporate value: An empirical analysis. *Journal of Financial Economics, 47*(1), 103–121.

Choi, A. H. (2018). Concentrated ownership and long-term shareholder value. *Harvard Business Law Review, 8*, 53.

Chung, K. H., & Zhang, H. (2011). Corporate governance and institutional ownership. *Journal of Financial and Quantitative Analysis, 46*(1), 247–273.

Claessens, S., Djankov, S., & Lang, L. H. (2000). The separation of ownership and control in East Asian corporations. *Journal of Financial Economics, 58*(1–2), 81–112.

Clarke, T. (2004). Theories of corporate governance. *The Philosophical Foundations of Corporate Governance, Oxon, 12*(4), 244–266.

Clarke, T. (2015). The transformation of corporate governance in emerging markets: Reform, convergence, and diversity. *Emerging Markets Finance and Trade, 51*(sup2), S25–S46.

Coffee, J. C., Jr. (2001). The rise of dispersed ownership: The roles of law and the state in the separation of ownership and control. *Yale Lj, 111*, 1.

Coffee, J. C. (2010). Dispersed ownership: The theories, the evidence, and the enduring tension between 'lumpers' and 'splitters'. *Columbia Law and Economics Working Paper*, (363).

Connelly, B. L., Hoskisson, R. E., Tihanyi, L., & Certo, S. T. (2010). Ownership as a form of corporate governance. *Journal of Management Studies, 47*(8), 1561–1589.

Cornett, M. M., Guo, L., Khaksari, S., & Tehranian, H. (2010). The impact of state ownership on performance differences in privately-owned versus state-owned banks: An international comparison. *Journal of Financial Intermediation, 19*(1), 74–94.

Cornett, M. M., Marcus, A. J., Saunders, A., & Tehranian, H. (2007). The impact of institutional ownership on corporate operating performance. *Journal of Banking & Finance, 31*(6), 1771–1794.

Correa-Mejía, D. A. (2022). Impact of corporate governance and institutional context on multilatinas' reporting quality. *Corporate Governance: The International Journal of Business in Society, 22*(6), 1259–1274.

Davies, P. (2015). Shareholders in the United Kingdom. In *Research handbook on shareholder power* (pp. 355–382). Edward Elgar Publishing.

del Carmen Briano-Turrent, G., & Rodríguez-Ariza, L. (2016). Corporate governance ratings on listed companies: An institutional perspective in Latin America. *European Journal of Management and Business Economics, 25*(2), 63–75.

Demsetz, H., & Lehn, K. (1985). The structure of corporate ownership: Causes and consequences. *Journal of Political Economy, 93*(6), 1155–1177.

Demsetz, H., & Villalonga, B. (2001). Ownership structure and corporate performance. *Journal of Corporate Finance, 7*(3), 209–233.

Ding, Y., Zhang, H., & Zhang, J. (2007). Private vs state ownership and earnings management: Evidence from Chinese listed companies. *Corporate Governance: An International Review, 15*(2), 223–238.

Donaldson, L., & Davis, J. H. (1991). Stewardship theory or agency theory: CEO governance and shareholder returns. *Australian Journal of Management, 16*(1), 49–64.

Douma, S., George, R., & Kabir, R. (2006). Foreign and domestic ownership, business groups, and firm performance: Evidence from a large emerging market. *Strategic Management Journal, 27*(7), 637–657.

Duggal, R., & Millar, J. A. (1999). Institutional ownership and firm performance: The case of bidder returns. *Journal of Corporate Finance, 5*(2), 103–117.

Dyck, A. (2000, April). Ownership structure, legal protections, and corporate governance. In *Annual World Bank Conference on Development Economics* (Vol. 8, pp. 291–330). World Bank.

Elvin, P., & Hamid, N. I. N. B. A. (2016). Ownership structure, corporate governance and firm performance. *International Journal of Economics and Financial Issues, 6*(3), 99–108.

Fama, E. F., & Jensen, M. C. (1983). Separation of ownership and control. *The Journal of Law and Economics, 26*(2), 301–325.

Fėlix, E. G. S., & David, D. S. T. (2019). Performance of family-owned firms: The impact of gender at the management level. *Journal of Family Business Management, 9*(2), 228–250.
Filatotchev, I., Lien, Y. C., & Piesse, J. (2005). Corporate governance and performance in publicly listed, family-controlled firms: Evidence from Taiwan. *Asia Pacific Journal of Management, 22*, 257–283.
Franks, J. (2020). Institutional ownership and governance. *Oxford Review of Economic Policy, 36*(2), 258–274.
Gao, L., & Kling, G. (2008). Corporate governance and tunneling: Empirical evidence from China. *Pacific-Basin Finance Journal, 16*(5), 591–605.
Gomez-Mejia, L. R., Cruz, C., Berrone, P., & De Castro, J. (2011). The bind that ties: Socioemotional wealth preservation in family firms. *The Academy of Management Annals, 5*(1), 653–707.
Greenaway, D., Guariglia, A., & Yu, Z. (2014). The more the better? Foreign ownership and corporate performance in China. *The European Journal of Finance, 20*(7–9), 681–702.
Gregory, H. J., & Simms, M. E. (1999, October). Corporate governance: What it is and why it matters. In *9th International Anti-Corruption Conference, Kuala Lumpur* (pp. 10–15).
He, C. X., Soh, W. N., Ong, T. S., Lau, W. T., & Zhong, B. (2022). Analysis of equity disputes in listed companies with dispersed ownership structure and protection of small and medium shareholders' interests. *Frontiers in Psychology, 13*, Article 857585.
Holderness, C. G., Kroszner, R. S., & Sheehan, D. P. (1996). *Were the good old days that good? The evolution of managerial stock ownership since the great depression* (No. 131). Working Paper.
Hooper, V., Sim, A. B., & Uppal, A. (2009). Governance and stock market performance. *Economic Systems, 33*(2), 93–116.
Iskandar, S. (2014). Shareholder types, their concentration and its effects on demutualized exchanges' operating and financial results-an empirical study. *Corporate Ownership & Control, 11*(4), 114–130
Khan, H. (2011, December). A literature review of corporate governance. In *International Conference on E-business, management and Economics* (Vol. 25, pp. 1–5). IACSIT Press.
Khanna, T., & Palepu, K. (2000). Is group affiliation profitable in emerging markets? An analysis of diversified Indian business groups. *The Journal of Finance, 55*(2), 867–891.
Kim, H. S., & Cho, K. S. (2021). The ownership of the largest family blockholders and Korean firm risk. *The Journal of Asian Finance, Economics and Business, 8*(3), 287–296.
Kirchmaier, T., & Grant, J. (2005). Corporate ownership structure and performance in Europe. *European Management Review, 2*(3), 231–245.
Koehn, D., & Ueng, J. (2005). Evaluating the evaluators: Should investors trust corporate governance metrics ratings? *Journal of Management & Governance, 9*, 111–128.
La Porta, R., Lopez-de-Silanes, F., & Shleifer, A. (1999). Corporate ownership around the world. *The Journal of Finance, 54*(2), 471–517.
Larcker, D., & Tayan, B. (2020). *Corporate governance matters*. FT Press.
Lemmon, M. L., & Lins, K. V. (2003). Ownership structure, corporate governance, and firm value: Evidence from the East Asian financial crisis. *The Journal of Finance, 58*(4), 1445–1468.
Liu, Q., & Tian, G. (2012). Controlling shareholder, expropriations and firm's leverage decision: Evidence from Chinese Non-tradable share reform. *Journal of Corporate Finance, 18*(4), 782–803.
Mallin, C. A. (2016). *Corporate governance*. Oxford University Press.
Manzaneque, M., Merino, E., & Priego, A. M. (2016). The role of institutional shareholders as owners and directors and the financial distress likelihood. Evidence from a concentrated ownership context. *European management journal, 34*(4), 439–451.
Meckling, W. H., & Jensen, M. C. (1976). Theory of the Firm. *Managerial Behavior, Agency Costs and Ownership Structure, 3*(4), 305–360.
Miller, D., & Le Breton-Miller, I. (2005). *Managing for the long run: Lessons in competitive advantage from great family businesses*. Harvard Business Press.
Miyajima, H., Hoda, T., & Ogawa, R. (2015). Does ownership really matter? The role of foreign investors in corporate governance in Japan. *IDEAS Working Paper Series from RePEc*.

Miyajima, K., Mohanty, M. S., & Chan, T. (2015b). Emerging market local currency bonds: Diversification and stability. *Emerging Markets Review, 22*, 126–139.
Mohanty, P. (2003). Institutional investors and corporate governance in India. *National Stock Exchange of India Research Initiative Paper*, (15).
Mollah, S., Al Farooque, O., & Karim, W. (2012). Ownership structure, corporate governance and firm performance: Evidence from an African emerging market. *Studies in Economics and Finance, 29*(4), 301–319.
Monks, R. A., & Minow, N. (2011). *Corporate governance*. John Wiley & Sons.
Morck, R. K. (Ed.). (2007). *Concentrated corporate ownership*. University of Chicago Press.
Morck, R., Shleifer, A., & Vishny, R. W. (1988). Management ownership and market valuation: An empirical analysis. *Journal of Financial Economics, 20*, 293–315.
Mülbert, P. O. (2013). Corporate governance of banks. In *Risk management and corporate governance* (pp. 243–281). Routledge.
Nuzula, N. F., & Lokuwaduge, C. S. D. S. (2017). Do ownership structures really matter? A study of companies listed on the Indonesia stock exchange. *Asia-Pacific Development Journal, 24*(1), 55–82.
Ozili, P. K., & Uadiale, O. (2017). Ownership concentration and bank profitability. *Future Business Journal, 3*(2), 159–171.
Pargendler, M. (2011). State ownership and corporate governance. *Fordham l. Rev., 80*, 2917.
Perotti, E. (2004). *State ownership: A residual role?* (Vol. 3407). World Bank Publications.
Pfeffer, J., & Salancik, G. R. (1978). *The external control.*
Ramady, M. A. (2010). The Saudi Arabian economy: Policies, achievements, and challenges.
Reyna, J. M. S. M., Vázquez, R. D., & Valdés, A. L. (2012). Corporate governance, ownership structure and performance in Mexico. *International Business Research, 5*(11), 12.
Rusmin, R., Tower, G., Achmad, T., & Neilson, J. (2011). Concentrated family ownership structures weakening corporate governance: A developing country story. *Corporate Ownership and Control, 8*(2), 95.
Saridakis, G., Lai, Y., Muñoz Torres, R. I., & Mohammed, A. M. (2018). Actual and intended growth in family firms and non-family-owned firms: Are they different? *Journal of Organizational Effectiveness: People and Performance, 5*(1), 2–21.
Short, H., & Keasey, K. (2005). Institutional shareholders and corporate governance in the UK. *Corporate Governance: Accountability, Enterprise and International Comparisons, Chichester: John Wiley and Sons*, 61–96.
Shubita, R., & Shubita, M. (2019). The impact of foreign ownership on corporate governance: Evidence from an emerging market. *Investment Management and Financial Innovations, 16*(2), 101–115.
Singal, M., & Singal, V. (2011). Concentrated ownership and firm performance: Does family control matter? *Strategic Entrepreneurship Journal, 5*(4), 373–396.
Singh, M., & Davidson, W. N., III. (2003). Agency costs, ownership structure and corporate governance mechanisms. *Journal of Banking & Finance, 27*(5), 793–816.
Sivathaasan, N. (2013). Foreign ownership, domestic ownership and capital structure: Special reference to manufacturing companies quoted on Colombo stock exchange in Sri Lanka. *European Journal of Business and Management, 5*(20), 35–41.
Teubal, M. (2004). Rise and collapse of neoliberalism in Argentina: The role of economic groups. *Journal of Developing Societies, 20*(3–4), 173–188.
Thomsen, S. (2004). Corporate values and corporate governance. *Corporate Governance: The International Journal of Business in Society, 4*(4), 29–46.
Tsionas, M. G., Merikas, A. G., & Merika, A. A. (2012). Concentrated ownership and corporate performance revisited: The case of shipping. *Transportation Research Part e: Logistics and Transportation Review, 48*(4), 843–852.
Tunay, K. B., & Yüksel, S. (2017). The relationship between corporate governance and foreign ownership of the banks in developing countries. *Contaduría y Administración, 62*(SPE5), 1627–1642.

Vives, X. (2000). Corporate governance: Does it matter. *Corporate Governance: Theoretical and Empirical perspectives*, 1–21.

Wahl, M. (2006). The ownership structure of corporations: Owners classification & typology. *EBS Review, 2*(21), 94–103.

Wang, J., & Wang, X. (2015). Benefits of foreign ownership: Evidence from foreign direct investment in China. *Journal of International Economics, 97*(2), 325–338.

Weiss, D. (2014). Internal controls in family-owned firms. *European Accounting Review, 23*(3), 463–482.

Welge, M. K., &Eulerich, M. (2014). *Corporate-governance-management*. Springer Fachmedien Wiesbaden.

Xu, X., & Wang, Y. (1997). *Ownership structure, corporate governance, and corporate performance: The case of Chinese stock companies* (Vol. 1794). World Bank Publications.

Xu, X., & Wang, Y. (1999). Ownership structure and corporate governance in Chinese stock companies. *China Economic Review, 10*(1), 75–98.

Yu, M. (2013). State ownership and firm performance: Empirical evidence from Chinese listed companies. *China Journal of Accounting Research, 6*(2), 75–87.

Yudaeva, K., Kozlov, K., Melentieva, N., & Ponomareva, N. (2003). Does foreign ownership matter? *The Russian Experience. Economics of Transition, 11*(3), 383–409.

Zhang, X., Yang, X., Strange, R., & Zhang, Q. (2017). Informed trading by foreign institutional investors as a constraint on: Evidence from China. *Corporate Governance: An International Review, 25*(4), 222–235.

Zhang, X., Yu, M., & Chen, G. (2020). Does mixed-ownership reform improve SOEs' innovation? Evidence from state ownership. *China Economic Review, 61*, Article 101450.

Chapter 5
The Role of the Board of Directors in Corporate Governance: A Case Study of Byju's

Kabirdoss Devi and V. Vardhini

Abstract Corporate Governance is weighted highly as the companies going globally and the stake holders of the companies increasing significantly. This chapter discusses about the board of directors in the governance structure of modern corporations by taking Byju's as a case study—a company having been at the forefront of the list of edutech companies. The impact of the board composition, diversity, and functioning on the strategic direction, risk management, and overall performance of the company is, therefore, examined with respect to Byju's. The chapter opens with a brief explication of the theoretical framework about corporate governance with respect to fiduciary duties and responsibilities of board members. The second is challenges which are specific to technology-driven firms like Byju's: leapfrogging market changes, keeping pace with regulatory compliance, and impetus for innovation. The focus falls on the composition structure of the board: executive versus non-executive directors, experience counting independence. It sits through a review of how the board structure has undergone strategic development ever since Byju's transitioned from just a start-up to a global edutech giant, overcoming problems arising from fundraising to international expansion and integration of acquisitions. There is also discussion on the role played by the board in developing the culture of compliance and ethical conduct and how it impacts the response of Byju's toward external scrutiny and criticism. The chapter has insights into how effective the board is with regard to aligning the interests of shareholders, employees, and customers. This would be done through a complete review of board meetings, committee structures, and decision-making processes at Byju's. Finally, there are lessons from the governance model of Byju's, and recommendations to companies, including those in the technology sector, on how to improve the effectiveness of boards toward long-term sustainability and success.

Keywords Byju's · CSR · Stakeholders · Governance · Board of directors

K. Devi (✉) · V. Vardhini
School of Management Studies, Vels Institute of Science Technology and Advanced Studies (VISTAS) Pallavaram, Chennai, India
e-mail: devikabirdoss@gmail.com

V. Kandpal et al. (eds.), *Developments in Corporate Governance*,
Approaches to Global Sustainability, Markets, and Governance,
https://doi.org/10.1007/978-981-96-6366-8_5

5.1 Introduction

Corporate governance assumes great importance in India if there has to be economic growth with investment and trust in the business environment. It would help to have transparency, accountability, and fairness in management and control of companies. Where family-owned business concerns abound in a country, good corporate governance helps strike a balance between the interests of majority and minority shareholders.

The risks drop, corruption comes down, and operational efficiency goes up. There is room for better decision-making by adding independence of directors and expertise in multifarious disciplines that bring in diversity into the decision-making process—very critical in the dynamic economic landscape that India is in.

These standards of governance have been enhanced multi-fold under the Companies Act, 2013, and SEBI guidelines. However, there are challenges to their implementation and enforcement. Improved corporate governance framework can help raise India's global competitiveness, attract higher inflows of foreign direct investments, and build conditions for long-term sustainable growth. The reason for corporate governance is to make available to the world a resilient, ethical, and successful business if India has to be a lead player for economic power.

While corporate governance typically works to the advantage of companies, it also presents challenges to owners. Theoretically, diminished control can prevent owners from doing what they want, because of oversight by boards and independent directors. Corporate governance, including structures, audit, and compliance measures, can be expensive to erect and maintain. Governance processes can be time-consuming, thus slowing the decision-making process and reducing the agility in high-speed markets.

Disclosure requirements may compel business owners to reveal proprietary business information that might benefit competitors. It also involves focusing too much on quarterly targets and not enough on long-term goals. Governance that is too strict may strip businesses of flexibility and hamper the degree of creativity that firms can propose to fully realize the idea of business practice or strategies.

Stricter scrutiny and more stringent accountability add to the legal exposure of owners. More stringent governance rules can scare away some executives, and holding onto the best talent becomes tougher. Power structures shake up in a family-run business. From this perspective, too much governance process may breed too much red tape that will make anything efficient totally unworkable. These disadvantages have to be balanced with the general benefits of good corporate governance.

5.2 Brief Over View of the Case

Byju's is an Indian multinational ed-tech company, founded in 2011 by Byju Raveendran and headquartered in Bangalore. The company offers personalized learning programs for students from kindergarten through 12th grade and for those preparing competitive exams like IIT-JEE, NEET, CAT, and IAS. Combining interactive videos, quizzes, and assessments, the Byju's app has achieved tremendous results in making learning fun and interesting.

The journey of the company from a small start-up to that of a global ed-tech giant is marked by its innovative use of technology combined with robust educational content. The acquisition of a string of platforms such as Osmo, WhiteHat Jr, and Aakash Educational Services led to rapid expansion for Byju's. But this came at a cost. Its shift to selling hardware like tablets and SD cards preloaded with content diluted its core focus on education. Failure to repay the $300 million loan in 2021 scared investors due to financial mismanagement. Further governance-related controversies in 2023 still incurred a loss of trust and proved that with this very competitive ed-tech space, growth must go with the core values of education.

5.3 About Byju's

Byju's is a multinational ed-tech company that offers engaging, personalized learning programs to students from kindergarten to the 12th grade and those appearing for competitive exams like IIT-JEE, NEET, CAT, and IAS. No other app is at par in comparison to interactive videos, quizzes, and assessments so that one learns through enjoyment.

Abhisheet Kumar (2024) in his article briefed that the case of Byju's showcases tremendous growth from a small, out-of-the-box idea into one of the largest ed-tech companies globally. The reason the firm succeeds is through the harmonious fusion between technology with strong educational content and styles of learning diversity. It grew into a global player by the acquisition of multiple education platforms such as Osmo, WhiteHat Jr, and Aakash Educational Services. This rapid growth has been framed with controversy over high valuations and aggressive marketing tactics, but it is still one of the very potent forces of change in education through technology and access to quality learning for millions of learners worldwide.

5.4 Fall of Byju

Rafsan Billah (2024) in his articles stated that Byju's was strong at first, having built its muscle in education services, but it was diluted with the introduction of hardware sales, like tablets and SD cards, pre-installed with educational content. This made more financial sense but diluted the very core essence of the brand and created questions in people's minds on what the priorities of the company were, alienating users seeking quality educational content. Byju's aggressive expansion has been at the root of serious money problems—cash flow mismanagement, heavy debt, and prominent loan defaults.

The company missed the on-schedule repayment in 2021 of a $300 million loan taken from a Singapore-based entity that had already triggered warnings about the financial health of the institution and shaken the confidence of investors. (Gupta, 2023) In 2023, governance and ethical concerns rocked Byju's, with the resignation of auditor Deloitte and the exit of three board members, marring its internal governance practices and reducing stakeholder confidence. Its aggressive marketing and sales-driven strategy, which propelled it to exponential growth, soon turned against it amid the rising public ire over their negative impact on customer trust.

The most critical mistake that the company did was to shift away from its core purpose of providing quality education. In the race to scale up and diversify at a fast pace, Byju's changed from enriching the learning experience to aggressive scaling and monetization, consequently affecting product quality and a loss of trust among their core audience: students and their parents. These challenges requires a strong strategies that are in line with core values, and extending to the fullest ability to the limit possible to overcome the competition in the EdTech sector.

5.5 Objectives of the Case Study

The objective of the case study is to-

1. Provide information on how Byju's transition from an educational services provider to a hardware seller has impacted the company's core values and brand identity, thereby assess the consequences of this shift.
2. Bring out the mistakes done through Byju's financial strategies that affected the cash flow management, excessive debt accumulation and loan defaults due to unplanned rapid expansion.
3. Rule out the reasons for auditor resignations and board member departures, to determine the company's internal governance challenges.
4. Analyze Byju's aggressive marketing and sales tactics which resulted in their initial success followed by subsequent public backlash and the consequences that affected the customer trust and brand reputation.
5. State the reasons and show how Byju's deviation from its primary goal of providing quality education changed the product quality and customer trust, particularly among students and parents.
6. Assess lessons learnt from Byju's experience as to the need to be in sync with the growth strategies and the core values and the downside of doing too much too fast within the environment of competition in the ed-tech sector.

7. Suggest recommendations for Byju's to restore with its core educational mission, improve financial management, enhance governance, and re-build stakeholder trust.

5.6 Problem Statement

Byju's, a leading ed-tech company, faced significant challenges as it shifted from their core educational services business to hardware sales. This resulted in brand dilution and user alienation. Governance issues highlighted by the resignation of auditor Deloitte and three board members also undermine stakeholder confidence. Aggressive marketing strategies generated public backlash and damaged customer trust and the company's reputation. This case study aims to explore these issues including aligning growth strategies with core values and the risk of over-expansion in the highly competitive ad tech sector.

5.7 Discussion

Byju's app was created by Think and Learn Pvt. Ltd., a company founded in 2011 by Byju Raveendran, Divya Gokulnath, and a group of students. Byju, originally an engineer, had been coaching students in mathematics since 2006. Initially, the company focused on providing online video-based learning programs for the K-12 segment and various competitive exams. In 2012, Byju's earned a spot in the Deloitte Technology Fast 50 India and Deloitte Technology Fast 500 Asia Pacific rankings and has maintained its presence there ever since.

5.8 Shift in Business Model

Gaurav Kumar and Padmini Muvel (2019) in 2015 studied about the Byju's e learning app among children. They made content analysis and confirmatory test to check on the effectiveness of Byju's e- learning app among children. They identified in their study that Byju concentrated in the four subjects mainly such as Mathematics, Chemistry, Physics, and Biology to start with for the classes from 4 through 12 inclusive of effective delivery for the competitive exams. Byju created the video lectures chapter wise for the effective learners understanding. They had become popular with the content made available in online portals, through solving queries on subject posted by the students online, quizzes and short notes made available for everyone considered as a strategic move attracting many students join the byjus.

Rafsan Billa (2024) in his article mentioned the shift of Byju from the core values and how the consequences affected its business. Byju's initially built its reputation and success on its innovative educational services, offering engaging, personalized learning programs through its app. However, a significant shift in strategy occurred when the company began focusing on hardware sales, including tablets and SD cards preloaded with educational content. While this move proved financially lucrative, it also diluted the essence of the Byju's brand, which was originally centered on delivering high-quality educational content and learning experiences.

Changes in hardware sales shifted Byju's attention and resources away from their core educational services. This has raised concerns about the company's priorities among users and stakeholders. This strategic focus created a connection with Byju's core user base where students and parents demanded reliable and effective educational content of hardware products. But it appears to be more focused on making money than on the company's basic mission of providing quality education. This brought huge tension on the stakeholders of Byju's at all level.

In corporate governance terms, this geographically based change of emphasis provoked basic questions about the overall strategic direction and decision-making. The shift in strategy implied inefficiency in monitoring by the board of directors and called for improvements in the organizational governance to better reflect organizational values. As for the board, they are charged with responsibility over strategic directions as well as ensuring that the company's focus is on its education mission put them under the spotlight. Also, the openness of the decision on such strategic changes and the ability of independent directors to assess the consequences of such changes in the long-term were questioned. This particular scenario highlighted the need to have a good corporate governance structure as a key for implementing growth strategies without straying off from the vision and Identity of the company, while at the same time ensuring that it always retains the trust of its' stakeholders.

5.9 Accounting Practices and Management of Funds

Before, Byju's used another way of recognizing revenues, and so had to practice Ind-AS 115 changes. By these standards, when one agrees to multiple year revenues that requires two, three or four years, the full amount cannot be recorded in the first year's financial statements but has to be spread over the term of the revenues contract. Based on the above analysis, it was detected by the auditor and used to make the change discussed in this figure. From the financial quarter ending September 2022, instead of recording streaming revenues fully in the usage term, Byju's started recognizing them over the usage term, in the contracts. Thus, this new accounting methodology of Hubble has reduced the capacity of Byju's to recognize the income upfront.

Moreover, earlier, Byju's used the funding capital to acquire domestic as well as international business companies and brands under it. Nevertheless, company's fast growth strategy which is associated with the identification of expanding territories has provoked investors' interest. Some of them have questioned Byju's fast expansion by deploying capital for international business and acquisitions and its feasibility and relevance. It was this extra attention which underlines the necessity to strengthen the provisions of corporate governance to prevent reckless expansion and focus on sustainable value creation.

From the lens of corporate governance, (Mishra et al., 2022) these problems signify the need to have proper disclosure of the company's finance and efficient strategies for governance. The requirement of accounting principles and policies to meet are important in order to provide a perfect picture of the financial state of the company and its revenues. Also, the problems revealed by investors about the accelerated rate of capital spending for acquisitions point to the need to board of directors to take stringent supervision to guarantee that the expansion plans will not compromise the overall organizational mission and that it will be manageable to deliver. Strong governance systems are especially important for Byju's to uphold its investors' trust while still keeping their financial and ethical integrity while the company expands.

Founded in 2011 by Byju Raveendran, Divya Gokulnath, and a group of students, Byju's secured seed funding from Aarin Capital in 2013. By 2019, the company had raised nearly $785 million (Thomas, 2018) from prominent investors like Sequoia Capital India, Chan Zuckerberg Initiative (CZI), and Tencent. By March 2018, (Vignesh et al., 2018) Byju's became a unicorn valued at ₹6,505 crore ($1 billion). It attained decacorn status in June 2020, valued at $10.5 billion following investment from Bond Capital, and reached a $22 billion (Singh, 2020) valuation by March 2022 after raising $800 million.

Despite its rapid expansion and strategic sponsorships, including the India national cricket team and Kerala Blasters FC, Byju's faced financial and governance challenges. The company shifted its revenue recognition practices in September 2022 to comply with Indian Accounting Standards (Ind-AS) 115, reducing its advance income bookings. Aggressive capital deployment for acquisitions and international growth drew investor skepticism.

In 2022, Byju's laid off nearly 4,000 employees amid a funding crunch and profitability push. The company faced litigation and grievances from various stakeholders. By December 2023, Byju's grappled with severe financial difficulties, leading to the sale of key assets to raise funds. Bankruptcy proceedings initiated by lenders in January 2024 resulted in Byju's U.S. division filing for Chapter 11 bankruptcy. A U.S. bankruptcy court penalized Byju's directors in May 2024 for contempt, and in July 2024, the National Company Law Tribunal appointed an interim insolvency resolution professional following unpaid dues allegations by BCCI.

The above table shows the (Sharma et al., 2022; Tyagi, 2021) data on the revenue and Profit & Loss in Crores. Byju's remarkably shown the growth and then the progressive losses due to financial turbulence (Table 5.1).

5.9.1 Governance and the Ethical Issues

Prosus is a leading global technology investor and BYJU's largest non-promoter shareholder. It has lambasted inadequate reporting and governance structures of the

Table 5.1 Revenue and profit and loss for the years 2018–2022

Year	Revenue (In crores)	Profits/Loss (In crores)
FY 2018	Δ490	Δ−29
FY 2019	Δ1376	Δ−8.82
FY 2020	Δ2381	∇−262.1
FY 2021	∇2280	∇−4558
FY 2022	Δ5298	∇−8245

Source The economics times and money control

ed-tech company. Peak XV Partners has also mentioned their inability to affect any improvement in compliance and governance norms in management of BYJU's as the reason for their director's resignation from the board. A milestone event in June 2023 was the resignation from BYJU's board by nominee directors of key investors Prosus, Peak XV Partners, and the Chan Zuckerberg Initiative.

Such isolated instances of governance lapses at BYJU's reflect trends of corporate mis-governance in various start-ups. This trend is aligned with the decrease of private capital inflows into the Indian start-up system in the backdrop of global financial market sentiments, which have remained quite muted over the past year.

Many remedies are suggested by the stakeholders to deal with such failures in governance. In-built ESG norms would help in pushing compliance and good practices. Recommendations from policymakers, industry bodies, and other major stakeholders said that self-regulation would act as a catalyst in bettering corporate governance. This is supported by the recently released Start-up 20 Engagement Group report under the Indian presidency of G20 and the Start-up Governance Playbook by IVCA.

Two recent reports underline that, in the case of start-ups, good governance structures require alignment between founders, investors, and the board, but need to evolve over time as start-ups mature. Adopting robust governance practices from the outset can foster a culture that allows for adaptive growth within a framework that encourages investor confidence and sustainable business practices as part of the start-up community.

5.9.2 Aggressive Sales Tactics and Advertisements

Some dodgy sales practices by WhiteHat Jr., (Bhushan, 2020) a wholly-owned subsidiary of Byju's, have been in the spotlight. Now, attention shifts more to wider problems with corporate governance. Following this, the Advertising Standards Council of India ordered WhiteHat Jr. to withdraw five TV ads for misleading content (Chen et al., 2021) and hard-sell tone. One famous case (Yahoo News, 2020) involved the invention of a child prodigy, "Wolf Gupta", who was supposedly offered very lucrative job offers, a story that proved to be a fabrication. The episode underscores the fact that good corporate governance needs to have ethical marketing practices as one of its elements.

In November 2020, WhiteHat Jr. filed a ₹20 million defamation suit against software engineer Pradeep Poonia (2021) for saying that the firm was making spurious claims and using (The Indian Express, 2021) unqualified teachers. The case was later withdrawn, which puts on tenterhooks the dichotomy between aggressive defense of corporate reputation and accountability. A data leakage (NDTV Gadgets 360, 2020) where over 200,000 users' personal information was compromised raises some serious concerns over data protection and privacy standards being followed, which constitutes an integral part of corporate governance.

The Department of Consumer Affairs expressed (Sinha, 2023) concern over aggressive sales and deceptive marketing strategies employed by EdTech companies, more precisely Byju's and its subsidiaries, in the meeting with India Edtech Consortium on 24 June 2022. DCA advised Byju's to extend cooperation in the process with ASCI so that these complaints can be properly redressed. Byju's has assured that it will get these issues resolved with a detailed action plan.

These incidents bring out the critical interplay between sales practices and corporate governance. Issues of ethical marketing, transparency, protection of customer data, and accountability go to the very heart of consumer trust and integrity in business. It will be able to re-build its reputation and ensure sustainable business practices if Byju's addresses (Outlook India, 2022) such governance issues.

5.9.3 Deviation from Primary Goal

The foundation stone of Byju is to provide quality education using technology. This was their core competency and strategy that differentiated Byju's from others. Aryan Gupta (2023) have discussed in his article on Byju's in 2023 that in the pursuit of rapid ascent and diversification the company started losing the sight of their goal and that's when the problem started when they began up-selling and cross-selling. Rafsan Billah (2024) in his case study on Byju's in 2024 presumed that Byju's deviated itself from its core competency of quality education to concentrating of selling hardware items to students and parents through ambitious and sometimes deceitful sales techniques. Due to the aggressive selling the Byju's started losing the trust of students and parents. Parents started to see these strategies as monetization strategies rather than for the educational purpose. Dr. Sinha (2023), Directorate of Academics, ICSI, has concluded in his case study that the take away from the Byju's case is the integral role of corporate governance while taking strategic decisions and to ascertain that the growth strategies are in align with the company's core competencies and also to maintain the ethical standards while cross-selling and up-selling.

Arora and Shah (2023) in a news article in 2023 concluded the learnings from Byju and that is you should not deviate from your core competency for the sake of cross-selling. Cross-selling is an important marketing strategy but it should not be at the cost of core competency. Customer should not take it as pushy selling techniques rather it should be viewed as complementary to the core competency. Company must understand that sustainable growth is more important than rapid growth. It is important to incorporate the governance principles so that the companies can smoothly sail through the complexities of growth while protecting the trust and integrity of the customers.

5.9.4 Risks of Overextension in the Competitive Ed-Tech Industry

Byju's became a pride in the Indian Start-up Ecosystem with its audacious dreams (Cornish et al., 2022), and strategic maneuvers. Not once but twice Byju Raveendran scored perfect score in CAT but decided not to study in IIM rather than teach millions of children around the world with their edutech company. Byju's have seen pinnacles of success in a very short span of time at the same time it has faced lots of strategic challenges. Pankaj Shrestha (2024) discussed in his article in 2024 that questions must be raised regarding the sustainability of the growth model whenever expenditures are surpassing the company's revenue.

For building a brand marketing plays a vital role. And Byju's have left no stones unturned when it is about marketing like getting sponsorship in Indian Cricket team and FIFA world cup and getting endorsements from world class celebrities like Shah Rukh Khan and Lionel Messi. But while doing an aggressive marketing, the company should not deviate from the sight of its core competency, on the principle which company was built. Aggressive sales techniques of Byju's where they have targeted parents to get course membership has led to have a bad reputation and lack of trust.

Vivek Goyal (2024) in 2024 stated in his article that Byju's have been operating in a landmine because Byju is an edutech company and in our country edutech industry is highly regulated and still evolving. There are regulations regarding data privacy and financial reporting etc. Working in an already complex regulatory framework in our country, Byju' expanded into international markets also that led to even more diverse regulatory environment. Divya Gokulnath, Byju Rveendran's wife and also Byju's board member once stated that "We are pioneers in Edutech, not pioneers in perfection". Sayan Chakraborty (2023) stated in his article in 2023 that poor governance and bad ethic always bring unmitigated damage to the company. Ronnie Screwvala, Chairman of UpGrad commented that it is crucial time when board has to understand their responsibilities that they are not just some art piece in the company's fund raiser but it is their duty to ask hard questions.

5.9.5 Why Board Members Failed to Detect Financial Mismanagement

In 2023–2024, BYJU's has faced severe scrutiny due to financial mismanagement and governance issues. Integrity of board members were questioned because of lack of transparency, delayed financial statements, and cash flow problems etc.

While investigating the financial downfall of BYJU's, it was identified that there was lack of transparency and that in BYJU's more than often the acquisition details are not fully disclosed to the directors. Meetings were convened on the short notice to approve major deals. This inadequate disclosure led the board's inability to meticulously assess and overview any major financial decisions that in turn led to unchecked managerial actions.

Company also lacked in appointing professional officers to effectively manage finances and compliance the government rules. This lack of management led to operational inefficiencies and also lack of rigorous financial control.

Significant delays were experienced in BYJU's for preparing and releasing financial statements. Timely assessment of company's financial health were significantly delayed due to such obstructions. These obstructions also hindered any proactive measures to address any emerging financial issues.

Another crucial factor in BYJU's downfall is the disruption in the board members. Due to the discord over internal control and business processes, many board members along with the major investors like Prosus Ventures and Peak XV Partners resigned. These significant withdrawals enfeeble BYJU board's oversight capacity and also disrupted the continuity in the governance.

BYJU's also failed to provide any necessary documents and the required support for the financial audits to their auditors like Deloitte and BDO therefore the auditors too resigned. This kind of insufficient response from the board indicated the board's lack of commitment to financial transparency and accountability.

The board showed poor oversight of the company's various aggressive acquisitions, inadequately assessing the strategic fit, valuation, and integration risks of multiple high-value acquisitions, including WhiteHat Jr ($300 million) and Aakash Educational Services ($950 million). All of these acquisitions were executed with little or no due diligence and even less post-merger integration planning, much to the detriment of the firms' firms from a value destruction and operational perspective.

The board also failed to properly question the sustainability of BYJU'S business model and rising losses. While it's possible they thought growth metrics were an end in itself, the board seemingly ignored glaring warning signals like significant customer acquisition costs, rising levels of customer complaints regarding sales practices, and poor unit economics. The company's aggressive revenue recognition practices and increasing losses should have warranted more serious board-level conversations about what was happening in terms of moving the business model forward and what strategic pivots might be necessary to get it there.

The board didn't pay adequate attention to cash flow management and working capital needs, a lapse that turned out to be particularly expensive. The breakneck growth and the acquisition binge resulted in significant working capital requirements and delayed collections at high operating costs further amp up liquidity pressures. As the company began to experience severe liquidity issues, and delays in suppliers and employee payment became apparent, the board failed to implement proper cash flow monitoring systems, and working capital controls. The board's already addressing issues like these were made worse by the board.

The BYJU'S start-up growth mind set mirrored a trend seen elsewhere in high-growth start-ups, namely the prioritization of aggressive expansion over rules-based governance. This resulted in a sort of myopia where the companies single-minded pursuit of growth and valuation took precedence over extremely worrying governance issues, and a culture where the idea of growth at all costs created an organization where there were no barriers to that growth, but that gave no thought to operational realities. This became clear from their aggressive sales practices, swift geographic expansion and ambitious acquisition strategy, all undertaken without use of adequate risk assessment or governance frameworks.

The pressure of investors for immediate returns was instrumental in sacrificing long-term sustainability. Each round had brought new investors willing to invest at ever-higher valuations, and those investors then sought more exponential growth to justify those valuations, thus creating a cycle where the company seemed to be forced to favor short-term growth numbers over sustainable business practices. This pressure from investors resulted in lofty revenue goals, accelerated scaling of operations, market share dominance, aggressive product road maps, and many more, often at the cost of financial prudence and operational stability.

The learning that is spread within the company is not even about operational efficiency, but about growth metrics. So, there is a core orientation with an attitude that the growth will continue. Performance was primarily assessed on metrics of user acquisition, market expansion and revenue growth, while important operational metrics such as unit economics, customer satisfaction and operational efficiency, were often considered secondary. This misplaced focus marginalized this need, causing wasteful use of resources and unsustainable cost structures as well as a lack of focus on creating strong internal systems and processes.

5.9.6 *Byju's Board Structure and Expertise in Risk Management*

One of the reasons for the failure of corporate governance in Byju's is the structure of the board and the expertise in risk management.

5.9.7 Board Structure

The Byju's Board composed of power concentrated among the Founders and resulted in the Founder-Centric Leadership. (Ray et al., 2024) Founder Byju Raveendran controlled decision-making with fewer checks and this power reduced the effective questioning or examination of key financial choices. Byju Raveendran did not allow the board members from various fields including finance, law and risk management thus resulting in the poor help in providing broader perspectives on potential risks. Because of dominant founder centric approach, the board initially focused on growth and scalability rather than risk management. This led to the insufficient financial transparency, acquisitions and compliance. Byju's authoritarian leadership style and the interventionist control of a small group of trusted players could have created an environment vulnerable to groupthink, where multiple viewpoints and dissenting opinions were not actively sought or even considered.

5.9.8 Board Independence

Although Byju's initial investors and board members accepted entities like Sequoia and Prosus, the board lacked substantial independent oversight. Independent board members usually play a crucial role in challenging management decisions, pointing out risks and confirming accurate financial reporting. (Vij et al., 2024) In Bjyu's instance, the comparatively smaller number of independent directors might have resulted in a lack of adequate scrutiny of management's behavior, especially as the firm decisions to go for high- growth stage. (Sinha, 2023) When large investors entered into Byju's board, they had their high priority on growth and return on investment. This attitude would have resulted in operational risks or mismanagement resulting in excessive emphasis on scaling instead of internal control strengthening.

5.9.9 Board Expertise

The board did not have members with major experience in financial oversight and risk management. Though the board comprised successful business persons and investors, there is no evident indication of experience in subjects like financial audits, compliance and risk management. (Mittal & Pianko, 2024) This shortage of experience and expertise resulted in making it more difficult for the board to spot warning signs of financial management, for example, deviations in revenue recognition, operation inefficiencies, or excessively aggressive acquisitions. Byju's board members seemed to consist predominantly of people who had excellent entrepreneurial and educational backgrounds but not necessarily in channelizing large- scale corporate risks. This minimal experience in managing financial or operational risks at the level of

Byju's operations seemed to have been a reason why company was unable to deal with issues like high leverage, unwanted acquisitions and aggressive growth.

5.9.10 Risk Management and Governance Failure

When the company's financial procedures raised red flags, the board did not move with speed to address them. This inaction permitted financial malpractice to exist for a substantial duration, eventually culminating in regulatory attention and investor loss of confidence. There was no properly defined risk management structure that would have been able to foresee and cushion the risks posed by Byju's high-growth strategies. Board members did not make it a priority to create auditing mechanisms or independent audits to maintain the financial well- being of the company as sustainable.

5.9.11 Effects of Poor Governance

Negligence in risk control has caused damage to Byju's reputation and investor trust. The company's value was reduced, and it came under scrutiny from regulators and investors meaning extreme governance failures. Resignations, (Kalra, 2024), and exit of senior board members such as Sequoia and Prosus are a loss of trust and misdirection in the company. These departures followed key governance breakdowns, further emphasizing the failure of the board to effectively manage risks.

5.9.12 How Good Governance Would Have Helped?

Byju's might have avoided its financial mismanagement woes by having tighter independent oversight so that board members with financial and risk management expertise were actively engaged in decision-making. Better focus on risk management by setting a strong foundation in the form of internal controls, compliance tests, and auditing of finances would have helped to identify risks beforehand. Moreover, regular independent audits and greater disclosure of finances may have alerted at an earlier time so that the board could move ahead before it became too late. Better governance here would have built accountability and investors' trust and facilitated long-term growth.

5.9.13 *Comparing Byju's Governance with Enron and WeWork*

Byju's, Enron and WeWork corporate governance failures share common threads of founder or executive dominance, inadequate board oversight and transparency shortcomings. Byju Raveendran, Byju's founder, exercised overbearing control of company decisions, raising questions about centralized power and inadequate oversight (Independent Directors Data Bank, 2024). Similarly, while Enron was not being led by its founder, the senior managers including Jeff Skilling and Ken Lay had too much authority, leading to unethical accounting and the subsequent collapse of the company (Columbia Law Scholarship, 2003). In WeWork, too much authority by co-founder Adam Neumann meant making reckless decisions that prioritized rapid expansion over sound business procedures, ultimately contributing to financial trouble (WeWork Governance Reports, 2022). These cases highlight the risks of unchecked executive power and the need for proper corporate governance mechanisms.

5.9.14 *Actionable Recommendation*

"One falls but doesn't rise again after the fall, that is failure". Likewise, Byju's had made some mistakes; now is the time to learn from those mistakes, correct them and again reach to the pinnacle. Prof. D Mukhopadhyay (2023) had stated in his article in Daily Excelsior in 2024 that Byju's must concentrate on its core competency and that is their high-quality educational content. It calls for them to invest in the curriculum development and pedagogical research in high-quality educational content.

In her article in 2024, Shreshtha Verma (2024) concluded that the skills of experienced teachers from the traditional teaching methods should be mixed with the technology to enhance the educational experience. More innovative and simulation-based learning needs to be brought in, such as gamified content and virtual labs, whereby students feel that the learning process is more interactive and engaging and another important aspect where Byju's has to work on is the transparency in financial reporting. Restoring faith among the stakeholders, involving customers and investors, is very vital. Byju's must represent the stakeholders clearly about the financial health of the company and must adhere to accounting standards; Byju's must not exaggerate their market projections. One can restore the credibility of Byju's by forecasting realistic growth and achievable goals.

Byju's lost its trust in aggressive marketing. Therefore, Byju's should prioritize ethical marketing practices keeping promotional activities honest, and reflect Byju's genuine commitment to education. It should also focus on teacher training programs and bring out recognition for outstanding educators to retain talent. Dona Das (2023) in her article in 2023 has put forward the fact that to drive maximum student outcome, Byju's has to offer personalized learning paths, which will help meet the different needs and styles of students. Byju's has to focus on user-centric approaches and, at the same time, reassure its commitment to educational excellence. Byju's has to address financial challenges by extensive restructuring.

Sunil Chajjar (2024) states in his article in 2024, "Only through transparent communication and governance reforms can Byju's, at this juncture, re-build stakeholder trust by fostering a culture of accountability that aligns with the expectations of the stakeholders".

From the Byju's case we can learn a valuable lesson for the other players in the Edutech ecosystem in terms of how they are able to deliver on both, the sustained and responsible growth.

5.9.15 Conclusion

Such aggressive expansion experiments of Byju indeed point to the bipolar potential of growth strategies for ed-tech businesses. At first branded as an array of new contents for educational purposes, the business faced stringent challenges when focus shifted to hardware revenue, other than the original mission and ethical framework established. After that point, overly tense marketing and financial blunders weakened stakeholder trust even more and introduced original massive corporate governance problems.

Byju's has to refocus its energy on the core objective: how the higher usage of the technology and usual forms of education are able reached its fundamental mission of delivering the quality education, for example, to invest in the curriculum and ethical advertisement. Shareholders can only be convinced through clear financial records and compliance to accounting best practices. Greater focus on individualized method called custom approach in combination with involving learners and parents might also help in enhancing learning achievements and the confidence of learners as well as parents.

The case of Byju's was one of a well-timed balanced organic growth and was backed by robust corporate governance. If a firm is to encourage sustainable prosperity, it is relevant that the company embrace a culture of professionalism in stewardship, contemporaneous information sharing, and accountability. It is only by means of these systems of expectations placed upon an organization that it may respond to set its course and navigate the convolutions of ed-tech en route to a more stable and confident future.

References

Arora, M., & Shah, A. (2023). Byju's crisis reflects the failure of all corporate governance pillars: Amarjit Chopra. *Economic Times CFO.* Retrieved from July

20, 2023, from https://cfo.economictimes.indiatimes.com/news/governance-risk-compliance/byjus-crisis-reflects-the-failure-of-all-corporate-governance-pillars-amarjit-chopra/101971610

Bhushan, R. (2020, October 28). Byju's WhiteHat Jr. agrees to withdraw ads after ASCI flags code violation. *Economic Times*. Retrieved from November 2, 2020.

Billah, R. (2024). *The rise and fall of Byju's: A comprehensive case study.* LinkedIn. https://www.linkedin.com/pulse/rise-fall-byjus-comprehensive-case-study-rafsan-billah-lx7bc/

Business Today. (2021, May 4). WhiteHat Jr drops Rs 20 cr defamation suit against critic Pradeep Poonia. *Business Today.* Retrieved May 18, 2022.

Chakraborty, S. (2023). Indian corporate governance comes under fire as Byju's stumbles. *Nikkei Asia*. Retrieved from July 3, 2023, from https://asia.nikkei.com/Spotlight/Market-Spotlight/Indian-corporate-governance-comes-under-fire-as-Byju-s-stumbles

Chen, B., Garcia, D. C., Caldas, A. E., Andrade, B., Nicholas, K., Dhesi, K., Costa, L., Huemer, S., & Shekhawat, V. (2021, May 23). *Youth economist compilation: For the youth by the youths.* Benjamin Chen. ISBN 979–8–5056–5091–2.

Cheng, Y.-J., & Doris, A. (2022, April 19). Why wework didn't work as planned: 4 lessons on corporate governance. Leadership & Management Strategy.8

Cornish, C., Singh, J., & Ruehl, M. (2022, October 3). How a teaching app feted by Silicon Valley was left chasing the Indian dream. *Financial Times*. Retrieved from October 8, 2022.

Daily Excelsior. (2023). Rise and fall of Byju's: A corporate governance failure. *Daily Excelsior.* Retrieved July 21, 2023.

Das, D. (2023). *Reviving Byju's: Navigating challenges and rebuilding the EdTech giant*. LinkedIn. https://www.linkedin.com/pulse/reviving-byjus-navigating-challenges-rebuilding-edtech-dona-das/

Gordon, J. N. (2003). *Governance failures of the Enron board and the new information order of Sarbanes-Oxley*. Columbia Law Scholarship Archive. https://scholarship.law.columbia.edu/cgi/viewcontent.cgi?article=4470&context=faculty_scholarship

Goyal, V. (2024, January). *Byju's—An example of risk management failure in EdTech*. GRM Institute. https://grm.institute/blog/byjus-a-case-of-risk-management-failure/

Gupta, A. (2023). *What were the key factors that contributed to the downfall of Byju's, a prominent EdTech company, and what lessons can be drawn from its decline?* LinkedIn. https://www.linkedin.com/pulse/what-were-key-factors-contributed-downfall-byjus-prominent-gupta/

Independent Directors Databank. (2024, July 4). *Corporate governance issues at Byju's*. Independent Directors Databank. https://www.independentdirectorsdatabank.in/newsletter/2024/07/4/1243

Jain, S. C. (2024). *Reflecting on Byju's: A case study of triumphs, trials, and lessons.* LinkedIn. https://www.linkedin.com/pulse/reflecting-byjus-case-study-triumphs-trials-lessons-chhajar-jain--sreuc/

Kalra, A. (2024, September 9). *Byju's auditor BDO resigns after start of bankruptcy proceedings, company says*. Reuters. https://www.reuters.com/technology/byjus-auditor-bdo-resigns-after-start-bankruptcy-proceedings-company-says-2024-09-07/

Kumar, A. (2024, January). Report indicates no fraud at Byju's, Govt says probe is still underway. *Business Standard*. https://www.business-standard.com/companies/news/probe-finds-no-evidence-of-fraud-at-byju-s-but-governance-failure-detected-124062600413_1.html

Kumar, G., & Muvel, P. (2019). Study of Byju's as an e-learning app among children. *Research Journey: International E-Research Journal.*

Mishra, D. (2022). Government airs concerns over complaints against EdTech companies. *The Economic Times*. Retrieved from July 19, 2022.

Mittal, A., & Pianko, D. (2024). *What investors should learn from the fall of EdTech unicorn Byju's.* Cornell SC Johnson College of Business. https://business.cornell.edu/hub/2024/07/01/what-investors-should-learn-from-fall-edtech-unicorn-byjus/

NDTV Gadgets 360. (2020, November 25). *WhiteHat Jr data exposed due to multiple vulnerabilities: Report*. NDTV Gadgets 360. Retrieved from May 18, 2022.

Outlook India. (2022, July 15). Govt expresses concern over alleged malpractices by BYJU's, other EdTech companies. *Outlook India.* Retrieved July 19, 2022.

Ray, S., Jain, K., Birru, P., & Mohata, R. (2024). A case study of BYJU's failure. *World Journal of Advanced Research and Reviews, 21(03)*, 674–689. https://doi.org/10.30574/wjarr.2024.21.3.0765

Sharma, S., & Mishra, D. (2022). Byju's losses swell to Rs 4,588 crore in delayed FY21 results. *The Economic Times.* Retrieved from September 16, 2022.

Shrestha, P. (2024). *Strategic insights from BYJU's journey.* LinkedIn. https://www.linkedin.com/pulse/strategic-insights-from-byjus-journey-pankaj-shrestha-i2fjc/

Singh, M. (2020, June 25). Mary Meeker's bond backs Indian online learning startup Byju's. *TechCrunch.* Retrieved from June 26, 2020.

Sinha, A. B. (2023). *From zenith to nadir—The case of BYJU's corporate governance fiasco.* Directorate of Academics, ICSI.

The Indian Express. (2021, May 6). WhiteHat Jr withdraws defamation case against software engineer. *The Indian Express.* Retrieved from May 18, 2022.

Thomas, P. M. (2018, March 29). *An entrepreneur at heart, Ranjan Pai is the new king of India's healthcare industry.* Moneycontrol.com. Retrieved May 4, 2019.

Tyagi, G. (2021, September 3). *Byju's consolidated revenue grew 82% in FY20 while losses jump 30X*. Entrackr. Retrieved from April 17, 2022.

Verma, S. (2024). *How corporate governance issues led Byju's fall? A lesson for startups.* TICE. Retrieved from July 1, 2024, from https://www.tice.news/tice-trending/corporate-governance-issues-that-led-byjus-fall-a-lesson-for-startups-4788038

Vignesh, J., & Chanchani, M. (2018, March 14). Here's how EdTech startup Byju's quietly turned into a unicorn company. *The Economic Times.* Retrieved from March 24, 2018.

Vij, S., Pandit, S., Kashish., & Saini, S. (2024, April 15). *BYJU's adventure: Rise and fall of Think & Learn Pvt. Ltd. SSRN*. https://ssrn.com/abstract=4880915

Yahoo News. (2020, October 12). *Who is Wolf Gupta? Byju's WhiteHat Jr latest marketing gimmick is making netizens furious, here's what we know so far about the 'fictitious' kid apparently working as AI researcher at Google.* Yahoo News. Retrieved from November 2, 2020.

Chapter 6
Ownership Structure and Their Impact on Corporate Governance

Khushboo Gulati, Seema Gupta, and C. P. Gupta

Abstract Investors are willing to invest in companies that are ethical and transparent in their dealings. Hence, the management's commitment to high governance standards is a magnet in attracting much-needed external funds for the company. Moreover, the theoretical viewpoint of well-known theories like the agency and the pecking order theory has also propounded that adopting a suitable governance mechanism helps to gain investors' confidence by ensuring transparency in management dealings. Although numerous studies have explored the effect of ownership structure on firm performance and corporate governance, research studies that consider the combined effect of corporate governance, ownership structure, and firm performance are rarely seen. Therefore, the present study examines the combined effect of ownership structure and good governance features (board diversity and multiple directorships) on firm performance. The study will make a novel contribution by studying the impact of foreign institutional investors and debt and equity in the ownership structure. The study's sample size comprises 500 BSE-listed firms whose performance is analyzed over ten years. The study found that board diversity and busyness led to a slight reduction in the significance level of the positive impact of equity on market performance. The study will provide valuable insights for policymakers and professionals in the field, keeping them engaged and interested in potential improvements to the institutional environment. Furthermore, the findings suggest that flexible rules should be applied to increase the inflow of equity participation by foreign institutions and domestic investors.

K. Gulati (✉)
School of Management and Commerce, Manav Rachna University, Faridabad, India
e-mail: khushboogulati55@gmail.com

S. Gupta
IIM-Sambalpur, Sambalpur, India
e-mail: seemaguptapd@gmail.com

C. P. Gupta
Lal Bahadur Shastri Institute of Management, Delhi, India
e-mail: guptacp@lbsim.ac.in

V. Kandpal et al. (eds.), *Developments in Corporate Governance*,
Approaches to Global Sustainability, Markets, and Governance,
https://doi.org/10.1007/978-981-96-6366-8_6

Keywords Corporate governance · Ownership structure · Performance · BSE firms · India

6.1 Introduction

The inculcation of good governance practices accelerates the growth of a business firm. The theoretical viewpoint of well-known theories like the agency and the pecking order theory has also propounded that adopting good governance mechanism helps to gain investors' confidence by ensuring transparency in management dealings. However, information asymmetry is relatively high in countries with inadequate governance norms. The unavailability of credible financial information affects the investors' decisions, resulting in a high cost of funds and hampering a firm's ownership structure.

Strong governance solves this problem of information asymmetry by reducing agency concerns and maximizing shareholder wealth (Shleifer & Vishny, 1997). When management is committed to high standards of transparency and accountability, it is easier for equity capital providers to keep a watch on their investments (Cohen et al., 2013). Additionally, an inflow of debt funds is also shown to reduce agency costs and improve firm performance (Ahmed et al., 2023; Grossman & Hart, 1982; Harris & Raviv, 1991; Williams, 2003) by enforcing the rules for managers to act in the best interest of shareholders (Myers, 1984) by limiting the amount of risk they take (Jensen & Meckling, 1976), the choice of investments (Ngatno & Youlianto, 2021), and the conditions under which they can go bankrupt (Grossman & Hart, 1982; Harris & Raviv, 1991; Jensen, 1986). The researchers (Dao & Ta, 2020; Jensen & Meckling, 1976; Myers & Majluf, 1984) have also reported that corporate finance policies effectively avoid agency conflicts.

Researchers in the recent past have examined the effect of ownership structure and governance variables like outside directors (Alqahtani et al., 2022; Jameson et al., 2014), the board size, CEO duality, and board meetings (Balagobei, 2017). However, studies on the board diversity index, board busyness, and ownership structure's effect on firm performance are rarely seen. Hence, the current research study contributes to the existing literature on ownership structure by examining the combined effect of board structure and ownership structure variables on firm performance. Furthermore, capital structure theory based heavily on large firms fails to explain the optimal debt-to-equity mix. Therefore, the choice of capital structure is an important issue for large and small companies. Thereby, a novel attempt has also been made to explore the effect of corporate governance and ownership structure on the performance of large, medium, and small-size firms by categorizing our sample into three categories based on market capitalization. Prior literature reveals that corporate governance impacts decisions regarding the ownership structure of the firms, and the effects can differ in developed and emerging countries (Khan et al., 2019; Pham & Nguyen, 2019; Zhou et al., 2021).

Therefore, the present study has been conducted in Indian settings for the following reasons: Firstly, in contrast to developed countries, India's corporate governance standards allow directors to hold several roles on the board of directors (Eulaiwi et al., 2016). Subsequently, a lack of time and effort on the part of busy directors hinders their ability to monitor effectively (Yasin et al., 2004), resulting in poor governance procedures and ineffective investment decisions (Chou & Feng, 2019; Jiraporn et al., 2012). Furthermore, it has been discovered that gender diversity and board independence influence the firm's ownership structure decision and help better monitor the business firm (Adams & Ferreira, 2009; Chen et al., 2016; Zalata et al., 2018).

This study provides useful guidelines for the corporate sector, financial institutions, shareholders, depositors, and investors as these can help companies to react effectively and efficiently to different economic conditions. Besides this, investors, regulators, and managers can get evocative insights, particularly who seek to improve their company's performance in the capital market through restructuring their ownership structure and board composition. The findings suggest that management adopting good governance practices will be able to raise more funds from FII and equity shares at a lower cost due to their high reputation for compliance with legal practices and procedures. Hence, it is advised that a company's management should aim to build a diversified board to reduce agency costs, design a low-cost ownership structure and thereby enhance firm performance.

The rest of the chapter is presented as follows: The second part discusses the extant literature. Then, the research methodology explains the variables, research model, and statistical technique employed for the analysis in the third section. Subsequently, the fourth section reports the panel data results, and the last section provides a brief chapter summary.

6.2 Literature Review

This section of the study synthesizes diverse aspects of the scholars' work in the context of the impact of ownership structure, corporate governance, and firm performance. This literature provided a strong background for framing hypotheses for the present investigation.

The ownership structure of a company is endogenously driven and differs between enterprises due to costs and benefits (Awan et al., 2016). Similarly, Chow et al. (2018) stated that in companies with higher levels of governance, there is a negative impact of macroeconomic uncertainty on ownership structure. Attig et al. (2021) reported that good governance has been found to lower a firm's equity cost and cost of debt (Funchal & Monte-Mor, 2016) for improving the firm's financing conditions. Chen et al. (2010) claimed good corporate governance is crucial for businesses that immediately need outside equity. Further, Martins et al. (2017) also proposed that in nations with good governance, businesses employ debt with shorter repayment periods that adequately protect debtholders. Managers frequently seek out less financial leverage when faced with the board of directors' effective corporate governance (Wen et al., 2002). Ashraf et al. (2017) noted that corporate governance and corporate funding decisions impact the performance of Pakistan's cement sector.

Additionally, apart from the external factors affecting the ownership structure of business. Chang et al. (2014) asserted that the firm's distinctive characteristics and conflicts of interest between its manager and shareholders impact the ownership structure (Jiraporn et al., 2012). Kieschnick and Moussawi (2018) revealed that managers permit their risk preferences to rule their firm's ownership structure decisions. Similarly, longer-tenured CEOs are more likely to emphasize their interests and be opportunistic when making strategic financial decisions, which results in agency costs for the company (Nadeem, 2020). Further, corporate charter restrictions and board composition also influence the ownership structure decisions (Kieschnick & Moussawi, 2018). Additionally, the board attributes such as CEO duality, insider ownership, board size, the presence of corporate governance committees, and multiple directorships (Andreou et al., 2013); age, gender, and education of the CEO (Tulung & Ramdani, 2018); management ownership; and institutional ownership (Javaid et al., 2023) are also found to be associated to ownership structure choice. Ghouma et al. (2018) reported that bondholders seem to care about the quality of the board's composition and the disclosure quality. On the other hand, Ngatno and Youlianto (2021) reported that only the board size can weaken the relationship between ownership structure and company performance. The gender diversity limits the impact of board structure on a company's financial choices (Zaid et al., 2020). Vijayakumaran and Vijayakumaran (2019) also indicated that the board structure variables do not impact a company's ownership structure choices.

Strebulaev and Yang (2013) found that dual-class organizations are initially more likely to be all-stock firms. However, as they become public corporations, they are more likely to use debt. In emerging countries, firms with a large-sized board use low leverage, whereas in developed countries, firms with small-sized boards use low leverage to enhance corporate performance (Kijkasiwat et al., 2022). Furthermore, the companies with smaller boards tend to use debt financing more sparingly, pursue riskier investment strategies, and experience more significant variability in performance (Huang et al., 2015). Kao et al. (2019) reported that ownership structure, block-holders' ownership, institutional ownership, foreign ownership family ownership, and director ownership (Rashid, 2020) are all positively related to firm value. Conversely, ownership by domestic financial institutions had a negative impact on the level of internationalization (Purkayastha et al., 2022). Abdullah and Tursoy (2021) revealed that a positive association between ownership structure and firm performance is the benefits of the tax shield and the lower costs of issuing debt compared to equity. Therefore, companies are focusing on debt financing instead of equity to achieve their financial as well as nonfinancial goals (Almosh et al., 2024). Conversely, Detthamrong et al. (2017) demonstrated that corporate governance is unrelated to financial leverage and firm performance. Taurus and Ayabei (2016) also found that leverage favorably correlates with director independence, whereas CEO duality and longevity negatively and significantly impact leverage.

The synthesis of extant literature indicate that researchers have explored the impact of corporate governance and ownership structure and ownership and structure and firm performance separately. However, the combined impact of corporate governance and ownership structure on firm performance has not be explored much. Therefore, the present study hypothesizes that:

H_1: The combined effect of corporate board structure and ownership structure affects the accounting performance of the BSE-listed sample firms.

H_2: The combined effect of corporate board structure and ownership structure affects the market value of the BSE-listed sample firms.

6.3 Research Methodology

This study section outlines the research framework, and the statistical techniques adopted to examine the combined effect of corporate governance and ownership structure on financial performance. We collected data for 2010–2019 for the 500 firms indexed on the Bombay Stock Exchange (BSE). Our sample firms comprise significant economic industries, representing nearly 93% of listed firms. Furthermore, due to their different regulations, 91 banking and financial services companies were not included. We excluded 139 firms with more than 20% missing data for the ownership structure variables. Additionally, four firms were declared sick and were removed. Therefore, we were left with the final sample size of 266 firms for the analysis. The data for ownership structure and firm-specific control variables have been collected from the capital line database and the company's annual reports.

6.3.1 Variables

See Table 6.1.

6.3.2 Statistical Technique

The fixed-effect panel data regression model has been used to capture the impact of corporate governance and ownership structure on the performance of BSE firms over time. The fixed-effect panel regression method has been used because it is possible to analyse the net impact on outcome variables of the predictors using the fixed-effect method since the time-invariant variables are removed. These time-invariant attributes are also presupposed to be unique to each individual and should not be correlated with any other individual characteristics. For example, there is no correlation between an entity's error term and its constant because each entity is unique and has a separate error term.

Table 6.1 List of variables with the abbreviations used in the study

Variables	Description	Label
Performance Variables Used as Dependent Variables		
Return on assets	Earnings before interest and tax/total assets	ROA
Return on equity	(Profit after tax- preference dividend)/ shareholders fund	ROE
Tobin's Q	(Market value of equity + book value of debt)/ total asset	Q
Corporate Board Structure and Ownership Structure Variables Are Used as Independent Variables		
Board Diversity Index		
Gender diversity	Blau value was calculated using two groups, i.e., male and female directors on board	
Board independence	Blau value was calculated using two groups, i.e., independent and non-independent directors on board	
Board Educational Qualifications	Blau value was calculated using five categories, i.e., graduation, post-graduation, technical, professional, and doctorate. These attributes have been assigned weights, i.e., 15, 17, 16, 18, 20 respectively, based on the total number of years of formal education (Ararat et al., 2015)	
Board Busyness		
Directorships	Average directorships are held by non-executive directors on the board of the other companies	
Ownership Structure Variables		
Domestic equity	Equity share capital, preference share capital, and reserve and surplus, excluding the equity investment by foreign institutional investors	
Foreign institutional investors	Equity investment by foreign institutional investors	
Debt	Long-term debt plus short-term debt/ equity	
Control Variables		
Research Expenditure	Research expenditure incurred by the firm/ total assets	R&D
Risk	The annualized standard deviation of monthly stock return	Risk
Market value to book value	The market value of equity/book value of equity	MVBV
Effective tax rate	Tax liability/ earnings before tax	ETR

Source Author's Compilation

6.3.3 Regression Models

$$\begin{aligned}\mathrm{ROA}_{\mathrm{it}} &= \beta_{\mathrm{i}} + \beta_1 \mathrm{Board\ Diversity}_{\mathrm{it}} + \beta_2 \mathrm{Directorships}_{\mathrm{it}} \\ &+ \beta_3 \mathrm{Domestic\ Equity} + \beta_4 \mathrm{Debt}_{\mathrm{it}} + \beta_5 \mathrm{FIIs} + \beta_6 \mathrm{MVBV} \\ &\beta_7 \mathrm{ETR} + \beta_8 \mathrm{R\&\ D}_{\mathrm{it}} + \beta_9 \mathrm{Risk}_{\mathrm{it}} + \varepsilon_{\mathrm{it}}\end{aligned} \quad (6.1)$$

$$\begin{aligned}\mathrm{ROE}_{\mathrm{it}} &= \beta_{\mathrm{i}} + \beta_1 \mathrm{Board\ Diversity}_{\mathrm{it}} + \beta_2 \mathrm{Directorships}_{\mathrm{it}} \\ &\beta_3 \mathrm{Domestic\ Equity} + \beta_4 \mathrm{Debt}_{\mathrm{it}} + \beta_5 \mathrm{FIIs} + \beta_6 \mathrm{MVBV} \\ &\beta_7 \mathrm{ETR} + \beta_8 \mathrm{R\&\ D}_{\mathrm{it}} + \beta_9 \mathrm{Risk}_{\mathrm{it}} + \varepsilon_{\mathrm{it}}\end{aligned} \quad (6.2)$$

$$\begin{aligned}\mathrm{Tobin's\ Q}_{\mathrm{it}} &= \beta_{\mathrm{i}} + \beta_1 \mathrm{Board\ Diversity}_{\mathrm{it}} + \beta_2 \mathrm{Directorships}_{\mathrm{it}} \\ &\beta_3 \mathrm{Domestic\ Equity} + \beta_4 \mathrm{Debt}_{\mathrm{it}} + \beta_5 \mathrm{FIIs} + \beta_6 \mathrm{ETR} \\ &+ \beta_7 \mathrm{R\&\ D}_{\mathrm{it}} + \beta_8 \mathrm{Risk}_{\mathrm{it}} + \varepsilon_{\mathrm{it}}\end{aligned} \quad (6.3)$$

ROA, ROE, and Tobin's Q are the dependent variables, β_i is the constant term, β (1,2,…..8) is the regressor's coefficient, and ε_{it} is the error term for the cross-section i at time t, respectively.

6.4 Analysis

The panel fixed-effect method results for the combined impact of corporate governance and ownership structure variables on the performance of BSE-listed firms and the disaggregate analysis of the sample firms are detailed in this part of the study.

Table 6.2 reports the summary statistics for corporate board composition, ownership structure, and firm-level control variables. The mean value for the multiple directorships in the sample firm is 5. Further, the data for directorships is positively skewed and leptokurtic, indicating considerable variation in the number of directorship positions held by the directors of the sample firms. However, the diversity index data is negatively skewed, indicating that the diversity index's mean value is less than the median value.

Similarly, the data for the inflow of funds from FII and debt is negatively skewed, presenting that the data is not normally distributed. Further, the data for debt is platykurtic. In contrast, it is leptokurtic for equity and FII, implying that sample firms are raising more equity. The inflow of FII funds will increase investors' risk, and consequently, they will demand higher returns. Further, it is highlighted that the effective tax rate for half of the sample firms is around twenty-seven percent, and less than twenty-five percent of the firms did not spend on research and development. The accounting performance measures (ROA) and (ROE) have a mean of 18.13% and 14.83%, highlighting that an average sample firm is earning a return of eighteen percent on assets. At the same time, shareholders of the sample Indian firms are

Table 6.2 Descriptive statistics of corporate governance and ownership structure variables

Variable	Mean	Std	Min	Max	P25	Median	P75
Diversity Index	2.487	0.256	1.196	3.193	2.234	2.511	2.648
Directorships	4.997	2.397	0	17.375	3.285	4.714	6.333
FII	0.687	0.966	−3.159	3.178	0.073	0.697	1.369
Domestic equity	3.296	0.644	1.58	5.606	2.856	3.214	3.631
Debt	0.736	0.851	−3.966	2.675	0.073	0.899	1.355
MVBV	4.0904	5.259	0.089	121.958	1.411	2.761	4.917
Effective tax rate	0.283	1.542	0	65.322	0.18	0.268	0.322
R&D	0.01	0.027	0	0.698	0	0.001	0.009
Risk	0.208	0.095	0.031	1.114	0.141	0.185	0.25
ROA	18.13	14.767	−23.754	147.051	8.971	15.395	24.073
ROE	14.838	24.59	−718.096	666.619	7.594	13.975	21.306
Tobin's Q	3.334	3.784	0.245	43.956	1.189	2.156	3.976

Source Author's Compilation

gaining a fifteen percent return on their investment. The mean value of 3.33 for Q suggests that the sample firms' asset value is less than the market value. These statistics are consistent with that of Raithatha and Haldar (2021).

The Pearson correlation for financial performance, ownership structure, corporate board diversity, and board busyness variables are depicted in Table 6.3. The correlation coefficient between the market-to-book value ratio and Tobin's Q is higher than 0.8, as these variables measure market performance. Therefore, to avoid the problem of multicollinearity, we have excluded the market-to-book value ratio as a control variable for measuring market performance. For the rest of the variables, the correlation value is less than 0.8, and the Variance Inflation Factor (VIF) values are less than 3.24, confirming the absence of multicollinearity in our data (Field, 2005).

The positive correlation coefficient of the diversity index, MVBV, and FII at a 1% significance level suggests that the presence of diversified board members increases the inflow of foreign equity in the company and, in turn, enhances the firm's market value. Additionally, larger firms with more busy directors prefer to raise funds from FII and debt rather than equity. On the other hand, the appointment of busy directors on the board adversely affects the firm's market value, as represented by the statistically significant correlation coefficient (−0.03). Moreover, increasing debt financing is negatively associated with the market value (−0.2401) and the research expenditure of the firm. Further, the alternate measures of financial performance are significantly correlated at a 1 percent level.

Table 6.4 displays the results of corporate board structure variables, i.e., diversity index, directorships, and ownership structure, on the financial performance of the

Table 6.3 Correlation matrix between variables used in the study

Variables	Diversity	Director	FII	Domestic	Debt	MVBV	ETR	Risk	R&D	ROA	ROE	Tobin's
Diversity	1											
Directorships	−0.0325*	1										
FII	0.1081**	0.0546**	1									
Domestic	−0.0018	0.0152	0.6977**	1								
Debt	−0.0317	0.1178**	0.3369**	0.4003**	1							
MVBV	0.1161**	–	−0.0178	–	–	1						
ETR	0.0231	−0.0012	−0.0082	−0.0132	0.0013	0.0023	1					
Risk	–	0.0359*	–	–	0.1178**	–	−0.018	1				
R&D	−0.0148	0.0288	–	−0.0286	–	0.0778**	−0.006	–	1			
ROA	0.014	–	–	–	–	0.5119**	−0.006	–	0.1223**	1		
ROE	0.0139	−0.0173	–	–	–	0.2815**	−0.01	–	0.0667**	0.5653**	1	
Torbin's	0.1435*	−0.0263	0.0119	–	–	0.8746**	0	–	0.1094**	0.6699**	0.3287**	1

Note ***,**, * indicates significant level at 1%, 5%, and 10% respectively
Source Author's Compilation

BSE firms. It has been observed that the presence of highly qualified directors, independent board members, and appointment of female directors significantly (0.01) improves the market performance. However, its favorable impact on accounting performance is non-significant {ROA ($\beta = -1.11$; ROE ($\beta = 0.49$)}. Further, outside directorships, the surrogate variable for board busyness, have an adverse effect on the firm's market valuation. However, busy directors favorably influence the accounting performance of the company.

These findings further reinforce that the FII is beneficial for improving the firm's financial performance since it considerably impacts the ROA and market value of the company. Besides this, debt funds are found to have a detrimental effect on financial performance as higher financial leverage decreases a firm's value by increasing bankruptcy risks; the results are in tune with the study conducted by (Shleifer & Vishny, 1997). In contrast to the hypothesis of pecking order theory, Mande et al.

Table 6.4 Panel fixed-effect results presenting corporate board and ownership structure effect on the firm performance sample firms for 2010–2019

Variables	Model 1	Model 2	Model 3
Independent Variables			
Diversity Index	−1.111 (−0.87)	0.493 (0.12)	0.663** (2.54)
Directorships	0.141 (0.86)	0.059 (0.14)	−0.027 (−0.59)
FII	1.552** (2.34)	0.766 (0.75)	0.736*** (4.07)
Equity	−5.704*** (−3.03)	−4.804 (−0.56)	1.245** (2.36)
Debt	−4.539*** (−5.55)	−2.684*** (−2.70)	−0.914*** (−3.00)
Control Variables			
MVBV	0.550** (2.17)	0.622 (1.55)	
Effective tax rate	−0.008 (−0.21)	−0.070 (−0.96)	−0.002 (−0.34)
R&D	9.861 (0.77)	16.573 (0.83)	2.579 (0.69)
Market Risk	5.428 (1.53)	12.576 (1.13)	2.211** (2.21)
Constant	37.776*** (5.62)	25.274 (1.05)	−2.602* (−1.65)
N	266		
R sq(within)	0.124	0.0173	0.1239
Rho	0.579	0.192	0.771

Note ***, **, * indicates significant level at 1%, 5% and 10% respectively
Source Author's Compilation

(2012) suggest that good governance lowers the cost of equity financing by reducing agency costs. The results report that the board's diversity and busyness led to a slight reduction in the significance level of the positive impact of equity on the market performance. In addition to this, except for the negative effect of the effective tax rate on the market performance, the findings for the control variables did not change while studying the combined effect of board and ownership structure variables.

6.4.1 Disaggregate Analysis Based on Large, Medium, and Small-Cap Companies

The disaggregate analysis is also performed to examine the joint impact of corporate board structure and ownership structure variables on the performance of large, medium, and small firms. The results sustained that the appointment of diversified board members considerably enhances the market performance of small-size firms. This is possibly due to the director's diverse knowledge of professional practices and experience. Contrary to this, we observed a significant inverse association between board diversity and ROA of the mid-size firms. The findings further reinforce that directors with multiple directorships adversely affect the market valuation for large firms only (Alqahtani et al., 2022). However, it is positively associated with the market performance of medium and small-sized firms (Table 6.5).

Raising funds from FIIs significantly enhances the shareholders' return in large firms with board diversity and multiple directorships. However, in confirming the findings of Kyereboah-Coleman and Bikepe (2006), suggesting that boards with outside directors employ more debt, this study also observed that the inflow of debt funds has more severe adverse effects on the market valuation of large-cap firms. On the other hand, appointing diverse board members reduces the adverse effects of debt funds on the market valuation of mid-cap firms. Further, in the case of small-cap firms, board diversity and busyness did not alter the effect of ownership structure on their performance (Table 6.6).

The inclusion of corporate board structure variables did not affect the results of the control variables for the large and the small-cap firms. However, a slight reduction in the adverse effects of the effective tax rate on the accounting performance of the mid-cap firms was witnessed (Table 6.7).

6.5 Conclusion

Good governance is a remedy for the problem of agency conflict. It further ensures transparency in the dealings of a business firm, thereby contributing to high performance. Investors are also willing to invest in companies that are ethical and transparent in their dealings. Hence, the management's commitment to high governance

Table 6.5 Panel fixed-effect results for the corporate board and ownership structure on the firm performance of large-cap firms for the years 2010–2019

Variables	Model 1	Model 2	Model 3
Independent Variables			
Diversity Index	−2.926 (−1.18)	−2.419 (−1.03)	1.129 (1.26)
Directorships	0.256 (0.96)	0.150 (0.51)	−0.194* (−1.71)
FII	2.422 (1.08)	4.065* (1.77)	−0.536 (−0.84)
Equity	−13.189*** (−4.11)	−19.724*** (−6.51)	1.228 (0.80)
Debt	−3.970*** (−2.81)	−1.662 (−1.48)	−0.990*** (−2.84)
Control Variables			
MVBV	1.227*** (5.61)	1.015*** (4.14)	
Effective tax rate	18.377*** (3.65)	14.565** (2.31)	0.999 (0.96)
R&D	79.706 (1.38)	13.117 (0.24)	24.773 (1.40)
Market Risk	1.409 (0.33)	−0.097 (−0.02)	0.533 (0.24)
Constant	67.470*** (6.38)	88.507*** (7.99)	−0.865 (−0.17)
N	75		
R sq(within)	0.2977	0.2805	0.105
Rho	0.627	0.654	0.849

Note ***, **, * indicates significant level at 1%, 5% and 10% respectively
Source Author's Compilation

standards is a magnet in attracting much-needed external funds for the company. Though research studies have examined the effect of ownership structure on firm performance and corporate governance on firm performance separately, very few studies have considered the combined effect of corporate governance and ownership structure on firm performance. This study attempts to bridge this gap in corporate governance and ownership structure literature.

The results suggest that the presence of highly qualified directors, independent board members, and appointment of female directors significantly (0.01) improves the market performance. Further, multiple directorships are held by the directors of the firm's market valuation. In contrast to the pecking order theory hypothesis and Mande et al. (2012) findings, good governance lowers performance. The results of this study report show that the board diversity and busyness led to a slight reduction in the significance level of the positive impact of equity on the market performance.

Table 6.6 Panel fixed-effect results for the corporate board and ownership structure on the firm performance of mid-cap firms for 2010–2019

Variables	Model 1	Model 2	Model 3
Independent Variables			
Diversity Index	−3.725* (−1.82)	2.794 (0.30)	0.151 (0.38)
Directorships	−0.178 (−0.77)	−0.389 (−0.45)	0.007 (0.09)
FII	1.282 (1.07)	−0.636 (−0.41)	1.195*** (3.03)
Equity	2.383 (0.61)	−7.133 (−0.44)	2.318*** (3.47)
Debt	−4.309*** (−3.82)	−2.253* (−1.77)	−1.054 (−1.61)
Control Variables			
MVBV	0.264* (1.94)	0.535 (1.11)	
Effective tax rate	−0.039 (−1.12)	−0.141* (−1.76)	−0.001 (−0.30)
R&D	7.335 (0.69)	13.240 (0.57)	0.045 (0.02)
Market Risk	19.127** (2.47)	20.387*** (2.73)	4.356** (2.21)
Constant	16.533 (1.28)	26.552 (0.78)	−5.341 (−2.70)
N	108		
R sq(within)	0.1224	0.0254	0.1861
Rho	0.544	0.136	0.697

Note ***, **, * indicates significant level at 1%, 5% and 10% respectively
Source Author's Compilation

The disaggregated analysis pinpointed that in large firms with board diversity and multiple directorships, raising funds from FIIs enhances the shareholders' return. Furthermore, in confirming the findings of Kyereboah-Coleman and Bikepe (2006), suggesting that boards with outside directors employ more debt, this study also observed that the inflow of debt funds has more severe adverse effects on the market valuation of large-cap firms. In comparison to the large and small-cap firms. The appointment of diverse board members reduces the adverse effects of debt funds on the market valuation of mid-cap firms. Further, in the case of small-cap firms, board diversity and busyness did not alter the effect of ownership structure on performance.

The study's findings will be useful to managers and upper management who wish to reform their internal system. The institutional environment in the country should be further strengthened so that companies are not dependent on only banks for their debt requirements. The flexible rules should be applied to increase the inflow of

Table 6.7 Panel fixed-effect results for the corporate board and ownership structure on the firm performance of small-cap firms for the years 2010–2019

Variables	Model 3a	Model 3b	Model 3c
Independent Variables			
Diversity Index	−1.549 (−0.86)	−1.382 (−0.36)	1.081*** (4.60)
Directorships	0.372 (1.26)	0.288 (0.58)	0.045 (1.27)
FII	0.113 (0.18)	0.9302 (0.84)	0.493*** (4.18)
Equity	−5.765*** (−2.88)	9.733 (0.62)	0.454 (0.71)
Debt	−2.876*** (−3.18)	−3.391 (−0.86)	−0.484** (−2.30)
Control Variables			
MVBV	2.062*** (6.90)	0.551 (0.36)	
Effective tax rate	7.124* (1.89)	15.340 (1.31)	0.702* (1.67)
R&D	−49.687 (−1.11)	35.649 (0.54)	13.002** (2.33)
Market Risk	0.322 (0.09)	15.945 (0.66)	1.076** (2.36)
Constant	30.232 (3.96)	−21.292 (−0.44)	−2.438 (−1.39)
N	83		
R sq(within)	0.1935	0.0181	0.1848
Rho	0.357	0.181	0.624

Note ***, **, * indicates significant level at 1%, 5% and 10% respectively
Source Author's Compilation

equity participation by foreign institutions and investors in the Indian market. This would encourage capital flow and reduce the cost of borrowing which ultimately enhance profitability of the companies. The inflow of funds from foreign investors will help in reducing the concentration of power in few hands resulting in protection of minority shareholder interest. The study further recommends that policymakers specify the minimum qualification for the board members as qualified directors would contribute to enhancing the company's performance. The minimum number of female and independent directors has been stated in the laws. However, their mere presence on the board will not improve the performance of the companies. On the contrary, it is suggested that their active presence with designated roles will contribute to better performance. Therefore, policies should be framed accordingly. The findings will aid the policymakers in further strengthening the norms for the inclusion of board members with diverse demographic backgrounds to increase firm performance. The appointment of professionally qualified directors will reduce information asymmetry and encourage investors to invest more in these companies.

Considering the implications of the study, the model can be replicated in other emerging economies, where the problem of principal-principal agency conflict is prevalent. A comparative study can assess the impact of board structure and financing choices of family and non-family firms. Research studies can also investigate the non-linear relationship between an independent and dependent variable that may exist.

References

Abdullah, H., & Tursoy, T. (2021). Capital structure and firm performance: Evidence of Germany under IFRS adoption. *Review of Managerial Science, 15*(2), 379–398.

Adams, R. B., & Ferreira, D. (2009). Women in the boardroom and their impact on governance and performance. *Journal of Financial Economics, 94*(2), 291–309.

Adjaoud, F., & Ben-Amar, W. (2010). Corporate governance and dividend policy: Shareholders' protection or expropriation? *Journal of Business Finance & Accounting, 37*(5–6), 648–667.

Ahmed, A. M., Nugraha, D. P., & Hágen, I. (2023). The relationship between capital structure and firm performance: The moderating role of agency cost. *Risks, 11*(6), 102.

Al Amosh, H., Khatib, S. F., Alkurdi, A., & Bazhair, A. H. (2024). Capital structure decisions and environmental, social and governance performance: Insights from Jordan. *Journal of Financial Reporting and Accounting, 22*(4), 972–989.

Alqahtani, J., Duong, L., Taylor, G., & Eulaiwi, B. (2022). Outside directors, firm life cycle, corporate financial decisions and firm performance. *Emerging Markets Review, 50*(1), 1–20.

Andreou, P. C., Ehrlich, D., & Louca, C. (2013, January). Managerial ability and firm performance: Evidence from the global financial crisis. In *European Financial Management Association, Annual Conference.*

Ashraf, M., Iqbal, T., & Tariq, S. (2017). Corporate governance and corporate financing decisions impact on firm performance a cement industry perspective of Pakistan. *International Journal of Sciences: Basic and Applied Research, 36*(3), 20–35.

Attig, N., El Ghoul, S., Guedhami, O., & Zheng, X. (2021). Dividends and economic policy uncertainty: International evidence. *Journal of Corporate Finance, 66*, Article 101785.

Awan, A. G., Abbas, M. G., & Scholar, M. S. (2016). Capital structure, corporate governance, and firm performance in Pakistan. *Journal of Poverty, Investment and Development, 21*(1), 9–23.

Balagobei, S. (2017). Impact of institutional ownership and individual ownership on dividend policy of listed plantation companies in Sri Lanka. *Asia Pacific Journal of Advanced Business and Social Studies, 3*(1), 329–335.

Chang, Y. K., Chou, R. K., & Huang, T. H. (2014). Corporate governance and the dynamics of capital structure: New evidence. *Journal of Banking & Finance, 48*, 374–385.

Chen, S., Ni, X., & Tong, J. Y. (2016). Gender diversity in the boardroom and risk management: A case of R&D investment. *Journal of Business Ethics, 136*(3), 599–621.

Chen, W. P., Chung, H., Hsu, T. L., & Wu, S. (2010). External financing needs, corporate governance, and firm value. *Corporate Governance: An International Review, 18*(3), 234–249.

Chou, T. K., & Feng, H. L. (2019). Multiple directorships and the value of cash holdings. *Review of Quantitative Finance and Accounting, 53*, 663–699.

Chow, Y., Muhammad, J., Bany-Ariffin, A., & Cheng, F. (2018). Macroeconomic uncertainty, corporate governance and corporate capital structure. *International Journal of Managerial Finance, 14*(3), 301–321.

Cohen, J. R., Hoitash, U., Krishnamoorthy, G., & Wright, A. M. (2013). The effect of audit committee industry expertise on monitoring the financial reporting process. *The Accounting Review, 89*(1), 243–273.
Dao, B. T. T., & Ta, T. D. N. (2020). A meta-analysis: Capital structure and firm performance. *Journal of Economics and Development, 22*(1), 111–129.
Detthamrong, U., Chancharat, N., & Vithessonthi, C. (2017). Corporate governance, capital structure and firm performance: Evidence from Thailand. *Research in International Business and Finance, 42*(1), 689–709.
Eulaiwi, B., Al-Hadi, A., Taylor, G., Al-Yahyaee, K. H., & Evans, J. (2016). Multiple directorships, family ownership and the board nomination committee: International evidence from the GCC. *Emerging Markets Review, 28*, 61–88.
Field, A. P. (2005). Is the meta-analysis of correlation coefficients accurate when population correlations vary? *Psychological Methods, 10*(4), 444–467.
Funchal, B., & Monte-Mor, D. S. (2016). Corporate governance and credit access in Brazil: The Sarbanes-Oxley act as a natural experiment. *Corporate Governance: An International Review, 24*(5), 528–547.
Ghouma, H., Ben-Nasr, H., & Yan, R. (2018). Corporate governance and cost of debt financing: Empirical evidence from Canada. *The Quarterly Review of Economics and Finance, 67*, 138–148.
Grossman, S. J., & Hart, O. D. (1982). *Corporate financial structure and managerial incentives the economics of information and uncertainty* (pp. 107–140). University of Chicago Press.
Harris, M., & Raviv, A. (1991). Capital structure and the informational role of debt. *The Journal of Finance, 45*(2), 321–349.
Huang, C. J., Wang, N. Y., & Wang, F. Y. (2015). How external corporate governance affects SEOs performance: Mediating effect of agency costs. *Journal of Business Economics and Finance, 4*(2).
Jameson, M., Prevost, A., & Puthenpurackal, J. (2014). Controlling shareholders, board structure, and firm performance: Evidence from India. *Journal of Corporate Finance, 27*(1), 1–20.
Javaid, A., Nazir, M. S., & Fatima, K. (2023). Impact of corporate governance on capital structure: Mediating role of cost of capital. *Journal of Economic and Administrative Sciences, 39*(4), 760–780.
Jensen, M. C. (1986). The agency costs of free cash flow: Corporate finance and takeovers. *American Economic Review, 76*(2), 323–329.
Jensen, M. C., & Meckling, W. (1976). Theory of the firm: Managerial behavior, agency costs and ownership structure. *Journal of Financial Economics, 3*(1), 305–360.
Jiraporn, P., Kim, J. C., Kim, Y. S., & Kitsabunnarat, P. (2012). Capital structure and corporate governance quality: Evidence from the institutional shareholder services (ISS). *International Review of Economics & Finance, 22*(1), 208–221.
Kao, M. F., Hodgkinson, L., & Jaafar, A. (2019). Ownership structure, board of directors and firm performance: Evidence from Taiwan. *Corporate Governance: The International Journal of Business in Society, 19*(1), 189–216.
Khan, S. N., Yaseen, M. N., Mustafa, F., & Abbasi, S. (2019). The interaction effect of financial leverage on the relationship between board attributes and firm performance; Evidence of non-financial listed companies of Pakistan. *Journal of Accounting and Finance in Emerging Economies, 5*(1), 115–122.
Kieschnick, R., & Moussawi, R. (2018). Firm age, corporate governance, and capital structure choices. *Journal of Corporate Finance, 48*, 597–614.
Kijkasiwat, P., Hussain, A., & Mumtaz, A. (2022). Corporate governance, firm performance and financial leverage across developed and emerging economies. *Risks, 10*(10), 185.
Kyereboah-Coleman, A., & Biekpe, N. (2006). Corporate governance and financing choices of firms: A panel data analysis. *South African Journal of Economics, 74*(4), 670–681.
Mande, V., Park, Y. K., & Son, M. (2012). Equity or debt financing: Does good governance matter? *Corporate Governance: An International Review, 20*(2), 195–211.

Martins, H. C., Schiehll, E., & Terra, P. R. S. (2017). Country-level governance quality, ownership concentration, and debt maturity: A comparative study of Brazil and Chile. *Corporate Governance: An International Review, 25*(4), 236–254.
Myers, S. C. (1984). The capital structure puzzle. *The Journal of Finance, 39*(1), 574–592.
Myers, S. C., & Majluf, N. S. (1984). Corporate financing and investment decisions when firms have information that investors do not have. *Journal of Financial Economics, 13*(1), 187–221.
Nadeem, M. (2020). Does board gender diversity influence voluntary disclosure of intellectual capital in initial public offering prospectuses? Evidence from China. *Corporate Governance: An International Review, 28*(2), 100–118.
Ngatno, A. E. P., & Youlianto, A. (2021). Moderating effects of corporate governance mechanism on the relation between capital structure and firm performance. *Cogent Business & Management, 8*(1), 1–22.
Pham, H. S. T., & Nguyen, D. T. (2019). The effects of corporate governance mechanisms on the financial leverage–profitability relation: Evidence from Vietnam. *Management Research Review, 43*(4), 387–409.
Purkayastha, S., Manolova, T. S., & Edelman, L. F. (2022). Innovation and internationalization in the Indian pharmaceutical sector: The moderating role of ownership structures. In *Research handbook on innovation in international business* (pp. 219–238). Edward Elgar Publishing.
Raithatha, M., & Haldar, A. (2021). Are internal governance mechanisms efficient? The case of a developing economy. *IIMB Management Review, 33*(3), 191–204.
Rashid, M. M. (2020). Ownership structure and firm performance: The mediating role of board characteristics. *Corporate Governance: The International Journal of Business in Society, 20*(4), 719–737.
Shleifer, A., & Vishny, R. W. (1997). A survey of corporate governance. *Journal of Finance, 52*(2), 737–783.
Strebulaev, I., & Yang, B. (2013). The mystery of zero-leverage firms. *Journal of Financial Economics, 109*(1), 1–23.
Tarus, D. K., & Ayabei, E. (2016). Board composition and capital structure: Evidence from Kenya. *Management Research Review, 39*(9), 1056–1079.
Tulung, J. E., & Ramdani, D. (2018). Independence, size and performance of the board: An emerging market research. *Corporate Ownership & Control, 15*(2).
Vijayakumaran, S., & Vijayakumaran, R. (2019). Corporate governance and capital structure decisions: Evidence from Chinese listed companies. *Journal of Asian Finance, Economics and Finance, 6*(3), 67–79.
Wen, Y., Rwegasira, K., & Bilderbeek, J. (2002). Corporate governance and capital structure decisions of the Chinese listed firms. *Corporate Governance: An International Review, 10*(2), 75–83.
Williams, R. J. (2003). Women on corporate boards of directors and their influence on corporate philanthrophy. *Journal of Business Ethics, 42*(1), 1–10.
Yasin, H., Shehab, I., & Saidi, N. (2004, June). Monitoring and overseeing disclosure. In *Paper presented at the corporate governance in MENA Countries: Improving transparency and disclosure, the second Middle East and North Africa regional corporate governance Forum,* Beirut (pp. 3–5).
Zaid, M., Wang, M., TF Abuhijleh, S., Issa, A., WA Saleh, M., & Ali, F. (2020). Corporate governance practices and capital structure decisions: The moderating effect of gender diversity. *Corporate Governance: The International Journal of Business in Society, 20*(5), 939–964.
Zalata, A. M., Tauringana, V., & Tingbani, I. (2018). Audit committee financial expertise, gender, and earnings management: Does gender of the financial expert matter? *International Review of Financial Analysis, 55*(1), 170–183.
Zhou, M., Li, K., & Chen, Z. (2021). Corporate governance quality and financial leverage: Evidence from China. *International Review of Financial Analysis, 73*, Article 101652.

Chapter 7
Lessons from the Corporate Governance Issues at BYJU's

Vijaya Kittu Manda and Lubza Nihar Khaliq

Abstract BYJU's is a major Indian Education Technology (EdTech). It is the poster boy of Indian startups and has set itself as an example of a successful startup globally. However, corporate governance issues hurt the company's image, eroded stakeholder trust, and impacted its regular operations. The company, which is now on a thin line of survival, came to glory for being the most preferred educational platform during the COVID-19 pandemic. Funding was more accessible as venture capitalists lined up to fund the company's future expansion. However, the company could not handle its situation well, leading to financial irregularities, inability to maintain transparency and disclosures, and took the blame for mismanagement. Board oversight and independence are often being questioned. Stakeholders' issues are not correctly handled, and shareholders' rights are often in trouble. A series of domestic and international court disputes and a spate of resignations at the senior leadership level have further hurt the company. As EdTech companies continue to grow and play a more critical role in education, adopting sound governance practices can help BYJU's to ensure their long-term success. This case on BYJU's highlights the importance of strong corporate governance and will be a classical lesson for EdTechs and any startup company globally.

Keywords Financial irregularities · Stakeholder engagement · Disclosure · Corporate reputation · Startup governance · Related party transactions · Board oversight · Founder dominance · Aggressive growth · Growth at any cost · Employee compensation issues

V. K. Manda (✉)
PBMEIT, Visakhapatnam, India
e-mail: vijaykittu@hotmail.com

L. N. Khaliq
Department of Finance, GITAM School of Business, GITAM (Deemed to Be University), Visakhapatnam, India
e-mail: lnihar@gitam.edu

V. Kandpal et al. (eds.), *Developments in Corporate Governance*,
Approaches to Global Sustainability, Markets, and Governance,
https://doi.org/10.1007/978-981-96-6366-8_7

7.1 Introduction

7.1.1 Significance of Corporate Governance

Corporate governance is a bundle of rules, practices, and processes that direct and control a company. It involves the relationships among the different stakeholders in the company, their interactions, and balancing their interests. Stakeholders of a company include the shareholders, the Board of Directors, the Management team, Customers, Suppliers, Financiers, the Government, the community, and other stakeholders. Good CG practices are necessary for the firm's long-term success. Companies that maintain good CG practices will be able to:

- Attract and retain investors.
- Improve the company's financial performance.
- Reduce the risk of fraud and corruption.
- Build a strong reputation for the company.

CG is important for all types of companies—Publicly traded companies, Large and complex companies, Highly regulated companies, Family-owned companies, Entrepreneurial firms, and even Startups. Figure 7.1 shows the various dimensions of Corporate Governance for a given company.

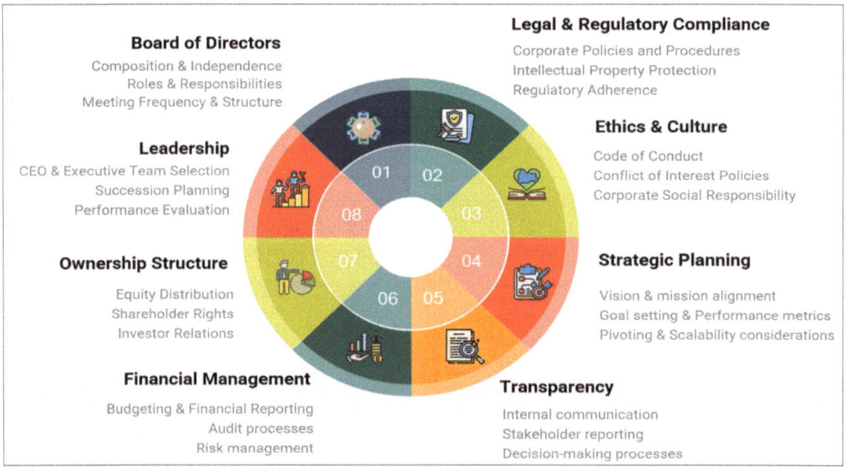

Fig. 7.1 Dimensions of Corporate Governance

7.1.2 *Corporate Governance in Startups*

Traditional studies on Corporate Governance focus on large and well-established corporations. CG in startups is not systematic and departs from norms set for large firms. The topic is a modern and emerging theme that has caught attention in the last decade as startup culture began to flourish globally. Several startups faced significant governance challenges; a few eventually failed and collapsed. The list includes WeWork (Wright & Nutt, 2021), Theranos, Zenefits, Uber, Wirecard, Luckin Coffee, Zynga, American Apparel, FTX cryptocurrency exchange (Manda & Nihar, 2023), and Juicero, among others. Understanding the unique corporate governance requirements of startups is essential. Corporate Governance gaps in startups arise because of internal factors such as:

1. Ownership structure
2. The Board of Directors and its composition (first layer of members from Founders and Management, second layer from Venture Capitalist Investors, and third layer of Independent Directors) (Fan, 2022)
3. Compliance requirements
4. High focus on innovation and product development
5. Nature of investors (venture capitalists).

7.1.3 *Corporate Governance in India*

Corporate governance in India is regulated by legislation and regulations (Manda & Khaliq, 2023). The three central legislations in this regard are:

1. The Companies Act, 2013
2. The Securities and Exchange Board of India (SEBI) Act, 1992, and
3. The Reserve Bank of India Act, 1934.

These laws and regulations set out the minimum standards of corporate governance that companies must comply with. Apart from these laws and regulations, companies can adopt several voluntary codes and guidelines to improve their corporate governance practices. Some popular mentions in this regard include the National Voluntary Guidelines on Corporate Governance (2017) and the Corporate Governance Code for Listed Companies (2019).

The Companies Act, 2013 is the source of company law in India. It is the updated version of the Companies Act 1956. It is an elaborate legislation with provisions for most companies in India. The legislation has specific provisions for corporate governance, transparency, and accountability. The legislation aligns well with Indian corporate law and makes it closer to global peers. An Amendment to the Companies Act, 2013, has further strengthened the law, especially the role of independent directors.

The Securities and Exchange Board of India (SEBI) is the Indian capital markets regulator. It is also a primary corporate governance regulator of listed companies.

SEBI has issued several regulations and guidelines on corporate governance. The SEBI Committee on Corporate Governance has formally defined "Corporate Governance." According to the formal definition, Corporate Governance is the "acceptance by management of the inalienable rights of shareholders as the true owners of the corporation and their role as trustees on behalf of the shareholders."

The Reserve Bank of India (RBI) is the Central Bank of India. The Reserve Bank of India (RBI) Act, 1934 gave powers to RBI to regulate the Indian financial system and promote corporate governance in the Indian financial system, especially among Banks and Non-Banking Financial Companies (NBFCs). These provisions include prudential regulations, supervisory powers, and Corporate governance guidelines. The RBI's corporate governance guidelines have helped to improve the corporate governance practices of banks and other financial institutions. Positive governance changes observed include decision-making, transparency, risk management, and supervision.

7.1.4 Research Themes on Corporate Governance Themes

Based on the literature review, the following research themes are identified as topics of academic and research interest:

1. Board diversity, composition, and effectiveness
2. Board independence and accountability
3. CEO and Executive compensation and pay-for-performance
4. Audit Committee Effectiveness
5. Corporate social responsibility (CSR)
6. CG and firm performance
7. Corporate transparency and disclosure practices
8. Risk Management and Internal Controls
9. CG and Sustainability (Csedő et al., 2022)
10. CG, Innovation, and Technology (Csedő et al., 2022)
11. Corporate governance in emerging markets
12. Whistleblower Protections and Internal Reporting
13. Shareholder Activism and Engagement
14. Corporate Governance and Ethics
15. Mergers and Acquisitions Governance

7.1.5 Overview of BYJU's and Its EdTech Industry Context

Educational Technology (EdTech) major BYJU's turned into a multinational brand during the COVID-19 lockdown times. It is one of the first companies to bring app-based learning to the international scene (Kovid & Bhati, 2024) by offering online learning platforms and educational resources for students of all ages. BYJU's offers a range of programs and services, including Online courses, Learning mobile applications for both the trainer and the learner, Test prep, Adaptive learning, and even offline classes. BYJU'S is a global major with over 150 million registered students. Apart from India, the company operates through subsidiaries in several other countries, such as the United States, the United Kingdom, South Africa, Australia, New Zealand, and the Middle East. BYJU's is a poster boy and icon for being one of the most successful startups. However, because of its poor governance practices, the once most sought-after company faces legal challenges, financial crises, and difficulties managing itself. This chapter examines some of the steps that led to the near collapse of the unicorn.

7.1.6 Case Methodology

This study follows the Case Study approach. The case concerns an Indian startup that rose to glory because of opportunities that favored it during the COVID-19 pandemic. Unfortunately, the glory was all lost because of the company's poor corporate governance. The study is based entirely on secondary data with verifiable inputs from reputed websites and portals, business newspapers and magazines, journals, books, and other case study repositories.

7.1.7 The Motivation Behind the Case Study

Studying BYJU's corporate governance lapses can give researchers and practitioners insight into corporate governance. Such cases can provide valuable lessons for current and future business leaders.

1. **Understanding Corporate Governance Failures**: BYJU's classic case highlights how weak corporate governance and compliance practices can lead to significant operational and financial challenges.
2. **Impact on Investors**: The lapses in governance have led to a loss of confidence among investors, with major investors like Prosus and Peak XV Partners leaving the Board and several of the investors almost writing off their investments.
3. **Crisis Management and Reputation Management**: BYJU's response to governance issues is not satisfactory to the stakeholders. Changes in auditors and board members appear to be a costly lesson in crisis management and organizational restructuring. How a company should deal with stakeholder management during turbulent times can be learned.
4. **Regulatory Insights**: The Ministry of Corporate Affairs investigation offers insights into how regulatory bodies address corporate governance issues and ensure compliance. The case further shows the complexities of adhering to regulations in India's evolving startup ecosystem.

5. **Financial Implications**: The lapses have significantly impacted BYJU's financial health, including a drastic reduction in its valuation.
6. **A case study in rapid growth challenges**: BYJU's has experienced enormous business growth quickly. Examining how BYJU's governance issues emerged during its explosive growth can provide insights into managing rapid scaling.
7. **EdTech sector implications**: As a prominent player in educational technology, BYJU's challenges highlight potential pitfalls and governance considerations for the broader EdTech industry.
8. **Financial Reporting and Transparency**: There is an apparent failure on the company's part to promptly disclose its financial reports and auditing issues. These lessons explain the importance of timely and accurate financial disclosures.

7.2 Background: BYJU's and Its Journey to Glory

7.2.1 About Byju Raveendran

Mr. Byju Raveendran hails from a small village in Kerala, India. He graduated as an engineer and joined a UK-based shipping company called Pan Ocean Shipping. During leisure time, he trained his friends to prepare for the Common Admission Test (CAT). CAT is the competitive entrance examination for admission into India's premium management school—the Indian Institute of Management (IIM). Raveendran himself wrote the exam and secured a 100 percentile! He appeared once again and achieved the same feat! Two years later, after he understood the approach to clearing the exam successfully and was motivated by the successful results of his friends in clearing CATS, he quit his job to start Byju's Classes in 2007. His coaching was so popular that there were instances where he delivered classes to packed auditoriums, and he was traveling across the country to train CAT aspirants. Many of his students went to the IIM and completed the Program. Their success stories motivated and inspired him to become a serious entrepreneur. In 2009, Byju Raveendran married Divya Gokulnath, a student in his CAT classes.

7.2.2 The Start of BYJU'S

Almost all startups have a humble beginning, and so did BYJU'S. BYJU's was founded by Byju Raveendran (CEO), his wife Divya Gokulnath, and his younger brother Riju Ravindran in November 2011. 'Think and Learn Pvt Ltd' (T&L) is the parent company of BYJU'S, headquartered in Bengaluru, India. The platform runs on a freemium model, offering simple services for free. A subscription fee is to be paid to access full features and content. The main aim of the business is to provide coaching through online video lectures for students of class 1 to class 12 and for people who are preparing for competitive exams. Initially, they started with 1000 students, and the company launched a series of online apps for various learners—*BYJU'S: The Learning App* (2015), *BYJU'S Math App for kids*, and *BYJU'S Parent Connect app* (2017). Within a short period, the company had products catering to the competitive and school examination needs of students in the K-12 (kindergarten to grade 12) segment and beyond. In 2018, the company reported having 15 million users, of which 900,000 are paid users. It extensively used social media and innovative methods to build customized products and offer personalized experiences (Ranjan, 2022). With venture capital funding available, the company went on organic and inorganic routes to increase the geographic spread and product offering (Tripathy & Devarapalli, 2021), eventually becoming India's first EdTech Decacorn in 2018. A Unicorn is a privately held startup valued at over $1 billion. So, technically, a Decacorn is ten times a Unicorn.

7.2.3 BYJU'S Path to Glory

The Unified District Information System for Education Plus (UDISE+) report on school education in India for 2021–22 says the number of enrolled students in formal school education from primary to higher secondary school was around 25.57 crores (PTI, 2022). BYJU's claims it had 11.5 crore students learning on its app as of December 2021. About 7 million students are its annual paid members (PTI, 2021). The COVID-19 pandemic was a blessing because it accelerated the adoption of online learning platforms, software, and content, which became the most sought-after for students worldwide. Parents were eager to purchase tablet PCs and subscriptions to the app. BYJU's took advantage by selling the bundle (of Tablet and app subscription) and even offered loans from NBFCs for those who cannot afford it. By June 2023, 150 million registered learners from 120 countries will be powered by 20,000+ teachers. The increased business attracted more investors, although the company did not make any profit. BYJU's succeeded in attracting marquee investors such as General Atlantic, Peak XV Partners (formerly Sequoia Capital India), Sofina, the Chan-Zuckerberg Initiative (CZI), Bond, Silver Lake, Netherlands-based Prosus Ventures (previously Naspers), Tencent (China), and Tiger Global to name a few. Investors poured in money, and the company peaked at its valuation of $22 billion in 2022. BYJU's used the funds to go on a mad acquisition spree to expand its depth and geographic research. However, the pace of acquisitions was faster than the integration with the company, leading to a later lack of acquisition synergies.

7.2.4 Downward Spiral

With COVID-19 lockdowns lifted and educational institutions opening up to offline classes, the demand for BYJU's fell. Also, as time progressed and Edtech platforms matured, The intended use of EdTech platforms by companies diverges increasingly from how these interfaces are utilized by students, teachers, families, and cultural communities. Further, because of standardization and scalability, there is little differentiation between various EdTech products (Bhatia et al., 2024). On the other hand, a financial crisis began to loom over the company. It made a few failed attempts to raise funds via the IPO route. Losses continued to mount, and investors began to be impatient. Globally, an increasing trend is observed wherein VCs are giving up more equity and control to founders. However, the startups are risky by not following regulations, scaling too quickly without synergies, and continuing to mount losses (Broughman & Wansley, 2023).

BYJU's experienced several deviations from normalcy. Mismanagement of resources (especially poor financial mismanagement), predatory business practices, concentration of decision-making with the founder CEO and his wife, delays in making statutory payments, poor treatment of employees and toxic work culture, compliance issues, inability to maintain proper accounting books and red flags in those reported, weak company board, and several such negative news led investors to write-off their investments bringing down its valuations. An economic crisis in 2023 forced the company to lay off 60% of its workforce (Basu & Majumdar, 2024) and even give up its prime office space. Soon, the company gave way to rivals like Unacademy, losing its position among the top e-learning websites (Seikh & Chatterjee, 2024). Additionally, the investors dragged the founders to courts in various jurisdictions for several disputes, some of which even questioned the company's ownership. Corporate Governance lapses are clearly at the core of the disgrace of the company image (Punita, 2024). The growth-at-all-costs attitude that the company took compromised its CG, eventually leading to its crisis, much like what is seen in another Indian startup—GoMechanic (Dhamija & Nayyar, 2024).

7.3 Corporate Governance Challenges and Controversies

7.3.1 Red Flags in Reporting and Compliance

Lack of transparency and disclosure always hurts investors and their sentiments. They are considered a reflection of trust by stakeholders. Corporations take this seriously and deploy additional resources to stay compliant and make timely disclosures. BYJU'S fell short on this count. It released its FY21 (2020–21) financials with an eighteen-month delay. It missed all four deadlines it set for itself to file its financial reports, raising red flags.

Moreover, the disclosures showed that the company made a US$555.46 million loss in FY21, nearly 20 times the adjusted US$28 million loss in FY20 (2019–20) (Nikhil, 2022). On August 25, 2022, the Indian Ministry of Corporate Affairs (MCA) asked the company to explain the delay in filing the returns. The company informed the stakeholders that the reds were due to the changes in accounting policies. The changes were incorporated as the audit firm Deloitte raised several red flags and refused to sign an unqualified report.

On August 25, 2022, the Indian Ministry of Corporate Affairs (MCA) asked the company to explain the delay in filing the returns and will also investigate corporate governance issues. However, Raveendran went on to say that the company's acquisitions were complex, leading to delays in getting audit reports of the companies. The CA Institute's Financial Reporting Review Board (FRRB) has taken up a speed-track probe into the alleged irregularities and accounting practices for review (Srivats, 2023b). The probe became necessary because of complaints from the Ministry of Corporate Affairs (MCA), the Serious Frauds Investigation Office (SFIO), and the CA Institute itself (Srivats, 2023a). FRRB took up the case *suo motu* over alleged governance concerns.

7.3.2 Auditor Resignation

Deloitte has audited books of T&L for over the past six years and was appointed auditor for five years—from April 1, 2020, to March 31, 2025. The financial results for March 31, 2022, were not filed until early July 2023. The extreme delays in providing financial data for audit requirements have forced Deloitte Haskins & Sells to announce their resignation as the statutory auditor for the firm on June 22, 2023, and with immediate effect. Deloitte wrote five letters reminding the need for information required for audit—four times in 2022 (September 30, November 5, November 12, December 24) and finally on March 29, 2023). It finally gave an unqualified report on BYJU'S internal financial controls in its FY21 audit report. Corporate Governance firm InGovern called the auditor's resignation "a very powerful signal."

BYJU'S quickly announced a replacement for Deloitte by appointing BDO (MSKA & Associates) as the statutory auditors for five years starting from the fiscal year 2022 and that the company has a "meticulously planned transition timeline." BDO audit will cover the parent T&L, all subsidiaries, and overall group consolidated results (Samidha & Digbijay, 2023a). BDO has experience working with multinational clientele, and their experience will help BYJU'S, considering the global acquisitions that the company has made. The company's Chief Financial Officer (CFO) informed the stakeholders that the audit of most subsidiaries had already been completed. On June 24, 2023, BYJU'S leadership team reportedly interacted with about 75 shareholders to assure and address the concerns regarding recent developments. The company said it is working closely with the new audit firm and will file audited FY22 results by September 2023 and results for March 2023 by December 2023 (Reuters, 2023).

7.3.3 Instability of the Board

The constitution of the Board is a crucial topic in corporate governance scholarship. Generally, the founder-centric model is underappreciated, while independent directors' roles are overstated, especially in the context of venture capitalists on Board (Fan, 2022).

In the context of BYJU's, the media reported that all three external directors of T&L submitted their resignations earlier in June 2023 and that their resignations were discussed in the shareholder meeting on June 19, 2023. Raveendran denied the claim and termed the news "entirely speculative" (Rajani & Sneha, 2023) but later confirmed the resignations and asked the members to reconsider their decision. The three external directors represent reputed financial institutions and are investors in the business. Hence, their exit is a serious concern and will be viewed critically by the stakeholders. With the external member resignations, only family members (Byju Raveendran himself, wife Divya Gokulnath, and brother Riju Raveendran) appear to be left on Board (DNA, 2023). With the resignations confirmed, the company began focusing on appointing new independent directors. The role of the independent directors is to oversee that there are no violations on the governance front. Resignation of non-executive directors often happens because of differences of views between them and the promoter (conflict of interest).

7.3.4 Delays in Statutory Deposits

T&L was held responsible for deducting the Provident Fund (PF) component but not depositing the deductions of several employees with the Employee Provident Fund Organisation (EPFO) accounts for several months. EPFO directs that the employers deposit the money deducted from the employees on or before the fifteenth of the next month, and delayed payments attract hefty penalties. It was also reported that there is a lack of discipline and consistency in making PF payments to its employees, and there were delayed payments (Parvathi, 2023a). Company sources said it had cleared payment of nearly US$142 million for the ten months from August 2022 to May 2023 and committed to pay the remaining arrears of US$415,000 in due course. The payment made the company avert a potential complaint the employees could make with the EPFO for the delayed payments (Yogima, 2023). EPFO data showed that BYJU had paid US$4.7 billion (Rs. 39.05 crores) between June 26 and June 29, 2023, from September 2022 to May 2023 (Parvathi, 2023b). Some media reports say that the company has deducted tax deduction at Source (TDS) but has not deposited the money into the government account—an act tantamount to a punishable offense.

7.3.5 *Foreign Exchange Violations*

On April 29, 2023, the Directorate of Enforcement (ED) conducted searches in several premises of T&L. The search was into alleged foreign exchange rules connecting investments received and the transfer of funds abroad. The searches were made under the Foreign Exchange Management Act (FEMA) provisions. BYJU'S brushed it off, saying the searches are a "routine inquiry." It was reported that BYJU'S attracted US$3.38 trillion in Foreign Direct Investment (FDI) and US$1.18 billion in Overseas Direct Investments (ODI) between 2011 and 2023. The ED reported the seizure of various incriminating documents and digital data. ED is cross-checking the genuineness of the figures given by the company with the bankers. ED had to investigate the case due to several complaints from various parties involved in these transactions. It is reported that the CEO has never appeared before the investigation, has been uncooperative, and has remained evasive (Salman, 2023). There are instances of oversight in completing specific processes/filing compliances related to inbound/outbound investment transactions (Rishabh, 2023).

7.3.6 *Inspection of Books and Probe*

In early July 2023, the Ministry of Corporate Affairs (MCA) called for an inspection of the company's books. Reports are that the Ministry is even considering involving the Serious Fraud Investigation Office (SFIO) to investigate auditor resignations and delayed filing of financial data, among others (Gireesh, 2023). The SFIO is a multidisciplinary probe team with experts from different departments and backgrounds. A relief here is that Deloitte Haskins and Sells, at the time of resignation as statutory auditor, did not mention "non-cooperation by management," "financial irregularities," or "fraud."

7.3.7 *Instability of the Top Management*

Atul Goel joined Byju in April 2023 but resigned in October 2023 and left back to his previous employer, Vedanta Limited, in less than seven months. He helped BYJU's with the long-delayed audit formalities for the financial year ending March 2022. Later, industry veteran Pradip Kanakia was appointed a senior advisor, while Nitin Golani was promoted to CFO (M. Singh, 2023). In mid-February 2024, four investors of Byju's filed a mismanagement and oppression suit against Raveendran in the NCLT, seeking the removal of Raveendran from the Board of directors.

BYJU's saga hit a new low after six key investors who hold a 32% stake in T&L, including Prosus NV and Peak XV Partners, voted to sack the founder CEO Byju Raveendran and family at an Extraordinary General Meeting (EGM) held in mid-February 2024. A small cohort of select shareholders attended the meeting, but Byju did not. The shareholder decision came after hours-long Zoom meetings often interrupted by unknown participants (Bloomberg, 2024). Byju rejected the resolutions, stating they were invalid and ineffective.

7.3.8 Spiral of Loans

With the disinterest of VC in their venture and a blocked road to raise funds through the equity route, BYJU'S started exploring the debt route. It hoped to take advantage of the low interest rate regime. It signed a credit agreement with lenders on November 24, 2021, for $1.2 billion as a Term Loan B (TLB). The TLB comes with an interest rate of 5.5% + LIBOR (0.20% at the time). It is a senior secured loan designated syndicated in the international loan market. It gives borrowers some years to repay the loan by initially making small annual payments. The lenders can sell the loan in the open secondary market (on the London Stock Exchange). However, when the loan is sold, especially in large quantities, hedge funds and high-risk-taking distress-loan buyers come in because the loan will be available cheaply. The instrument carries a high default risk. JPMorgan and Morgan Stanley were loan agents, and Glas Trust Company managed the lenders. Unfortunately, the interest rate market took a surprising turn in 2022, and the LIBOR rate changed to 5%, making the once cheap loan turn expensive and costing an interest close to 10.5%. Investors who brought the loan requested repayment.

One of the reasons for taking the loan was to acquire a US EdTech company named 2U for US$1 billion, but the deal did not go through. Ideally, BYJU'S would have returned the unused loan money, but the low interest rate perhaps made it to keep the money. To repay the existing loans, the company contemplated raising further loans. BYJU'S planned to present its subsidiaries Great Learning, Tangible Play (Osmo), Whitehat Jr, and the parent T&L as guarantors for the loan. Its request to RBI seeking permission for WhiteHat Jr. to be a guarantor for the loan was rejected. However, BYJU'S had relief as the lenders extended the repayment deadline to October 8, 2022, with 0.125% on the outstanding loan. As the company failed to present its audited financial reports on time in October 2022, it lost its opportunity to take WhiteHat Jr. as a guarantor.

In October 2022, it took a loan of $250 million from existing investors. The company failed to meet loan requirements, and lenders were worried because the loan was trading at a 30% discount in the debt market. In December 2022, creditors of TLB asked for a partial payment considering the debt market developments. In February 2023, lenders gave a counter proposal for an immediate cash payment of US$240 million, paying an exit fee of US$150 million and increasing the interest rate by 700 basis points. On March 3, 2023, lenders issued a notice of default of dues worth US$1.26 billion. They removed existing directors of BYJU'S Alpha and named a new director. In the same month, BYJU offered to pay a higher TLB rate to negotiate debt financing. In April 2023, the lenders sought up to $200 million with higher interest as a restructuring rider.

7.3.9 Losing the Trust of Finance Companies

Further adding to the existing financial woes, starting April 2023, Indian NBFCs such as Avanse Financial Services, Aditya Birla Finance, and Fullerton India suspended their loans to the companies. They informed the education finance intermediary Propelld about the same. The quantum of such loans stood at US$365 million during the COVID years of 2020–21. RBI's First Loss Default Guarantees (FLDG) policy, announced in September 2022, had BYJU'S in a tight spot. According to this, the Lending Service Provider (LSP) (BYJU'S in this case) must compensate the Regulated Entity (such as a bank or an NBFC) for the loss due to default or up to a certain threshold of the loan portfolio.

BYJU'S has lost the support of its lenders, and its cash flows continued to get tighter as the revenues were also falling. The finance crunch forced BYJU'S to use the last and most dangerous route to source funds—parent company turning guarantor—and funding from its balance sheet so that it can continue to offer financial product features such as BYJU'S Assure, BYJU'S Advantage, and BYJU'S Direct to keep the revenue stream going. These products are the lifeline for the company, as they form 65% of orders, and customers can pay the course fee in EMI at a 0% interest rate. RBI, with the information on the governance, financial, and operational problems of the business, asked the NBFCs (some of them include SMFG India Credit Company, Avanse, InCred, IIFL, Aditya Birla Finance, Credit Saison, IIFL, among others) to report the RBI on their exposure to the EdTech sector (Pratik, 2023).

7.3.10 New Loans to Clear Old Loans!

In May 2023, BYJU'S took a Rs. 2000 crore ($250 million) loan via structured credit transactions in debt rounds from New York-based Davidson Kempner Capital Management. The three-year loan facility is given against future cash flows that accrue upon listing its Aakash subsidiary. The rate of interest is at an exorbitant 12% p.a! The repayments will be made from Aakash's future cash flows. Media sources revealed that the loan release would be in tranches contingent upon the company complying with its promises. A Condition Precedent (CP) of the loan is that the company needs to file its financial accounts. Until now, Rs. 800 crores, representing 40% of the total committed loan, was released in two tranches. In the wake of developments in June 2023, Davidson Kempner is reconsidering the release of the remaining loan (Samidha & Digbijay, 2023b). Delays in inflows of cash or loan cancellations will mean trouble for the company.

In addition, lenders dragged BYJU'S to Delaware court after a $500 million sanction as part of the loan went missing. As a part of this, a bitter legal battle arose. Because of defaulting on payment, lenders have put a claim on BYJU'S Alpha—a US subsidiary company of BYJU'S.

Additionally, Google and Facebook stopped running online advertisements of BYJU'S after the company did not settle dues for invoices for the advertisement campaigns run for Think & Learn's India entities, including Aakash Educational Services Ltd (AESL). The quantum of dues is not public. The creditors accused founder Raveendran of violating their debt contracts by refusing to give them financial details about the three units, known as "alleged debtors," now that they face bankruptcy. BYJU's refused to make some interest payments, making the dispute even more bitter. Overall, the parent company is on a thin thread and is trying to save its subsidiary.

7.3.11 Failed Merger with Aakash

In April 2021, BYJU's acquired Aakash Institute in a $950 million deal that comprised a 70% cash component and a 30% equity component. The acquisition is believed to be the biggest acquisition in this space. There was a share swap deal between the promoters of Aakash—the Chaudhry family and private equity fund Blackstone. The share swap deal hit a roadblock after the Chaudhry family did not show interest in taking equity, citing governance problems. Byju's have sent a legal notice for their resistance to share swap.

BYJU'S struggled to make payments to Blackstone during the acquisition process of Aakash. Raveendran had to buy some stake in his personal capacity to keep the majority stake in Aakash. T&L holds 73%, and Raveendran holds a 27% stake in Aakash (Madhav, 2023). On June 5, 2023, T&L announced plans to take its test preparatory subsidiary—Aakash Educational Services, to IPO. The Board of BYJU'S has given its nod, and the IPO will likely be in mid-2024. BYJU'S claims Aakash has annual revenues of Rs. 3000 crores and is likely to touch the Rs. 4000 crores mark by 2024 (as per a LinkedIn post by Divya Gokulnath). Market experts say the valuation of Aakash itself could be at Rs. 17,000 crores (US$2 billion). Aakash is undoubtedly a cash cow, but this is perhaps the last cash cow left that can get BYJU's out of the financial troubles, and the company can use the money to meet the financial pressures of T&L.

In November 2023, investor Ranjan Pai emerged as the white knight after he invested close to $200 million in Aakash and helped Byju clear its debt and interest to Davidson Kempner. With this, he bailed out a 27% stake held by the fund as collateral. Pai was also reported to be in talks with other minority shareholders. By converting his debt to equity, Pai emerged as the biggest investor in Aakash. Reports in February 2024 suggested that Ranjan Pai was holding a 40% stake in the company and had given an additional Rs. 250 crores as working capital to run the daily operations of T&L and to pay salaries for which Byju has pledged his personal stake (Mishra, 2024). These steps make him the largest investor in Aakash.

Prosus has sent a legal notice over converting $250–300 million (Rs 2072–2486 crore) debt into equity at Aakash. A Bengaluru civil court, however, gave a reprieve for BYJU's, for which the lenders are seeking an ex-parte injunction on the transaction (Digbijay, 2024). The court also added that the jurisdiction of the case was not correct. Meanwhile, Mr. Pai moved the Competition Commission of India (CCI) for its nod to the deal. Exhibit 5 shows the ownership structure and the value erosion at Aakash.

In March 2023, the Aakash and Think & Learn merger application was withdrawn. The withdrawal was a pre-planned and mutually agreed process (Jyoti, 2024). After this, Aakash distanced itself from BYJU's. There was an apparent reduction in the presence of BYJU'S name or logo in the marketing campaigns (Anees, 2024).

7.4 Corporate Governance Theories and Byju's Case Analysis

7.4.1 Importance of Theoretical Underpinning

Theoretical connections play a critical role in research. Researchers can use the essential framework the theories provide to interpret their findings within the broader context of existing knowledge (Grant & Osanloo, 2014). They enable the researchers to build upon previous work and contribute meaningfully to their field. In case studies specifically, theoretical frameworks serve as critical lenses. Case analysts can use them to analyze and make sense of complex, real-world happenings (Ridder, 2017). The theories allow them to move beyond mere description to deeper understanding. Theories sometimes give clear explanations of the observed patterns and relationships. Strong theoretical connections also enhance the applicability of case study findings. Researchers can thus draw insights that may apply to similar situations or contexts beyond the specific case being studied. This allows the case analyst to avoid the five misunderstandings about case study research (Flyvbjerg, 2006). For example, the learnings from the present case study on BYJU's can probably be applied to similar companies in the EdTech domain or startups. After all, studies show that startups have a high failure rate (Pollman, 2023), and corporate governance is undoubtedly a top reason for this (Pollman, 2020).

7.4.2 Academic Theory and BYJU's Case

The corporate governance failures at BYJU's can be critically examined through three prominent governance theories:

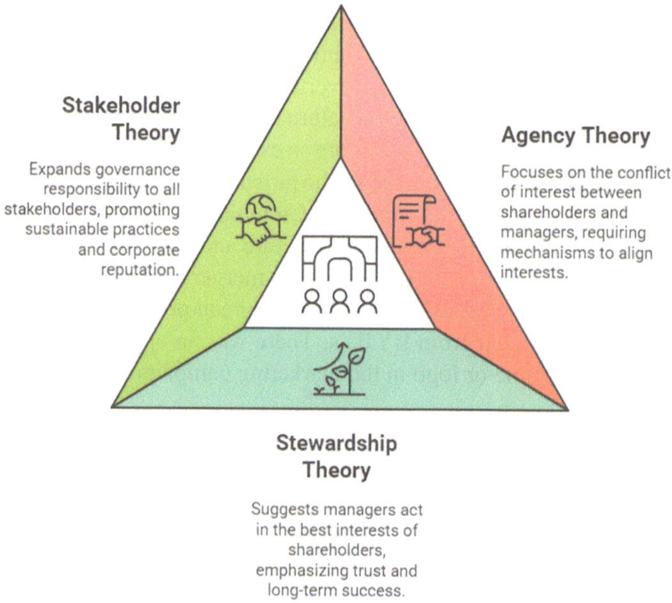

Fig. 7.2 Key theories that explain corporate governance

a. Agency Theory
b. Stewardship Theory
c. Stakeholder Theory.

These frameworks provide deeper insights into the company's governance lapses and their impact on stakeholders. Figure 7.2 gives a glimpse of the three key corporate governance theories.

7.4.3 *Agency Theory and BYJU's Case*

Agency theory was primarily developed by Michael C. Jensen and William H. Meckling in their seminal 1976 paper (Jensen & Meckling, 1976). This theory explores the relationship between principals (shareholders) and agents (company executives) in business settings. It highlights the conflicts that arise when agents do not perfectly align their actions with the best interests of the principals, leading to agency costs. While Jensen and Meckling formalized the theory, earlier work by Stephen Ross in 1973 (Ross, 1973) and Barry Mitnick in 1975 (Mitnick, 1975) also contributed to the conceptual foundation of agency theory. Figure 7.3 gives a glimpse of Agency theory.

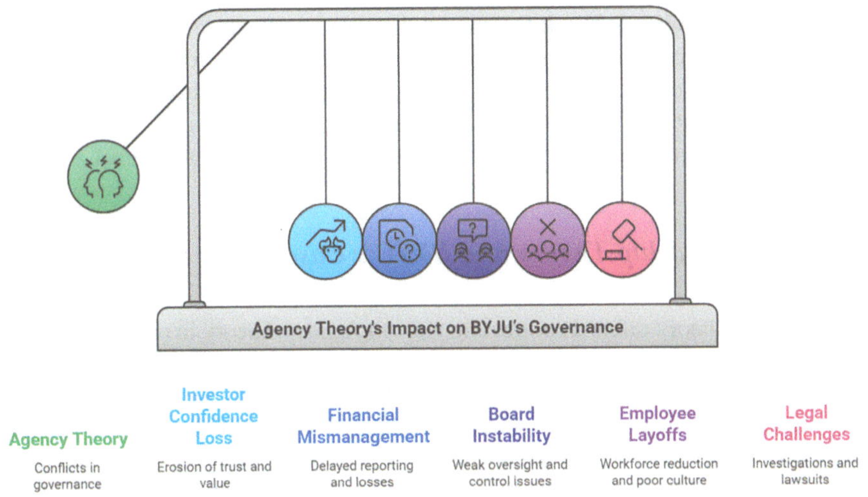

Fig. 7.3 Agency theory in corporate governance

With relation to BYJU's, the theory highlights:

1. **Shareholder vs. Management Conflict**: Agency Theory highlights conflicts when managers prioritize their self-interest over shareholder value. In the case of BYJU's, the excessive control held by the founder and the board's lack of independence created a classical principal-agent problem.
2. **Founder-Centric Decision-Making**: The CEO, Byju Raveendran, exercised disproportionate control over strategic decisions, sidelining investor concerns. The lack of transparency in financial reporting and the non-disclosure of critical information to investors indicate an agency problem. The resignation of major investors from the board, such as Prosus NV and Peak XV Partners, signals a lack of oversight, raising concerns about how managerial discretion has compromised investor confidence.
3. **Financial Mismanagement & Lack of Disclosure**: Delayed financial reporting (such as the 18-month delay in FY21 results) and the eventual revelation of significant losses indicate misalignment between the interests of the founders and shareholders. The resignation of Deloitte as an auditor over financial irregularities further exemplifies governance failures due to management's reluctance to comply with audit requirements.
4. **Excessive Risk-Taking & Growth-At-All-Costs Mentality**: The aggressive expansion strategy, including numerous high-value acquisitions (e.g., Aakash Institute, WhiteHat Jr., Great Learning), was executed without proper risk assessment or ensuring synergy between acquisitions. The $1.2 billion Term Loan B (TLB) was another instance where managerial risk appetite exceeded reasonable financial prudence, leading to a debt spiral that worsened BYJU's financial stability.

5. **Board Instability & Weak Oversight**: The exodus of independent directors and investor-backed board members reflects a fundamental agency issue, where the board failed to function as a check on management power. Instead of independent oversight, the dominance of founder-family members on the board further exemplifies weak governance and entrenchment of managerial control.

7.4.4 Stewardship Theory and BYJU's Case

Stewardship Theory offers a contrasting view to Agency Theory in corporate governance. While Agency Theory assumes executives may act in self-interest, Stewardship Theory suggests that managers are inherently trustworthy stewards who prioritize organizational goals over personal gain (Lehrer & Segal, 2020). According to this perspective, executives derive satisfaction from successfully serving the organization and its stakeholders, making external monitoring less necessary (Xu & Hu, 2024). The theory suggests that when executives identify strongly with the organization, they align their self-concept with organizational success, leading to decision-making that naturally benefits the company (Aitken & Von Treuer, 2021). This perspective has implications for governance structures, often recommending less control-oriented approaches and more empowering, collaborative relationships between boards and management (Kang et al., 2023). With relation to BYJU's, the theory highlights the application of Stewardship Theory. Figure 7.4 dipicts stewardship theory with relation to corporate governance.

In BYJU's case, the excessive concentration of power within the founder's family resulted in governance failures rather than responsible stewardship.

1. **Founder's Vision vs. Reality**: The theory suggests that founder-led organizations can perform well when stewards act in the company's best interests. However, BYJU's missteps, such as prioritizing rapid growth over financial sustainability, indicate a failure to balance expansion and profitability. Despite being a deca-unicorn valued at $22 billion (2022), the company was unable to generate sustainable revenue, forcing major layoffs and cost-cutting measures.

Fig. 7.4 Stewardship theory and corporate governance

2. **Lack of Professionalized Leadership**: Unlike well-governed companies where professional management is entrusted with day-to-day operations, BYJU's leadership remained heavily concentrated in the founder's hands. The CEO's refusal to step down despite investor pushback showcases resistance to corporate governance reforms, contradicting the core tenets of Stewardship Theory.
3. **Mismanagement of Financial Resources**: Instead of acting as responsible stewards, BYJU's leadership took excessive loans (e.g., Davidson Kempner $250 M loan) without ensuring repayment feasibility. The failed merger with Aakash Institute and the ongoing ownership disputes further illustrate the downside of excessive founder control without external oversight.
4. **Resistance to External Influence**: Investors tried to push for governance improvements, but the leadership's reluctance to accept external guidance (such as replacing the CEO) signals a key failure in stewardship-driven governance.

7.4.5 Stakeholder Theory and BYJU's Case

Stakeholder Theory, developed by R. Edward Freeman in the 1980s, fundamentally challenges the traditional shareholder primacy model by advocating for a more inclusive approach to business management (Freeman et al., 2018). According to this theory, corporations should serve the interests of all parties affected by or affecting the organization, not just those with financial stakes (Parmar et al., 2010). With relation to BYJU's, the theory highlights stakeholder impact. Figure 7.5 pictures Stakeholder Theory and Corporate Governance.

There is a profound impact on Investors, Employees, and Customers BYJU's corporate governance failures severely impacted multiple stakeholders.

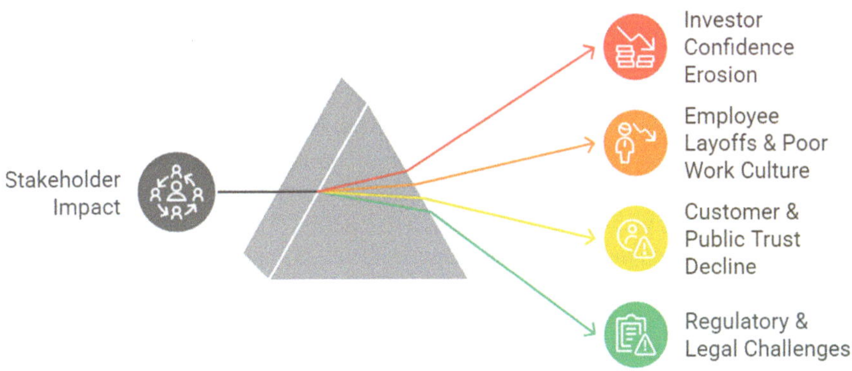

Fig. 7.5 Stakeholder theory and Corporate Governance

1. **Investor Confidence Erosion**: Investors such as Prosus and Peak XV Partners were forced to write-off their investments due to the company's financial instability. The stakeholder trust erosion led to declining valuations, with BYJU's market capitalization plummeting from $22 billion to around $5 billion in 2024.
2. **Employee Layoffs & Poor Work Culture**: About 60% workforce reduction in 2023–24 indicates poor internal governance and financial mismanagement. Reports of delayed employee salaries, non-payment of provident funds (PF), and toxic workplace culture highlight a failure in stakeholder-centric governance.
3. **Customer & Public Trust Decline**: The aggressive sales tactics used to push parents into purchasing subscriptions via NBFC-backed loans backfired, leading to complaints and legal scrutiny. Educational quality suffered, and customer dissatisfaction led to a decline in the company's brand reputation.
4. **Regulatory & Legal Challenges**: Investigations by India's Ministry of Corporate Affairs (MCA) and the Enforcement Directorate (ED) over foreign exchange violations show that regulatory authorities were concerned about governance lapses. Multiple lawsuits from investors and creditors (such as the Delaware court case on missing loan funds) further illustrate the consequences of poor corporate governance.

7.5 Learning Outcomes

The BYJU's episode has several learning outcomes. The company once again reminded of the need for good corporate governance for startups. Companies using the aggressive growth strategy appear to be taking on too much debt and making poor acquisitions. Failure to disclose its financial results promptly led to concerns from investors and regulators.

BYJU's failure to properly manage its risks led to the company facing several challenges, including financial distress and legal disputes. The delays in financial reporting, auditor resignations, and allegations of financial irregularities underscore the importance of maintaining proper financial records and stakeholder transparency.

The case highlights the importance of independent directors in corporate governance. BYJU's lack of independent directors contributed to the company's poor governance practices and financial irregularities. The signs of weak corporate governance structures are visible—the resignation of external directors and the dominance of family members on the Board. An independent and diverse board is crucial. Excessive control by the founder-led to a lack of checks and balances. Startups need to professionalize management as they grow. Startups need robust compliance mechanisms as they scale. BYJU faced several compliance issues, such as delays in statutory payments and foreign exchange violations.

Proper communication and stakeholder support are critical at times of crisis. Unfortunately, BYJU has strained relationships with investors, lenders, employees, and regulators. The case shows how poor handling of issues can spiral into more significant crises. Having proper crisis management protocols is essential. The case study provides lessons for EdTech companies. It stresses the importance of corporate governance, transparency, and stakeholder engagement in ensuring the long-term. The global EdTech market size was $2.8 billion in 2020. It is projected to grow to $10.4 billion by 2025. This statistic shows that BYJU's still has scope to fix its issues and make a comeback (Kovid & Bhati, 2024).

7.6 Scope for Future Research

The BYJU's story is ongoing as the company battles for survival and tries its best to stay afloat. Researchers from Corporate Governance and outside the area can find several dimensions of studies derived from the ongoing developments. A brief list of aspects that can be further studied includes:

1. A financial perspective on how the financial valuations shot up and eroded so quickly requires a thorough study, primarily because the valuations are done by experts from some of the finest venture capitalist investors.
2. The senior secured loan taken by the company, missing out on repaying, the expensive loan from Davidson Kempner, skipping interest payments, and the fight for BYJU's Alpha over ownership and missing funds can be studied in detail.
3. The demand of large shareholders asking the founders to step down requires a study by leadership research enthusiasts.
4. The failed merger of BYJU's and Aakash can be another case study that can give further potential insights. Further, the reasons for both companies' failure to raise funds through IPO can be examined.

7.7 Conclusion

The BYJU's case is a cautionary tale for rapidly growing startups. It gives valuable lessons for the EdTech sector. It highlights the importance of robust corporate governance practices, financial transparency, and responsible growth strategies. The company rose to be a Decacorn but faced multiple crises. Startups need to professionalize their management, establish robust compliance mechanisms, and engage effectively with stakeholders. The dominance of founders and lack of independent directors on the Board reflect the poor governance practices. These issues ultimately lead to the downfall. The case also underscores the importance of timely financial reporting, auditor independence, and transparency in financial dealings. As the EdTech market grows, startups must prioritize good corporate governance to ensure long-term success and sustainability.

References

Aitken, K., & Von Treuer, K. (2021). Leadership behaviours that foster organisational identification during change. *Journal of Organizational Change Management, 34*(2), 311–326. https://doi.org/10.1108/JOCM-01-2020-0029

Anees, H. (2024, March 14). As brand Byju's takes a hit, Aakash tries to distance itself. *Financial Express.* https://www.financialexpress.com/jobs-career/education-as-brand-byjus-takes-a-hit-aakash-tries-to-distance-itself-3424504/

Basu, R., & Majumdar, A. (2024). The economic crisis faced by one of the biggest Ed-Tech companies in the world: The case study on Byju's layoff. In K. L. Tennin & S. Ray (Eds.), *Advances in finance, accounting, and economics* (pp. 161–168). IGI Global. https://doi.org/10.4018/979-8-3693-1544-6.ch008

Bhatia, K. V., Arora, P., & Gupta, S. (2024). Edtech platforms from below: A family ethnography of marginalized communities and their digital learning post-pandemic. *Learning, Media and Technology, 1–15.* https://doi.org/10.1080/17439884.2024.2328693

Bloomberg. (2024, February 23). *Byju's investors vote to oust CEO in hours-long zoom call crashed by staff*. NDTV. https://www.ndtv.com/india-news/byjus-investors-vote-to-remove-ceo-byju-raveendran-from-troubled-edtech-startup-5113687

Broughman, B. J., & Wansley, M. (2023). Risk-seeking governance. *Vanderbilt Law Review, 76*(5), 1299–1378.

Csedő, Z., Magyari, J., & Zavarkó, M. (2022). Dynamic corporate governance, innovation, and sustainability: Post-Covid period. *Sustainability, 14*(6), 3189. https://doi.org/10.3390/su14063189

Dhamija, S., & Nayyar, R. (2024). Corporate governance mishap in a startup: A case of GoMechanic. *Emerald Emerging Markets Case Studies, 14*(1), 1–25. https://doi.org/10.1108/EEMCS-05-2023-0179

Digbijay, M. (2024, February 26). Court relief for Byju's in TLB lenders' suit. *Economic Times.* https://economictimes.indiatimes.com/tech/startups/court-relief-for-byjus-in-tlb-lenders-suit/articleshow/107437846.cms

DNA. (2023, June 23). *Byju Raveendran's Rs 67000 crore giant in trouble? Reports claim only founder family left on Byju's board, auditor quits.* DNA. https://www.dnaindia.com/business/report-byju-raveendran-s-rs-67000-crore-giant-in-trouble-reports-claim-only-founder-family-left-on-board-auditor-3048786

Fan, J. S. (2022). The landscape of startup corporate governance in the founder-friendly era. *New York University Journal of Law & Business, 18*(2), 317–389.

Flyvbjerg, B. (2006). Five misunderstandings about case-study research. *Qualitative Inquiry, 12*(2), 219–245. https://doi.org/10.1177/1077800405284363

Freeman, R. E., Harrison, J. S., & Zyglidopoulos, S. (2018). *Stakeholder theory: Concepts and strategies* (1st ed.). Cambridge University Press. https://doi.org/10.1017/9781108539500

Gireesh, C. P. (2023, July 8). Govt weighs SFIO probe after auditor exit at Byju's. *Mint.*

Grant, C., & Osanloo, A. (2014). Understanding, selecting, and integrating a theoretical framework in dissertation research: Creating the blueprint for your "house." *Administrative Issues Journal Education Practice and Research, 4*(2), 15. https://doi.org/10.5929/2014.4.2.9

Jensen, M. C., & Meckling, W. H. (1976). Theory of the firm: Managerial behavior, agency costs and ownership structure. *Journal of Financial Economics, 3*(4), 305–360. https://doi.org/10.1016/0304-405X(76)90026-X

Jyoti, B. (2024, March 12). Aakash and Byju's merger application withdrawn. *Business Line.* https://www.thehindubusinessline.com/news/education/aakash-and-byjus-merger-application-withdrawn/article67943729.ece

Kang, Y., Peng, J., & Nie, Q. (2023). Peer reaction to manager stewardship behavior: Crediting or stigmatizing the behavior? *Journal of Business Ethics, 183*(2), 453–474. https://doi.org/10.1007/s10551-022-05069-2

Kovid, R. K., & Bhati, B. (2024). Evolution of online learning platforms during and post pandemic. In J. Singh & V. Kumar (Eds.), *Technology-mediated learning during the pandemic* (1st ed., pp. 15–32). Routledge India. https://doi.org/10.4324/9781003491712-2

Lehrer, M., & Segal, L. (2020). The stewardship organization: Essential characteristics and conditions of feasibility. *American Journal of Business, 35*(3/4), 175–190. https://doi.org/10.1108/AJB-04-2020-0046

Madhav, C. (2023, June 23). *Amid crises, Byju Raveendran addresses Aakash deal, family investments.* https://thearcweb.com/article/crises-byju-raveendran-divya-gokulnath-aakash-family-office-lpTEnLjUPRTbjmkj

Manda, V. K., & Khaliq, L. N. (2023). Corporate governance failure and the regulatory fight against it: An examination of select instances from India. In R. Hasan (Ed.), *Advances in logistics, operations, and management science* (pp. 24–54). IGI Global. https://doi.org/10.4018/978-1-6684-9867-5.ch002

Manda, V. K., & Nihar, L. K. (2023). Lessons from the FTX cryptocurrency exchange collapse. In S. Saluja, D. Kulshrestha, & S. Sharma (Eds.), *Advances in business strategy and competitive advantage* (pp. 19–36). IGI Global. https://doi.org/10.4018/978-1-6684-8488-3.ch002

Mishra, D. (2024, February 25). White knight in control? Ranjan Pai seen setting course at Aakash. *Economic Times.* https://economictimes.indiatimes.com/tech/startups/white-knight-in-control-ranjan-pai-seen-setting-course-at-aakash/articleshow/107973719.cms

Mitnick, B. M. (1975). The theory of agency: The policing "paradox" and regulatory behavior. *Public Choice, 24*(1), 27–42. https://doi.org/10.1007/BF01718413

Nikhil, P. (2022, September 14). Byju's logs in surprise fall in FY21 revenue, losses widen 20 times. *Money Control.* https://www.moneycontrol.com/news/business/startup/byjus-posts-surprise-fall-in-fy21-revenue-losses-surge-17-times-9177991.html

Parmar, B. L., Freeman, R. E., Harrison, J. S., Wicks, A. C., Purnell, L., & De Colle, S. (2010). Stakeholder theory: The state of the art. *Academy of Management Annals, 4*(1), 403–445. https://doi.org/10.5465/19416520.2010.495581

Parvathi, B. (2023a, June 26). Byju's has not paid PF money for months, shows EPFO data. *Financial Express.*

Parvathi, B. (2023b, July 7). Byju's denies PF delays, yet pays Rs. 39 crore dues last month. *Business Line.*

Pollman, E. (2020, January 14). *Startup governance.* Harvard Law School Forum on Corporate Governance. https://corpgov.law.harvard.edu/2020/01/14/startup-governance

Pollman, E. (2023). *Startup failure* (Working Paper No. Working Paper N° 729/2023; p. 50). University of Pennsylvania and ECGI. https://www.ecgi.global/sites/default/files/working_papers/documents/startupfailure.pdf

Pratik. (2023, June 26). Crisis at Buju's plays up elenders' asset quality problem. *Economic Times.*

PTI. (2021, December 8). Byju's acquires Austria's GeoGebra. *The Times of India.* https://timesofindia.indiatimes.com/business/india-business/byjus-acquires-austrias-geogebra/articleshow/88163527.cms

PTI. (2022, November 3). More than 20,000 schools closed in India during 2020–21; number of teachers declined by 1.95%: Ministry of Education. *The Hindu.* https://www.thehindu.com/news/national/more-than-20000-schools-closed-in-india-during-2020-21-number-of-teachers-declined-by-195-ministry-of-education/article66090905.ece

Punita, R. (2024, March). *Start-up governance—Is it time to put an end to 'fake it till you make it'?* Artha, IIM Calcutta. https://www.iimcal.ac.in/FinLab/email-template9/pdf/Punita-R.pdf

Ranjan, R. (2022). Product innovation and personalization via social media: Learnings from the phenomenal success of BYJUs. In M. Pejic-Bach & Ç. Doğru (Eds.), *Advances in logistics, operations, and management science* (pp. 288–302). IGI Global. https://doi.org/10.4018/978-1-7998-7793-6.ch012

Rajani, R., & Sneha, S. (2023, June 23). Blow to Byju's as auditor Deloitte, 3 directors resign. *Mint.* https://www.livemint.com/companies/start-ups/blow-to-byju-s-as-auditor-deloitte-3-directors-resign-11687458700828.html

Reuters. (2023, June 26). Will file '22 financials by Sept: Byju's to investors. *Business Standard.*
Ridder, H.-G. (2017). The theory contribution of case study research designs. *Business Research, 10*(2), 281–305. https://doi.org/10.1007/s40685-017-0045-z
Rishabh, B. (2023, June 23). Byju's: How troubles have mounted for the edtech giant over the past year. *BQ Prime.* https://www.bqprime.com/business/byjus-how-troubles-have-mounted-for-the-edtech-giant-over-the-past-year
Ross, S. A. (1973). The economic theory of agency: The principal's problem. *American Economic Review, 63*(2), 134–139.
Salman, S. (2023, April 30). ED searches Byju's premises. *Financial Express.*
Samidha, S., & Digbijay, M. (2023a, June 23). Three key directors, Auditor Deloitte resign from Byju's. *Economic Times.*
Samidha, S., & Digbijay, M. (2023b, June 29). Byju's is yet to get full Rs. 2k crore pledged from Davidson Kempner. *Economic Times.*
Seikh, M. R., & Chatterjee, P. (2024). Evaluation and selection of E-learning websites using intuitionistic fuzzy confidence level based Dombi aggregation operators with unknown weight information. *Applied Soft Computing, 163*, Article 111850. https://doi.org/10.1016/j.asoc.2024.111850
Singh, M. (2023, October 24). *Byju's CFO quits in six months amid delayed accounts.* Tech Crunch. https://techcrunch.com/2023/10/23/byjus-cfo-quits-in-six-months-delayed-accounts/
Srivats, K. (2023a, March 10). CA Institute confident of completing probe on BYJU's this year, says Talati. *Business Line.*
Srivats, K. (2023b, June 24). Review of Byju's accounting practices to conclude in 2 months'. *Business Line.*
Tripathy, S., & Devarapalli, S. (2021). Emerging trend set by a start-ups on Indian online education system: A case of Byju's. *Journal of Public Affairs, 21*(1), Article e2128. https://doi.org/10.1002/pa.2128
Wright, S. L., & Nutt, S. (2021). The Wework company: A case about a failed IPO, Unicorn startup. *Journal of Finance Case Research, 19*(1), 81–87.
Xu, Y., & Hu, W. (2024). Linking servant leadership to employees' knowledge sharing: The role of thriving at work and organizational identification. *Current Psychology, 43*(40), 31517–31529. https://doi.org/10.1007/s12144-024-06735-6
Yogima, S. (2023, June 28). Byju's staff to get their PF dues as co transfers Rs. 123 cr to EPFO. *Economic Times.*

Chapter 8
The Dimensions of Family Business with Effective Corporate Governance—A Worm's Eye View

Srihari Jwalapuram, K. S. Anand Sudarshan, and Nella Bhargava

Abstract Family businesses are pivotal to the global economy, contributing significantly to job creation and innovation. However, their unique dynamics necessitate a distinct approach to corporate governance to ensure sustainability and long-term performance. The dimensions of family business and their interplay with effective corporate governance practices include family involvement, governance structure, succession planning, socioemotional wealth, and stakeholder engagement. The level of family engagement often shapes governance frameworks, leading to the necessity for clear delineation of roles among family and non-family members. Effective corporate governance in this context promotes transparency, accountability, and balanced decision-making, fostering trust among stakeholders. The governance structure within family businesses varies significantly compared to non-family firms. It typically encompasses family councils, advisory boards, and formalized governance policies. This structure is critical in aligning family interests with business objectives and mitigating potential conflicts. Strong governance frameworks empower family businesses to navigate challenges while preserving family legacy and values. Succession planning is another vital dimension that impacts both family business continuity and governance effectiveness. The lack of a structured succession plan can lead to uncertainties that jeopardize business stability. This research underscores the importance of proactive succession strategies that not only prepare the next generation of leaders but also reinforce governance mechanisms, ensuring a seamless transition and continuity in governance practices.

Keywords Family business · Corporate governance · Transition · Sustainability · Long-term planning

S. Jwalapuram (✉)
Bharatiya Engineering Science Technology Innovation University, Bengaluru, India
e-mail: harijwalapuram@gmail.com

K. S. A. Sudarshan
Dhanwantari Academy for Management Studies, Bengaluru, India

N. Bhargava
Acharya Institute of Graduate Studies, Bengaluru, India

V. Kandpal et al. (eds.), *Developments in Corporate Governance*,
Approaches to Global Sustainability, Markets, and Governance,
https://doi.org/10.1007/978-981-96-6366-8_8

8.1 Introduction

Family businesses represent a vital segment of the global economy, often characterized by unique governance challenges and opportunities. These enterprises blend family values with business objectives, which can significantly influence their strategic direction, operational practices, and overall success. Effective corporate governance is essential for these businesses, as it helps navigate the complexities inherent in familial relationships while promoting transparency, accountability, and long-term sustainability. In a global context, the dynamics of family business governance vary widely, influenced by cultural, economic, and regulatory factors. Research has shown that effective governance practices are critical for addressing issues such as succession planning, conflict resolution, and stakeholder engagement, particularly in diverse environments. The interplay between family interests and professional management can enhance innovation and resilience when managed thoughtfully.

Recent studies underscore the importance of adapting governance frameworks to local contexts while embracing global best practices. For instance (Gómez-Mejía et al., 2021) emphasizes the role of socioemotional wealth in family firms and its impact on governance decisions, while (Miller et al., 2022) explores the effects of internationalization on governance practices in family businesses. Additionally (Sharma & Chrisman, 2020) discusses the implications of generational transitions and the importance of strategic planning in governance. With regards to succession planning ensuring a smooth transition of leadership across generations is vital. Research highlights that early and strategic planning can mitigate conflicts and sustain business continuity (Miller et al., 2022). As a consequence of conflict resolution, family dynamics often lead to conflicts that can impact business decisions. Establishing formal governance structures, such as boards of directors, can help manage disputes and align interests (Bammens et al., 2021). About stakeholder engagement, family businesses must balance the needs of various stakeholders, including family members, employees, and external partners. Effective governance frameworks can facilitate communication and decision-making processes that consider these diverse interests (Astrachan & Shanker, 2020). Ethical considerations of family firms often emphasize values and ethics, which can influence governance practices. Integrating ethical considerations into governance can enhance trust and long-term performance (Dyer & Whetten, 2022). Globalization and adaptability as family businesses expand internationally, they face varying regulatory and cultural landscapes. Understanding local contexts while adhering to global governance standards is essential for sustained success (Gómez-Mejía et al., 2021). The transition of leadership is critical for sustainability. Research indicates that structured succession planning enhances resilience and minimizes conflict during generational shifts (Smith et al., 2024). Family conflicts can disrupt business operations. Establishing clear governance frameworks, such as advisory boards, can facilitate effective dispute resolution and ensure alignment of interests (Jones & Williams, 2024). Engaging various stakeholders, including family members and employees, is essential. Effective governance practices promote inclusive decision-making, leading to better business outcomes (Garcia et al., 2024). Ethical governance of family firms often prioritizes ethical practices. Integrating ethics into governance frameworks can strengthen trust and improve overall performance (Kim & Lee, 2024). Global adaptability of family businesses expand globally, they must navigate diverse regulatory and cultural landscapes. A flexible governance approach that respects local contexts while adhering to global best practices is crucial (Chen & Zhao, 2024).

By synthesizing these perspectives, this paper aims

- To contribute to the growing body of knowledge on effective corporate governance in family businesses, offering practical insights for practitioners and researchers alike.
- To examine the dimensions of corporate governance within family businesses across different regions, highlighting best practices and strategies that have emerged globally.
- To provide insights into how family firms can align their governance structures with both familial aspirations and business objectives, thereby ensuring their competitiveness and longevity.

8.2 Literature Review

Socioemotional wealth (SEW) serves as a cornerstone for understanding family businesses. It encompasses the non-financial aspects that families prioritize, such as family reputation, emotional attachment, and long-term legacy. Effective corporate governance recognizes and incorporates SEW considerations into decision-making processes, aligning business strategies with family values and objectives.

- Stakeholder engagement is essential for family businesses' resilience and adaptability. By fostering strong relationships with various stakeholders—including employees, customers, suppliers, and the community—family businesses can enhance their governance frameworks. This engagement not only strengthens trust and loyalty but also informs strategic decisions, thereby supporting sustainable business practices.
- The interplay between these dimensions and effective corporate governance ultimately shapes the performance and longevity of family businesses. The study reveals that family businesses that adopt comprehensive governance frameworks are better positioned to face market fluctuations and internal challenges. These frameworks facilitate the balancing of family interests with professional management practices, ensuring both business success and family harmony.

The study of family firms is essential as many growing entrepreneurial ventures evolve into family-run businesses. Given that these businesses rely on a significant portion of job creation, it's crucial to understand their dynamics. This review was selected to explore their impact and contribution to the economy, given their vital role in employment generation and business growth.

Table 8.1 Comparison of Tata group and reliance industries (2023)

Metric	Tata Group	Reliance Industries
Revenue (2023)	Approx. $100 billion	Approx. $225 billion
Market Cap (2023)	Approx. $150 billion	Approx. $200 billion
Key Sectors	Automotive, IT, Steel, Consumer Goods	Petrochemicals, Telecom, Retail
Number of Employees	750,000+	250,000+
Global Presence	100+ countries	100+ countries
Notable Subsidiaries	Tata Motors, Tata Consultancy Services	Reliance Jio, Reliance Retail

The Table 8.1 is self-evident with reference to the power of family-run businesses over decades in India.

8.2.1 *Succession as a Good Gesture*

The younger son feels pressured by his older brother's presence. Family business, insufficient exploration of ownership and board governance in unlisted firms, and an overemphasis on financial metrics in studies of listed family businesses. Further, it suggested succession planning yields always (Azila-Gbettor et al., 2022). For family businesses to grow and endure across generations, they must establish robust internal control and governance structures, particularly in first and second-generation firms. The corporate board is a vital governance mechanism for family firms, significantly influencing their performance. Relationships, revealing that board attributes impact performance differently based on cultural and legal contexts. It improves governance in family firms and suggests that managers should consider these contextual factors when structuring their boards. Agency conflict between active family owners involved in management and passive family owners in unlisted Spanish family firms. Using agency theory, we found that ownership concentration and governance mechanisms enhance performance, particularly in later-generation firms. Direct control by passive owners also positively impacts performance, while board presence shows no significant effect.

Identified a direct link between corporate governance mechanisms and family business performance, covering innovation, internationalization, auditing, and failure risk (Teixeira & Carvalho, 2024). The evolution of corporate governance in family business centered around centralized decision-making, addressing key governance challenges, and embracing family governance models and modern trends. Succession planning eventually leads to diversity and thereby innovation (Nakpodia, 2024). Found succession planning results in proper responsibility and self-leadership (Maroga et al., 2024). Family firms significantly contribute to global economies by generating wealth and creating employment. It examines the role of entrepreneurship in succession planning and decision-making. Key drivers of knowledge transfer for effective succession in family businesses include trust, organizational culture, competitive advantage, technological advancements, and new knowledge sources. In the future value creation, internationalization, and business innovation will be crucial in facilitating knowledge-sharing between family and non-family members during succession (Luong et al., 2024). While family businesses recognize the importance of sustainability, awareness of sustainable processes and procedures remains low. Furthermore, sustainability efforts are not managed or implemented in a systematic manner, indicating the need for more structured approaches (Jamil et al., 2024).

8.2.2 Ownership Structure and Firm Performance

Predicts firm performance, highlighting that family firms foster a culture that enhances human capital and overall performance. Distinguishing family businesses from others by exploring the unique family-business interface and identifying key characteristics—such as demographics, strategies, continuity, capital, and values—that contribute to their success. In family businesses, gender bias often excludes daughters from succession. Societal attitudes and family dynamics contribute to this exclusion, rendering daughters' abilities overlooked. However, daughters can ascend to leadership in special situations, such as lacking male heirs or during critical transitions, often achieving competitive advantages (Wang & Miao, 2010). Business restructuring is vital for the longevity of family firms, yet research in these areas has largely been underdeveloped. Exploration of how "grey literature" can enhance the relevance and impact of management and organization studies with respect to corporate governance in family firms (Adams et al., 2017).

Families often prefer shutting down satellite businesses over selling them, driven by identity factors, asset recycling goals, and intentions to restart the business later. Family business managers can take leverage to understand how portfolios can adapt to decline and maintain long-term entrepreneurship. Shutting down a satellite firm rather than selling it serves as an effective turnaround strategy, helping preserve family identity while ensuring the continuity of the family business portfolio (Akhter et al., 2016). While cultural long-term orientation does not moderate the negative effect of family-intensive governance on performance, it reveals a mediating effect, suggesting that long-term orientation decreases performance by increasing family involvement in governance (Pinelli et al., 2024). Presents an organizational identity-based explanation for why family firms prioritize non-financial goals. The trans-generational sustainability intentions and the firm's ability to enhance family identity increase the importance of identity alignment and corporate reputation, hence it drives family firms to pursue non-financial goals benefiting non-family stakeholders, with reinforcing feedback loops supporting these processes (Zellweger et al., 2013).

8.2.3 Family Firms and High Technology Mergers and Acquisitions

Investigated whether family firms engage in value-creating high-tech M&A Family ownership positively influences abnormal returns during announcements, though this effect diminishes at higher ownership levels (André et al., 2014). Existing research offers conflicting views on whether family firms perform better or worse in mergers and acquisitions (M&As) than non-family firms (Palm et al., 2024). Family firms, with ownership concentrated among family members, influence their internationalization of decisions. Indian family firms contribute to the mixed gambling perspective by highlighting these dynamics in internationalization strategies (Fuad et al., 2024). Family-owned investment companies effectively balance the hard and soft aspects of the M&A process by leveraging their entrepreneurial experience to apply soft values. Coupled with external management's focus on hard values, they achieve equilibrium (Fagrell & Gunnarsson, 2024). From a mixed gamble perspective, we argue that family firms (FFs) with varying family shareholder structures face distinct trade-offs between current and prospective financial and socioemotional wealth (SEW), influencing their international acquisition decisions (Calabrò et al., 2024). Application of socioemotional wealth theory to explore how non-family shareholder governance influences family firms' digital transformation (Du et al., 2024). Family businesses often begin with co-founders from the same family, which helps retain ownership (Song et al., 2024).

8.2.4 Founding-Family Ownership, Corporate Diversification, and Firm Leverage

Single-family offices (SFOs) manage trillions of assets globally, highlighting their importance as cohesive wealth management tools. Despite their growing significance, research on SFOs is still limited, especially regarding the capital structure of the firms they own. Analyzing a hand-collected sample of SFO-owned firms in the DACH region, we find that SFO-owned firms have a higher long-term debt ratio compared to family-owned firms, indicating adherence to trade-off theory, similar to private equity firms. This trend is more pronounced in SFOs that have sold their original family business. In contrast, family-owned firms exhibit more conservative financial decision-making, aligning with the pecking order theory (Block et al., 2024). This study examines the relationship between family governance and innovativeness in family firms, focusing on how family commitment and involvement of the founding generation moderate this relationship. Drawing on agency theory, it is analyzed that family firms across various countries and industries using moderated regression analysis. Our results reveal that strong family commitment and involvement of the founding generation enhance the link between governance and innovativeness. Furthermore, when founding family members are absent, firms with high family commitment experience greater improvements in governance-innovation relationships than those with weaker commitment. This research adds to agency theory by identifying important boundary conditions in family firms (Dibrell et al., 2024).

Research on entrepreneurship in family firms has surged over the past two decades, yet it remains fragmented due to diverse theoretical perspectives and contexts. This study addresses this gap through a bibliometric analysis, integrating co-citation analysis and bibliographic coupling to reveal the field's intellectual foundations and current research discourses. Through published studies, identifying four key foundations: socioemotional wealth (SEW), entrepreneurial orientation, family-embedded resources, and agency theory. Current research is categorized into seven main discourses, including entrepreneurial motivation, gender and success, and family firm internationalization. Our findings highlight developmental trends and identify future research opportunities within the field (Anwar et al., 2024). Family firms constitute two-thirds of all businesses globally, making absorptive capacity (AC) crucial for acquiring and utilizing external knowledge for value creation. This highlights that family influence leads to ambiguous effects on AC integration. While family members can enhance knowledge integration, they may also isolate firms from external knowledge sources. Most AC research in family firms remains conceptual, with few empirical studies relying on simplistic constructs. We advocate for future empirical research to explore multidimensional constructs of AC and family influence in greater depth (Pütz & Werner, 2024). The case analysis indicates that entrepreneurial family (EF) members need to acquire distinct knowledge types at various generational stages to foster new ventures, specifically technical knowledge in the second generation and business knowledge in the third. Trust among generations and role separation facilitated these transitions, empowering younger generations to pursue entrepreneurial opportunities (Capolupo et al., 2024).

Knowledge networks in regional clusters are crucial for innovation and local development, with family firms playing a significant role in creating business opportunities and inter-organizational networks. However, their impact on knowledge (Ghinoi et al., 2024) transfer compared to non-family firms remains unclear. This paper uses Exponential Random Graph Models (ERGMs) to analyze network data from the Parabiago footwear cluster in Italy. We identify unique network strategies based on various types of knowledge exchange: technological, market, and managerial. Our findings show that family firms are more adept at expanding networks than non-family firms, often connecting with each other, indicating a homophilous networking approach.

Conversely, non-family firms do not exhibit the same tendency. These insights suggest that ownership structure influences knowledge exchange, guiding managers and policymakers in developing effective cluster strategies. Corporations with promising core businesses tend to concentrate on these areas, often neglecting their internal corporate ventures (ICVs) and affecting their performance. Corporate parents, reveals that attractive core businesses lead to greater planning autonomy for ICV managers (Covin et al., 2024), enhancing performance. Additionally, the experience of venture managers and product similarity within the corporation moderate these effects. The research highlights that corporations reliant on ICV success due to weak core business prospects may mismanage them, as direct managerial involvement increases. Autonomy in venture planning is crucial to prevent unrealistic expectations for ICVs and applies to other entrepreneurial settings emphasizing experimentation and learning. Many family firms derive their property rights from the privatization of state-owned enterprises, especially in emerging markets. While the advantages of this ownership are well-documented, disadvantages exist. We introduce the concept of ownership origin liability, suggesting it challenges the legitimacy of privatized family firms, reducing their owners' desire for family and trans-generational control. However, corporate philanthropy can alleviate these negative impacts. Our empirical analysis, based on a national survey of Chinese family-owned firms, supports these findings (Jin & Hu, 2024). Accelerating innovation and launching new businesses have become priorities for firms, supported by various corporate venturing (CV) modes. However, CV initiatives compete for organizational resources and management support, especially during crises. The study during the COVID-19 crisis reveals how external challenges impact their internal legitimacy. We identify four legitimacy-seeking mechanisms—adjusting, sensing, focusing, and mobilizing—that help CV units secure resources and support from stakeholders. These mechanisms highlight the dynamic nature of legitimation efforts, influenced by perceived organizational distress and pre-crisis legitimacy, enhancing our understanding of CV's role in resource allocation (Weniger & Jarchow, 2024). In the context of open innovation, understanding learning dynamics is crucial for corporate success. Firms reveal an inverted U-shaped relationship between technology sourcing ambidexterity and innovation, indicating that excessive ambidexterity can hinder organizational learning. Organizational slack alleviates these constraints, enabling resource orchestration and risk-taking beyond existing competencies (Anokhin et al., 2024). Corporate entrepreneurship (CE) plays a vital role in entrepreneurship research, impacting firms' survival and growth. These systematic literature reviews examine top-tier journals between 1985 and 2023. It highlights a lack of focus on context in CE processes and suggests a research agenda addressing various contextual factors.

8.2.5 Entrepreneurship in Family vs. Non–Family Firms

Family firms typically have concentrated ownership and significant socioemotional investments, indicating a greater need to hedge compared to non-family firms. However, empirical evidence has been inconclusive (Abeysekera et al., 2024). The impact of family ownership, management, and prior transaction experience on long-term stock market performance following acquisitions and divestments in Japan. Family ownership positively influences stock performance (Song et al., 2024). Examined the impact of family ownership and control on firm performance, considering gender diversity as a moderating factor (Amin et al., 2024). Family ownership positively influences firm performance, a relationship strengthened by female board members Family firms operate within institutional frameworks that influence their governance and performance (Al-Haddad et al., 2024).

8.2.6 Family-Controlled Firm and Corporate Venturing

Research indicates that nepotism in public family firms can lead to agency costs due to mismatches in cash flow and voting rights. It is further demonstrated that institutional ownership is crucial in mitigating these agency costs within public family firms. Family ownership adds value only when the founder is CEO or serves as Chairman with a hired CEO.

8.2.7 Good Corporate Governance Structures: A Must for Family Businesses

Significance of socioemotional wealth (SEW) in family firm takeovers and shows that incumbent management holds power in building ideal corporate governance (Campbell & Jerzemowska, 2017). Using behavioral and stakeholder theories, family firms pursue family-centered non-economic goals that influence their behaviors. Family influence partially mediates the relationship between family involvement and the adoption of these goals (Chrisman et al., 2012). Evaluating family firm effectiveness relies heavily on the measurement of goals and performance outputs. The green innovation differences between family and non-family firms, focusing on business type and firm age as moderators affecting green patents (Francesco et al., 2023). The Companies Act emphasizes the separation of ownership and management, involving various stakeholders such as banks, the Board of Directors, employees, suppliers, customers, and regulators. Family firms and business groups are crucial in emerging economies. Sustainability drivers can initiate change in corporate governance models while identifying barriers can aid in developing strategies for overcoming obstacles and institutionalizing sustainability efforts. Corporate governance and firm performance in economies with dispersed stock ownership and active corporate control markets. Aiming to clarify the relationship between generation and innovation in family firms, examining how family influence varies across generations. It tests whether family management and the intention to transfer control affect this relationship. Long-term orientation (LTO) refers to the tendency to focus on long-term impacts of decisions, a common trait in family businesses. Suggesting LTO positively influences innovativeness, proactiveness, and autonomy, but negatively affects risk-taking and competitive aggressiveness. Business Model Innovation (BMI), while environmental dynamism and new digital technologies have driven many firms, including family businesses, to innovate their business models moderates this effect.

8.2.8 Entrepreneurship and Family Firms

Adopted a generational perspective to explore entrepreneurial orientation (EO) in family firms, analyzing how internal and external factors influence EO across different generations. Additionally, non-family managers positively impact EO only in third-generation firms and beyond, with non-family investors significantly influencing EO in later generations (Cruz & Nordqvist, 2012). Growth is crucial for a business's long-term success, yet the impact of family influence on firm growth is often overlooked. Significant economic impact on growth rates and its implications are identified.

8.2.9 Family Business Journals and Their Contribution

The bibliometric analysis reveals that family business journals have gained increasing recognition over the years, evident from rising citation counts and low self-citation rates. The US is the leading country in this field, with prominent authors like Frank W. Kellermanns and Joseph H. Astrachan. European countries also perform well, with Germany, Spain, and the UK among the top ten, but authors in these countries have yet to fully realize their potential for productivity and impact.

8.2.10 Trust and Reputation in Family Businesses

Family businesses are known for their trustworthiness and reputation, which significantly influence their long-term financial success (Chaudhary et al., 2021). While many business-owning families (BOFs) engage in community social efforts through both family foundations and corporate social responsibility (CSR) programs, the connection between these activities is unclear (Cruz et al., 2024). Globally, U.S. public family firms with private foundations, has a positive link between foundation contributions and CSR activities (Chaudhary et al., 2021) Socioemotional wealth in family businesses, its impact on trust and reputation, differences from non-family firms, and strategies for its creation and preservation (Ramadani, 2024). Presents a theoretical framework for supply chain management (SCM) that aligns with previous research on family business characteristics and resilience strategies adopting product enhancements and digitization to improve supply chain processes. Laissez-faire leadership (LFL) and Family businesses (FB) both affective and cognitive trust are vital for problem-solving in FBs, while cognitive trust dominates in NFBs, bridging leadership and conflict management literature.

8.2.11 Family Business and Corporate Governance in Pan-India (Global)

The role of block holders and their governance mechanisms within France's complex ownership structures. Additionally, family businesses homogeneity reduces conflicts, emphasizing the importance of control contestability in governance (Ducassy & Guyot, 2017). Identifying the corporate controller is crucial for discussions on corporate governance within firms and markets. The shift from a national to a global economy has impacted corporate operations in various ways. Different corporate governance models highlight the need for clear objectives and responsibilities (Deb & Dube, 2017). Concentrated ownership is often viewed as inefficient due to risks of minority expropriation linked to unequal control and cash-flow rights. Director stock ownership is strongly and consistently linked to future corporate performance, making it significant for public policymakers and long-term investors (Bhagat & Bolton, 2019). Corporate governance in Indian financial institutions remains a concern, with fraud, weak disclosures, and rising non-performing assets raising alarms (Goyal & Gulati, 2025). Governance codes, with a particular focus on corporate governance codes for listed companies, stewardship codes for institutional investors, and best-practice codes for proxy advisors.

8.2.12 Socioemotional Wealth Theory—An Invisible Phenomenon

Socioemotional wealth is a distinctive aspect of family businesses that shapes their strategic decisions (Chang & Mubarik, 2021) by generating non-financial benefits that influence investment choices (Diéguez-Soto et al., 2024; Hernández-Perlines et al., 2019; Souder et al., 2017) identify SEW as a key factor differentiating family firms, explaining their diversity in characteristics, behaviors, and interests (Berrone et al., 2012). Conflicting arguments (Davila et al., 2023) on the socioemotional wealth (SEW)-family firm performance link. The managerial decisions mediate this relationship and how SEW dimensions moderate it. A positive SEW-performance link, showing this pursuit doesn't harm financial utility, with governance and stakeholder strategies offsetting negative effects. Examining numerous studies referencing SEW in recent years (Jiang et al., 2018) it is concerning that a clear consensus on its definition remains elusive. While many agree that SEW reflects the "affective endowments of family owners" (Gomez-Mejia et al., 2011), its sufficiency remains debatable. Suggested that due to the challenges of directly measuring socioemotional wealth (SEW), an alternative construct is necessary. It is socioemotional wealth importance scale (SEWi), which assesses the significance attributed to SEW rather than its level or accumulation. Social mandates. SEW remains a relatively recent (Smajić et al., 2023) concept, underscores the positive emotional wellbeing of a family business is directly linked to the success/growth of the enterprise.

The following are few cases in the recent past with reference to social-emotional wealth and family business firms

Risk Aversion and Long-Term Orientation of (Gómez-Mejía et al., 2007). The Ford family has maintained control over Ford Motor Company for over a century, prioritizing stability and long-term growth. During the 2008 financial crisis, Ford chose to take private loans instead of accepting a government bailout, unlike General Motors and Chrysler. This decision reflected their commitment to preserving family control and legacy, a core aspect of SEW. Stakeholder and Employee Relationships (Berrone, 2012). Tata Group, a multinational family-owned conglomerate, strongly prioritizes social responsibility. During the 2008 Mumbai attacks, Tata Hotels continued paying salaries to affected employees and their families, reinforcing SEW-driven commitment to social capital and corporate reputation. Governance Conflicts Due to SEW (Chrisman et al., 2005) after the death of founder Dhirubhai Ambani, his sons Mukesh and Anil Ambani engaged in a power struggle over Reliance Industries. Their strong emotional attachment to control led to a bitter feud, resulting in the division of the business. SEW can sometimes lead to governance conflicts in family firms. Resistance to Change and Innovation (Gómez-Mejía et al., 2010). Cargill, one of the world's largest family-owned businesses, has historically been slow to adopt disruptive technologies and strategic alliances. The family's desire to retain control and protect SEW has sometimes led to resistance to change, even when market conditions demanded innovation. Philanthropy and Community Engagement (Zellweger et al., 2013). Mars Inc., a privately held family business, actively engages in sustainability and ethical sourcing initiatives. The long-term focus on environmental and social impact aligns with SEW principles, emphasizing reputation preservation and legacy over short-term financial returns. Influence of Female Family Directors on Innovation (Bauweraerts et al., 2022) the presence of female family members on the board of family firms affects innovation activities. More female family directors can enhance innovation, but this effect is moderated by SEW factors. Specifically, strong family control and influence can weaken the positive impact of female directors on innovation, while a strong family identification with the firm can enhance it. Socioemotional Wealth and Initial Public Offerings (IPOs) (Berrone et al., 2022) through Initial Public Offerings (IPOs) Family firms with high SEW are more cautious about going public due to concerns over losing control and diluting family influence. However, when they do choose to go public, they often implement mechanisms to retain control, such as dual-class shares, to protect their socioemotional wealth. SEW and Financial Decision-Making During the COVID-19 Pandemic (Hernández-Perlines et al., 2022) how family firms managed their socioemotional wealth during the COVID-19 pandemic. Family firms with abundant slack resources were more likely to prioritize SEW preservation, such as maintaining employee welfare and community support, even at the expense of short-term financial performance. This behavior underscores the importance of SEW in guiding family firms' responses during crises. Leadership Decisions Influenced by SEW (Subramanian, 2022) the Tata Group, a prominent Indian family-owned conglomerate, dismissed its non-family chairman, Mr. Cyrus Mistry the decision was driven more by the family's desire to preserve SEW—maintaining family control and upholding legacy—than by financial performance considerations. This underscores how SEW can influence leadership choices in family firms. SEW as a Driver of Business Performance (Palalić & Smajić, 2022) examined two family businesses in Bosnia and Herzegovina, exploring the relationship between SEW, transformational leadership, and business performance. SEW positively influences business performance when mediated by transformational leadership, highlighting the importance of leadership styles in leveraging SEW for organizational success. Impact of CEO Tenure and Generational Influence on SEW (Åberg et al., 2024). SEW affects family firm performance, considering the moderating roles of CEO tenure and the presence of millennial CEOs. The longer CEO tenures may strengthen the positive effects of SEW on firm performance, while millennial CEOs might introduce new perspectives that alter traditional SEW dynamics.

8.2.13 Agency Theory—A Quintessential for Family Business

Agency relationships as contracts under which principals engage agents to perform services on their behalf, delegating decision-making authority. An introduction of the concept of agency costs, which include monitoring expenditures by the principal, bonding expenditures by the agent, and residual loss. The agency costs arise from the separation of ownership and control and can influence a firm's capital structure (Jensen & Meckling, 1976). The separation of ownership and control in large firms, managerial, labor markets and competition can mitigate agency problems. The role of internal and external mechanisms in aligning the interests of managers and shareholders (Fama, 1980). A comprehensive review of agency theory, discussing its assumptions, empirical support, and applications. The two streams within agency theory: positivist and principal-agent, and examines mechanisms to address agency problems, such as outcome-based contracts and information systems (Eisenhardt, 1989). The separation of decision-making and risk-bearing functions in organizations. Likewise separating ownership and control can lead to agency problems but also propose that organizational structures, such as boards of directors, can serve as effective control mechanisms (Fama & Jensen, 1983). Ross introduces the principal-agent problem, analyzing how principals can design incentive structures to align agents' actions with their interests along with the role of information asymmetry and risk-sharing in agency relationships (Ross, 1973). The agency theory addresses issues stemming from the delegation of decision-making authority, emphasizing the importance of contract design to mitigate agency problems and associated costs (Hendrastuti & Harahap, 2023). How agency theory explains conflicts arising from differing risk preferences between principals and agents, particularly in the context of corporate governance and corruption (Al-Faryan, 2024).The agency and altruism in family firms, arguing that family dynamics, particularly altruism, can exacerbate agency problems. Models (Schulze et al., 2003). The application of agency theory in understanding family firm behavior, potentially causing harm to the organization (Madison et al., 2016a). The head of a family-owned firm may prioritize the welfare of certain family members within the organization (Kallmuenzer, 2015). How firm size influences the relationship between family ownership and performance, suggesting that size plays a crucial role in shaping governance structures and outcomes in family businesses (Galavotti et al., 2025). Revisiting the separation between ownership and control, reviewing studies on agency theory and governance, and examining their implications for corruption within organizations (Al-Faryan, 2024). The family firms emphasize ownership concentration, board size, and gender diversity, which significantly impact financial performance (Mashele et al., 2024). Drawing on agency and stewardship theories, the presence of a family protocol influences governance structures (Mansouri et al., 2024) the strengths and limitations of various corporate governance approaches in understanding family business phenomena, advocating for an integrated perspective (Jayantilal et al., 2024).

8.2.14 Through the Lens of Stewardship Theory

The above Table 8.2 explains the key differences between stewardship theory and agency theory

Stewardship theory suggests that individuals within an organization are not primarily driven by self-interest, as proposed by principal-agent theory. Instead, they are inherently motivated to contribute to the organization's success (Abels, 2023). The differing interests of principals (board chairs) and agents (CEOs) suggest that these two sets of non-profit actors frequently operate from different perspectives, potentially affecting the governance of their organizations (Bernstein et al., 2016). Stewardship theory is a popular alternative to agency theory for studying family firm governance, a widely recognized alternative to agency theory in examining family firm governance (Chrisman, 2019; Madison et al., 2016a, 2016b). According to (Davis et al., 1997), stewardship theory posits that individuals are driven by the desire to fulfill higher-order needs through pro-organizational actions, leading them to naturally align their interests with those of the organization and its principals. They argue that even when interests are not fully aligned, stewards prioritize cooperation over defection (Davis et al., 1997; Donaldson & Davis, 1991). Stewardship, a key

Table 8.2 Comparison with Agency Theory

Feature	Stewardship theory	Agency theory
Manager's Motivation	Intrinsic (loyalty, commitment)	Extrinsic (self-interest, financial incentives)
Governance Approach	Trust, empowerment	Monitoring, control mechanisms
Decision-Making	Long-term focus, collective goals	Short-term, individualistic
Manager's Role	Steward (guardian of the company's wellbeing)	Agent (may act in self-interest)

governance practice, has diverse interpretations that carries out the efficiency, and productivity of the family business by default (Martin & Ishihara, 2024). Stewardship application provided room for much activist funds during Japan's second wave of activism after the implementation of Stewardship and Corporate Governance Codes (Miyachi & Takeda, 2021).

The foundations of stewardship theory largely stem from its underlying model of human nature, which contrasts with the model of human behavior in agency theory. (Hernandez, 2012), especially because it assumes that self-interest always leads to opportunistic behavior. However (Jensen & Meckling, 1976), states that "If both parties to the [agency] relationship are utility maximizers, there is good reason to believe that the agent will not always act in the best interests of the principal". This suggests that agency theory does not portray an extreme view of human nature but rather recognizes that opportunism may or may not occur (Davis et al., 1997) Similarly acknowledge this point in their development of stewardship theory.

In work from the early 1970s, later published in 1994 (Jensen & Meckling, 1994), reject the economic model of man often cited by stewardship theorists (Davis et al., 1997) favoring the Resourceful, Evaluative, Maximizing Model (REMM) of human behavior. The model states that people care about a wide range of things, not just money, and they make trade-offs to maximize satisfaction based on these preferences, demonstrating resourcefulness. Also suggests that self-interest and other interests are not mutually exclusive, as individuals respond to a variety of incentives, not just monetary ones (Jensen & Meckling, 1994).

While stewardship theory critiques agency theory for presenting an unrealistic portrayal of human behavior, it can also be criticized for its own extreme assumptions. As stewardship theorists suggest (Davis et al., 1997; Hernandez, 2012) factors such as management philosophies, individual psychological mechanisms (e.g., reward motivations, identification, use of power), and situational mechanisms (e.g., cultural differences) influence these behaviors. Furthermore, how individuals behave can shape the organizational environment, for better or worse. Examples of such behaviors include altruism and bifurcation bias in family firms (Schulze et al., 2001). Although stewardship is intrinsically built over the years (voluntary) and based on the "comply or explain" principle, it provides an important reference point for assessing the quality of family firms' corporate governance (Esposito De Falco, 2024).

8.2.14.1 The Italian and Chinese Style of Stewardship

Globally speaking, Italy's legal and self-regulatory framework for asset manager and asset owner stewardship focuses on the interplay between hard and soft law. It explores five key dimensions and advocates for a nuanced soft law approach that aligns with national governance, ESG integration, debt holder stewardship, collaboration, and enforcement strategies (Corgatelli, 2024).

China proposed corporate governance reforms, prompting an analysis of their financial impact. The board structure changes using a dynamic panel. Removing board size limits and supervisory boards but oppose eliminating main boards for small firms (Famba et al., 2024).

8.3 Discussion

Succession planning aims to reach the family firm's objectives quickly and productively. Entrepreneurship out of family firms propels the finest motives in corporate governance. Family-controlled firms significantly hand over the largest share to their future generations. Whereas in non-family firms, entrepreneurship fails to materialise efficiently. Social economic theory denotes family-owned businesses prioritize non-financial goals—such as family identity, legacy, and social responsibility—over purely financial performance, while agency theory In family businesses, where owners often serve as managers, these conflicts might be minimized. This alignment fosters trust-based governance models, emphasizing collective goals over individual ones. Family businesses may thus adopt participative governance structures that empower managers and family members to act in the organization's best interest. The trust factor builds reputation, but the building up of trust comes out of stewardship theory practices.

8.4 Conclusion

Understanding the dimensions of family business in conjunction with effective corporate governance is vital for fostering resilience and sustainable growth. This research contributes to the literature by providing a multidimensional framework that elucidates the intricate relationships between family dynamics and governance practices, ultimately offering valuable insights for practitioners and scholars alike. By embracing these dimensions, family businesses can navigate complexities and thrive in an increasingly competitive landscape, securing their legacy for future generations.

8.5 Limitations of the Study and Future Research Directions

This review paper was carried out in a time-bound manner. Hence, there is a wider scope still remaining in this area of research. The gem of the research, we could've included, but we didn't due to its obsoleteness. Future research directions could include assessing cross-cultural comparisons, succession planning, and governance board dynamics in family firms' governance and innovation, sustainability and corporate social responsibility, governance in family firms during crises, digital transformation, and governance, gender dynamics in family business governance, family governance structures and firm performance, and comparative governance models.

References

Abels, C. M. (2023). Stewardship theory. In R. A. List, H. K. Anheier, & S. Toepler (Eds.), *International encyclopaedia of civil society*. Springer, Cham. https://doi.org/10.1007/978-3-319-99675-2_9556-1

Abeysekera, A., Brockman, P., & Fernando, C. S., & Salas, J. M. (2024). Undiversified shareholders, socioemotional wealth, and corporate hedging. *Evidence from family firms.* https://ssrn.com/abstract=4774066 https://doi.org/10.2139/ssrn.4774066

Åberg, C., Botero, I. C., & Rondi, E. (2024). Socioemotional wealth and family firm performance: The moderating role of CEO tenure and millennial CEO. *British Journal of Management*. https://doi.org/10.1111/1467-8551.12818

Adams, R. J., Smart, P., & Huff, A. S. (2017). Shades of grey: Guidelines for working with the grey literature in systematic reviews for management and organizational studies. *International Journal of Management Reviews, 19*, 432–454. https://doi.org/10.1111/ijmr.12102

Akhter, N., Sieger, P., & Chirico, F. (2016). If we can't have it, then no one should: Shutting down versus selling in family business portfolios. *Strategic Entrepreneurship Journal, 10*, 371–394. https://doi.org/10.1002/sej.1237

Al-Faryan, M. A. S. (2024). Agency theory, corporate governance and corruption: An integrative literature review approach. *Cogent Social Sciences, 10*(1). https://doi.org/10.1080/23311886.2024.2337893

Al-Faryan, M. A. S., & Shil, N. C. (2024). A regulatory appraisal of the Saudi stock exchange: A study based on secondary data. *Journal of Governance & Regulation, 13*(3), 27–45. https://doi.org/10.22495/jgrv13i3art3

Al-Haddad, L. M., Saidat, Z., Seaman, C., & Gerged, A. M. (2024). Does capital structure matter? Evidence from family-owned firms in Jordan. *Journal of Family Business Management, 14*(1), 64–76. https://doi.org/10.1108/JFBM-09-2022-0115

Amin, A., Ali, R., Rehman, R. U., & Hasan, M. (2024). Family ownership, control, and firm performance: Does gender diversity matter? *Eurasian Business Review, 14*, 501–525. https://doi.org/10.1007/s40821-024-00256-9

André, P., Ben-Amar, W., & Saadi, S. (2014). Family firms and high technology mergers & acquisitions. *Journal of Management Governance, 18*, 129–158. https://doi.org/10.1007/s10997-012-9221-x

Anokhin, S. A., Hess, M., & Wincent, J. (2024). Technology sourcing ambidexterity in corporate venture capital: Limitations of learning from open innovation. *Small Business Economics*. https://doi.org/10.1007/s11187-024-00900-8

Anwar, M., Clauss, T., & Meyer, N. (2024). Entrepreneurship in family firms: An updated bibliometric overview. *Review of Managerial Science, 18*, 539–575. https://doi.org/10.1007/s11846-023-00650-z

Astrachan, J. H., & Shanker, M. C. (2020). Family businesses and their stakeholders: A global perspective. *Journal of Family Business Strategy, 11*(3), Article 100346. https://doi.org/10.1016/j.jfbs.2020.100346

Azila-Gbettor, E. M., Honyenuga, B. Q., Blomme, R. J., & Kil, A. (2022). Review of differences between listed and unlisted family business research trends: A corporate governance context. *Journal of Family Business Management, 12*(4), 853–875. https://doi.org/10.1108/JFBM-03-2020-0023

Bammens, Y., Voordeckers, W., & Van Gils, A. (2021). Governance in family businesses: The role of the family board of directors. *Family Business Review, 34*(1), 21–37. https://doi.org/10.1177/0894486520959627

Bauweraerts, J., Rondi, E., Rovelli, P., De Massis, A., & Sciascia, S. (2022, September 8). *Women can help family firms innovate—But socioemotional factors matter*. FamilyBusiness.org. Retrieved February 6, 2025, from https://familybusiness.org/content/women-can-help-family-firms-innovate-but-socioemotional-factors

Bernstein, R., Buse, K., & Bilimoria, D. (2016). Revisiting agency and stewardship theories. *Nonprofit Management and Leadership, 26*, 489–498. https://doi.org/10.1002/nml.21199

Berrone, P., Cruz, C., & Gomez-Mejia, L. R. (2012). Socioemotional wealth in family firms: Theoretical dimensions, assessment approaches, and agenda for future research. *Family Business Review, 25*(3), 258–279.

Berrone, P., Cruz, C., & Gomez-Mejia, L. R. (2022). Why and how do family firms go public? A socioemotional wealth perspective. *Family Business Review*. https://doi.org/10.1177/08944865241273380

Berrone, P., Cruz, C., Gomez-Mejia, L. R., & Larraza-Kintana, M. (2010). Socioemotional wealth and corporate responses to institutional pressures: Do family-controlled firms pollute less? *Administrative Science Quarterly, 55*(1), 82–113. https://doi.org/10.2189/asqu.2010.55.1.82

Bhagat, S., & Bolton, B. (2019). Corporate governance and firm performance: The sequel. *Journal of Corporate Finance, 58*, 142–168. https://doi.org/10.1016/j.jcorpfin.2019.04.006

Blanco-Mazagatos, V., de Quevedo-Puente, E., & Delgado-García, J. B. (2016). How agency conflict between family managers and family owners affects performance in wholly family-owned firms: A generational perspective. *Journal of Family Business Strategy, 7*(3), 167–177. https://doi.org/10.1016/j.jfbs.2016.07.003

Block, J., Fathollahi, R., & Eroglu, O. (2024). Capital structure of single family office-owned firms. *Journal of Family Business Strategy, 15*(3), Article 100596.

Calabrò, A., Torchia, M., Quarato, F., Valentino, A., Cambrea, D. R., & Lohe, F.-W. (2024). Family firms' shareholder structure and international acquisitions: A differentiated socioemotional wealth approach. *Family Business Review, 37*(2), 244–266. https://doi.org/10.1177/08944865231205847

Campbell, K., & Jerzemowska, M. (2017). Contested takeovers of family firms and socioemotional wealth: A case study. *Baltic Journal of Management, 12*(4), 447–463. https://doi.org/10.1108/BJM-12-2016-0277

Capolupo, P., Messeni Petruzzelli, A., & Ardito, L. (2024). A knowledge-based perspective on transgenerational entrepreneurship: Unveiling knowledge dynamics across generations in family firms. *Journal of Knowledge Management, 28*(5), 1444–1467. https://doi.org/10.1108/JKM-05-2023-0451

Cha, J., & Chrisman, J. J. (2012). Family involvement, family influence, and family-centered non-economic goals in small firms. *Entrepreneurship Theory and Practice, 36*(2), 267–293. https://doi.org/10.1111/j.1540-6520.2010.00407.x

Chang, A. A., & Mubarik, M. S. (2021). Where socioemotional wealth (sew) stands today? *A review into the past, present and the future.*

Chaudhary, S., Dhir, A., Ferraris, A., & Bertoldi, B. (2021). Trust and reputation in family businesses: A systematic literature review of past achievements and future promises. *Journal of Business Research, 137*, 143–161. https://doi.org/10.1016/j.jbusres.2021.07.052

Chen, X., & Zhao, Y. (2024). Navigating global markets: Governance challenges and strategies in family firms. *International Journal of Family Business, 6*(1), 22–41. https://doi.org/10.1108/IJFB-202

Chrisman, J. J. (2019). Stewardship theory: Realism, relevance, and family firm governance. *Entrepreneurship Theory and Practice, 43*(6), 1051–1066. https://doi.org/10.1177/1042258719838472

Chrisman, J. J., Chua, J. H., Pearson, A. W., & Barnett, T. (2012). Family involvement, family influence, and family–centered non–economic goals in small firms. *Entrepreneurship Theory and Practice, 36*(2), 267–293. https://doi.org/10.1111/j.1540-6520.2010.00407.x
Chrisman, J. J., Chua, J. H., & Sharma, P. (2005). Trends and directions in the development of a strategic management theory of the family firm. *Entrepreneurship Theory and Practice, 29*(5), 555–576.
Corgatelli, M. (2024). Reconstructing the framework of institutional investor stewardship in Italy: Synergies between hard and soft law. *European Business Organization Law Review, 25*, 533–570. https://doi.org/10.1007/s40804-024-00315-8
Covin, J. G., Garrett, R. P., Bouncken, R. B., Ratzmann, M., & Muhammad, M. (2024). Core business prospects and the management of internal corporate ventures. *Small Business Economics, 62*, 435–459. https://doi.org/10.1007/s11187-023-00824-9
Cruz, C., Milanov, H., & Klein, J. (2024). It's a family affair: A case for consistency in family foundation giving and family firm community CSR activity. *Journal of Business Ethics, 191* , 633–649. https://doi.org/10.1007/s10551-023-05424-x
Cruz, C., & Nordqvist, M. (2012). Entrepreneurial orientation in family firms: A generational perspective. *Small Business Economics, 38*, 33–49.
Davila, J., Duran, P., Gómez-Mejía, L., & Sanchez-Bueno, M. J. (2023). Socioemotional wealth and family firm performance: A meta-analytic integration. *Journal of Family Business Strategy, 14*(2), Article 100536. https://doi.org/10.1016/j.jfbs.2022.100536
Davis, J. H., Schoorman, F. D., & Donaldson, L. (1997). Toward a stewardship theory of management. *Academy of Management Review, 22*(1), 20–47.
Deb, S., & Dube, I. (2017). Corporate governance disclosure for complex ownership structure in India. *Indian Journal of Corporate Governance, 10*(2), 143–175. https://doi.org/10.1177/0974686217730937
Dibrell, C., Gamble, J. R., Clinton, E., & Sherlock, C. (2024). Family firm innovativeness: An investigation of family governance, commitment, and generation involvement. *Entrepreneurship & Regional Development, 1–22*. https://doi.org/10.1080/08985626.2024.2338171
Diéguez-Soto, J., Campos-Valenzuela, M., Callejón-Gil, Á. M., & Aldeanueva-Fernández, I. (2024). Family firm heterogeneity on CSR approach: A socio-emotional (SEW) perspective. *BRQ Business Research Quarterly, 27*(4), 443–466.
Donaldson, L., & Davis, J. H. (1991). Stewardship theory or agency theory: CEO governance and shareholder returns. *Australian Journal of Management, 16*(1), 49–64.
Du, S., Ma, L., Li, Z., & Ma, C. (2024). Non-family shareholder governance and the digital transformation of family firms: Evidence from China. *Corporate Governance: An International Review, 32*(1), 89–115. https://doi.org/10.1111/corg.12517
Ducassy, I., & Guyot, A. (2017). Complex ownership structures, corporate governance and firm performance: The French context. *Research in International Business and Finance, 39*(A), 291–306. https://doi.org/10.1016/j.ribaf.2016.07.019
Dyer, W. G., & Whetten, D. A. (2022). Family business governance and firm performance: An ethical perspective. *Entrepreneurship Theory and Practice, 46*(4), 759–788. https://doi.org/10.1177/1042258721992472
Eisenhardt, K. M. (1989). Agency theory: An assessment and review. *Academy of Management Review, 14*(1), 57–74. https://doi.org/10.5465/amr.1989.4279003
Esposito De Falco, S. (2024). Codes of governance, stewardship and proxy advisors. In *Corporate governance*. Palgrave Macmillan, Cham. https://doi.org/10.1007/978-3-031-74089-3_11
Fagrell, O., & Gunnarsson, N. (2024). *M&A within family-owned investment companies: A qualitative study on the juggling of hard and soft values in the pre-acquisition process* (Dissertation). https://urn.kb.se/resolve?urn=urn:nbn:se:hj:diva-64456
Fama, E. F. (1980). Agency problems and the theory of the firm. *Journal of Political Economy, 88*(2), 288–307. https://doi.org/10.1086/260866
Fama, E. F., & Jensen, M. C. (1983). Separation of ownership and control. *Journal of Law and Economics, 26*(2), 301–325. https://doi.org/10.1086/467037

Famba, T., Chiriseri, L., Chituku-Dzimiro, G., & Kurauone, O. (2024). Two-tier corporate board structure, corporate governance reforms and financial performance of firms listed on the China stock market. *SN Business & Economics, 4*, 130. https://doi.org/10.1007/s43546-024-00729-z

Francesco, A., Pupo, V., & Mannarino, L. (2023). *Green innovation and firm heterogeneity*. Beyond Dichotomies between family and non-family firms. https://doi.org/10.2139/ssrn.4869397

Fuad, M., Thakur, V., Pattnaik, C., & Jain, R. (2024). Family firms and the mixed gamble perspective in cross-border acquisitions: A study of Indian firms. *International Business Review, 33*(1), Article 102205.

Galavotti, I., D'Este, C., & Cerrato, D. A. (2025). Size-based contingency approach to family firms' performance: The role of family power. *Journal of Management and Governance*. https://doi.org/10.1007/s10997-025-09736-0

Garcia, M., Chen, Y., & Patel, R. (2024). Stakeholder engagement in family businesses: Effective governance practices. *Entrepreneurship Theory and Practice, 48*(1), 89–112. https://doi.org/10.1177/10422587231204567

Ghinoi, S., De Vita, R., Steiner, B., & Sinatra, A. (2024). Family firm network strategies in regional clusters: Evidence from Italy. *Small Business Economics, 62*, 87–103. https://doi.org/10.1007/s11187-023-00755-5

Gómez-Mejia, L. R., Cruz, C., Berrone, P., & De Castro, J. (2011). The bind that ties: Socioemotional wealth preservation in family firms. *The Academy of Management Annals, 5*(1), 653–707.

Gómez-Mejía, L. R., Haynes, K. T., Núñez-Nickel, M., Jacobson, K. J., & Moyano-Fuentes, J. (2007). Socioemotional wealth and business risks in family-controlled firms: Evidence from Spanish olive oil mills. *Administrative Science Quarterly, 52*(1), 106–137.

Gómez-Mejía, L. R., Haynes, K. T., Núñez-Nickel, M., Jacobson, K. J. L., & Moyano-Fuentes, J. (2021). Socioemotional wealth and business decisions: The importance of family business governance. *Family Business Review, 34*(3), 286–302. https://doi.org/10.1177/08944865211003824

Gómez-Mejía, L. R., Makri, M., & Kintana, M. L. (2010). Diversification decisions in family-controlled firms. *Journal of Management Studies, 47*(2), 223–252.

Goyal, B., & Gulati, R. (2025). A nexus between corporate governance and performance: Evidence from the Indian insurance industry. *International Journal of Disclosure Governance*. https://doi.org/10.1057/s41310-024-00283-y

Gupta, P., & Chauhan, S. (2023). Dynamics of family business succession: A literature review and future research agenda. *Family Business Review, 36*(1), 4–28. https://doi.org/10.1177/08944865221126706

Hassan, L. M., & O'Sullivan, T. (2019). The role of family firms in sustainable development: A systematic literature review. *International Journal of Entrepreneurial Behavior & Research, 25*(4), 715–731. https://doi.org/10.1108/IJEBR-09-2018-0442

Hendrastuti, R., & Harahap, R. (2023). Agency theory: Review of the theory and current research. *Jurnal Akuntansi Aktual*. https://doi.org/10.17977/um004v10i12023p085

Hernandez, M. (2012). Toward an understanding of the psychology of stewardship. Academy of Management Review, 37(2), 172–193

Hernández-Perlines, F., Ariza-Montes, A., & Araya-Castillo, L. (2022). Socioemotional wealth of family firms during the COVID-19 pandemic: The role of slack resources. *Journal of Family Business Management*. https://www.researchgate.net/publication/370486057_Socioemotional_wealth_of_family_firms_during_the_COVID-19_pandemic_the_role_of_slack_resources

Hernández-Perlines, F., Moreno-García, J., & Yáñez-Araque, B. (2019). The influence of socioemotional wealth in the entrepreneurial orientation of family businesses. *International Entrepreneurship and Management Journal, 15*, 523–544.

Jamil, M., Stephens, S., & Md Fadzil, A. F. (2024). Sustainability in family business settings: A strategic entrepreneurship perspective. *Journal of Family Business Management*, Vol. ahead-of-print No. ahead-of-print. https://doi.org/10.1108/JFBM-01-2024-0001

Jayantilal, S., Jorge, S. F., & Alcarva, P. (2024). Unravelling the determinants of family firms' governance: The family protocol. *Journal of Family Business Management, 14*(4), 829–847. https://doi.org/10.1108/JFBM-10-2023-0190

Jensen, M. C., & Meckling, W. H. (1976). Theory of the firm: Managerial behavior, agency costs and ownership structure. *Journal of Financial Economics, 3*(4), 305–360. https://doi.org/10.1016/0304-405X(76)90026-X

Jensen, M. C., & Meckling, W. H. (1994). The nature of man. *Journal of Applied Corporate Finance, 7*(2), 4–19.

Jiang, D. S., Kellermanns, F. W., Munyon, T. P., & Morris, M. L. (2018). More than meets the eye: A review and future directions for the social psychology of socioemotional wealth. *Family Business Review, 31*(1), 125–157.

Jin, R., & Hu, H. W. (2024). Liability of ownership origin, corporate philanthropy, and desire for control in Chinese family firms. *Entrepreneurship Theory and Practice, 48*(3), 763–787. https://doi.org/10.1177/10422587231226111

Jones, L., & Williams, S. (2024). Conflict resolution strategies in family firms: Governance mechanisms that work. *Journal of Family Business Strategy, 15*(2), 145–162. https://doi.org/10.1016/j.jfbs.2023.100467

Kallmuenzer, A. (2015). Agency theory and the family business. *ResearchGate.*

Kim, H., & Lee, J. (2024). Ethical governance in family businesses: A framework for enhancing trust and performance. *Journal of Business Ethics, 179*(1), 97–116. https://doi.org/10.1007/s10551-023-05176-4

Kraiczy, N. D., & Hack, A. (2018). Business models in family firms: A systematic review and research agenda. *Family Business Review, 31*(1), 38–54. https://doi.org/10.1177/0894486516675597

Litz, R. A. (1995). The family business: Toward definitional clarity. *Academy of Management Review, 20*(1), 75–101. https://doi.org/10.5465/amr.1995.9503271991

Luong, T.-T., Dang, C.-M., & Tran, Q. N. (2024). Drivers of knowledge transfer for succession in family business: A perspective article. *Journal of Family Business Management, 14*(4), 813–822. https://doi.org/10.1108/JFBM-09-2023-0159

Madison, K., Holt, D. T., Kellermanns, F. W., & Ranft, A. L. (2016a). Viewing family firm behavior and governance through the lens of agency and stewardship theories. *Family Business Review, 29*(1), 65–93. https://doi.org/10.1177/0894486515594292

Madison, K., Holt, D., Kellermanns, F., & Ranft, A. (2016b). *Stewardship theory and family firm governance.*

Mansouri, A., Chafai, M., & Moufdi, N. (2024). Family business governance: An integrated approach using agency, stewardship, and social capital theories. *Corporate & Business Strategy Review, 5*(1), 108–116. https://doi.org/10.22495/cbsrv5i1art11

Maroga, R., Schultz, C., & Smit, P. (2024). Succession planning mediates self-leadership and turnover intention in a state-owned enterprise. *SA Journal of Human Resource Management, 22*. https://doi.org/10.4102/sajhrm.v22i0.2304

Martin, C., & Ishihara, H. (2024). Establishing a typology for stewardship: A nexus of opportunity for organisational and environmental management. *SNF, 32*, 1. https://doi.org/10.1007/s00550-024-00543-z

Mashele, A., Mouton, M., & Pelcher, L. (2024). Corporate governance and financial performance: Family firms vs. non-family firms. *Journal of Risk and Financial Management, 17*(10), 444. https://doi.org/10.3390/jrfm17100444

Miller, D., & Le Breton-Miller, I. (2006). Family governance and firm performance: Agency, stewardship, and capabilities. *Family Business Review, 19*(1), 73–87. https://doi.org/10.1111/j.1741-6248.2006.00061.x

Miller, D., & Le Breton-Miller, I. (2014). Configurations of family business strategies: A dual process model. *Strategic Management Journal, 35*(5), 645–664. https://doi.org/10.1002/smj.2096

Miller, D., Le Breton-Miller, I., & Lester, R. H. (2022). Internationalization and governance in family firms: A global perspective. *Entrepreneurship Theory and Practice, 46*(2), 255–284. https://doi.org/10.1177/1042258720985624

Miyachi, H., & Takeda, F. (2021). Hedge fund activism in Japan under the stewardship and corporate governance codes. *International Advances in Economic Research, 27*, 119–130. https://doi.org/10.1007/s11294-021-09824-5

Morck, R., & Yeung, B. (2011). The effect of ownership structure on firm performance: Evidence from China. *Journal of Corporate Finance, 17*(5), 1034–1055. https://doi.org/10.1016/j.jcorpfin.2011.05.004

Nakpodia, F. (2024). Corporate governance and family business: A perspective article. *Journal of Family Business Management, 14*(4), 808–812. https://doi.org/10.1108/JFBM-10-2023-0198

Palalić, R., & Smajić, H. (2022). Socioemotional wealth (SEW) as the driver of business performance in family businesses in Bosnia and Herzegovina: The mediating role of transformational leadership. *Journal of Family Business Management, 12*(4), 1043–1064. https://doi.org/10.1108/JFBM-07-2021-0067

Palm, M., Kraft, P. S., & Kammerlander, N. (2024). Family firms, M&A strategies, and M&A performance: A meta-analysis. *Journal of Management, 50*(7), 2818–2849. https://doi.org/10.1177/01492063231178027

Pérez, C., & Sánchez, P. (2017). The role of family influence in the firm performance of family businesses: A systematic review. *Family Business Review, 30*(2), 144–161. https://doi.org/10.1177/0894486516687513

Pérez, C., & Sanchez, P. (2018). The impact of family involvement on the firm's innovation behavior: A meta-analysis. *Family Business Review, 31*(4), 537–558. https://doi.org/10.1177/0894486518763385

Pinelli, M., Debellis, F., & De Massis, A. (2024). Long-term orientation, family-intensive governance arrangements, and firm performance: An institutional economics perspective. *Small Business Economics, 63*, 731–754. https://doi.org/10.1007/s11187-024-00877-4

Pütz, L., & Werner, A. (2024). Absorptive capacity in family firms: A systematic literature review. *Review of Managerial Science, 18*, 577–632. https://doi.org/10.1007/s11846-023-00651-y

Ross, S. A. (1973). The economic theory of agency: The principal's problem. *American Economic Review, 63*(2), 134–139. https://www.jstor.org/stable/1817064

Schulze, W. S., Lubatkin, M. H., & Dino, R. N. (2001). Altruism, agency, and the competitiveness of family firms. *Managerial and Decision Economics, 22*(4), 317–331.

Schulze, W. S., Lubatkin, M. H., & Dino, R. N. (2003). Toward a theory of agency and altruism in family firms. *Journal of Business Venturing, 18*(4), 473–490. https://doi.org/10.1016/S0883-9026(03)00054-5

Sharma, P., & Chrisman, J. J. (2020). Toward a multi-theoretical framework of family business governance. *Journal of Family Business Strategy, 11*(2), Article 100375. https://doi.org/10.1016/j.jfbs.2019.100375

Smajić, H., Palalić, R., & Ahmad, N. (2023). Future perspective of socioemotional wealth (SEW) in family businesses. *Journal of Family Business Management, 13*(4), 923–954. https://doi.org/10.1108/JFBM-05-2022-0070

Smith, J., Johnson, R., & Taylor, A. (2024). The role of succession planning in family business governance: A global perspective. *Family Business Review, 37*(1), 34–56. https://doi.org/10.1177/08944865231004567

Song, B., & Schwienbacher, A. (2024). How do family founders help novice entrepreneurs to develop their firms? *Small Business Economics, 63*, 781–804. https://doi.org/10.1007/s11187-024-00879-2

Song, T., Kubo, Y., & Kutsuna, K. (2024, March 1). *Family firms, acquisitions and divestments, and long-term stock market performance: The moderating effect of prior experience.* https://ssrn.com/abstract=4787235 or https://doi.org/10.2139/ssrn.4787235

Souder, D., Zaheer, A., Sapienza, H., & Ranucci, R. (2017). How family influence, socioemotional wealth, and competitive conditions shape new technology adoption. *Strategic Management Journal, 38*(9), 1774–1790.

Subramanian, S. (2022). Importance of socioemotional wealth in Indian family business group: The case of Tata Group. *Paradigm*, 26(2), 138–154. https://ideas.repec.org/a/sae/padigm/v26y2022i2p138-154.html

Teixeira, J. F., & Carvalho, A. O. (2024). Corporate governance in SMEs: A systematic literature review and future research. *Corporate Governance, 24*(2), 303–326. https://doi.org/10.1108/CG-04-2023-0135

Ting, Y. S., & Sweeney, B. (2020). The impact of family ownership on firm performance: A review of the literature and future research directions. *Journal of Family Business Management, 10*(2), 163–178. https://doi.org/10.1108/JFBM-04-2019-0057

Verbeke, A., & Kano, L. (2012). The dynamic interplay between the resource-based view and the institution-based view. *Global Strategy Journal, 2*(4), 361–379.

Wang, Y., & Miao, Q. (2021). Family firm performance: A review and future research directions. *Family Business Review, 34*(2), 159–187. https://doi.org/10.1177/0894486520951116

Weniger, S., & Jarchow, S. (2024). Corporate venturing in times of crisis: Securing resources through legitimacy. *Review of Managerial Science*. https://doi.org/10.1007/s11846-024-00728-2

Zellweger, T. M., Nason, R. S., & Nordqvist, M. (2012). From longevity of firms to transgenerational entrepreneurship: Exploring the unexplored. *Family Business Review, 25*(2), 136–155. https://doi.org/10.1177/0894486510387450

Zellweger, T. M., Nason, R. S., Nordqvist, M., & Brush, C. G. (2013). Why do family firms strive for nonfinancial goals? An organizational identity perspective. *Entrepreneurship Theory and Practice, 37*(2), 229–248. https://doi.org/10.1111/j.1540-6520.2011.00466.x

Part II
Sustainable Governance and CSR (Corporate Social Responsibility)

Chapter 9
Employee Engagement in CSR: The Role of Green HRM in Building a Sustainable Culture

Deeksha, Ranjit Kumar Mukherji, and Riya Sharma

Abstract Today, information-level business strategy is significantly focusing on environmental and corporate social responsibility. In this context, one of the most efficient concepts of engaging in the mentioned principles is indeed green human resource management (GHRM). It involves appending elements to existing HR practices to foster the development of competent and environmentally friendly human capital. Using Green Human Resource Management is a potent way to promote these values (GHRM). This approach incorporates elements into HR practices to develop a staff that is highly skilled, productive, as well as environmentally conscious. GHRM plays a critical role in developing an atmosphere by giving workers an awareness of accountability for sustainability. It elaborates on a situation where every adult can speak about his sustainable goal at the workplace. As the time one joins the organization, professionals are immersed in an environmentally friendly environment. These are considered to be green employees and are the ones hired through green recruitment. This category is continually trained on methods of sustainability, including energy saving and wastage reduction, hence giving the prerogative to practice change within organizational premises and in their own lives. GHRM also provides incentives to workers, as the latter are motivated to participate in sustainability initiatives due to appreciation of their work. Organizational commitment towards sustainability enhances this culture whereby leaders mirror top-notch, greener practices and elaborate on environmental objectives. Conducting clear and open communication makes all people aware of the organization's efforts towards sustainability and creates awareness of the common goal in place. Thus, GHRM turns sustainable development from the company's policy into the individual responsibility of every worker. Since sustainability becomes a part of corporate culture and organizational behavior, GHRM significantly contributes to creating a sense of the company's and employees' responsibility towards the environment.

Deeksha · R. K. Mukherji (✉) · R. Sharma
Graphic Era (Deemed to Be University), Dehradun, Uttarakhand, India
e-mail: rkmmukh@gmail.com

V. Kandpal et al. (eds.), *Developments in Corporate Governance*,
Approaches to Global Sustainability, Markets, and Governance,
https://doi.org/10.1007/978-981-96-6366-8_9

Keywords Green human resource management · Sustainability · Corporate social responsibility · Green recruitment · Wastage reduction · Employee accountability

9.1 Introduction

In the present business environment, there is growing interest in the relationship between CSR and GHRM. Interestingly, the opinion among the firm is to understand that they cannot succeed by focusing solely on economic value, but they must also consider their social and ecological value. The last is the involvement of the employees in the company's CSR projects, as it ensures that it becomes a sustainable culture both for the company and the rest of society. Through the incorporation of GHRM, organizations and the needed human capital will be in a position to participate in the subsequent CSR activities, hence having the much-needed sustainable culture in their operations. The term Green HRM describes the prospective involvement of environmental management in personnel management strategies such as employment, training and development, performance evaluation, and other employee relation strategies (Renwick et al., 2013). These practices not only support the organization's environmental goals but also enhance employee motivation and commitment to sustainability. When employees see that their organization is genuinely committed to environmental responsibility, they are more likely to adopt eco-friendly behaviors and participate in CSR initiatives (Zoogah, 2011).

Specifically, in the modern business environment, CSR and GHRM practices constitute a significant business phenomenon. Businesses are finally waking up from their self-centered sleep driven by pure financial results; they now have to care about society and the environment. Implementation of a CSR program is incomplete without the engagement of employees because it also creates awareness about sustainable development in an organization and the larger society. As it was illustrated throughout this paper, the integration of GHRM practices serves as a critical factor in managing CSR engagement and development in organizations, leading to the creation of a sustainable organizational culture. Green HRM is defined as the process of implementing environmental management into the human resource management function of an organization in aspects such as employment practice, training, performance evaluation, and management of Human resources (Renwick et al., 2013). The above practices not only serve to improve the organization's environmental objectives but also serve to engage and increase the level of commitment of the employees to the objectives of sustainable development. When employees perceive that their organization is serious about CSR, they are likely to exhibit a change in behavior and support CSR.

The role of GHRM in promoting employee engagement in CSR is multifaceted. Firstly, it involves educating and training employees on the importance of sustainability and how they can contribute. These may involve online lessons, conferences, as well as events that highlight environmentally friendly methods along with difficulties with the environment. Secondly, GHRM practices aim to embed sustainability into the company's culture by aligning organizational values with environmental goals. This alignment can be achieved through sustainable leadership, green policies, and incentives that reward eco-friendly behaviors (Jackson et al., 2011). Moreover, fostering a sustainable culture through GHRM requires effective communication and collaboration. Employees need to feel that their contributions to CSR are valued and that they are part of a collective effort to make a positive impact. Regular updates on CSR achievements, feedback mechanisms, and employee recognition programs are essential for maintaining high levels of engagement and motivation (Jabbour & Santos, 2008).

The company, along with everyone involved, benefits from such an all-encompassing method of long-term viability, which opens up the path towards a more accountable and environmentally friendly tomorrow. Thus, the involvement of GHRM in the process of promotion of employee engagement in CSR has several aspects. Firstly, it entails creating awareness among the employees on the subject of sustainability as well as creating capability in the same. These may include lessons conducted online and/or conferences and events that showcase green approaches as well as challenges about the environment. Secondly, GHRM practices seek to make sustainability part and parcel of the company's culture by checking on some organizational values towards environmental objectives. It can be created by stable leadership supported by green policies and by using incentives for environmentally friendly behaviors (Jackson et al., 2011).

Also, amplifying GHRM to promote and sustain the culture of the elements must involve good communication and cooperation. Employees have to believe that CSR initiatives in their firms are important and that everyone is a stakeholder in the change. It is necessary to have regard to the method of ongoing reporting of the CSR achievements and the feedback outstanding both to the employees and to the organization (Jabbour & Santos, 2008). Thus, the integration of CSR and GHRM is critical in establishing the sustainability of the organization's culture. Subsequently, green HR is highly effective because CSR practices involve the engagement of employees in environmental conservation, and the process leads to the satisfaction of individual and organizational objectives. Both the company and everyone related to the business profit from such an inclusive approach towards long-term sustainability, which clears the path for a more responsible and less damaging tomorrow.

9.1.1 Literature Review

In the last couple of years, there has been increasing concern about GHRM's relative position in fostering environmental sustainability and staff participation in CSR. This work reviews the available literature on these subjects, highlighting major findings while at the same time making remarks on sectors that seem to require more research.

9.1.1.1 The Concept of CSR and Employee Engagement

The commitment of a business to operate SRI, contribute to economic development, and improve the quality of life of its personnel, neighbors, and society, in general, is called CSR (Carroll, 1991). It is essential to employ CSR for the improvement of organizational engagement because it matches the common goals of the employee and the organization. Aguinis and Glavas (2012) pointed out that where employees participate in CSR, a positive relationship exists between job satisfaction, organizational commitment, and job performance.

9.1.1.2 The Importance of Employee Engagement in CSR

The involvement of people in organizations as far as the implementation of sustainability practices is concerned is central to the subject of CSR. Employees who are engaged will be more committed to supporting and contributing to the CSR activities of their organization, and this will result in better organizational performance (Glavas, 2016). The literature also supports the fact that when the employees receive the signal of the organizational commitment to social and environmental issues, job satisfaction and organizational commitment are likely to go high (Brammer et al., 2007). This involvement results in a virtuous cycle since the backing of CSR also solidifies the stand of the organization on sustainability.

9.1.1.3 Green HRM: General and Specific Linking of Sustainability to HRM Practice

Green HRM is the concept of incorporating environmental considerations into ordinary practices, including, but not limited to, recruitment, training, performance evaluation, and handling employee relations. According to Renwick et al., (2013) Green HRM is the application of HR policies to the responsible use of resources by an organization. The effort of applying human resource regulations in an organization to the optimum consumption of the company's resources is commonly referred to as green HRM.

Green HRM is essentially a subset of Green Operations and involves integrating environmental management into standard HR practices, including hiring, training, organizing, motivating, rewarding, and communicating with employees. Renwick, Redman, and Maguire, in their paper, define Green HRM simply as the integration of HR policies aimed at the sustainable use of resources by organizations. The act of applying HR regulations aimed at increasing businesses' resource savings is called "Green HRM". Besides promoting the organization's environmental mission, this approach also boosts employees' participation by making them embrace environmental causes. For example, green recruitment strategies incorporate the principles of sustainable human resource management because they add value to the environment by identifying and targeting sustainable candidates to join the firm as a way of aligning them with the organizational goals (Hronová & Špaček, 2021; Jamil et al., 2023). Likewise, green training regimes inform the employees about some sustainable measures that can be implemented in some organizational processes, and the employees are challenged to implement them in their day-to-day operations (Jose Chiappetta Jabbour, 2011).

9.1.1.4 Training and Development for Sustainability

One critical component of GHRM is providing employees with training and development opportunities focused on sustainability. Jabbour and Santos (2008) emphasize the importance of educating employees about environmental issues and sustainable practices. Training programs can include workshops, seminars, and e-learning modules that equip employees with the knowledge and skills needed to contribute to the company's CSR objectives.

9.1.1.5 Sustainable Leadership and Organizational Culture

Leadership plays a vital role in embedding sustainability within the organizational culture. Jackson et al., (2011) argue that sustainable leadership involves promoting green values, setting environmental goals, and encouraging eco-friendly behaviors among employees. Leaders who demonstrate a commitment to sustainability can inspire employees to participate in CSR initiatives and adopt green practices in their daily work.

9.1.1.6 Sustainability and Leadership and Management of Cultural Change

It is the leadership's responsibility to foster a culture of sustainability in the organization. CEOs with environmental goals and perspectives who act accordingly can encourage CSR employee activities (Ehnert et al., 2016). Sustainable leadership raises awareness of the organization's sustainability objectives, ensures that all the required resources are provided to complete these objectives, and rewards employees for their CSR efforts.

In addition, more importantly, sustainability must become a part of the organizational culture, which means that the organization has to make a continuous effort to bring its values in line with the established goals and objectives regarding the environment (Sroufe, 2017). There are different ways of achieving this alignment, including the company's mission statement, establishing organizational policies, and the disclosure of information on CSR goals, as well as social performance and progress (Chong, 2009; Im et al., 2017; Min et al., 2024).

9.1.1.7 Employee Motivation and Engagement

Prior research has well established a connection between adopted GHRM practice and the level of employee motivation. In particular, Zoogah (2011) stated that it is crucial if the employee has a perception that the organization is sincere while practicing environmentally sustainable conduct. Other factors, such as offering incentives, awards, and recognition to employees for the CSR initiatives they undertake, can also encourage employees to embrace more CSR activities.

9.1.1.8 Building a Sustainable Culture

The process of transforming a culture into a sustainable one requires an organization to have environmental values. Sustainable alignment can be brought about through the various policies and practices at organizations with respect to sustainability. The authors of the publication Renwick et al., (2013) mention that, firstly, communication and collaboration create a sustainable culture. If the organization's CSR activities and policies are to remain effective in motivating its employees, periodic reports of the organization's CSR activities, modes of receiving feedback, and other means of rewarding high performers should be made available.

9.1.2 Developing a Common Goal

Having a clear and shared vision means that there must be integration of the overall organization sustainability picture to the overall picture of the employee morale and required dreams. Such a process is very important because it enables the formation of a corporate-wide culture towards the support of CSR activities. They serve as a compass that narrates a common sustainability story so that all the people in the organization are headed in the same direction.

9.1.2.1 Key Elements of Creating a Shared Vision

Alignment of Goals: This is because people tend to be happy whenever their employer's objectives match their culture, religion, or other beliefs. It creates solidarity and group propulsion towards the accomplishment of sustainability objectives.

Communication and Transparency: It is necessary to explain the vision of the organization connected with sustainable development. It gets on the shoulders of leaders to communicate the vision in the simplest, and most encouraging manner. Publicizing goals, and process assists in encouraging employees' trust as well as clarifying drivers of achievement and highlighting to the employees the achievement's implication to the organization. The blogging and frequent meetings, where information is shared can help sustain this vision.

Involvement in Strategy Formulation: Getting the employees involved in the making of the sustainability strategies is a necessary procedure that lets their opinions be considered. This approach also helps in making these strategies more useful and effective since it increases the engagement of employees in the process. According to the theory, one can see that the more the employees believe they have an interest in the results of the programs or policies the higher the level of their interest.

Role Modeling by Leaders: Leaders' role in building a shared vision is vital, and they should personally model the vision. Through the organizational leader's actions, much can be done to encourage employees to adopt sustainable practices. This role modeling develops a corporate culture that puts pressure on organizations' employees to adopt sustainable practices in their daily lives.

Cerco and Wooldridge suggest that the level of shared vision in organizations affects employee engagement and motivation. For instance, working with Kim et al. (2019), the results showed that increased essential compatibility with organizational sustainability initiatives contributed to higher levels of these respondents' organizational commitment to CSR practices and their job satisfaction and performance. Furthermore, Mitchell et al. (1997) pointed out the necessity of stakeholder involvement in the strategic decision-making process stabilizing the CSR activities' effectiveness and legitimacy.

Through engaging in the process of developing the shared vision it is always possible to create a unified and enthusiastic workforce aimed at the achievement of sustainability aims. Such efforts help not only in the augmentation of organizational performance but also in the creation of societal welfare and continued support to Green HRM for the development and sustenance of a green culture within organizational boundaries.

9.1.2.2 Making Sustainability Personal: How Green HRM Interacts with the Employee Party

Pretend that you are going to work one day, and you enter the office, and you have the feeling that everything you do from that precise moment will be for the good of the planet, or at least for the good of the office's sustainability. This is the crux of the modern sustainability movement, and this is where GHRM comes in, helping to build this bridge. GHRM guarantees that sustainability is not an empty word for the company, but rather a value that each of the staff is aware of.

Personal Connection Through Engagement: Green human resource management practice involves getting the employees involved When the managers design the strategies to implement the sustainability policy, they should consider the following.

This approach enables employees to understand that what they do triggers the environment, hence incorporating sustainability in practice. For instance, when a company comes up with paperless policies or even encourages working from home to minimize its carbon footprint, the workers feel they are working to make the world a better place (Renwick et al., 2013).

Training and Development: Training courses in environmental consciousness or sustainable development are also provided in GHRM. Visualize yourself attending the workshops in which one gets to know how to save energy, how to dispose of waste etc., and sustainable ways of sourcing the products. These training sessions do not only entail providing knowledge as to how various green practices can be implemented but also provide a sense of responsibility and empowerment at the workplace. The employees trained in such manners are likely to conduct themselves sustainably both at the workplace and in their day-to-day activities (Jose Chiappetta Jabbour, 2011).

Recognition and Rewards: The other important thing to do to make sustainability personal is to ensure that there is a provision for rewarding the employees in their endeavors towards making sustainable changes. Imagine getting an award, or a bonus for an idea that would help reduce the use of plastics in the office environment. Such recognition helps to reaffirm sustainability and encourages the employees to engage in activities that will achieve organizational goals regarding the environment. Studies have shown that when companies acknowledge and reward sustainable practices, employees are more committed and enthusiastic about participating in green initiatives (Govindarajulu & Daily, 2004).

Creating a Sustainable Culture: Green HRM fosters an organizational culture that makes sustainability an organizational value. Perhaps there is no better job to go into than a workplace that is committed to reducing its carbon footprint from the top management to the new employees. In this way, these common commitments promote the sense of belonging to a group and even the group is worth fighting for. The employee feels a higher meaning or work commitment when sustainability is embedded in the business strategy or mission statement (Daily et al., 2008).

Communication and Involvement: The fact remains that communication is erudite to individualizing sustainability. GHRM ensures that employees are kept informed about the company's sustainability goals and progress. Regular updates, newsletters, and meetings that highlight green initiatives and achievements make employees feel involved and valued. When employees are aware of the bigger picture and their role in it, they are more likely to take ownership of sustainability efforts.

A vital component of personalizing sustainability is efficient interaction. GHRM makes sure that workers are aware of the business's performance towards its objectives for sustainability. This may include being updated from time to time, given periodicals, or even being briefed on evergreen environmentally friendly programs and projects. Organizational members are more willing to be held accountable for their environment because they see the picture from the bigger perspective and how they fit into that picture (Milliman & Clair, 1996).

Real-Life Example: An example can be a company that introduces the green commute policy, in which employees are asked to take cars together, bike, or use public transport. This way, employees who take part in this program cut their carbon footprint and develop a feeling of togetherness with the other employees. For this, the company offers some incentives such as half fares on transportation and bike stands at the workplace. Thus, the employees get the sense that their choices affect their private lives and thus, the employees get more engorged with the company's sustainability agenda.

9.1.2.3 Challenges and Future Research Directions

Nonetheless, the following challenges can still be observed in both of these research areas, which are GHRM and CSR. Another is the cross-cultural application of GHRM practices, as the practices may work in one region while the opposite is true in another (Jabbour & de Sousa Jabbour, 2016). There is still so much more that can be done to understand the contextual factors that may present a threat to GHRM and what can be done to mitigate the threats.

One of the future research avenues is understanding how GHRM affects organizational performance in the long run. These reviews show that GHRM can positively impact employee engagement and support CSR objectives; more time design research is required to investigate the impact of GHRM on the long-term sustainable and financial performance (Tang et al., 2018).

9.1.2.4 Green Work Practices

Sustainable workplace practices refer to all the measures and activities that are employed by organizations in their day-to-day functioning to reduce their ecological footprint. They are part of Green HRM and contribute to the process of the implementation of sustainability goals across the organization. Not only does the act of observing green work practices and participating in the process increase employee commitment to those goals, but it also affects job satisfaction and organizational loyalty.

Key Aspects of Green Work Practices

Reducing Waste: Discounting wasteful practices is a core part of green work practice, and this entails developing policies in this respect. This entails encouraging the use of products that can be recycled, encouraging record systems to reduce the use of paper, and encouraging recycling programs. This way, organizations not only have an impact on reducing their contribution to garbage production but also on improving the working environment.

Energy Conservation: Organizations can do many things to support sustainability when it comes to energy conservation. This comprises installing efficient lighting and appliances, encouraging employees not to leave lights and other appliances on when not in use, and efficiently working the heating and cooling systems. Energy saving also saves the organization's operational expenses and, at the same time, shows society the organization's efforts in the fight against climate change.

Sustainable Procurement: Another is the selection of environmentally friendly goods and services in the workplace, which is also another crucial element of working environmentalism. This means choosing suppliers that meet environmentally responsible standards, employing packaging materials that are bio-degradable or recyclable, and using energy-efficient technologies. Sustainable procurement makes it possible for an organization to dance its sustainability mission through its supply chain.

Promoting Green Transportation: It is also important to suggest to staff members that they share a car or bicycle and use public transportation or electric cars. This will reduce the organization's overall carbon footprint. To encourage such measures, incentives such as the passing of cheap short distances, the provision of bike stations, and other services in support of green commutes could come in handy.

Eco-friendly Office Design: Sustainable design of work environments can also make the premises more environmentally friendly. This comprises avoiding the use of toxic materials, using natural light, and using plants in the building to enhance air quality and openness, which reduces the use of artificial light and air conditioning. Sustainable office design is also beneficial for the green target and the employees of an organization and their effectiveness.

Green IT Practices: Green IT, on the other hand, entails the proper usage of technology to the lowest possibility of energy consumption and the least amount of electronic gadgets or materials that can be disposed of. This also entails employing effective control of power consumption in servers, promoting the use of cloud computing to reduce the use of physical infrastructure, and writing policies on e-waste management. Green IT strategies make certain that the use of technology conforms to the green objectives of the organization.

Studies establish the role of green work practices in promoting the performance of the organization and the level of commitment from the workforce. According to (Daily & Huang, 2001) study, green practices at the workplace make the employees satisfied with their jobs and increase their commitment towards sustainability. Further, Renwick et al., (2013) pointed out that to reach a sustainable future, it is required to pay attention to green work practices and noted that they enhance organizational image and competitive position.

It can be achieved through advocacy of a corporate environmental corporate through the workplace procedures of sustainable use. Such procedures do not only act as measures towards environmental conservation but also enhance worker pride towards work and, hence, participation.

9.1.2.5 Building a Supportive Culture

Cultivation of a sustainable culture entails fostering a strong and healthy culture that will support sustainability at every level of the organization. An organizational culture spearheads the support of the employees on sustainability, which results in a culture that supports sustainability, where such practices are encouraged and promoted.

Key Elements of Building a Supportive Culture:

Leadership Commitment: The management strategy applied here is to find a supportive atmosphere that embraces everyone. Practitioners who regularly exercise sustainability through their decisions and managerial courses set a good example for their personnel. This engagement can be manifested at the strategic level of organizational goals, resources, and dollars and at the individual level through patronage of sustainability projects. Had the leaders in the various organizations been keen on the support of green practices, they would have assisted in renewing focus on their significance within organizations.

Employee Involvement: Promoting sustainability within the company is crucial, and the staff must be included in the process. They welcome ideas of green participation, engage them in developing and implementing sustainability ideas, and acknowledge them for their effort. Employees can get engaged by being registered in green teams or committees and attending sustainable workshops and brainstorming meetings.

Open Communication: People need to know what is being done and what is intended for the sustainability of the organization and its operations. Updates and meetings keep the employees informed of the goals of the organization's sustainable practices, the effects of the work done, and any new plans. Opportunities for feedback and suggestions provide a means for the employees to be listened to and encouraged to express their opinions.

Recognition and Rewards: When employees find themselves in a company that practices sustainability, commending and rewarding them would ensure that they help maintain sustainability. These may be categorized based on the structure of the rewards as in/formal recognition programs and in/tangible incentives. Organizational achievements ought to be always recognized whether they are major or minor, this is because doing so lays the foundations for sustainability.

Supportive Policies: The other area that requires commitment towards encouraging green work practices is the adoption of policies that facilitate such culture. This includes offering support to sustainable projects in the organization, developing policies on behavior that support the environment, and education on environmentally responsible behavior. Policies in this context guarantee that workers are equipped with proper and sufficient resources and information for their efficiency towards the accomplishment of sustainable initiatives.

Integration into Organizational Values: Thus, the first condition of making the activity supportive of sustainability is the provision of solutions that will integrate sustainable development concepts into the organization's value proposition. In other words, sustainability is integrated into the organizational culture and the concept is just an everyday working norm. This integration makes sure that every employee, ranging from the top management downwards to the common junior employee, appreciates sustainability and its importance in his or her working environment.

Continuous Improvement: A culture of continuous improvement involves regularly assessing and refining sustainability practices. Encouraging employees to identify areas for enhancement and implement new ideas fosters a dynamic and evolving approach to sustainability. This ensures that the organization is in a continuous improvement activity in addressing the environment and remains very committed to green values.

Employers can create an environment of employment where sustainability is not only a project but a principle of the employee experience by creating an environment that supports it. This culture shift also leads to a good work environment for the employees and increases the productivity of the firm. This continuous improvement mindset helps organizations stay ahead of environmental challenges and maintain a strong commitment to green practices. Research underscores the importance of a supportive culture in achieving sustainability goals. Organizations with a strong culture of sustainability, driven by leadership commitment and employee engagement, experience higher levels of environmental performance and employee satisfaction (Sypniewska et al., 2023; Yang et al., 2024). Additionally, Eisenbeiss et al. (2008) highlighted that supportive organizational cultures are essential for the successful implementation of green practices and for fostering innovation. Organizations can cultivate a supportive atmosphere to establish an atmosphere of employment where sustainability is no longer merely a project but a core component of employee engagement. This culture change fosters a productive and enjoyable job atmosphere and improves the organization's efficiency.

9.1.2.6 Inclusive Decision-Making

The concept of inclusion-making choices is all about giving an added and broader range of opinions on viewpoints while in the decision-making process. The inclusion of viewpoints in making choices viable for sustainable CSR efforts is what the concept of inclusion-making choices is all about. This method also ensures diverse points of view are considered and enhances observance of the staff members' commitment and accountability.

Key Aspects of Inclusive Decision-Making:

Engaging Diverse Perspectives: It is the case that the involvement of a large number of points of view in the decision-making allows revealing the opportunities and the possible problems that the decision-maker might not have thought of in case he or she worked only in the narrow frame of reference. This diversity can be departmental in terms of hierarchy and in terms of demographics. By bringing together varied experiences and ideas, organizations can develop more robust and innovative sustainability strategies.

Encouraging Participation: Creating opportunities for employees to contribute their ideas and opinions ensures that everyone has a voice in shaping the organization's sustainability efforts. Accessible discussion boards, discussion groups, inquiries, and discussions about ideas may all assist in accomplishing goals. Employee involvement within the corporation's environmental objectives is increased as well as their feeling of importance is enhanced when involvement is encouraged.

Transparent Processes: Being honest and forthcoming about how decisions are made increases confidence and reliability. Transparent discussion of the norms and techniques employed for making choices facilitates employee comprehension of how decisions are reached while ensuring that their points of view have been taken into consideration.

Recognizing Contributions: Acknowledging and appreciating the contributions of employees who participate in decision-making processes reinforces their importance and encourages continued involvement. Medals with honors are examples of official recognition; verbal compliments, as well as open acknowledgment, are examples of casual acknowledgment.

9.2 Conclusion and Practical Implications

Synthesizing GHRM with CSR is a formidable strategy for developing a company environment that will endure. Organizations are under pressure to adopt green procedures and processes; hence, staff should be engaged in CSR activities such as GHRM. The above approach not only helps attain a happy, welcoming, and driven workplace but also helps attain sustainable goals.

It is necessary to involve the staff in CSR programs as major changes inside firms can be generated this way. Employees who are part of the CSR process build the organization's values, which gives purpose and meaning to the work, thus, boosting happiness at work and engagement. It has been found that employees who think that the organization they work for is socially and environmentally responsible are more likely to engage in behaviors that are receptacles for sustainable development—a boon for the business and the planet. The use of environmentally friendly human resources management is essential to encouraging such types of interaction. Through the integration of ecological variables into HR operations, including recruitment, instruction, performance, evaluation, and relations with employees, companies may develop staff who are proficient, competent, and dedicated to sustainability.

Therefore, the positive interaction between CSR and GHRM is essential for the creation of a responsible organizational culture. There is a direct and positive correlation between green HR practices on unions and employees on one side and efficient performance, increased satisfaction, loyalty, and organizational commitment from the employees' end on the other. As businesses continue to navigate the complexities of sustainability, the integration of CSR and green human resources management continues to be an essential tactic for attaining long-term achievement and contributing towards the betterment of both the planet and the community.

Every corporation is always looking for ways through which they can enhance sustainability efforts with regard to training and development programs put in place to provide the necessary tools for sustainable contributions to the company's CSR goals. Moreover, stable, well-communicated, and practiced sustainable leadership strengthens the notion of environmental responsibility from the position of the company's leadership, thus making it a part of the company's culture with the help of widely communicated and practiced CSR goals.

Thus, to save, the organization has to develop a sustainable culture of environmental goals with reference to organizational values as a whole. Activities such as offering incentives and recognition to employees for committing to environmentally friendly practices, involving employees in green projects, and informing them of the company's CSR performances are other crucial elements of this approach. Through organizational commitment, sustainability becomes an organizational culture, and employees are encouraged to participate fully in CSR activities and continually optimize the company's environmental impact.

Finally, they support the view that both CSR and GHRM are vital for creating the specific context of a sustainable organization. Effective implementation of green HR practices not only assists the organization in meeting its environmental goals but also increases employee satisfaction, commitment, and organizational performance. Thus, incorporating CSR and GHRM remains a viable and crucial strategy for organizations in the pursuit of sustainable endeavors and the enhancement of the quality of life for the world and people. Implementing green performance metrics involves setting up measurable indicators to track and evaluate an organization's environmental impact across various aspects of its operations, like energy consumption, waste generation, water usage, and material sourcing. The goal is to drive sustainable practices and continuous improvement towards a reduced environmental footprint.

References

Aguinis, H., & Glavas, A. (2012). What we know and don't know about corporate social responsibility a review and research agenda. *Journal of Management, 38*, 932–968. https://doi.org/10.1177/0149206311436079

Brammer, S., Millington, A., & Rayton, B. (2007). The contribution of corporate social responsibility to organizational commitment. *The International Journal of Human Resource Management, 18*(10), 1701–1719. https://doi.org/10.1080/09585190701570866

Carroll, A. B. (1991). The pyramid of corporate social responsibility: Toward the moral management of organizational stakeholders. *Business Horizons, 34*(4), 39–48. https://doi.org/10.1016/0007-6813(91)90005-G

Chong, M. (2009). Employee participation in CSR and corporate identity: Insights from a disaster-response program in the Asia-Pacific. *Corporate Reputation Review, 12*(2), 106–119. https://doi.org/10.1057/crr.2009.8

Daily, B. F., Bishop, J. W., & Govindarajulu, N. (2008). A conceptual model for organizational citizenship behavior directed toward the environment. *Business & Society, 48*(2), 243–256. https://doi.org/10.1177/0007650308315439

Daily, B. F., & Huang, S. (2001). Achieving sustainability through attention to human resource factors in environmental management. *International Journal of Operations & Production Management, 21*(12), 1539–1552. https://doi.org/10.1108/01443570110410892

Ehnert, I., Parsa, S., Roper, I., Wagner, M., & Muller-Camen, M. (2016). Reporting on sustainability and HRM: A comparative study of sustainability reporting practices by the world's largest companies. *The International Journal of Human Resource Management, 27*(1), 88–108. https://doi.org/10.1080/09585192.2015.1024157

Eisenbeiss, S. A., van Knippenberg, D., & Boerner, S. (2008). Transformational leadership and team innovation: Integrating team climate principles. *Journal of Applied Psychology, 93*(6), 1438–1446. https://doi.org/10.1037/a0012716

Glavas, A. (2016). February). Corporate social responsibility and organizational psychology: An integrative review. *Frontiers in Psychology, 7*, 1–13. https://doi.org/10.3389/fpsyg.2016.00144

Govindarajulu, N., & Daily, B. F. (2004). Motivating employees for environmental improvement. *Industrial Management & Data Systems, 104*(4), 364–372. https://doi.org/10.1108/02635570410530775

Hronová, Š., & Špaček, M. (2021). Sustainable HRM practices in corporate reporting. *Economies, 9*(2). https://doi.org/10.3390/economies9020075

Im, S., Chung, Y. W., & Yang, J. Y. (2017). Employees' participation in corporate social responsibility and organizational outcomes: The moderating role of person–CSR Fit. In *Sustainability, 9*(1). https://doi.org/10.3390/su9010028

Jabbour, J. C. C. (2011). How green are HRM practices, organizational culture, learning and teamwork? A Brazilian study. *Industrial and Commercial Training, 43*(2), 98–105. https://doi.org/10.1108/00197851111108926

Jabbour, C. J. C., & de Sousa Jabbour, A. B. L. (2016). Green human resource management and green supply chain management: Linking two emerging agendas. *Journal of Cleaner Production, 112*, 1824–1833. https://doi.org/10.1016/j.jclepro.2015.01.052

Jabbour, C. J. C., & Santos, F. C. A. (2008). The central role of human resource management in the search for sustainable organizations. *The International Journal of Human Resource Management, 19*(12), 2133–2154. https://doi.org/10.1080/09585190802479389

Jackson, S. E., Renwick, D. W. S., Jabbour, C. J. C., & Muller-Camen, M. (2011). State-of-the-art and future directions for green human resource management: Introduction to the special issue. *German Journal of Human Resource Management, 25*(2), 99–116. https://doi.org/10.1177/239700221102500203

Jamil, S., Zaman, S. I., Kayikci, Y., & Khan, S. A. (2023). The role of green recruitment on organizational sustainability performance: A study within the context of green human resource management. In *Sustainability, 15*(21). https://doi.org/10.3390/su152115567

Milliman, J., & Clair, J. (1996). Best environmental HRM practices in the US. In *Greening people* (p. 25). Routledge. https://doi.org/10.4324/9781351283045-2/best-environmental-hrm-practices-us-john-milliman-judith-clair

Min, Y. A., Hao, M., Yang, X., Ling, D. Y., & Yuan, J. S. (2024). June). How employee corporate social responsibility participation promotes pro-environmental behavior. *Frontiers in Earth Science, 12*, 1–11. https://doi.org/10.3389/feart.2024.1393386

Mitchell, R. K., Agle, B. R., & Wood, D. J. (1997). Toward a theory of stakeholder identification and salience: Defining the principle of who and what really counts. *The Academy of Management Review, 22*(4), 853–886. https://doi.org/10.2307/259247

Renwick, D. W. S., Redman, T., & Maguire, S. (2013). Green human resource management: A review and research agenda. *International Journal of Management Reviews, 15*(1), 1–14. https://doi.org/10.1111/j.1468-2370.2011.00328.x

Sroufe, R. (2017). Integration and organizational change towards sustainability. *Journal of Cleaner Production, 162*, 315–329. https://doi.org/10.1016/j.jclepro.2017.05.180

Sypniewska, B., Baran, M., & Kłos, M. (2023). Work engagement and employee satisfaction in the practice of sustainable human resource management—Based on the study of Polish employees. *International Entrepreneurship and Management Journal, 19*(3), 1069–1100. https://doi.org/10.1007/s11365-023-00834-9

Tang, G., Chen, Y., Jiang, Y., Paillé, P., & Jia, J. (2018). Green human resource management practices: Scale development and validity. *Asia Pacific Journal of Human Resources, 56*(1), 31–55. https://doi.org/10.1111/1744-7941.12147

Yang, Y., Din, A. U., Mohi Ud Din, Q., & Khan, I. U. (2024). Green leadership in manufacturing industry: Unveiling the green revolution's impact on organizational performance. *Heliyon, 10*(6), e27831. https://doi.org/10.1016/j.heliyon.2024.e27831

Zoogah, D. B. (2011). The dynamics of green HRM behaviors: A cognitive social information processing approach. *German Journal of Human Resource Management, 25*(2), 117–139. https://doi.org/10.1177/239700221102500204

Chapter 10
Sociological Relevance of Corporate Social Responsibility and Sustainable Governance

Sai Aanchal

Abstract Sociology is the study of society and social institutions like economic institutions. Corporate organizations are integral aspects of economic institutions functioning in human societies around the world. In contemporary times, especially with an emphasis on the Sustainable Development Goals (SDGs), corporates are committed to operate in an economically, socially, and environmentally sustainable manner through functionally effective Corporate Social Responsibility (CSR) initiatives and sustainable governance. CSR and sustainable governance are closely intertwined in the field of sociology, as both emphasize the need for businesses to balance economic, environmental, and social considerations to achieve long-term sustainability and positive societal impact. The present study makes an attempt to understand the phenomena of CSR and sustainable governance by undertaking a thematic review of the literature. Further, sociological theories have been employed to study the sociological relevance of CSR and sustainable governance by highlighting its impact on the tribal, rural, and urban societies in India.

Keywords Sociology · Society · Economic institutions · Corporate social responsibility · CSR · Sustainable governance · Sustainable Development Goals · SDG

10.1 Introduction

Sociology is intricately related to the economics because economic institutions are socially constructed, and are significant to the subject matter of contemporary economic sociology (Rizza, 2006). Corporate organizations are integral aspects of economic institutions functioning in human societies around the world. In contemporary times, especially with an emphasis on the Sustainable Development Goals

Sai Aanchal (✉)
Department of Sociology, SGGS College, Chandigarh, India
e-mail: sggssaiaanchal@gmail.com

V. Kandpal et al. (eds.), *Developments in Corporate Governance*,
Approaches to Global Sustainability, Markets, and Governance,
https://doi.org/10.1007/978-981-96-6366-8_10

(SDGs), corporates are committed to operate in an economically, socially, and environmentally sustainable manner through functionally effective Corporate Social Responsibility (CSR) initiatives and sustainable governance. CSR has become a prominent concept in business, encompassing a corporate's self-regulated adherence to ethical standards and societal norms (Chakraborty, 2015; Josan, 2015). It is not easy to define CSR, considering it is an all-expansive umbrella term for diverse and intricate collaborations between various stakeholders, i.e., corporates, non-governmental organizations, government organizations, human societies, and communities (Redmond, 2005). CSR is not a hollow term or a catch-all fashionable on paper term to tick-mark the corporate obligations but CSR is a definite rather conscious societal responsibility of the corporates towards the human societies in which they operate, in a way mandating for these corporates to give back to the human societies whose individuals are their social capital and functionally productive human resources (Khan et al., 2013). The major objectives of CSR are to address myriad issues with emphasis on social and human development, contribution to promoting education and skill development in the targeted areas for poverty alleviation, consistent efforts to resolve environmental problems, and preservation of natural resources and biodiversity (Dobers & Halme, 2009; Renouard & Ezvan, 2018). The economists and social activists who defend the relevance of CSR are of the opinion that the myriad issues addressed by the corporations through their CSR projects add financial value to the corporations besides building their positive image (Carroll & Shabana, 2010; Okpara & Idowu, 2013; Schreck, 2011). There are many corporations that have adopted accountable CSR practices in an attempt to be accountable to their vision, business ethics, and fundamental values that are core to their foundational principles (Kong et al., 2020; Tarabella & Burchi, 2013). Thus, CSR is imperative for more transparent, accountable, and socio-economically fruitful practices that bring a definite social impact both for human societies and the natural environment, without which societies cannot survive.

Sustainable governance is not merely a composite but a panoptic concept that is significant for safeguarding the sustainable interests of human societies and dimensions that form the basis for those human societies. Sustainable governance incorporates social, economic, and ecological dimensions for the prospicient welfare of human societies. It entails both scripted and spontaneous oral regulations, which colligate institutions and norms governing those institutions with ecological citizenship (Bosselmann et al., 2008; Ekardt, 2019). On the one hand, we have environmental governance where the emphasis is on ecological citizenship, i.e., environmentally responsible behavior of humans; on the other hand, we have sustainable governance that emphasizes both sustainable preservation of the environment and promoting human development (Eyong, 2003; Roy, 2016). Research studies have substantiated that efficacious sustainable governance adds productive value to sustainable development in developing and developed nations (Güney, 2017; Kardos, 2012; Stojanović et al., 2016). It is important to understand that the major components of governance for sustainability and sustainable development incorporate transition management, reflexive governance, and adaptive management (Kemp et al., 2005; Sol et al., 2018). It is veritably challenging to achieve sustainable governance irrespective of a vast multitude of pacts, conventions, and global initiatives (Abbott, 2012; Bressers & Rosenbaum, 2003). Thus, it is important for the nations to emphasize on sustainable governance for the conservation of natural resources and welfare of human societies (Bleischwitz & Bringezu, 2008). Sustainable governance is fundamental to human societies signifying constructive social change that merges innovation with policy implementation (Folke et al., 2005).

CSR and sustainable governance are closely intertwined in sociology, as both emphasize the need for businesses to balance economic, environmental, and social considerations to achieve long-term sustainability and positive societal impact. This paper elaborates a discussion on the various factors that highlight strong bases and requisite conditioning that impact CSR and sustainable governance with an emphasis on the theoretical implications of CSR and sustainable governance in sociology. It focuses on three specific themes, and finally, through conclusion, the paper presents implications for future research in CSR and sustainable governance. The paper is structured under three themes: CSR in Sociology, Sustainable Governance in sociology, and Theoretical Implications of CSR and Sustainable Governance in Sociology. These themes will be discussed in the following sections.

10.2 CSR in Sociology

Corporate Social Responsibility (CSR) has become a significant area of sociological inquiry, offering insights into the complex relationship between business and society (Singh & Misra, 2022). Sociological perspectives can enhance our understanding of CSR's diversity and dynamics across various levels of analysis (Athanasopoulou & Selsky, 2015; Strumińska-Kutra, 2013). However, CSR is not without criticism, with some scholars arguing that it represents market-based ethics that commodifies humanitarian values and perpetuates capitalist structures (Daily, 2017; Fleming & Jones, 2012; Karaca Akbaş, 2012; Schneider, 2020). The concept of CSR is characterized by competing agendas, including professional, political, activist, and corporate interests, which can hinder sustainable change (Allen & Craig, 2016; Dawkins, 2021; Timms, 2017). Many multinational corporations increasingly engage in community development and social reporting, and CSR communication through corporate websites has become essential for stakeholder dialogue (Amaladoss & Manohar, 2013; Mahapatra, 2019; Morsing & Schultz, 2006). While CSR reporting practices are more prevalent in developed countries, there is a growing interest in understanding the impact and effectiveness of CSR in developing nations (Jamali & Karam, 2018; Jamali & Mirshak, 2007; Mahapatra, 2019; Rahman Belal & Momin, 2009).

CSR in sociology has been studied in the following sections under the following three main themes:

2.1. CSR for Tribal Society
2.2. CSR for Rural Society
2.3. CSR for Urban Society

10.2.1 CSR for Tribal Society

CSR initiatives have shown positive impacts on tribal communities in India. Tata Steel's efforts in Jharkhand have focused on creating employability for *Primitive Tribal Groups* (PTGs) like Birhor and Savar (Agarwal et al., 2013). According to Rahman and Singh (2019), the CSR initiatives of the Tata Group have played a significant role in the financial empowerment of tribals. Tata Group has funded CSR projects for the advancement of agricultural measures that are making tribal communities more self-reliant in states such as Chhattisgarh, Jharkhand, and Odisha. These CSR projects also supported tribal handicrafts and provided skill development opportunities for tribal artisans (Kumari et al., 2020).

Uchoi and Singh (2020) found that the CSR initiatives of the Tripura Forest Development & Plantation Corporation Limited (TFDPC) have a positive impact on the lives of tribals in Tripura state, i.e., improved the quality of tribal livelihoods by improving the income-earning opportunities of the tribals. A study in Odisha found that the Rourkela Steel Plant of Steel Authority has promoted the development and sustainable livelihoods of tribal people by rendering education and healthcare services and further ensuring their empowerment in all terms (Acharya & Patnaik, 2018). Similar findings were reported by the study of researchers Uma and Uma (2021) which stated that CSR initiatives by corporates have played a substantial role in the socio-economic development of tribal societies, and this result has brought immense benefits to not just the tribal population but also the corporates involved. However, it is important to note that there are consistent challenges in ensuring that CSR programs are in line with the requisite needs of the tribal society, as pressure groups wielding socio-economic and political powers often use CSR projects as a shield for causing damage to the marginalized tribals (Majumdar & Saini, 2013). Irrespective of several such barriers, CSR has remained a significant business agenda for Indian corporations across diverse sectors, reinstating the determination of corporations for the welfare of tribal communities (Uchoi & Singh, 2020).

10.2.2 CSR for Rural Society

CSR initiatives in Indian rural societies have documented a substantial, incontrovertible impact on rural development. Research studies have highlighted that CSR initiatives bring positive gains for both corporates and rural societies, irrespective of the corporate strength or the business sphere in which that corporation operates (Das, 2015; Pradhan & Ranjan, 2011).

It is important to note that corporations, irrespective of the sector in which they operate, implement CSR initiatives by placing crucial value on the population of rural societies by perceiving the rural population as their significant stakeholder (Joshi et al., 2023; Kale, 2020; Sarkar & Pingle, 2018). CSR initiatives primarily focus on the socio-economic development of rural societies by addressing socio-economic issues like unemployment and generating employment through skill development, infrastructure development, and healthcare (Hussain, 2014).

Narwal (2007) conducted an extensive study on 33 public–private sector banks in Chandigarh and Northern Haryana with the sole objective of examining their CSR activities aimed at rural development. Findings reflected how these banks focused on sustainable growth, education, healthcare, and creating awareness of environmental issues through their efficient marketing strategies. Similar findings were reported by the study of Gautam et al. (2023), which established that consistent financial support to the CSR initiatives for rural development has significantly contributed to the poverty reduction and sustainable development of Indian states by improving the quality of education, healthcare, and environment.

Research studies have also substantiated that corporate policies for CSR disbursement require thorough modifications for the implementation of efficient activities and initiatives for rural development (Bendell et al., 2011; Deshmukh & Gupta, 2020; Frynas, 2005). Since 2014, Maharashtra has been the highest beneficiary of CSR disbursement; a district-wise study revealed the need for improved allocation of resources based on the Human Development Index and rural population components (Deshmukh & Gupta, 2020). Overall, analysis by various studies has only reinstated that CSR initiatives by corporations have significantly rather positively supplemented the attempts made by the government for rural development in India (Gupta & Gupta, 2019; Moharana, 2013).

10.2.3 CSR for Urban Society

CSR has emerged as a crucial concept in India, integrating social, cultural, and environmental aspects into corporate operations (D Cunha et al., 2019). It benefits society by improving living standards and helps companies enhance their goodwill (Aggarwal, 2014). Wikantiyoso and Suhartono (2018) examined the positive potency and problems of private sector involvement in the urban development processes, especially the maintenance of urban green open spaces, and developed a model of private sector involvement policy in the revivification of urban green open spaces through CSR programs. CSR activities in rural and urban India span across various sectors, including healthcare, education, and skill development (Singh, 2018a).

Indian corporates, through their CSR initiatives, are playing a prominent role in the urban development of India through constructive collaborations with the government. One can witness a win-win agenda being fostered through the public–private partnerships that are helping in building smart cities in India while emphasizing the social and environmental sustainability of the cities (Jain & Gupta, 2016; Singh et al., 2024). A smart city is one that applies technology to improve the quality of life for its residents, with key elements including a smart economy, mobility, governance, sustainable living, aware people, and a sustainable environment (Jain & Gupta, 2016).

Jain and Gupta (2016) emphasized that CSR can play a significant role in addressing urban challenges, creating financially sustainable infrastructure, and collaborating with various stakeholders to create liveable and inclusive cities. While CSR has become mandatory for corporations to contribute a portion of their profits towards social welfare (Aggarwal, 2014), its impact on social upliftment is often limited due to challenges such as inadequate implementation and insufficient stakeholder engagement (Singh, 2018a). Some Indian corporates mistakenly believe that merely complying with laws fulfills their social responsibility (Deswal & Raghav, 2013). To maximize CSR's effectiveness, corporates should adopt a strategic approach, prioritizing stakeholder engagement, impact assessment, and transparency (Singh, 2018a).

10.3 Sustainable Governance in Sociology

Sustainable governance in sociology encompasses complex interactions between various stakeholders and institutions to address social, economic, and environmental challenges. Telecentres in Kerala demonstrate the importance of managing relationships among government, private sector, and community actors for long-term sustainability (Madon, 2005). The European Union recognizes sustainable development as a major governance challenge requiring significant changes to existing practices (Meadowcroft et al., 2005). A dynamic, responsive model of governance-in-action emphasizes the role of multiple institutions and actors in sustaining effective governance, as illustrated in pharmaceutical governance (Fox & Ward, 2008). Deliberative democracy and governance are proposed as potential solutions for Western societies to address conflicts arising from the unequal distribution of benefits, costs, and risks associated with sustainable development. Implementing institutional reforms towards deliberation may enable societies to achieve environment-centered sustainable development with significant economic and social benefits (Lehner, 2022).

10.3.1 Sustainable Governance in Tribal Society

Sustainable governance in tribal societies is closely linked to their traditional practices and indigenous knowledge systems, which foster a harmonious relationship with nature (Mollick & Singh, 2022; Sahoo, 2024). These practices contribute to sustainable development by promoting environmental protection and social stability (Sahoo, 2024).

Tribal communities often possess unique folk knowledge related to resource management, agriculture, and traditional medicine, which supports their sustainable way of life (Mollick & Singh, 2022). Effective governance for sustainable development in indigenous communities requires building on local control and cultural strengths (Dodson & Smith, 2003). It is crucial to understand that key principles for good governance include self-determination and development, as outlined in constitutional provisions like India's Sixth Schedule (Prabhu, 1993). Recognizing and incorporating these traditional practices and governance structures into broader development strategies can lead to more sustainable outcomes for tribal societies while preserving their distinct cultural identities (Dodson & Smith, 2003; Sahoo, 2024).

Singh (2018b) emphasized that civil society organizations and non-governmental organizations, in partnership with state governments, can constructively aim for inclusive governance in tribal communities by focusing on economic, social, and political initiatives that empower them while leading to social cohesion and integration of the tribals.

Mehta (2024) studied the role played by sustainable governance in tribal-occupied southern Rajasthan and found that systematic planning and implementation of ecotourism would socially empower the tribals, improve their standard of living, and make them culturally aware. It would also improve the infrastructure and enhance the tribals' community mobilization and participation.

10.3.2 Sustainable Governance in Rural Society

Sustainable governance in rural societies involves complex interactions between local communities, governments, and other stakeholders. Social capital plays a crucial role in fostering innovation and leveraging local resources for sustainable development (Go et al., 2013). Local governance partnerships have emerged as important mechanisms for promoting rural development, particularly in the EU context, though challenges remain in balancing efficiency and democratic representation (Cawley, 2009).

Sustainable governance must be inseparable from rural development projects because they involve rural communities, business organizations, and varied tiers of government (Secco & Burlando, 2017). Research studies that compare different nations often ignore the relevance of sustainable governance arrangements in rural development projects (Shand, 2016).

Liu and Yang (2019) found that in China, rural communities are closely working with government social networks to promote sustainable rural development; besides that, Chinese leaders in action have taken up a vital role in the efficacious use of governance measures. Jain (2004) studied the relevance of good governance for fueling sustainable development in India. Rist et al. (2007) studied the transition of Indian rural societies from sustainable resource management to sustainable governance of natural resources, ascribing this significant shift to communicative actions and social learning processes for rural development. Behnassi and Yaya (2011) found in their study that land resource governance was important for Indian rural development, especially in terms of improving rural livelihoods and assuring food security. The findings of their study showed the relevance of sustainable governance initiatives for rural populations, forest-dependent rural communities, small landholder farmers, and pastoralists.

Naik (2011) studied sustainable business models for e-governance-embedded rural telecentres (EGERT) in India. These rural telecentres were backed by information and communication technology (ICT) and aimed at promoting inclusive growth in rural societies by offering private and public services such as financial inclusion, employment, health, education, and access to relevant certificates for the residents of rural areas in India.

10.3.3 Sustainable Governance in Urban Society

Sustainable governance in the urban societies in India must be an inseparable aspect of the institutions to safeguard the present and futuristic verve of the concerned stakeholders—urban residents, urban local governments, urban non-governmental organizations, and urban businesses (Smith & Weik, 2012). Sustainable governance in urban societies has become the need of the hour, with the principal emphasis being the creation of an equilibrium between ecological constancy and socio-economic functionalities with an emphasis on socio-economic equities and equalities. Sustainable governance incorporates strategies aimed at quashing the wastage of urban resources to target the problems created by urban wastage, provide efficient provisions for sanitation and cleanliness, and extenuate pollution problems (Lieberherr-Gardiol, 2008). There are several obstructions caused in the thriving execution of sustainable governance strategies, which call for addressing issues of latent and manifest oppositions by the urban institutions, gaps in conceptual clarity of the sustainable governance within the urban institutions, and overall transformation of traditional approaches to the urban governance (Bridges, 2017; Smith & Wiek, 2012).

Thus, efficient, sustainable governance strategies comprise forwarding resolutions by integration of innovations with operations backed by data, collaborative justified solutions, and consistent valuation of the performance through methods recommended by multiple stakeholders involved in the process. The transition towards sustainability demands a paradigm shift in urban governance, moving beyond minimizing environmental losses to maximizing ecological gains and fostering self-sustaining development that improves both human and planetary health (Tan, 2023). Public participation is an important prerequisite for good governance in the urban areas of India, and it can ensure sustainable development (Menon & Hartz-Karp, 2019).

Ahluwalia (2019) highlighted the strengthening of urban governance in India and institutional reforms for planned urbanization and renewal of urban infrastructure. In the context of the renewal of urban infrastructure, state governments must play a significant role in furthering strategic leadership for transparency of fund allocation, decentralization of decision-making powers, and nurturing an environment that enables the efficient and timely functioning of institutional and legislative reforms.

Desai et al. (2019) supported the cause of sustainable governance in urban India for achieving Sustainable Development Goals (SDGs). Governments must ensure that there are revised policies besides constructive governance and regulatory mechanisms for sustainable cities. It is important to understand that good governance can keep in check the growth of unsustainable cities that are marked by health hazards, environmental pollution, misery, and urban decay for their ever-increasing population. Sustainable governance must address the provision of environmental services that reduce carbon footprint.

10.4 Theoretical Implications of CSR and Sustainable Governance in Sociology

Timms (2017) identifies four competing CSR agendas: professional, political, activist, and corporate, viewing CSR as a battleground of rival pressures and power relations. Melé (2007) critically reviews four CSR theories: Corporate Social Performance, Shareholder Value Theory, Stakeholder Theory, and Corporate Citizenship, highlighting the need for a more comprehensive approach. Wehmeier and Schultz (2011) emphasize organizational sociology's focus on interactions between organizations, society, and individuals in CSR contexts. Frynas and Yamahaki (2016) survey CSR theories, categorizing them into external drivers (e.g., stakeholder theory) and internal drivers (e.g., resource-based view). They note the dominance of external driver theories in CSR research and advocate for multi-theory studies and increased focus on internal dynamics and individual-level analysis to enhance CSR scholarship.

10.4.1 Functionalist Perspective

The functionalist perspective in sociology views society as an interconnected system where each element contributes to overall stability and order (Chugh, 2011). The functionalist perspective derives its basis from the intellectual thoughts of Europe, and it places a significant emphasis on the unified social system of shared values, social coherence, and integration (Garada, 2013). Talcott Parsons was a renowned sociologist who consistently advocated for the functionalist perspective. Parsons used a functionalist perspective to study the social order and doctor–patient relationship in the context of medicine (Bissell et al., 2002).

A section of scholars not ideologically aligned with the functionalist perspective have challenged its relevance for CSR by arguing that CSR has evolved beyond the definitive realms of the functionalist perspective, and today, CSR can be studied through the lens of varied theoretical perspectives. Gond et al. (2010) purported a broad-based model taking a shift from functionalism, proposing four substitute approaches for CSR: social regulation, power relationship, cultural product, and socio-cognitive construction. This multifarious interpretation of CSR lays more emphasis on its functional purpose in the development of communities. Ismail (2009) established through a study that CSR must incorporate ethics that favor the welfare of societies and communities without any inherent or hidden biases or self-vested interests. Indian spirituality has consistently focused on the development of a society where everyone serves their fellowmen and the societies in which they live without any hidden agenda or bias. Thus, CSR in India can be seen through the lens of ancient Indian traditions and doctrine like Kautilya's Arthashashtra, which stirs the inner soul and conscience of those aiming for change, especially if it is of value to those leaders who want relevant, sustainable change (Muniapan & Dass, 2008).

We can also see the impact that institutions have on the choices of individuals in social settings. Brammer et al. (2012) also observed that institutional theory can prove to be more significant in understanding the relevance of CSR because various forms of CSR in human societies are the result of socio-economic conditions, polity, and history of those human societies in which they operate. In this context, there can be a contrast rather than conflict between voluntary CSR engagement and CSR, which is mandatory because it is socially binding and non-negotiable for the legal setup of the corporates. Thus, CSR must be considered important for economic sustainable governance (Brammer et al., 2012).

Frostenson (2013) clearly made a distinction between CSR, which is ideologically pushed forward as part of a corporate agenda for seeking the legitimacy of external parties, and CSR, which is a fundamental inseparable aspect of business because it aims at focusing on internal business principles that drive the transparency and social responsibilities of the corporates. This differentiation between the two types of CSR—one that is forced and one that is natural underlines the variations in the ethics and business ideological process that legitimatize CSR.

We can see perspectives on CSR that entail the economic realm of CSR functionality, focusing on welfare mechanisms. However, this is not excluded from public–private policies and consumer markets in which the corporations operate (Kitzmueller & Shimshack, 2012).

The functionalist perspective on sustainable governance calls for addressing environmental concerns by employing integrative and adaptative mechanisms. Sustainable governance must aim for sustainable development that strikes a perfect equilibrium between the decisive processes of multiple stakeholders and enhances the quality of human lives by focusing on environmental sustainability (van Zeijl-Rozema et al., 2008). The functionalist perspective in contemporary times also postulates a transformative shift from traditional environmental governance to value-based sustainable governance, promoting human development and focusing on preserving the environment and natural resources (Roy, 2016).

Effective sustainable governance involves multiple actors, including governments, civil society, and businesses, operating across different scales and sectors (Delmas & Young, 2009). However, implementing sustainable development in Western democracies faces obstacles, requiring strategic governance and adaptation of existing political structures (Lafferty, 2006). The debate on governance for sustainable development can be clarified by explicitly defining the perspective on sustainability and the mode of governance pursued to avoid miscommunication and mismatches in approach (van Zeijl-Rozema et al., 2008).

10.4.2 Marxist Perspective

The Marxist perspective in sociology offers a critical approach to understanding social phenomena, emphasizing class conflict and historical materialism. In law, Marxist theory has evolved from legal economism to more nuanced views of law-society relationships, exploring politics, ideology, and history (Spitzer, 1983). Archeology has benefited from Marxist theory, providing long-term historical perspectives to enrich the discipline and related social sciences (Spriggs, 1984). Despite perceived incompatibility, Marxism and positivism share a commitment to scientific inquiry, with Marxist criticisms of positivism rooted in scientific principles (York & Clark, 2006). Historically, Marxism and sociology emerged as distinct approaches, with Marx focusing on political economy and class conflict, while early sociologists like Comte emphasized different aspects of social analysis (Swingewood, 1970). These perspectives continue to influence various fields within sociology, offering unique insights into social structures, change, and institutions. The Marxist perspective on Corporate Social Responsibility (CSR) offers a critical view of its role in capitalism. CSR is seen as a spectacle that commodifies humanitarian values and confronts the working class with each other (Karaca Akbaş, 2012). It is criticized as a self-defeating oxymoron by new Marxist thinkers, who challenge its coherence and effectiveness (Toft, 2015). In the construction industry, CSR initiatives for worker health, safety, and well-being are viewed as commodification that ultimately benefits companies rather than workers (Sherratt & Sherratt, 2017).

The political perspective of CSR is noted for its conceptual novelty but criticized for ambiguities and oversights. It is suggested that CSR should be seen as a form of globalization rather than its consequence and that Western multinational corporations likely engage in CSR for instrumental reasons (Whelan, 2012). The critical approaches towards CSR call for more transparency and accountability in CSR, stating that CSR for corporates must not be a tick-mark thing to fulfill mandatory social responsibility imposed by legal authorities or the state. The Marxist perspective proposes climacteric perceptiveness for sustainable governance. We know that Marx was critical of exploitative capitalism, and he emphasized the establishment of a social order that is sustainable. Marx in a way proposed sustainability forwarding the vision of communist society (Foster, 1997). Marxist perspectives distinguish between the prerequisite goals for sustainability and exploitative capitalist systems that hamper the achievement of those sustainability goals. Marxist perspective also argues that contemporary models of CSR subtly proliferate inequalities irrespective of assurances of social inclusion (Weis & White, 2020).

Marxist approach to ecology postulates that it is of utmost importance to take into consideration the ecological crisis on planet Earth, and this can only be addressed by ensuring that market forces are subordinated to the multiple-level democratic planning for sustainable governance (Albert, 2020). There has also been a trend to address the issues of sustainable governance by incorporating radical class-based approaches with a rational intent to tackle the gaps between what is on paper and what actually needs to be addressed, keeping in mind societal Deutz (2014) called for development of a radical perspective on environmental justice with an emphasis on class systems and dynamics involved in the discourse of sustainability, because environmental approaches often are backed by mindless competition for seeking further fundings rather than ensuring equity. As a result, most of the green jobs add value to the lives of only elite workers and not labor classes.

10.4.3 Social Exchange Theory

George Homans formulated social exchange theory to contradict the Marxist perspective and Durkheimian functionalism. Homas was critiqued for ignoring social structural factors and psychological reductionism (Wardani, 2016). In sociology, social exchange theory is often employed to examine the types and relevance of social interactions among the various types of human relationships, i.e., personal, professional, and familial relationships. In contemporary times, scholars from diverse fields are applying it to study routine social interactions in terms of the exchange of words, gestures, services, and resources that could either yield rewards or lead to punishments (Barbalet, 2017). Social exchange theory furnishes how social interactions shape, change, and alter power dynamics, social cohesion, and collective consciousness within social institutions and social structures (Cook & Hahn, 2021; Turner, 2021). The approach underlines that humans evaluate profits and costs even for every social interaction, determining the rewards that they may gain in exchange for their interaction (Inuzuka, 1974). However, exchange theory has its own limitations (Inuzuka, 1974).

In the context of CSR, exchange theory is relevant because it highlights the significance of employee engagement and organizational relationships. CSR can initiate co-operative efforts between suppliers and firms through social network resources, reputational risk management, and trust building among different stakeholders of CSR (Peters & Caro, 2013). Social exchange processes as a part of CSR initiatives are known to shape, change, and evolve the behavior and attitude of the employees, further impacting the performance of the organization, improving social settings in which those CSR initiatives operate, and addressing environmental concerns (Gond et al., 2010). It is important to note that not all employees in all corporates are actively engaged with CSR. Their engagement with CSR would vary depending on their interests and participation, which is often influenced by the extent to which CSR is culturally embedded within the corporate agendas in alignment with the objectives of the businesses run by these corporates and the communication on CSR by the leadership (Slack et al., 2015).

Midttun (2005) stated that it is important to realign the interests of civil societies, businesses, and government in the context of sustainable governance. In contemporary times, social governance models that are oriented towards CSR have moved away from (neo)liberal and Keynesian welfare state models, with a clear predilection for market forces that are relative to social governance. Thus, in alignment with social exchange theory, we can observe intricate interdependence between CSR, behavioral participation of the stakeholders, and exchange processes. This interdependence can impact the exchange processes of the corporate organizations both externally and internally in all aspects, i.e., social, political, cultural, economic, and environmental.

Social exchange theory substantiates the relevance of sustainable governance in all terms. It is observed that in corporations, social exchange processes are fundamental aspects of sustainable governance and the manner in which the members within the corporation coordinate to achieve the set targets to achieve sustainable governance (Jussila et al., 2012). Similarly, in web3 governance, exchange theory emphasizes the bargaining processes in the entrepreneurial setup, revealing value through pseudonymous, composable, and permissionless governance (Allen et al., 2023). In business ethics, exchange theory emphasizes sustainable governance and building trust through stakeholder relationships that bring mutual benefits for each stakeholder involved and social responsibility (Suhendi et al., 2024). The social exchange theory to governance highlights the significance of value-based reciprocal exchange, which requires social interactions and adding consistent value for furthering sustainable corporate and organization systems across diverse spheres, from traditional businesses to decentralized systems within nations.

10.5 Conclusion and Implications for Future Research

While conducting a thematic review of the literature, it was found that CSR and sustainable governance have been researched in the context of the backward and somewhat advanced areas of developing countries. However, it was important to study both CSR and sustainable governance in the realm of sociology with a focus on sociological theoretical perspectives. Furthermore, the context of research undertaken must clearly specify socio-political improvements, but most of the studies focused only on the ecological and financial aspects of CSR and sustainable governance.

It is noteworthy that the meanings associated with corporate social responsibility (CSR) and sustainable governance differ between Western nations and those in South Asia and the Middle East. The development and coverage of CSR and sustainable governance initiatives have been influenced by a myriad of sociocultural, economic, and political circumstances. This phenomenon across the world seems to have either a positive or negative impact in terms of the potential to transform the environment and human societies. Thus, suggested implications for future research can take up an in-depth study of the phenomena of CSR and sustainable governance, with the following constructive agenda:

i. It is important to consider empirical research in traditional, developing, and developed nations and their evolving thoughts on CSR and sustainable governance in tribal, rural, and urban societies. This research must be undertaken in the context of climate change, ecological conservation, and sociocultural repercussions.
ii. It is more than relevant to examine contemporary sociocultural issues in the context of CSR and sustainable governance, with an emphasis on ethnic/racial/gender/national identities, values and attitudes inherent in a community, secularism, rights to social justice, pertinent information pertaining to good governance, and press freedom.
iii. It is also important to note that there is a lack of qualitative longitudinal studies that have examined the processes of CSR and sustainable governance in tribal, rural, and urban societies in developing and developed countries in all relevant contexts, with a focus on sociocultural, economic, political, and environmental consequences and also beyond these specified repercussions.
iv. There is also extensive scope for future interdisciplinary research by incorporating multi-disciplinary perspectives from sociology, economics, environmental science, political science, law, and economics, especially in the context of contemporary issues, new media, information communication technologies, and AI (artificial intelligence).

References

Abbott, K. W. (2012). Engaging the public and the private in global sustainability governance. *International Affairs, 88*(3), 543–564.

Acharya, J., & Patnaik, S. N. (2018). Corporate social responsibility in community development and sustainability: Rourkela Steel Plant, a unit of SAIL, India. *Asian Journal of Business Ethics, 7*, 53–79. https://doi.org/10.1007/s13520-017-0079-5

Aggarwal, M. (2014). Impact of corporate social responsibility. *Asian Journal of Research in Business Economics and Management, 4*(7), 94–98.

Agarwal, R., Pruthi, M., Lodhi, P., & Kumar, P. (2013). Sustainable livelihood options for primitive tribal groups in Jharkhand: CSR initiatives by Tata Steel. *AIMA Journal of Management & Research, 7*(1/4).

Ahluwalia, I. J. (2019). Urban governance in India. *Journal of Urban Affairs, 41*(1), 83–102.

Albert, M. J. (2020). Capitalism and earth system governance: An ecological marxist approach. *Global Environmental Politics, 20*(2), 37–56.

Allen, D. W., Berg, C., Lane, A. M., MacDonald, T., & Potts, J. (2023). The exchange theory of web3 governance. *Kyklos, 76*(4), 659–675.

Allen, M. W., & Craig, C. A. (2016). Rethinking corporate social responsibility in the age of climate change: A communication perspective. *International Journal of Corporate Social Responsibility, 1*(1), 1–11.

Amaladoss, M. X., & Manohar, H. L. (2013). Communicating corporate social responsibility—A case of CSR communication in emerging economies. *Corporate Social Responsibility and Environmental Management, 20*(2), 65–80.
Athanasopoulou, A., & Selsky, J. W. (2015). The social context of corporate social responsibility: Enriching research with multiple perspectives and multiple levels. *Business & Society, 54*(3), 322–364.
Barbalet, J. (2017). Social exchange theory. In B. S. Turner (Ed.), *The Wiley-Blackwell encyclopedia of social theory* (pp. 1–11). Wiley-Blackwell.
Behnassi, M., & Yaya, S. (2011). Land resource governance from a sustainability and rural development perspective. In M. Behnassi, S. Shahid, & J. D'Silva (Eds.), *Sustainable agricultural development* (pp. 3–23). Springer.
Bendell, J., Miller, A., & Wortmann, K. (2011). Public policies for scaling corporate responsibility standards: Expanding collaborative governance for sustainable development. *Sustainability Accounting, Management and Policy Journal, 2*(2), 263–293.
Bissell, P., Traulsen, J. M., & Haugbølle, L. S. (2002). Sociological theory and pharmacy practice research:(2) An introduction to functionalist sociology: Talcott Parsons' concept of the "sick role." *International Journal of Pharmacy Practice, 10*(1), 60–68.
Bleischwitz, R., & Bringezu, S. (2008). Global governance for sustainable resource management. *Minerals & Energy-Raw Materials Report, 23*(2), 84–101.
Bosselmann, K., Engel, R., & Taylor, P. (2008). *Governance for sustainability: Issues, challenges, successes* (No. 70). IUCN.
Brammer, S., Jackson, G., & Matten, D. (2012). Corporate social responsibility and institutional theory: New perspectives on private governance. *Socio-Economic Review, 10*(1), 3–28.
Bressers, H. T., & Rosenbaum, W. A. (Eds.). (2003). *Achieving sustainable development: the challenge of governance across social scales*. Bloomsbury Publishing.
Bridges, A. (2017). The role of institutions in sustainable urban governance. *Natural Resources Forum, 40*(4), 169–179.
Carroll, A. B., & Shabana, K. M. (2010). The business case for corporate social responsibility: A review of concepts, research and practice. *International Journal of Management Reviews, 12*(1), 85–105.
Cawley, M. (2009). Local governance and sustainable rural development: Ireland's experience in an EU context. *Journal for Geography, 4*(1), 53–64.
Chakraborty, U. K. (2015). Developments in the concept of corporate social responsibility (CSR). *The Researchers' International Research Journal, 1*(1), 23–45.
Chugh, D. (2011). Understanding sociological perspectives. *Research Journal of Humanities and Social Sciences, 2*(4), 215–217.
Cook, K. S., & Hahn, M. (2021). Social exchange theory: Current status and future directions. In S. Abrutyn & K. McCaffree (Eds.), *Theoretical sociology* (pp. 179–205). Routledge.
D Cunha, S. I., Kapil, R., & Kumar, M. (2019). Corporate social responsibility in India: A reality check. *Journal of Management (JOM), 6*(2), 261–274.
Daily, L. A. (2017). *Ethical capitalism, commodities, and the consumerist gaze* (Doctoral dissertation, George Mason University).
Das, S. K. (2015). Corporate social responsibility initiatives taken for rural development in India. *Sumedha Journal of Management, 4*(3), 4–17.
Dawkins, C. E. (2021). An agonistic notion of political CSR: Melding activism and deliberation. *Journal of Business Ethics, 170*(1), 5–19.
Delmas, M. A., & Young, O. R. (Eds.). (2009). *Governance for the environment: New perspectives*. Cambridge University Press.
Desai, B. H., Sidhu, B. K., & Kumari, N. (2019). Sustainable cities in India: A governance challenge. In H. Ginzky, E. Dooley, I. L. Heuser, E. Kasimbazi, T. Markus, & T. Qin (Eds.), *International yearbook of soil law and policy 2018* (pp. 29–54). Springer Nature.

Deshmukh, H. A., & Gupta, R. (2020, February 24). *Mandatory corporate social responsibility: Impact on rural development sector of Maharashtra*. https://ssrn.com/abstract=3779108 or https://doi.org/10.2139/ssrn.3779108

Deswal, P., & Raghav, N. (2013). Corporate social responsibility: A relationship between business organizations and the society. *OIDA International Journal of Sustainable Development, 6*(11), 37–44.

Deutz, P. (2014). A class-based analysis of sustainable development: Developing a radical perspective on environmental justice. *Sustainable Development, 22*(4), 243–252.

Dobers, P., & Halme, M. (2009). Corporate social responsibility and developing countries. *Corporate Social Responsibility and Environmental Management, 16*(5), 237–249.

Dodson, M., & Smith, D. E. (2003). *Governance for sustainable development: Strategic issues and principles for Indigenous Australian communities*. Centre for Aboriginal Economic Policy Research (CAEPR), The Australian National University.

Ekardt, F. (2019). *Sustainability: Transformation, governance, ethics, law*. Springer.

Eyong, C. T. (2003). Poverty eradication and sustainable development in Cameroon. *Journal of Sustainable Development in Africa, 5*(2), 30–58.

Fleming, P., & Jones, M. V. (2012). *The end of corporate social responsibility: Crisis and critique*. Sage.

Folke, C., Hahn, T., Olsson, P., & Norberg, J. (2005). Adaptive governance of social-ecological systems. *Annual Review of Environment and Resources, 30*(1), 441–473.

Foster, J. B. (1997). The crisis of the earth: Marx's theory of ecological sustainability as a nature-imposed necessity for human production. *Organization & Environment, 10*(3), 278–295.

Fox, N. J., & Ward, K. J. (2008). What governs governance, and how does it evolve? The sociology of governance-in-action. *The British Journal of Sociology, 59*(3), 519–538.

Frostenson, M. (2013). CSR travels into the organization. In M. Jutterström & P. Norberg (Eds.), *CSR as a management idea* (pp. 126–140). Edward Elgar Publishing.

Frynas, J. G. (2005). The false developmental promise of corporate social responsibility: Evidence from multinational oil companies. *International Affairs, 81*(3), 581–598.

Frynas, J. G., & Yamahaki, C. (2016). Corporate social responsibility: Review and roadmap of theoretical perspectives. *Business Ethics: A European Review, 25*(3), 258–285.

Garada, R. (2013). Beyond structural-functional perspective: A critical heart searching for dalit, tribal and gender in Indian sociology. *IOSR Journal of Humanities and Social Science (IOSR-JHSS), 16*(6), 7–18.

Gautam, R. S., Bhimavarapu, V. M., Rastogi, S., Kappal, J. M., Patole, H., & Pushp, A. (2023). Corporate social responsibility funding and its impact on India's sustainable development: Using the poverty score as a moderator. *Journal of Risk and Financial Management, 16*(2), 1–12.

Go, F. M., Trunfio, M., & Lucia, M. D. (2013). Social capital and governance for sustainable rural development. *Studies in Agricultural Economics, 115*(2), 104–110.

Gond, J. P., El-Akremi, A., Igalens, J., & Swaen, V. (2010). Corporate social responsibility influence on employees. *International Center for Corporate Social Responsibility, 54*(4), 1–47.

Gupta, O., & Gupta, K. (2019). CSR in India: The companies act and its implications. *Journal of Contemporary Management Research, 13*(1), 1–23.

Güney, T. (2017). Governance and sustainable development: How effective is governance? *The Journal of International Trade & Economic Development, 26*(3), 316–335.

Hussain, S. (2014). Corporate social responsibility: Initiatives and its impact on socio economic development of rural areas in India. *Asian Journal of Multidisciplinary Studies, 2*(4), 102–107.

Inuzuka, S. (1974). The current and nature of the exchange theory. *Japanese Sociological Review, 24*(4), 2–18.

Ismail, M. (2009). Corporate social responsibility and its role in community development: An international perspective. *Journal of International Social Research, 2*(9), 199–209.

Jain, R. B. (2004). Good governance for sustainable development: Challenges and strategies in India. *Indian Journal of Public Administration, 50*(1), 50–71.

Jain, M., & Gupta, P. (2016). Exploring the role of CSR activities in accomplishing urban amenities in smart cities. *Asia Pacific Journal of Research, 1*(41), 2347–4793.
Jamali, D., & Karam, C. (2018). Corporate social responsibility in developing countries as an emerging field of study. *International Journal of Management Reviews, 20*(1), 32–61.
Jamali, D., & Mirshak, R. (2007). Corporate social responsibility (CSR): Theory and practice in a developing country context. *Journal of Business Ethics, 72*, 243–262.
Josan, G. (2015). Corporate social responsibility or irresponsibility. *Academicia: An International Multidisciplinary Research Journal, 5*(5), 270–282.
Joshi, G., Sharma, D., Kunte, M., & Shikalgar, S. (2023). Was CSR in our minds? The relevance of CSR in the vision and mission of Indian companies through the lens of ownership. *Social Responsibility Journal, 19*(9), 1787–1807.
Jussila, I., Goel, S., & Tuominen, P. (2012). Governance of co-operative organizations: A social exchange perspective. *Business and Management Research, 1*(2), 14–25.
Kale, S. S. (2020). From company town to company village: CSR and the management of rural aspirations in eastern India's extractive economies. *The Journal of Peasant Studies, 47*(6), 1211–1232.
Karaca Akbaş, E. (2012). *A sociological study of corporate social responsibility: A marxist perspective* (Master's thesis, Middle East Technical University).
Kardos, M. (2012). The reflection of good governance in sustainable development strategies. *Procedia-Social and Behavioral Sciences, 58*, 1166–1173.
Kemp, R., Parto, S., & Gibson, R. B. (2005). Governance for sustainable development: Moving from theory to practice. *International Journal of Sustainable Development, 8*(1/2), 12–30.
Khan, A., Muttakin, M. B., & Siddiqui, J. (2013). Corporate governance and corporate social responsibility disclosures: Evidence from an emerging economy. *Journal of Business Ethics, 114*, 207–223. https://doi.org/10.1007/s10551-012-1336-0
Kitzmueller, M., & Shimshack, J. (2012). Economic perspectives on corporate social responsibility. *Journal of Economic Literature, 50*(1), 51–84.
Kong, Y., Antwi-Adjei, A., & Bawuah, J. (2020). A systematic review of the business case for corporate social responsibility and firm performance. *Corporate Social Responsibility and Environmental Management, 27*(2), 444–454.
Kumari, G., Ratnesh, M., & Eguruze, E. S. (2020). Role of CSR in supporting tribal handicrafts of Jharkhand. *Journal of Critical Reviews, 7*(4), 57–61.
Lafferty, W. M. (Ed.). (2006). *Governance for sustainable development: The challenge of adapting form to function*. Edward Elgar Publishing.
Lehner, F. (2022). *Deliberative governance for sustainable development: An innovative solution for environment, economy and society* (1st ed.). Routledge.
Lieberherr-Gardiol, F. (2008). Urban sustainability and governance: Issues for the twenty-first century. *International Social Science Journal, 59*(193–194), 331–342.
Liu, Y., & Yang, W. (2019). Leadership and governance tools for village sustainable development in China. *Sustainability, 11*(20), 1–17.
Madon, S. (2005). Governance lessons from the experience of telecentres in Kerala. *European Journal of Information Systems, 14*, 401–416.
Mahapatra, M. (2019). Multinationals, CSR and social reporting: A sociological analysis. *International Journal of Economics and Management Studies, 6*(6), 41–45.
Majumdar, S., & Saini, G. K. (2013). Perceptions of corporate social responsibility: A study of the social segments in the Indian state of Goa. *South Asian Journal of Management, 20*(1), 139–150.
Meadowcroft, J., Farrell, K. N., & Spangenberg, J. H. (2005). Developing a framework for sustainability governance in the European Union. *International Journal of Sustainable Development, 8*, 3–11.
Mehta, S. (2024). Adaptation policy and governance towards tribal communities of southern Rajasthan, India. In K. K. Sharma, S. Sharma, V. K. Pandey, & R. Singh (Eds.), *Climate change and human adaptation in India: Sustainability and public policy* (pp. 207–221). Springer. https://doi.org/10.1007/978-3-031-55821-4_15

Melé, D. (2007). Responsabilidad social de la empresa: una revisión crítica a las principales teorías. *Ekonomiaz: Revista vasca de economía, 65*(2), 50–67.
Menon, S., & Hartz-Karp, J. (2019). Institutional innovations in public participation for improved local governance and urban sustainability in India. *Sustainable Earth, 2*(1), 1–19.
Midttun, A. (2005). Realigning business, government and civil society: Emerging embedded relational governance beyond the (neo) liberal and welfare state models. *Corporate Governance, 5*(3), 159–174.
Moharana, S. (2013). Corporate social responsibility: A study of selected public sector banks in India. *IOSR Journal of Business and Management, 15*(4), 1–9.
Mollick, F., & Singh, J. N. (2022). Folk knowledge, sustainable development and tribal culture: An anthropological perspective. *Studies of Tribes and Tribals, 19–20*(1/2), 1–6.
Morsing, M., & Schultz, M. (2006). Corporate social responsibility communication: Stakeholder information, response and involvement strategies. *Business Ethics: A European Review, 15*(4), 323–338.
Muniapan, B., & Dass, M. (2008). Corporate social responsibility: A philosophical approach from an ancient Indian perspective. *International Journal of Indian Culture and Business Management, 1*(4), 408–420.
Naik, G. (2011). Designing a sustainable business model for e-governance embedded rural telecentres (EGERT) in India. *IIMB Management Review, 23*(2), 110–121.
Narwal, M. (2007). CSR initiatives of Indian banking industry. *Social Responsibility Journal, 3*(4), 49–60.
Okpara, J. O., & Idowu, S. O. (2013). Corporate social responsibility: A review of the concept and analysis of the business case for corporate social responsibility in the twenty-first century. In J. Okpara & S. Idowu (Eds.), *Corporate social responsibility* (pp. 3–15). Springer. https://doi.org/10.1007/978-3-642-40975-2_1
Peters, R., & Caro, C. A. (2013). Promoting cooperation between corporate social responsibility and inter-organizational relationships. *Journal of Business & Economics Research, 11*(10), 417–430.
Prabhu, P. (1993). Sustainable tribal development. *Indian Journal of Public Administration, 39*(3), 479–487.
Pradhan, S., & Ranjan, A. (2011). Corporate social responsibility in rural development sector: Evidences from India. *School of Doctoral Studies (European Union) Journal, 2*, 139–147.
Rahman, H., & Singh, R. (2019). An overview of CSR taken by Tata group. *International Journal of Development Research, 9*(1), 25485–25489.
Rahman Belal, A., & Momin, M. (2009). Corporate social reporting (CSR) in emerging economies: A review and future direction. *Accounting in Emerging Economies, 9*, 119–143.
Redmond, P. (2005). Corporate social responsibility: An overview. *Reform, 87*, 7–11. https://doi.org/10.3316/ielapa.200602376
Renouard, C., & Ezvan, C. (2018). Corporate social responsibility towards human development: A capabilities framework. *Business Ethics: A European Review, 27*(2), 144–155.
Roy, S. K. (2016). The principle of sustainable development, human rights and good governance. *Contemporary Indigenous and Constitutional Issues, 3*(2), 200–220.
Rist, S., Chidambaranathan, M., Escobar, C., Wiesmann, U., & Zimmermann, A. (2007). Moving from sustainable management to sustainable governance of natural resources: The role of social learning processes in rural India, Bolivia and Mali. *Journal of Rural Studies, 23*(1), 23–37.
Rizza, R. (2006). The relationship between economics and sociology: The contribution of economic sociology, setting out from the problem of embeddedness. *International Review of Sociology, 16*(1), 31–48.
Sahoo, P. (2024). Practices of Indigenous People are leading towards sustainability. *International Journal for Multidisciplinary Research, 6*(1), 1–7.
Sarkar, S., & Pingle, S. (2018). CSR governance addressing social issues: A critical review. *South Asian Journal of Management, 25*(1), 1–28.
Schneider, A. (2020). Bound to fail? Exploring the systemic pathologies of CSR and their implications for CSR research. *Business & Society, 59*(7), 1303–1338.

Schreck, P. (2011). Reviewing the business case for corporate social responsibility: New evidence and analysis. *Journal of Business Ethics, 103*, 167–188.

Secco, L., & Burlando, C. (2017). Social capital, network governance and social innovation: Towards a new paradigm? In E. Pisani, G. Franceschetti, L. Secco, & A. Christoforou (Eds.), *Social capital and local development: From theory to empirics* (pp. 83–105). Springer Nature.

Shand, R. (2016). *The governance of sustainable rural renewal: A comparative global perspective*. Routledge.

Sherratt, F., & Sherratt, S. (2017, September). The road to hell: Worker health, safety and well-being within UK corporate social responsibility practices. In *Proceedings of the 33rd Annual Association of Researchers in Construction Management (ARCOM) Conference* (pp. 4–6).

Singh, K. P. (2018a). An analytical review of recent CSR practices and their role in social upliftment in India. *Psychology and Education, 55*(1), 361–368.

Singh, V. (2018b). Sustaining development through social capital among tribes: Analysing state-civil society interaction in inclusive governance. *Journal of Rural and Industrial Development, 6*(2), 36–42.

Singh, K., Kolar, P., Nanduri, S., Ram, V. S., & Kumar, D. (2024). CSR and smart city progress: A proposed model for urban development. *Letters in Spatial and Resource Sciences, 17*(1), 1–18.

Singh, K., & Misra, M. (2022). The evolving path of CSR: Toward business and society relationship. *Journal of Economic and Administrative Sciences, 38*(2), 304–332.

Slack, R. E., Corlett, S., & Morris, R. (2015). Exploring employee engagement with (corporate) social responsibility: A social exchange perspective on organisational participation. *Journal of Business Ethics, 127*(3), 537–548.

Smith, R., & Wiek, A. (2012). Achievements and opportunities in initiating governance for urban sustainability. *Environment and Planning C: Government and Policy, 30*(3), 429–447.

Sol, J., Van der Wal, M. M., Beers, P. J., & Wals, A. E. (2018). Reframing the future: The role of reflexivity in governance networks in sustainability transitions. *Environmental Education Research, 24*(9), 1383–1405.

Spitzer, S. (1983). Marxist perspectives in the sociology of law. *Annual Review of Sociology, 9*(1), 103–124.

Spriggs, M. (Ed.). (1984). *Marxist perspectives in archaeology*. Cambridge University Press.

Stojanović, I., Ateljević, J., & Stević, R. S. (2016). Good governance as a tool of sustainable development. *European Journal of Sustainable Development, 5*(4), 558.

Strumińska-Kutra, M. (2013). Socjologiczny wkład w debatę o CSR. *Roczniki Nauk Społecznych, 41*(2), 67–91.

Suhendi, Devi, N. A., & Syahidah, N. (2024). Exchange theory in business ethics: Building trust and business sustainability. *International Journal of Integrative Research, 2*(2), 115–130.

Swingewood, A. (1970). Comte, Marx and political economy. *The Sociological Review, 18*(3), 335–349.

Tan, Y. X. (2023). Sustainable cities: The role of governance for greater innovations. *International Journal on Smart and Sustainable Cities, 1*(01), 1–18.

Tarabella, A., & Burchi, B. (2013). Systematic review of the business case for CSR. *International Journal of Economic Practices and Theories, 3*(1), 10–28.

Timms, J. (2017). A sociological approach to the problem of competing CSR agendas. In S. Vertigans & S. O. Idowu (Eds.), *Corporate social responsibility: Academic insights and impacts* (pp. 103–118). Springer.

Toft, K. H. (2015). Liberal CSR and new Marxist criticism. In S. Idowu, C. Frederiksen, A. Mermod, & M. Nielsen (Eds.), *Corporate social responsibility and governance: Theory and practice* (pp. 303–316). Springer. https://doi.org/10.1007/978-3-319-10909-1_16

Turner, J. (2021). Karen S. Cook and Michael Hahn on exchange theory. In S. Abrutyn & K. McCaffree (Eds.), *Theoretical sociology* (pp. 206–208). Routledge.

Uchoi, O., & Singh, R. (2020). Self-reliance of tribal through corporate social responsibility (CSR): A case study in Tripura. *Indian Journal of Extension Education, 56*(1), 96–99.

Uma, E., & Uma, K. (2021). Corporate social responsibility in India—An overview. *Academicia: An International Multidisciplinary Research Journal, 11*(5), 906–912.

van Zeijl-Rozema, A., Cörvers, R., Kemp, R., & Martens, P. (2008). Governance for sustainable development: A framework. *Sustainable Development, 16*(6), 410–421.

Wardani, W. (2016). Membedah Teori Sosiologi: Teori Pertukaran (Exchange theory) George Caspar Homans. *Studia Insania, 4*(1), 19–38.

Wehmeier, S., & Schultz, F. (2011). Communication and corporate social responsibility: A storytelling perspective. In Ø. Ihlen, J. Bartlett, & S. May (Eds.), *The handbook of communication and corporate social responsibility* (pp. 467–488). Wiley.

Weis, V. V., & White, R. (2020). A Marxist perspective on the 2030 agenda for sustainable development. In J. Blaustein, K. Fitz-Gibbon, N. W. Pino, & R. White (Eds.), *The Emerald handbook of crime, justice and sustainable development* (pp. 63–83). Emerald Publishing Limited.

Whelan, G. (2012). The political perspective of corporate social responsibility: A critical research agenda. *Business Ethics Quarterly, 22*(4), 709–737.

Wikantiyoso, R., & Suhartono, T. (2018). The role of CSR in the revitalization of urban open space for better sustainable urban development. *International Review for Spatial Planning and Sustainable Development, 6*(4), 5–20.

York, R., & Clark, B. (2006). Marxism, positivism, and scientific sociology: Social gravity and historicity. *The Sociological Quarterly, 47*(3), 425–450.

Chapter 11
Corporate Social Responsibility Reporting in India

Shelly

Abstract There has, in recent times, been a meteoric rise in comprehending Corporate Social Responsibility ubiquitously, and how it may generate value for the business. CSR is perceived as a mechanism through which a company balances economic, environmental, and social objectives. In the Indian context, CSR has covered a very long journey, right from being primarily an outcome of philanthropy in an ancient era to the time when, in 2014, it was mandated for specific Indian companies. The Companies Act, 2013 emphasises corporate philanthropy rather than strategic CSR for Indian companies. The mandated disclosures on Corporate Social Responsibility, foster transparency and accountability by the Indian Companies. These disclosures intend to make sure that companies fulfil their societal obligations thereby promoting sustainable development. This study attempts to examine the extent of corporate social responsibility reporting practices under a mandatory regime. For this purpose, Annual Reports of Top Indian Companies by Average Market Capitalization, for the FY23 are perused. The CSR Disclosure index for this study is prepared according to the activities set out in Schedule VII of the Companies Act, 2013. The paper uses the technique of Content Analysis to estimate the extent of disclosure by Indian companies. The findings of the paper reveal the mean CSR disclosure of 44% only, in spite of operating under the mandatory regime of CSR rulings. The CSR initiative of 'Eradicating hunger, poverty and malnutrition, promoting health care including preventive health' is the most disclosed area, followed by 'Promotion of education' and 'Enhancing vocation skills, especially among children, women, elderly, and the differently abled and livelihood enhancement projects'.

Keywords CSR · Schedule VII · Reporting · Content Analysis · India

---

Shelly (✉)
Department of Management, Global Group of Institutes, Amritsar, India
e-mail: shelly.aug@gmail.com

V. Kandpal et al. (eds.), *Developments in Corporate Governance*,
Approaches to Global Sustainability, Markets, and Governance,
https://doi.org/10.1007/978-981-96-6366-8_11

11.1 Introduction

The world is witness to many occasions where companies did a tremendous job in combating issues of health, education and environment by virtue of their CSR interventions. The most recent example of the COVID-19 pandemic bears testimony to how many companies rose above their financial imperatives and contributed substantially to the social well-being of communities where they are operating without any mandatory legislation. The contemporary approach of CSR is based upon the thought that since businesses operate in society, they have what it takes to make a significant contribution to goals and aspirations of those living in the society (The World Bank Group, 2004). No doubt, it is fast emerging as a mechanism through which a business balances its economic, environmental, and social objectives ('Triple-Bottom-Line-Approach"), and parallelly, it seeks to address the expectations of stakeholders (Singh, 2018; UNIDO, 2022). While slogging through strategic choices, businesses try to ascertain the impact of their strategic alternatives on people, the planet, and profit. Following this, CSR reporting is becoming an aspect by dint of which CSR performance data is now becoming a more and more important part of companies' annual reports (Vartiak, 2016). Transparency in communicating performance data fosters trust and enhances credibility among its stakeholders.

11.2 Evolution of CSR in India

11.2.1 # Philanthropic Phase of CSR

In the Indian context, the roots of social responsibility can be traced to ancient times, having its references found in the Vedas; for example, Atharva Veda preaches that one should get possession of wealth with a hundred hands but should distribute it with ten times more hands. In its inceptive form, CSR was merely considered as an outcome of philanthropy and was influenced by religion, culture, and family values. However, the approach to CSR changed under colonial rule. Under the influence of Western industrialization leading industrialists like Tatas, Birlas, and Godrej pioneered the concept of 'CSR motivated business', of course with commercial motive, in supporting efforts for nation-building. In a way, our nation's fight for independence marked the second phase of CSR evolution (Ghosh & Mukerjee, 2020).

The following stage was marked by the arrival of PSUs as the government was striving to provide socially just economic growth. With the public sector being perceived as the driving force behind nation-building, the private sector suffered a setback. However, the public sector could achieve only limited success, which shifted the pendulum of expectations from the public sector to the private sector for contributing actively to the socio-economic development of India (Sharma, 2009). The year 1991, which ushered in an era of LPG (Liberalisation, Privatisation, and Globalisation) in India, witnessed a big push for structural and policy reforms to survive the internal economic crisis and make India a global competitive force. Indian economy witnessed an increased momentum of growth under this wave. Since global economies were more focused on labour and environmental issues, Indian companies needed to show some level of sensitivity towards the same. As a result, they started integrating CSR into coherent and sustainable business strategies and, henceforth, showed more willingness to contribute towards social responsibility (Sharma, 2009).

11.2.2 # Legal Phase of CSR

The Ministry of Corporate Affairs, GOI issued the Voluntary Corporate Social Responsibility Guidelines in 2009, which articulated that CSR activities are purely voluntary in nature (MCA, 2009), followed by the National Voluntary Guidelines on Social, Environmental & Economic Responsibilities of Business in July 2011, which specifically stated that businesses need to be responsible in their conduct towards the society so that every effort of theirs lead to sustainable growth and economic development. These guidelines urged businesses to embrace a "triple bottom line" and laid down nine principles along with the core elements (to actualize principles) so that businesses sincerely foster inclusive economic growth (MCA, 2011).

The transition to a regulated reporting regime dawned in 2012, with the Securities & Exchange Board of India (SEBI) making it legally compulsory for the top 100 listed companies by Mkt Cap to file a Business Responsibility Report (BRR) along with their Annual Report. Later, this mandatory requirement was specified for the top 500 companies in 2015 and eventually for the top 1000 listed companies in 2019. The inception of new reporting requirements on ESG (Environment, Social, and Governance) parameters, known by the name of Business Responsibility & Sustainability Report (BRSR), took the place of the Business Responsibility Report. The purpose behind bringing this new report was to manifest linkages between the financial results of a firm with its environmental, social, and governance performance. The BRSR, applicable on the top 1000 listed companies by Mkt Cap, may be followed voluntarily for FY 22 and subsequently would be mandatory with effect from FY23.

The watershed moment in the evolution of CSR came about when Section 135 of the Companies Act, 2013 was enacted, and it legally mandated Corporate Social Responsibility for certain categories of companies. Thus, companies with a net worth of Rs. five hundred crores or more, OR with a turnover of Rs. one thousand crores or more, OR with a net profit of Rs. five crores or more, during the immediately preceding FY, will be under legal compulsion to spend at least 2% of their average net profits for the three immediately preceding financial years.

These companies will be required to constitute a CSR committee of three or more directors, one of whom at least shall be an independent director. The CSR Committee will perform the following functions, namely:

a. Formulating and recommending to the board for its approval, a CSR Policy elaborating the initiatives to be undertaken by the company as per activities enumerated in Schedule VII to the Companies Act, 2013. The activities mentioned therein range from basic life amenities to social and environmental concerns.

b. Recommending the amount of CSR expenditure (for subsequent approval by board).
c. Monitoring CSR policy from time to time.

The board's report shall include an Annual Report on CSR activities, giving information such as a brief outline of CSR policy, composition of the CSR committee, CSR amount spent or unspent for the financial year, and details of CSR projects, among other pertinent details. The board's report shall further state the reasons thereof if CSR funds remain unspent, and this unspent amount shall be transferred to a Fund specified in Schedule VII within six months from the completion of the financial year. However, if an unspent amount is pertinent to an ongoing project, it shall be transferred to a Special Bank Account opened with the name Unspent Corporate Social Responsibility Account in this regard. It is further articulated that the amount so transferred to a bank account would continue to be utilised for CSR obligations, spanning across a period of three financial years, or it would be transferred to a Fund specified in Schedule VII within thirty days after the third financial year ends. The breach of all these regulations would invite a penalty of an amount twice the funds required to be transferred (Fund specified in Schedule VII or the Unspent Corporate Social Responsibility Account, as per the circumstances) or Rs. one crore, whichever is less. Likewise, every defaulting officer of a company shall be penalised with one-tenth of the amount required to be so transferred or Rs. two lakhs, whichever is less (MCA, 2022).

The year 2020 turned out to be a cursed year, probably among the worst in the history of mankind, with the COVID-19 pandemic disrupting lives across all nations and negatively affecting global economic growth. The health infrastructure of many countries, including India, took a beating in combating this ferocious and deadly pandemic. In March 2020, GOI announced that any amount spent on creating health infrastructure for COVID-19 care would be tantamount to CSR expenditure (ET Online, 2020). In January 2021, the definition of Corporate Social Responsibility was revised, and a few activities were omitted from its scope, which is enumerated (MCA, 2021):

Political Contributions
#Activities performed in the normal course of business
Activities which benefit employees of the company only or fulfil any statutory requirement
Any activity performed in a country other than India unless it relates to coaching sports personnel who are either representing either at domestic level or on a global scale.
Activities sponsored by companies for marketing purposes.
An activity undertaken to fulfil any statutory obligation in India

On the basis of the above discussion, it can be said that, indeed, Corporate social responsibility in Indian context has had an eventful journey which started during ancient times. The socio-economic development of any country is no longer the responsibility of the administration, especially in a country like ours which tussles with many challenges like growing population, poverty, unemployment, inequality, lack of education, gender imparity, environmental degradation, and many others. Certain Indian companies were doing well for the betterment of our society but a need was felt to invoke accountability, transparency, mobilise resources for holistic development, and bring Indian corporate sector at par with global trends of sustainability. Thus, the government of India decided to mandate CSR for companies to attain the inclusive growth and foster the competitiveness of our country at the global level.

Against the backdrop of aforementioned discussion, the purpose of this study is to discern CSR disclosure patterns of the most valued companies in India which come within ambit of Section 135 to the Indian Companies Act, 2013. It moves with the following specific objectives:

1. Estimate the extent of CSR Disclosure made by the most valued companies in India.
2. Examine the theme-wise CSR Disclosure made by the most valued companies in India.

11.3 Literature Review

The roots of analysing CSR disclosure can be traced to 1970s, when Abbott and Monsen (1979), taking a sample of Fortune 500 Companies, studied CSR disclosure practices over the years 1971–75. Their study revealed that disclosure had increased to 85.7% in 1975, from 51.4% in 1971. Environment matters were found to be the most reported category in the annual reports. Likewise, Kelly (1981) reported an increasing trend of CSR disclosure in Australia, over a period of study 1969–78. However, he observed that large corporations disclosed more environmental, energy and products information than their smaller counterparts. Andrew et al. (1989) observed a disclosure of 26%, in developing nations of Malaysia and Singapore. Guthrie and Parker (1990) analysed the disclosure practices in Western countries. They found high disclosure of 98% in the U.S., followed by the UK with 85%, and Australia with 56%. Fukukawa and Moon (2004), in their study on top 50 Japanese firms by annual turnover, revealed that 90% companies disclosed their CSR initiatives. In Malaysia, Ahmada et al. (2003) reported a disclosure extent of 84% for top 98 companies by market capitalization. Cecil (2008) found an increasing trend in voluntary CSR reporting in the U.S., over a period 1991–2006. The study revealed that the number of corporates issuing CSR reports grew from 2 in 1991 to 154 in 2001, and that in the year 2006, this number increased to 230 companies. Bashtovaya (2014) compared CSR disclosures of top ten companies in the oil and gas industry in the U.S. and Russia. Their findings revealed that Russian companies reported more expansively about issues of employees and consumers than their American counterparts. However, American companies delved deep into environmental issues, a concern somewhat overlooked by Russian companies. Bhatia and Chander (2014), in studying disclosure of SENSEX companies in India for the year 2009–2010, reported that the company-wise mean disclosure is only 31% whereas the theme-wise mean disclosure is 40.32%. It was also reported that the category of Others (inclusive of CSR Intent; Awards & Recognition; Sustainability Report) was the most disclosed theme, with 'Environment' and 'Community Involvement' concerns at the second and third spots respectively. However, Maqbool and Zamir (2019) reported a mean disclosure of 66% by SENSEX Companies for the year 2016–2017. Their most reported themes were education, health, and energy conservation. Bhatia and Dhawan (2018) compared the CSRD extent of the voluntary era with that of the mandatory regime in India. They reported an average of 36.84% in voluntary times, and an average score of 47.62% under mandatory rulings. Nair et al. (2019) reported an average score of CSR disclosure of 49.6% for 369 firms listed on BSE, for the period 2014 to 2017. They further investigated if CSR disclosure could improve financial transparency under a mandatory regime in India. They suggested, on the basis of their results, that CSR disclosure improves financial transparency. They also stated that the retail ownership bolsters the relation of disclosure with financial transparency. Santoso et al. (2022), taking a sample of 261 listed companies in Indonesia, for the period 2013–2019, estimated an extent of CSRD of 15.3%. They also argued that just because companies disclose CSR practices, it should never be construed that the same companies would also be ethical in their business practices.

11.4 Database and Research Methodology

11.4.1 Sample of the Study

Top BT 20 companies by average market capitalization (market capitalization refers to stock price multiplied by the number of outstanding shares) were considered for this study in the first instance, out of which five Banking and Finance companies were excluded since there is lot of variation in the expectations of CSR from the companies in the financial sector as put forth by Maqbool and Zamir (2019). Besides, LIC was excluded since it does not fall under the purview of Section 135. In this way, the final sample turned out to be of 14 companies. The disclosure data was studied for the FY23. This year was chosen since it is representative of the most recent period for which the requisite data were available for all sample companies.

11.4.2 Data Source

The Annual Reports for the FY 2022–2023 have been perused for extracting relevant CSR data for analysis.

11.4.3 Technique of Content Analysis

The Annual Reports were examined as per the items set out in Schedule VII of the Companies Act, 2013, using the technique of Content Analysis. Under Content Analysis, scoring is done on a dichotomous basis, since dichotomy is the mere level of measure that may be attained for each category (Abbot & Monsen, 1979). The score of '1' is assigned if disclosure of attribute under analysis is made, otherwise score '0' is assigned.

11.4.4 Data Collection Instrument

The Corporate Social Responsibility Disclosure index for this study is prepared according to the activities enumerated under Schedule VII of the Companies Act, 2013. The activities are given under 12 clauses but for smooth and an in-depth analysis, these have been broken down into 18 statements. The codes are given to represent activities for the sake of convenience, since they are large statements. The maximum score that can be attained by any company is 18. Table 11.1 presents the Corporate Disclosure Index, used in the present study.

11.5 Results and Discussion

11.5.1 Company-Wise Disclosure

The results of company-wise disclosure are presented in Table 11.2, a perusal of which reveals the score range of 17% to 72%. The top scorers include Reliance Industries Ltd. (72%), HCL Technologies (72%), Larsen & Toubro (50%), and Adani Enterprises (50%). The companies at the bottom are Asian Paints (17%), Bharti Airtel (22%), and Maruti Suzuki Ltd. (28%). In other words, Reliance Industries Ltd. and HCL Technologies Ltd. stand together at rank 1. Even, rank 4 is shared by three companies, i.e., TCS, Larsen & Toubro, and Adani Enterprises, with a disclosure score of 50%, each. Likewise, three companies, Sun Pharma Ltd, Avenue SuperMarts, and Titan Company Ltd. share rank 9, with 33% disclosure, each. The overall average score of 14 companies is 44%. When compared to earlier studies in the Indian context, the average score is higher than that reported by Bhatia and Chander (2014), but lower than those reported by Bhatia and Dhawan (2018) and Maqbool and Zamir (2019). However, when the results of this study are compared with those of earlier studies, one fact which cannot be overlooked is that the present analysis of companies was done at that point of time when certain activities do not qualify now for CSR requirements of companies, namely, political contributions, activities done in normal course of business, activities which benefit company employees only or

Table 11.1 CSR disclosure index

Code	Disclosure items	Total score
A1	Eradicating hunger, poverty, and malnutrition, promoting health care including preventive health	1
A2	Improving sanitation facilities and safe drinking water	1
A3	Promotion of education	1
A4	Enhancing vocation skills, especially among children, women, elderly, and the differently abled and livelihood enhancement projects	1
A5	Empowering women; promoting gender equality; setting up homes and hostels for women and orphans	1
A6	Old age homes; day care centres; facilities for senior citizens	1
A7	Measures for reducing inequalities faced by marginalised groups	1
A8	Ensuring environmental sustainability and ecological balance; conservation of natural resources; maintaining the quality of soil, air and water	1
A9	Protecting flora and fauna; animal welfare; agro forestry	1
A10	Protection of national heritage, art, and culture including restoration of buildings and sites of historical importance and works of art; setting up public libraries; promotion and development of traditional arts and handicrafts	1
A11	Promoting the welfare of veterans, widows, and dependents of armed forces	1
A12	Training to promote rural sports, nationally recognised sports Paralympic sports and Olympic sports	1
A13	Contribution to the PM's National Relief Fund or PM's Citizen Assistance and Relief in Emergency Situations Fund (PM CARES Fund) or any other fund set up by the Central Government for socio-economic development and relief and welfare of the scheduled caste, tribes, other backward classes, minorities and women	1
A14	Contribution to incubators or research and development projects in the field of science, technology, engineering and medicine, funded by the Government at the central or state level or PSU	1
A15	Contributions to public-funded Universities; Indian Institute of Technology (IITs); National Laboratories and autonomous bodies established under Department of Atomic Energy (DAE); Department of Biotechnology (DBT); Department of Science and Technology (DST); Department of Pharmaceuticals; Ministry of Ayurveda, Yoga, and Naturopathy, Unani, Siddha and Homoeopathy (AYUSH); Ministry of Electronics and Information Technology, and other bodies, namely Defense Research and Development Organisation (DRDO); Indian Council of Agricultural Research (ICAR); Indian Council of Medical Research (ICMR), and Council of Scientific and Industrial Research (CSIR), engaged in conducting research in science, technology, engineering, and medicine aimed at promoting Sustainable Development Goals (SDGs)	1
A16	Rural development projects	1
A17	Slum area development	1
A18	Disaster management	1

Source Retrieved as such from MCA (2022)

fulfil any statutory obligation for that matter, and activities undertaken outside India unless related to coaching of sports personnel who are representing either at the domestic level or internationally. Further, the sample size is smaller for the present study than the compared studies, which may have contributed to the difference in the mean score of CSR disclosure.

11.5.2 Theme-Wise Disclosure

Table 11.3 shows the theme-wise disclosure of companies. The most reported CSR activity is A1, which represents 'Eradicating hunger, poverty and malnutrition, promoting health care including preventive health', with a score of 93%.

The second spot is shared by two initiatives, i.e., A3 and A4, which represent activities 'Promotion of education' and 'Enhancing vocation skills, especially among children, women, elderly, and the differently abled and livelihood enhancement projects', respectively. The score for these two is 86%, each. At the bottom, there is CSR intervention of A13 representing 'Contribution to the PM's National Relief Fund or PM's Citizen Assistance and Relief in Emergency Situations Fund (PM CARES Fund) or any other fund set up by the Central Government for socio economic development and welfare of the schedule castes, tribes, OBCs, minorities and women'. It has score of zero, as no company has disclosed this initiative. The second last spot is shared by three interventions of A11, A14, and A17, which represent 'Promoting welfare of veterans, widows and dependents of armed forces'; 'Contribution to incubators or research and development projects in the field of science, technology, engineering and medicine, funded by Government at central or state level or PSU' and 'Slum Area Development', respectively. The disclosure for all these three CSR interventions is 7%, each.

On the basis of the above results, it is heartening to observe that the top Indian companies have given preference to initiatives like eradication of poverty and malnutrition, promotion of education, and enhancing vocational skills among youth, women and specially-abled people, although the company-wise average disclosure level is 44% only. The top scorers include Reliance Industries Ltd. (72%), HCL Technologies (72%), Larsen & Toubro (50%), and Adani Enterprises (50%). The companies at the bottom are Asian Paints (17%), Bharti Airtel (22%), and Maruti Suzuki Ltd. (28%).

11.6 Findings and Conclusion

Corporate social responsibility is a concept that has been contemplated since time immemorial.

However, in the current scenario, it is fast emerging as a mechanism by virtue of which companies attain their economic, environmental, and social imperatives.

Table 11.2 Company-wise CSR disclosure

Companies	A1	A2	A3	A4	A5	A6	A7	A8	A9	A10	A11	A12	A13	A14	A15	A16	A17	A18	Reported score	CSRD %	Rank
RIL	✓	✓	✓	✓	✓	✓	✓	✓	✓	✓		✓				✓		✓	13	72	1
TCS	✓	✓	✓	✓	✓		✓	✓								✓		✓	9	50	4
HUL	✓		✓	✓	✓		✓	✓				✓				✓			8	44	7
INFOSYS	✓	✓	✓	✓	✓	✓	✓			✓	✓				✓	✓		✓	12	67	3
ITC	✓		✓	✓	✓		✓	✓	✓							✓			8	44	7
BHARTI AIRTEL			✓		✓		✓							✓					4	22	13
LARSEN & TOUBRO	✓	✓	✓	✓				✓		✓		✓			✓	✓			9	50	4
ADANI ENTERPRISE	✓	✓	✓	✓			✓	✓	✓			✓						✓	9	50	4
HCL TECHNOLOGIES LTD	✓	✓	✓	✓	✓		✓	✓	✓	✓		✓				✓	✓	✓	13	72	1
ASIAN PAINTS	✓	✓		✓															3	17	14
MARUTI SUZUKI LTD	✓		✓	✓												✓		✓	5	28	12
SUN PHARMA LTD	✓	✓	✓					✓								✓		✓	6	33	9
AVENUE SUPERMARTS	✓		✓	✓			✓	✓							✓				6	33	9
TITAN COMPANY LTD	✓	✓		✓			✓			✓						✓			6	33	9

Source Author's Calculation

Table 11.3 Theme-wise CSR Disclosure

Activity code	Total score	CSRD %
A1	13	93
A2	9	64
A3	12	86
A4	12	86
A5	7	50
A6	2	14
A7	10	71
A8	9	64
A9	4	29
A10	5	36
A11	1	7
A12	5	36
A13	0	0
A14	1	7
A15	3	21
A16	10	71
A17	1	7
A18	7	50

Source Author's Calculations

In the same vein, stakeholders want to see that their companies care sincerely for society and that they put in their whole-hearted efforts to foster an inclusive growth of the society that they are part of. In the Indian context, CSR has progressed a long way, right from being primarily an outcome of philanthropy in an ancient era to the time when in 2014, it was mandated for specific Indian companies. The present study examined the CSR disclosure practices of the most valued companies in India, under an era of legal requirement, and found an average disclosure level of 44%. This score may not be very appreciative since the sample comprised companies which have a high market capitalization and are among representatives of the Indian economy. On second thoughts, this score may be considered satisfactory given a very small sample size which is fourteen, in question. The results also reveal that the Indian companies have prioritised the eradication of poverty and malnutrition, the promotion of education, and strengthening vocational skills among youth, women and specially-abled people, which is a very promising sight to behold. However, on the flip side, we lag behind in terms of contribution to incubators or research and development projects in the field of science, technology, engineering, and medicine, which is not healthy for any society. In recent times, India, with a growing appetite for innovation, is remarkably transforming into a hub of startups. Now, incubators which help early-stage businesses in translating their ideas into viable products/services need financial support of corporates by means of CSR interventions. So,

Indian Corporates should strategically align with projects having the potential for national impact, and for this, a platform should be set up that may help identify such projects.

11.7 Implications of the Study

There are certain significant implications associated with the findings of the present study for the companies, authorities, and stakeholders with regard to undertaking CSR practices:

- The Indian Companies should align CSR policy with their strategic intent, which would strengthen their social credibility as well as enhance competitive advantage, at the same time.
- Linking executive compensation to CSR performance can raise the level of CSR innovation. A robust internal incentive system may spur more indulgence in R & D and thereby bring up more innovation.
- The companies, with turnover running into hundreds of crores may find penalty amounting to a few lakhs rupees, a peanut to them. Therefore, a penalty worth burning a hole in their pocket may serve the purpose.
- Even for the defaulting officers, disqualification from holding their present positions can act as a strong deterrent against violations of CSR norms.
- Companies may be recognised for their CSR contributions, be it through tax incentives or other facilities which can facilitate ease of doing business.
- A special independent audit team may be constituted by the concerned Ministry for enhanced scrutiny of CSR expenditures.
- Companies should actively engage stakeholders in their planning and implementation of CSR Initiatives, which would send out good signals to customers and investors, thereby leaving a good taste in their mouth.

References

Abbott, W. F., & Monsen, R. J. (1979). On the measurement of corporate social responsibility: Self-reported disclosures as a method of measuring corporate social involvement. *Academy of Management Journal, 22*(3), 501–515.

Ahmada, N. N. B. N., Sulaimanb, M. B., & Siswantoroc, D. (2003). Corporate social responsibility disclosure in Malaysia: An analysis of annual reports of KLSE listed companies. *IIUM Journal of Economics and Management, 11*(1), 1–37.

Andrew, B. H., Gul, F. A., Guthrie, J. E., & Teoh, H. Y. (1989). A note on corporate social disclosure practices in developing countries: The case of Malaysia and Singapore. *The British Accounting Review, 21*(4), 371–376.

Bashtovaya, V. (2014). CSR reporting in the United States and Russia. *Social Responsibility Journal, 10*(1), 68–84.

Bhatia, A., & Chander, S. (2014). Corporate social responsibility disclosure by SENSEX companies in India. *Management and Labour Studies, 39*(1), 1–17.

Bhatia, A., & Dhawan, A. (2018). Extent of corporate social responsibility reporting: India's transition from voluntary to mandatory regime. *International Journal of Sustainable Strategic Management, 6*(1), 73–99.

Cecil, L. (2008). Corporate social responsibility reporting in the United States. *McNair Scholars Research Journal, 1*(1), 43–52, Article 6. https://commons.emich.edu/mcnair/vol1/iss1/6

ET Online. (2020). What Covid-19 contributions get qualified as CSR expenditure. Here's the complete list. *Economic Times*. https://economictimes.indiatimes.com/news/economy/policy/what-covid-19-contributions-get-qualified-as-csr-expenditure-heres-the-complete-list/articleshow/75093908.cms?from=mdr

Fukukawa, K., & Moon, J. (2004). A Japanese model of corporate social responsibility? A study of online reporting. *Journal of Corporate Citizenship, 16*, 45–59.

Ghosh, S., & Mukerjee, S. (2020). The relevance of spirituality and corporate social responsibility in management education: Insights from classical Indian wisdom. *Philosophy of Management, 19*(4), 469–497.

Guthrie, J., & Parker, L. (1990). Corporate social disclosure practice: A comparative international analysis. *Advance in Public Interest Accounting, 3*(2), 159–175.

Kelly, G. J. (1981). Australian social responsibility disclosure: Some insights into contemporary measurement. *Accounting and Finance, 21*(2), 97–107.

Maqbool, S., & Zamir, M. N. (2019). Corporate social responsibility disclosure by SENSEX companies in India. *Management and Labour Studies*, 1–15.

MCA. (2009). *Voluntary Guidelines on Corporate Social Responsibility, 2009*. Ministry of Corporate Affairs, Government of India. https://www.mca.gov.in/Ministry/latestnews/CSR_Voluntary_Guidelines_24dec2009.pdf

MCA. (2011). *National Voluntary Guidelines on Social, Environmental and Economical Responsibilities of Business, 2011*. Ministry of Corporate Affairs, Government of India. https://www.mca.gov.in/Ministry/latestnews/National_Voluntary_Guidelines_2011_12jul2011.pdf

MCA. (2021). *Circular, Amendment in Rules to Companies Act 2013*. https://www.mca.gov.in/Ministry/pdf/CSRAmendmentRules_23012021.pdf

MCA. (2022). *Companies Act, 2013*. https://www.mca.gov.in/content/dam/mca/pdf/CompaniesAct2013.pdf

Nair, R., Muttakin, M., Khan, A., Subramaniam, N., & Somnath, V. S. (2019). Corporate social responsibility disclosure and financial transparency: Evidence from India. *Pacific-Basin Finance Journal, 56*, 330–351.

Santoso, E. B., Basuki, & Isnalita. (2022). The paradox of corporate social responsibility disclosure in developing countries: Evidence from Indonesia. *Academy of Strategic Management Journal, 21*(3), 1–18.

SEBI. (2012). *Circular on Business Responsibility Reports*. https://www.sebi.gov.in/sebi_data/attachdocs/1344915990072.pdf

SEBI. (2021a). *Circular on Business Responsibility and Sustainability Reporting by listed entities*. https://www.sebi.gov.in/media/press-releases/may-2021/sebi-issues-circular-on-business-responsibility-and-sustainability-reporting-by-listed-entities-_50097.html

SEBI. (2021b). *Meeting file on Business Responsibility and Sustainability Reporting by listed entities*. https://www.sebi.gov.in/sebi_data/meetingfiles/apr-2021/1619067265752_1.pdf

Sharma, S. G. (2009). Corporate social responsibility in India: An overview. *The International Lawyer, 43*(4), 1515–1533. https://scholar.smu.edu/cgi/viewcontent.cgi?article=1317&context=til

Singh, A. K. (2018). Corporate social responsibility. *Anveshna: Search for Knowledge, 8*(2), 2–20. https://doi.org/10.23872/aj%2F2018%2Fv8%2Fi2%2F180670

The World Bank Group. (2004). *Public sector roles in strengthening corporate social responsibility: Taking Stock*. https://documents1.worldbank.org/curated/en/548301468313740636/pdf/346560CSR1Taking1Stock.pdf

UNIDO. (2022). *What is CSR?* https://www.unido.org/our-focus-advancing-economic-competitiveness-competitive-trade-capacities-and-corporate-responsibility-corporate-social-responsibility-market-integration/what-csr

Vartiak, L. (2016). CSR reporting of companies on a global scale. *Procedia Economics and Finance, 39*, 176–183.

Chapter 12
CSR and Job Seekers' Organizational Attractiveness: Reframing Relationships Through the Lens of Signaling and Stakeholder Theories

Esrafil Ali and Biswajit Satpathy

Abstract Drawing on signaling and stakeholder theory, the aim of this paper is to investigate the impact of CSR on job seekers' organizational attractiveness, taking Western Odisha as a case. This study uses a qualitative approach to understand the CSR and HR professionals' perceptions about CSR. Purposive and convenient sampling were used to select both the sample frame. The interview notes were further analyzed using thematic and matrix analysis. The data was treated in the Nvivo software. The findings support the two concepts of signaling and stakeholder theory, as most respondents believe that the organizations that develop CSR competencies based on their internal and external CSR practices, are more likely to attract potential job seekers. This study combines the theoretical viewpoints of signaling and stakeholders theory to understand the perceptions of industry professionals about CSR and its impact on the organizational attractiveness of job seekers. Very limited studies have attempted to find such relationships. This study has a noble contribution as it empirically found out how CSR can be a strategy for attracting job seekers.

Keywords Corporate social responsibility · Signaling theory · Stakeholder theory · Job seekers' organizational attractiveness

12.1 Introduction

Today, organizations face many challenges in attracting and retaining qualified job seekers for business operations. At the same time, it is also observed that organizations are implementing various innovative strategies to cope with this problem. CSR

E. Ali (✉)
School of Business Management, SVKM's NMIMS (Deemed to Be University), Indore, India
e-mail: esrafil.ali@nmims.edu

B. Satpathy
Department of Business Administration, Sambalpur University, Burla, Odisha, India

V. Kandpal et al. (eds.), *Developments in Corporate Governance*,
Approaches to Global Sustainability, Markets, and Governance,
https://doi.org/10.1007/978-981-96-6366-8_12

has been used as an intervention to create organizational attractiveness resulting in the attraction of prospective employees to organizations. Therefore, CSR of an organization is not restricted to social activities but has now aligned with business strategies. Today, organizations are using CSR as an intervention to attract and retain the best workforce. Various studies have shown how companies develop their HR strategies to recruit the best employees. The individuals prefer to work in organizations with a socially valued culture. One of the surveys conducted by McKinsey (2009) revealed that the qualified workforce would be the inevitable resource for organizations in the next two decades. The study also showed that "ability" was the only factor promoting sustainable business, innovation, research, and development. This results in sustainable business practices. The corporate leaders' concern is to attract these talents to organizations for increased productivity and growing business. CSR empowers an organization to be competitive in attracting and retaining the best talent for business practices.

In today's competitive world, CSR is one of the crucial factors which receives more attention. Previous literature on the connotation of CSR and job seekers' attitudes focuses on various ways to understand their relationships. For example, Kumar et al. (2021) viewed a significant association between CSR and the branding of organizations. Employer Brand, Perceived Benefits, and Perceived Career Development significantly and positively influenced potential candidates' intention to apply to organizations (Gandasari et al., 2024). Further, CSR also ensured job seekers' intention to stay in an organization longer. The study by Agnihotri and Bhattacharya (2022) examined that CSR fit influenced the organizational attractiveness among job applicants. Likewise, Oliveira et al. (2022) emphasized a positive relationship and direct effect between corporate volunteering and employer attractiveness. The job applicant's volunteer experience had a positive influence on CSR. Job applicants' self-motives were crucial in attracting top socially responsible organizations (Choi et al., 2020). The internal and external CSR positively impact job seekers' perception of employer prosocial orientation (Song et al., 2024).

In a related study, Choi et al. (2021) proposed that CSR increased the motivation of job applicants to apply for the job. It was also established that the job seekers' choice was influenced by their calling on applications for the job. Moreover, Ronda et al. (2020) studied that salary and ethics of an organization were influencing factors for job choice decisions. Further, it was also revealed that the preference of job applicants in choosing a job had influenced the level of organizational attractiveness. Similarly, Ouyang et al. (2021) showed that the negative publicity of CSR influenced the job pursuit negatively for employment via organizational attractiveness. On the contrary, George et al. (2020) argued that the CSR dimensions indirectly affected corporate reputation, trust, and identification, influencing CSR behavior. Accordingly, Junior et al. (2020) identified that philanthropic, ethical, legal, and economic was the order of the highest importance. Eger et al. (2019) found that the business students valued social factors in attractiveness and employer branding.

Presently, India's private and public sector organizations are facing various challenges in promoting CSR as an intervention to attract and retain qualified job seekers, where getting employment is a mammoth task. Though these organizations are taking active steps to redefine and foster a strong business-society relationship to gain competitive advantages, they fail to get a qualified workforce in the long run. Due to the Companies Act 2013, organizations are becoming aware of the CSR mandate and spending as per the mandate rules and regulations. Therefore, it is paramount for organizations to identify CSR strategies and action plans to continue finding potential job seekers. This will enable organizations to develop CSR action plans and programs to create an effective tool for attracting job seekers. This timely study attempts to understand on what CSR magnitudes are valued most by job seekers, ensuring competitive advantage. Specifically, the book chapter's primary objective is to understand the perceptions among CSR and HR professionals toward the influence of CSR in attracting job seekers and to draw out some key underlying issues for future CSR agenda. The study employs the perspectives of signaling theory and stakeholder theory as a theoretical foundation.

12.2 Literature Review

12.2.1 CSR and Job Seekers' Organizational Attractiveness

In a study, Turker (2009) found that CSR and organizational commitments were highly correlated. According to Behrend et al. (2009), posting environmental messages on an organization's website attracted job seekers to apply for jobs. It was also contended that organizational reputations were built if the organization published positive news about the society and environment. On the contrary, Das (2009) argued that CSR was not only meant for philanthropic activities. It was also focusing the business interest of an organization. According to Gomes and Neves (2010), the individual context factors were important in organizational attractiveness. The study also revealed that prior job experience and advertisement significantly impacted organizational attractiveness.

Further, Strobel et al. (2010) found that individual-level ethical behavior could significantly impact the organizational attractiveness of individuals. Additionally, the organization's CSR and social development investments helped build its corporate reputation and people's image. According to Du et al. (2010), CSR communication was critical for an organization in their research. It had significant effects on stakeholders. For Ali et al. (2010), there was a significant association between CSR and employee commitment. On a related note, Sharp and Zaidman (2010) conducted a study on using CSR to build a reputation strategically. It was reported that CSR was a substantive strategic activity for an organization. Likewise, Bakanauskiene et al. (2011) researched and reported that the best workplace was an essential attribute in the employer branding and attraction toward the job. Moreover, Dilling (2011) studied the stakeholders' perception of CSR and found that the age of an organization, reporting and publication of CSR, cultural diversity, and community people's involvement significantly impacted the perception of CSR of stakeholders. The finding also showed that the publication of CSR facts and figures and annual sustainability reports were very influential in organizational attraction and brand building. In their study, Irena et al. (2011), they revealed that the factor "best place to work" contributed to a deeper understanding of the employer images. Yin and Zhang (2012) undertook a study about the institutional dynamics and CSR from China's perspective. The study claimed that CSR in China was still in preliminary stages and yet to be explored. CSR in China focused more on ethical aspects.

The work environment that supports innovation, sustainable career development, and competitive compensation policies are the main factors in talent management (Kusuma & Sentosa, 2024).

Remišová and Búciová (2012) attempted to measure the CSR toward employees. The findings stated that the corporations that implemented CSR seriously were called responsible organizations rather than those that lacked clarity on the issue. Angelo et al. (2012) conducted a study on strategic CSR based on a Brazilian Case study. They found that aligning HR dimensions was inevitable to implement CSR strategically. According to Glavas and Godwin (2013), employees' perceptions of an organization's CSR were more important than the organization's reality. Perceived CSR helped identify the organizational identification and brand. In a study, Kaur and Dubey (2014) found a significant impact on job seekers' organizational attractiveness at the initial stage. It was also proposed that the effects of social media or social networking were very high on employee perceptions in terms of organizational attractiveness. Also, another study by Farooq et al. (2014), wherein the relationship between corporate trust, organizational identification, and affective organizational commitments were examined, concluded that CSR perceptions were strong indicators of affective organizational obligations. Zweers (2015) contended that the pro-environmental CSR information in recruitment and job seekers' organizational attractiveness had a significant positive relationship. For Santhosh and Baral (2015), leadership and top management support were crucial and vital in effective CSR initiatives for Indian firms. According to Sohn et al. (2015), the information about CSR was of great importance while they went for employment.

Furthermore, Rahman et al. (2016) investigated the connection between job satisfaction and OC in CSR. The study revealed that there was a positive association between these variables. In their study, Marques-Mendes and Santos (2022) found that CSR maturity involved obedience, negation, efficiency, strategic innovations, integration, and transformation. In the same line of thought, Famiyeh et al. (2016) contended that CSR activities had a significant positive relationship with organizational reputations. According to Gaudencio et al. (2017), the perception of CSR among employees helped to predict the attitude and behavior through trust and belief. Likewise, Sommer et al. (2017) found that organizations with more innovative products and cultures were the critical influencers for employees' attractiveness. Where Bohlmann et al. (2018) proposed that CSR's economic, social, and environmental performance positively affected its attractiveness. Held and Bader (2018) researched the impact of images on organizational attractiveness. The finding indicated that individuals were more attracted and inclined toward developed countries' organizations than developing countries.

Stefańska (2018) empirically investigated the relationships between organizations and the attitude of employees in terms of CSR. The paper examined that the organization's perceptions by employees were conditional based not on what actions were taken by organizations but rather on what initiatives organizations took in implementing CSR awareness and promotion. For Kim and Han (2019), CSR perceptions significantly influenced organizational compliance. Research by Zhang et al., (2019) revealed that corporate reputations substantially influenced CSR and customer satisfaction relationships. Likewise, Carlini et al. (2019) demonstrated a clear connection between CSR reputation and brand image. The study revealed that CSR provided internal and external signaling to the organization's brand image. CSR information and communication were also crucial in building their reputations.

Salanţă et al. (2019) argued that the responsibilities to employees, the customer, and the environment had a weaker impact on the overall CSR strategy. Garg and Gupta (2020) studied how CSR expenditure and firms' performance helped build the CSR reputation of the organization. The study claimed that firms who complied with mandatory CSR expenditures experienced low performance in social issues areas. Also, Caputo (2021) studied the CSR approach toward market expectations and how it helped in brand building. The research revealed that CSR had become an organization's attention to promoting strategies. It was also based on the perceptions of actors that influence a company's CSR. For Mahmud et al. (2023), perceived CSR on societal behavior positively influenced and impacted. The factor also had an association with CSR engagement and CSR positivity. Lee (2021) noted that the internal CSR of an organization had a positive impact on collaborative employee relationships. It was also reported that it increased organizational citizenship behavior. In a research, Huang (2022) noticed that person-job fit feedback had a more significant influence on organizational attractiveness. Individuals receiving high fit feedback were found to be more attracted to organizations. CSR promotes ethical and sustainable business practices that benefit not only the company but also its stakeholders and the environment (Mg, 2024).

The foregoing discussion demonstrates that the relationship between CSR and JSOA has been studied from several perspectives in many countries. However, a significant gap in existing conceptualizations is the absence of studies in Indian cultural settings. All considerable investigations have been done outside India to the best of our knowledge and developing countries like India are not much explored in these terms. As private business organizations in India face many challenges in attracting and retaining potential job seekers, this research on CSR and the JSOA relationship would provide valuable insights on how to strategically implement CSR in attracting job seekers in India. Therefore, the study aims to address this acknowledged gap by answering the following research questions:

Research Questions (RQ) for CSR Professionals:

1. Does your organization consider CSR as a strategic management policy? If yes, how?
2. How and where your organization is strategizing CSR as a competitive advantage? Please state a few examples from your organization.

3. How does CSR as a competitive advantage attract job seekers to your organization? Please elaborate on this.
4. How does the Internal CSR (such as Employee development, job satisfaction, work safety, welfare, etc.) and External CSR (such as local community development, protecting natural environment, consumer satisfaction) influence job seekers while applying for a job? Please state a few occurrences of your company.

Research Questions (RQ) for HR Professionals:

5. According to you, what is the importance of CSR in recruitment?
6. Do you believe that CSR information should be included while recruitment advertising of a company? If yes. Why?
7. While recruitment, how does your company communicate CSR externally toward prospective employees for a job?
8. Do you believe that CSR can be an important factor in job choice for job seekers? If yes, why?
9. How does the Internal CSR (such as Employee development, job satisfaction, work safety, welfare, etc.) and External CSR (such as local community development, protecting the natural environment, consumer satisfaction) influence job seekers while applying for a job? Please state a few occurrences of your company.

12.2.2 Signaling Theory

Spence (1973) proposed that job seekers need to observe the organization's CSR initiatives and activities to get quality information and sustainable business functions. In this study, the signaler's (organizations) voluntary CSR actions may signal (CSR practices) positive attributes of organizations to the receivers (job seekers). Figure 12.1 captures the perspectives of the theory for this study.

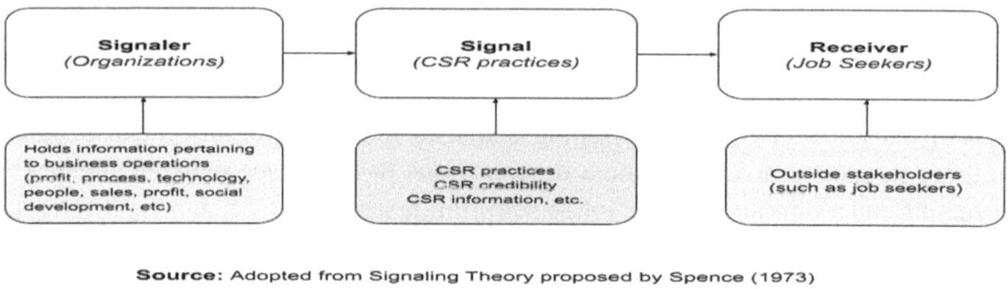

Fig. 12.1 Perspectives of signaling theory in this study. *Source* Author's own figure. Adopted from Signaling Theory proposed by Spence (1973)

12.2.3 Stakeholder Theory

Freeman (2010) proposed the stakeholder theory. Following the deployment of this theory, the study explores making a case for the stakeholder approach to organizational attractiveness, interpreting the perceptions of organizations' CSR and HR professionals. Figure 12.2 shows the perspectives of this theory. Based on the conception of the above two theories, Fig. 12.3 demonstrates the proposed framework of the study.

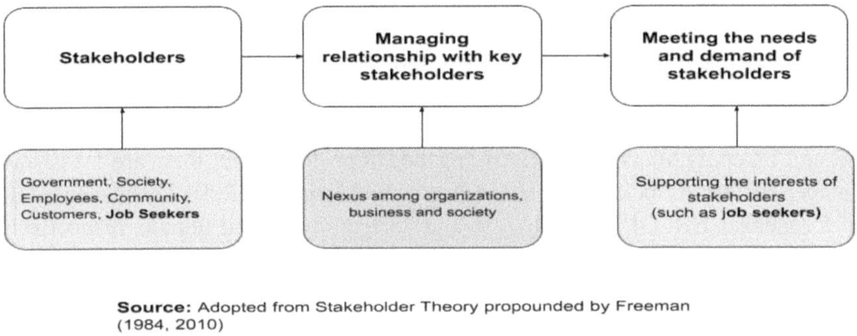

Fig. 12.2 Perspectives of stakeholder theory in this study. *Source* Author's own figure. Adopted from Stakeholder Theory propounded by Freeman (2010)

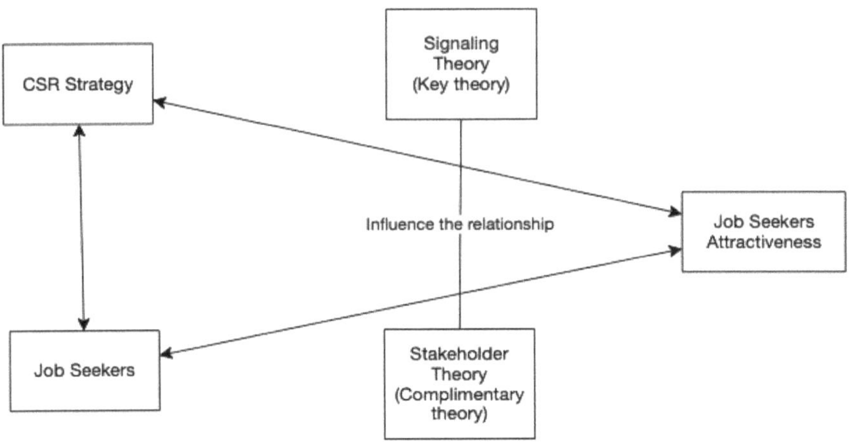

Fig. 12.3 Research framework of the study. *Source* Author's own figure

12.3 Method

The study adopts a qualitative approach to understanding the viewpoints of CSR and HR professionals about CSR and its importance in JSOA. The sample for the study is obtained from the CSR and HR professionals of 6 selected companies in Western Odisha. Applying purposive sampling, only *private limited companies* of Western Odisha who are strongly committed to CSR activities and give due importance to the organization's CSR reputation are selected. They are ACC Limited, Hindalco, JSW Steel, L&T, TRL, and Vedanta. The strategy to choose only private companies is to reach only those populations that match the research objectives' requirements and the study's needs. A total of 6 CSR professionals and 15 HR professionals are selected and interviewed for the study responses using convenient sampling. The selected CSR and HR professionals are from different levels and positions. Of the CSR professionals, there are 01 Head CSR, 2 Assistant Manager CSR, 2 Manager CSR and 1 Deputy Manager CSR. Among the HR professionals, there are 4 Assistant Manager HR, 01 Associate GM HR, 4 Associate Manager HR, 01 Dy Manager HR, and 5 Manager HR. Of the total, 72% and 28% are male and female, respectively. Also, those professionals with at least two years of work experience in the sector are selected. They are experienced and familiar with industry CSR/HR practices and deployed as key officers in their organizations. The summary of the respondent's profile is shown in Table 12.1.

Being a qualitative study, semi-structured interviews are conducted to ascertain the responses. The first part centers on the open-ended questions for CSR professionals to map the research objectives. The second part concentrates on the interview questions (open-ended) to address the "why" and "how" of CSR aspects by HR Professionals. The interview question's reliability, validity, and authenticity are duly verified and tested using expert opinion and feedback from academicians engaged in the same/allied field of study. To give more accessibility and flexibility to the interview process, they adopt various modes for the convenience of the respondents. Therefore, 5 telephonic, 6 face-to-face, and 10 online interviews (zoom/google meet) are conducted. The note of the interviews is taken for record and analysis.

The interview responses are collected and transcribed into text data. The data are then analyzed via a thematic and matrix analysis approach (Fig. 12.4 captures the steps followed in thematic analysis). The data are treated in Nvivo software for coding and analysis purposes.

Table 12.1 Respondents' profile of the study

S. No.	Respondents code	Company	Designation	Department	Experience in years	Gender
1	CSRP01	ACC Limited	Head CSR	CSR	18	Male
2	CSRP02	Hindalco	Assistant Manager CSR	CSR	9	Male
3	CSRP03	JSW Steel	Manager CSR	CSR	6	Female
4	CSRP04	L&T	Assistant Manager CSR	CSR	5	Male
5	CSRP05	TRL	Manager CSR	CSR	9	Male
6	CSRP06	Vedanta Limited	Deputy Manager CSR	CSR	12	Female
7	HRP01	ACC Ltd	Assistant Manager HR	HRD	8	Male
8	HRP02	ACC Ltd	Assistant Manager HR	HR	8	Male
9	HRP03	Hindalco	Assistant Manager HR	L&D	9	Male
10	HRP04	Hindalco	Manager HR	HR	7	Male
11	HRP05	JSW Steel	Assistant Manager HR	HR	9	Male
12	HRP06	JSW Steel	Deputy Manager HRD	HRD	4	Female
13	HRP07	JSW Steel	Manager HR	HRD	2.3	Male
14	HRP08	L&T	Associate Manager HR	HR	9	Female
15	HRP09	L&T	Assistant Manager HR	HR	2.5	Male
16	HRP10	L&T	Associate Manager HR	HR	4	Male
17	HRP11	TRL	Manager HR	HRD	2.5	Male
18	HRP12	TRL	Manager HR	HR	9	Male
19	HRP13	Vedanta limited	Assistant General Manager HR	HR	8	Male
20	HRP14	Vedanta limited	Associate Manager HR	HR	15	Female
21	HRP15	Vedanta limited	Associate Manager HR	HR	2	Female

CSRP = Corporate Social Responsibility Professional, HRP = Human Resource Professional
Source Research Data

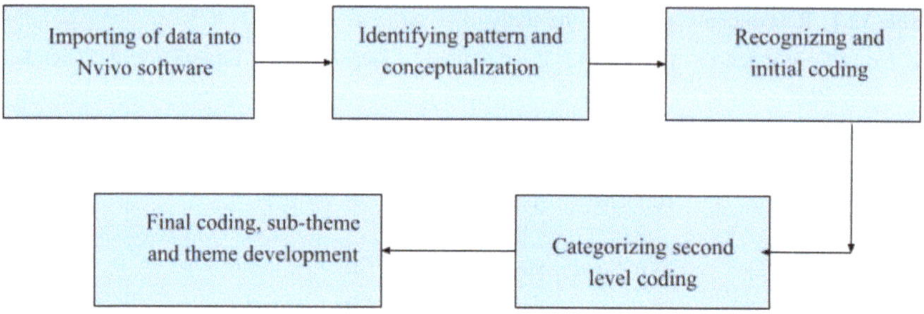

Fig. 12.4 Process involved in thematic analysis for the study. *Source* Author's own figure

12.4 Results and Discussion

12.4.1 Thematic Analysis—Perspectives of CSR Professionals

CSR and Job Seekers' Organizational Attractiveness*: Perceived by the CSR Professionals*

A total of four themes emerged through thematic analysis and further contended to interpret the participants' responses regarding CSR: applying CSR as a competitive advantage, inclusion of CSR in management policy, role of internal and external CSR in attracting job seekers, and ways *to attract job seekers through CSR*. Table 12.2 shows the theme and sub-theme developed from survey questions and responses recorded.

RQ1. Does your organization consider CSR as a strategic management policy? If yes, how?

Theme 1: Inclusion of CSR in Management Policy

It is widely established that today CSR is not only a philanthropic activity but a part of corporate and business strategy. The following responses validate these views:

> "CSR is an integral part of our overall organization development, and it's growth story" (CSRP01)
> "It is strategized at the highest level by involving the ManCom members" (CSRP01)
> "Yes. CSR has been included as an important component in the business strategy of the company" (CSRP03)
> "We use CSR in recruitment, brand building, promotions, sustainable business environment" (CSRP03)

Table 12.2 Theme and sub-theme developed from research questions

No	Focus	Survey questions	Theme	Sub-theme
1	CSR	Does your organization consider CSR as a strategic management policy? If yes, how?	• Inclusion of CSR in the management Policy • Applying CSR as a competitive advantage	• Brand building • CSR is important for policy, Role of external CSR • Role of internal CSR • Building positive image through CSR activities
2	CSR	How and where your organization is strategizing CSR as a competitive advantage? Please state a few examples from your organization		
3	CSR	How does CSR as a competitive advantage attract job seekers to your organization? Please elaborate on this	• Ways to attract job seekers through CSR • Role of internal and external CSR in attracting job seekers	
4	CSR	How does the Internal CSR (such as Employee development, job satisfaction, work safety, welfare etc.) and External CSR (such as local community development, protecting natural environment, consumer satisfaction) influence job seekers while applying for a job? Please state a few occurrences of your company		
5	HR	According to you, what is the importance of CSR in recruitment?	• Strategic use of CSR for recruitment • Relevance of CSR information in recruitment process	• CSR information requirement • Various methods used for CSR Information • CSR Information not Relevant • CSR Information Relevant • Building CSR Reputation • Various CSR Initiatives
6	HR	Do you believe that CSR information should be included while recruitment advertising of a company? If yes. Why?		
7	HR	While recruitment, how does your company communicate CSR externally toward prospective employees for a job?	• Methods of CSR communication during recruitment • CSR as an important factor for job seekers	• Job seekers give importance to brand • Vivid perspectives of job seekers toward CSR • External CSR Credibility • Internal CSR Credibility
8	HR	Do you believe that CSR can be an important factor of job choice for job seekers? If yes, why?		
9	HR	How does the Internal CSR (such as Employee development, job satisfaction, work safety, welfare etc.) and External CSR (such as local community development, protecting natural environment, consumer satisfaction) influence job seekers while applying for a job? Please state a few occurrences of your company	• Influence of internal and external CSR on job seekers	

Source Research Data

RQ2. How and where your organization is strategizing CSR as a competitive advantage? Please state a few examples from your organization.

Theme 2: Applying CSR as a Competitive Advantage

Most professionals believe that CSR practices of organizations help brand building and use strategically to gain a competitive advantage in the market. They provided several aspects of how CSR is used as a competitive advantage. For example, CSRP03 points out, CSR is always a positive message for stakeholders and helps improve the reputation and credibility of organizations:

> "We used to deploy CSR benefits for brand building and social development" (CSRP03)
> "Hence, it's always an attractive way for improving the brand value/image of the organization to attract best talent" (CSRP03)
>
> Similarly, CSRP04 sees CSR as a benefit to stakeholders:
>
> "CSR helps fulfil stakeholders' demands and build images for the company" (CSRP04)

RQ3. How does CSR as a competitive advantage attract job seekers to your organization? Please elaborate on this.

Theme 3: Ways to Attract Job Seekers Through CSR

When the respondents are asked about the various methods to use CSR for attracting job seekers, they described multiple ways. It is pertinent to note that organizations focus more on their internal and external CSR to leverage the benefits and influence job seekers' perceptions of the job. For instance:

> "Committed to take challenging tasks, are always given an opportunity" (CSRP01)
> "Internal CSR in our company benefits employees, which gives a positive image in the eyes of applicants who apply for the job" (CSRP02)

RQ4. How does the Internal CSR (such as employee development, job satisfaction, work safety, welfare, etc.) and External CSR (such as local community development, protecting natural environment, consumer satisfaction) influence job seekers while applying for a job? Please state a few occurrences of your company.

Theme 4: Role of Internal and External CSR in Attracting Job Seekers

It is believed that the organizations engaged in socially responsible activities for external stakeholders result in a high CSR reputation and brand building. Further, the organizations' internal and external CSR practices influence job seekers' perceptions of their attractiveness as an employer. In this context, when the respondents are questioned as to how the internal and external CSR affects the perception of job seekers while they choose an organization, there are varied responses; some of them are:

> "External CSR—making a balance with the local society always leads to growth of organization. There are many organizations who has to close their work due to local disturbance" (CSRP05)
> "Good CSR practices in our company attract talents to join our company. Many employees have praised our CSR practices during recruitment" (CSRP06)

12.4.2 Matrix Analysis—Perspectives of CSR Professionals

Tables 12.3 and 12.4 present the matrix analysis of the responses recorded from interviews with CSR professionals. The respondents are asked to share their perceptions about CSR on:

- CSR as strategic factor
- CSR as competitive advantages
- CSR as a factor to attract the job seekers
- Role of internal and external CSR in influencing job seekers

Table 12.3 Matrix analysis: theme and sub-theme identified for research questions and responses received from CSR professionals

Theme and sub-theme	Brand building	CSR is important for policy	Role of External CSR	Role of Internal CSR	Building positive image	Through CSR activities
Applying CSR as a Competitive Advantages	6	0	0	0	0	0
Inclusion of CSR in Management Policy	0	6	0	0	0	0
Role of Internal and External CSR in attracting job seekers	0	0	10	5	0	0
Ways to attract job seekers through CSR	0	0	0	0	5	2
Total	6	6	10	5	5	2

Source Research Data

Table 12.4 Matrix analysis—sub-theme and theme identified from CSR professional per interview responses

No	Respondents	Applying CSR as a competitive advantages	Inclusion of CSR in management policy	Role of internal and external CSR in attracting job seekers	Ways to attract job seekers through CSR	Total
1	CSRP01	0	2	1	3	6
2	CSRP02	0	0	2	2	4
3	CSRP03	2	2	3	0	7
4	CSRP04	1	1	1	2	5
5	CSRP05	1	1	4	0	6
6	CSRP06	2	0	4	0	6
	Total	6	6	15	7	34

Source Research Data

The matrix analysis indicates that the "*Role of internal and external CSR in attracting job seekers*" is the most discussed (being shown as 15) theme. All the respondents interviewed comprehend that internal and external CSR activities are pervasive characteristics positively related to job seekers' organizational attractiveness. Research evidence also shows that an organization's internal and external CSR positively affects the job seekers (Carlini et al., 2019; Lee, 2021; Ouakouak & Ouedraogo, 2019). The second most talked-about theme is "*Ways to attract job seekers through CSR.*" The majority of the respondents agree that CSR has a link to the competitive advantage, which helps to gain the best and most qualified talents in organizations. In line with the agreement by a few researchers (Gomes & Neves, 2010), most of the respondents' internal and external CSR reporting of an organization is inevitable and sufficient to promote CSR image and attract the job seekers.

12.4.3 Thematic Analysis—Perspectives of HR Professionals

CSR and Job Seekers' Organizational Attractiveness: *Perceived by the HR Professionals*

RQ5. According to you, what is the importance of CSR in recruitment?

Theme 1: Strategic Use of CSR for Recruitment

Magbool et al. (2016) indicated that those organizations that practice a high level of CSR could attract more talented employees. In the same line, Presley et al. (2018) reported that CSR was essential in attracting potential employees. Wang (2013) examined that Corporate Reputation, CSR, and Job seekers' perceptions were closely related factors. According to Turban and Greening (1996), firms having higher CSP had a more positive reputation. Albinger and Freeman (2000) conducted studies and found that corporate social performance was positively associated with organizational attractiveness for job seekers.

From the respondent's perspective, CSR is an essential factor in the recruitment process of an organization. Here are some excerpts that support the statement:

> "It helps bring understanding of the organisation amongst potential job seekers" (IIRP06)
> "As he is going to take a job in that place, awareness of CSR of concern areas" (HRP07)

RQ6. Do you believe that CSR information should be included while recruitment advertising of a company? If yes. Why?

Theme 2: Relevance of CSR Information in Recruitment Process

Job seekers need CSR information about an organization to make rational decisions. It is pertinent to note that the message embedded in an organization's recruitment advertisement increases the applicant's willingness to pursue employment. The majority of the respondents believe that CSR information affects the job applicants' following job choice intention. The respondent's statement that validates the above views are:

> "It creates faith and sense of respect among candidates" (HRP01)
> "Yes, it can attract the individual with the audiology mix with CSR thoughts" (HRP05)

RQ7. While recruitment, how does your company communicate CSR externally towards prospective employees for a job?

Theme 3: Methods of CSR Communication During Recruitment

As revealed in the literature about CSR, it is observed that organizations use various traditional and contemporary methods of recruitment messages that affect the potential applicants applying. Today, the internet provides a common medium for CSR communication. From the interview analysis, the respondents have stated different methods of CSR communication that are adopted in their organizations during the recruitment process.

RQ8. Do you believe that CSR can be an important factor in job choice for job seekers? If yes, Why?

Theme 4: CSR as an Important Factor for Job Seekers

When asked whether CSR is an essential factor for job choice decisions by job seekers while applying for jobs, nearly all the respondents agree that CSR plays a vital role in job choice. They consider CSR an influential factor for job seekers when they make decisions. This can be inferred from the statements made by the respondents, such as:

> "Yes, I strongly believe CSR can be an important factor in choosing a job" (HRP05)
> "Yes, it can be an important factor as it helps create differentiation in the market/competitors by brand value that it brings" (HRP06)

RQ9. How does the Internal CSR (such as Employee development, job satisfaction, work safety, welfare, etc.) and External CSR (such as local community development, protecting natural environment, consumer satisfaction) influence job seekers while applying for a job? Please state a few occurrences of your company.

Theme 5: Influence of Internal and External CSR on Job Seekers

Past studies have emphasized that the organization's internal and external CSR positively influences job seekers. The respondents are asked about the various methods the organizations adopt to affect the job seekers using internal and external CSR messages. Each respondent has a different answer about the same. For example, *free education, employee engagement, women empowerment, job satisfaction, infrastructure support,* etc. As remarked by HRP07:

> "Our CSR activities are very strong; we offer free education to students having merits. We are giving infrastructure support, sanitation support, etc., to different localities" (HRP07)
>
> HRP08 points out that:
>
> "Yes, it does. It motivates job seekers to join in hands with the good cause" (HRP08)
>
> For HRP11:
>
> "Internal and external CSR activities of any company signal positive image to job seekers and help build CSR reputations" (HRP11)

12.4.4 Matrix Analysis—Perspectives of HR Professionals

The matrix analysis for HR professional responses is exhibited in Tables 12.5 and 12.6. The matrix analysis observes that "*Influence of internal and external CSR on job seekers*" is the most significant and influencing theme out of all the identified themes (23 in the matrix table). Additionally, the themes "*CSR as an important factor for job seekers,*" "*Methods of CSR communication during recruitment,*" and "*Strategic use of CSR for recruitment*" are also found to be relevant in terms of explaining the research objectives and purpose of the study. The matrix analysis provides insights into how the respondents have discussed the various themes. Overall, the theme

"*Influence of internal and external CSR on job seekers*" increases the attention of HR professionals. The study findings are consistent with similar research about CSR and job seekers (Hameed et al., 2016; Jia et al., 2019; Low et al., 2017). The analysis demonstrates a more significant influence of internal and external CSR in improving the perceived reputation of the organization among job seekers. The underlying reason for this could be the contribution of effective CSR communication to job seekers. In essence, the job seekers may perceive that if an organization can take care of the demand and interests of its internal and external stakeholders, then it is reputable, worthy, prestigious, and can facilitate better treatment for its potential employees.

12.5 Conclusion

The paper has two conclusive remarks. *First*, the interviews conducted in the study reveal that most CSR professionals believe CSR is being widely used as a business strategy to profit from society. They further comprehend that CSR reputation and brand image are always leveraged to attract the best talent in organizations. Therefore, the study results are consistent with the earlier findings which stress that CSR provides a competitive advantage to organizations (Fombrun & Shanley, 1990). In a similar line, the majority of respondents are of the view that organizations may develop competitive advantages by encouraging CSR practices. Drawing on stakeholders' theory (Freeman, 2010), it is perceived that organizations who contribute more to stakeholders' benefits directly affect the social image of that organization and help them gain competitive advantages (Greening & Gray, 1994; Wood, 1991). Consistent with the above perspectives that CSR plays a crucial role in brand building, the findings from this study also reveal that organizations' CSR influences job seekers for attraction. Organizations adopt various methods for attracting job seekers. From the respondents' perception, it is noted that the techniques such as equal employment opportunities, extending welfare to the community, involvement in society, CSR awareness campaigns, building brand image/value, etc., are practiced in their organizations. In addition to the above findings, the study also makes important observations of internal and external CSR and their influence on job seekers' decisions. Organizations that develop CSR competencies based on their internal and external CSR practices are more likely to attract potential job seekers. This validates other research findings also (Hameed et al., 2016; Jia et al., 2019).

Second, the study is also carried out to understand the views and perceptions of HR professionals about CSR and JSOA. From the thematic analysis, it is found that respondents believe CSR is an important factor in the recruitment process of an organization. Many Indian organizations rope in the CSR's social, philanthropic, and environmental initiatives in their recruitment advertisement. Earlier research has also demonstrated the significant effect of organizational CSR on job seekers' attraction (Turban & Greening, 1996). The other important finding shows that the respondents indicate many innovative ways that organizations practice, including

Table 12.5 Matrix Analysis: Theme and sub-theme identified for research questions and responses received from HR professionals

Theme and sub-theme	Job seekers give importance to brand	Vivid perspectives of job seekers toward CSR	External CSR credibility	Internal CSR credibility	CSR information requirement	Various methods used for CSR information	CSR information not relevant	CSR information relevant	Building CSR reputation	Various CSR initiatives
CSR as an important factor for Job Seekers	4	10	0	1	0	0	0	0	0	0
Influence of internal and external CSR on Job Seekers	1	0	18	5	0	0	0	0	0	0
Methods of CSR Communication during Recruitment	0	0	0	0	3	11	0	0	0	0
Relevance of CSR information in recruitment process	0	0	0	0	0	0	2	9	0	0
Strategic use of CSR for recruitment	0	0	0	0	0	0	0	0	7	7
Total	5	10	18	6	3	11	2	9	7	7

Source Research Data

Table 12.6 Matrix Analysis—sub-theme and theme identified from HR professional per interview responses

No.	Respondents	CSR as an important factor for job seekers	Influence of internal and external CSR on job seekers	Methods of CSR communication during recruitment	Relevance of CSR information in recruitment process	Strategic use of CSR for recruitment	Total
1	HRP01	0	1	1	1	1	4
2	HRP02	0	1	2	0	1	4
3	HRP03	1	2	0	0	2	5
4	HRP04	0	1	0	0	2	3
5	HRP05	3	2	1	0	0	6
6	HRP06	2	0	0	0	1	3
7	HRP07	1	3	1	0	1	6
8	HRP08	2	1	1	3	0	7
9	HRP09	2	1	0	0	1	4
10	HRP10	1	2	1	0	2	6
11	HRP11	0	2	1	2	1	6
12	HRP12	0	3	1	1	1	6
13	HRP13	1	3	2	2	0	8
14	HRP14	0	1	2	1	0	4
15	HRP15	1	0	1	1	1	4
	Total	14	23	14	11	14	76

Source Research Data

CSR information in the recruitment process. They are—by mentioning terms and conditions and benefits in the offer letter, willingness to work in all sectors, campus presentations, job postings on LinkedIn, website publication, the release of CSR reports, and information sharing of CSR initiatives through in-house publications thorough organization's annual report. Thus, the results suggest that organizations should spend more on CSR publicity and be concerned about attracting prospective job seekers. Consistent with the signaling theory, which proposes that prospective job seekers interpret organizations based on their information, our findings also indicate that CSR provides a positive signal to job seekers for job choices.

12.6 Practical Implications

The study provides various practical implications for organizations (corporate leaders, HR/CSR professionals), and government (policy makers/regulators). Organization should emphasize on CSR-driven efforts such as equal employment opportunities, safe working conditions, work-life balance, employee engagement plan, development of employees, and social benefits, that would positively impact the mindset and behavior of job seekers,. Further, organizations should integrate the CSR disclosures in their recruitment policies during job advertisement and recruitment process. This results in positive encouragement among job seekers to apply for jobs in larger numbers. In addition, it is also recommended that organization should make their best efforts in building positive CSR images, highlighting the CSR agenda, CSR spending, CSR communication and reporting, and CSR initiatives while job positing to grab a competitive advantages to attracting best talent. Moreover, organizations can also implement a generative-AI-based feedback mechanism (e.g. creating CSR-based personalized feedback link while submitting forms) to collect the responses from job seekers on CSR image and its influence on their motivation to apply for job. This would help organization to know the HR perceptions of recruitment and convergence ratio of job seekers. This way, organizations can to review, revise, and analyze the CSR baseline and benchmark studies for job seekers attraction.

The dynamic role of government (policy makers/regulators) is also imperative and crucial where it is necessary that government take active part in triggering positive CSR behavior in job seeking population. For this, government should propose a policy where the private organizations, and local/state/central government organizations can jointly contribute toward CSR-driven hiring processes. Also, government should ensure that the CSR central rules and policies are to be embraced into the recruitment and hiring process of organizations. This would tailor CSR reputation to job seekers with high organization-job fit perceptions toward organizations.

12.7 Research Limits and Scope for Future Research

The study is grounded on the viewpoints of the signaling and stakeholder theory. However, future research may be undertaken based on other CSR theories and frameworks. The second limitation could be using the subject sample across companies located in Western Odisha. The inference drawn from such a limited sample may pose challenges in complete generalization. However, this study may be extended by taking another geographic boundary. Third, in the survey portion, the private companies were only covered. Public companies are not included in the sample survey. Therefore, academicians and researchers may analyze the relationship between CSR and job seekers in future, prioritizing the public sector. Lastly, this study has adopted a qualitative approach to understanding the perceptions which may not be generalizable to the target population. Thus, there is a scope for extending this research by applying the quantitative method to gain a better interpretation on the subject matter.

References

Agnihotri, A., & Bhattacharya, S. (2022). CSR fit and organizational attractiveness for job applicants. *International Journal of Organizational Analysis, 30*(6), 1712–1727.

Albinger, H. S., & Freeman, S. J. (2000). Corporate social performance and attractiveness as an employer to different job seeking populations. *Journal of Business Ethics, 28*(3), 243–253.

Ali, I., Rehman, K. U., Ali, S. I., Yousaf, J., & Zia, M. (2010). Corporate social responsibility influences, employee commitment and organizational performance. *African Journal of Business Management, 4*(13), 2796–2801.

Angelo, F. D., Bartocci Liboni Amui, L., Ferreira Caldana, A. C., & Jose Chiappetta Jabbour, C. (2012). Towards a strategic CSR: A Brazilian case study. *Business Strategy Series, 13*(5), 224–238.

Bakanauskiene, I., Bendaraviciene, R., Krikstolaitis, R., & Lydeka, Z. (2011). Discovering an employer branding: Identifying dimensions of employer's attractiveness in university. *Management of Organizations: Systematic Research, 59*(1), 7–22.

Behrend, T. S., Baker, B. A., & Thompson, L. F. (2009). Effects of pro-environmental recruiting messages: The role of organizational reputation. *Journal of Business and Psychology, 24*(3), 341–350.

Bohlmann, C., Krumbholz, L., & Zacher, H. (2018). The triple bottom line and organizational attractiveness ratings: The role of pro-environmental attitude. *Corporate Social Responsibility and Environmental Management, 25*(5), 912–919.

Caputo, F. (2021). Towards a holistic view of corporate social responsibility. The antecedent role of information asymmetry and cognitive distance. *Kybernetes, 50*(3), 639–655.

Carlini, J., Grace, D., France, C., & Lo Iacono, J. (2019). The corporate social responsibility (CSR) employer brand process: An integrative review and comprehensive model. *Journal of Marketing Management, 35*(1–2), 182–205.

Choi, B. K., Moon, H. K., & Joo, Y. R. (2020). Who is attracted to socially responsible organizations? Roles of job applicants' volunteer experience and motives. *Baltic Journal of Management, 15*(5), 817–836.

Choi, J., Park, Y., & Sohn, Y. W. (2021). Corporate social responsibility and job seekers' application intention: A mediated moderation model of calling and value congruence. *Career Development International, 26*(1), 65–82.

Das, S. C. (2009). Status and direction of corporate social responsibility in Indian perspective: An exploratory study. *Social Responsibility Journal, 5*(1), 34–47.

Dilling, P. F. (2011). Stakeholder perception of corporate social responsibility. *International Journal of Management and Marketing Research, 4*(2), 23–34.

Du, S., Bhattacharya, C. B., & Sen, S. (2010). Maximizing business returns to corporate social responsibility (CSR): The role of CSR communication. *International Journal of Management Reviews, 12*(1), 8–19.

Eger, L., Mičík, M., Gangur, M., & Řehoř, P. (2019). Employer branding: Exploring attractiveness dimensions in a multicultural context. *Technological and Economic Development of Economy, 25*(3), 519–541.

Famiyeh, S., Kwarteng, A., & Dadzie, S. A. (2016). Corporate social responsibility and reputation: Some empirical perspectives. *Journal of Global Responsibility, 7*(2), 258–274.

Farooq, O., Payaud, M., Merunka, D., & Valette-Florence, P. (2014). The impact of corporate social responsibility on organizational commitment: Exploring multiple mediation mechanisms. *Journal of Business Ethics, 125*(4), 563–580.
Fombrun, C., & Shanley, M. (1990). What's in a name? Reputation building and corporate strategy. *Academy of Management Journal, 33*(2), 233–258.
Freeman, R. E. (2010). *Strategic management: A stakeholder approach.* Cambridge University Press.
Gandasari, D., Tjahjana, D., Dwidienawati, D., & Ichsan, M. (2024). How to attract talents? The role of CSR, employer brand, benefits and career development. *Cogent Business & Management, 11*(1), 2323774.
Garg, A., & Gupta, P. K. (2020). Mandatory CSR expenditure and firm performance: Evidence from India. *South Asian Journal of Business Studies, 9*(2), 235–249.
Gaudencio, P., Coelho, A., & Ribeiro, N. (2017). The role of trust in corporate social responsibility and worker relationships. *Journal of Management Development, 36*(4), 478–492.
George, N. A., Aboobaker, N., & Edward, M. (2020). Corporate social responsibility and organizational commitment: Effects of CSR attitude, organizational trust and identification. *Society and Business Review, 15*(3), 255–272.
Glavas, A., & Godwin, L. N. (2013). Is the perception of 'goodness' good enough? Exploring the relationship between perceived corporate social responsibility and employee organizational identification. *Journal of Business Ethics, 114*(1), 15–27.
Gomes, D. R., & Neves, J. G. (2010). Do applicants' prior experiences influence organizational attractiveness prediction? *Management Research: Journal of the Iberoamerican Academy of Management, 8*(3), 203–220.
Greening, D. W., & Gray, B. (1994). Testing a model of organizational response to social and political issues. *Academy of Management Journal, 37*(3), 467–498.
Hameed, I., Riaz, Z., Arain, G. A., & Farooq, O. (2016). How do internal and external CSR affect employees' organizational identification? A perspective from the group engagement model. *Frontiers in Psychology, 7*, 788.
Kim, H. L., & Han, J. (2019). Do employees in a "good" company comply better with information security policy? A corporate social responsibility perspective. *Information Technology & People, 32*(4), 858–875.
Held, K., & Bader, B. (2018). The influence of images on organizational attractiveness: Comparing Chinese, Russian and US companies in Germany. *The International Journal of Human Resource Management, 29*(3), 510–548.
Huang, J. C. (2022). Effects of person-organization fit objective feedback and subjective perception on organizational attractiveness in online recruitment. *Personnel Review, 51*(4), 1262–1276.
Irena et al. (2011). They revealed that the factor "best place to work" contributed to a deeper understanding of the employer images
Jia, Y., Yan, J., Liu, T., & Huang, J. (2019). How does internal and external CSR affect employees' work engagement? Exploring multiple mediation mechanisms and boundary conditions. *International Journal of Environmental Research and Public Health, 16*(14), 2476.
Kaur, T., & Dubey, R. (2014). Employee reviews on company independent sites and its impact on organizational attractiveness: Role of information realism, person-environment fit and source credibility framework. *Business: Theory and Practice, 15*(4), 390–397.
Kumar, V., Jain, S., & Singh, A. K. (2021). A nomological network study of employer branding with CSR, motivation and intention to stay. *Asia-Pacific Journal of Business Administration, 13*(4), 553–573.
Kusuma, R. P., & Sentosa, B. M. (2024). Talent management in the knowledge economy: Strategies for attracting and retaining top talent. *Indonesia Journal of Engineering and Education Technology (IJEET), 2*(2), 408–414.
Lee, J. Y. (2021). Corporate Social Responsibility. In P. M. Pedersen (Ed.), *Encyclopaedia of sport management* (pp. 106–109). Edward Elgar Publishing.

Low, M. P., Ong, S. F., & Tan, P. M. (2017). Would internal corporate social responsibility make a difference in professional service industry employees' turnover intention? A two-stage approach using PLS-SEM. *Global Business & Management Research, 9*(1).
Magbool, M. A. H. bin, Amran, A., Nejati, M., & Jayaraman, K. (2016). Corporate sustainable business practices and talent attraction. *Sustainability Accounting, Management and Policy Journal, 7*(4), 539–559.
Mahmud, A., Ding, D., & Ali, Z. (2023). An investigation of employee perception of micro-corporate social responsibility and societal behavior: A moderated-mediated model. *International Journal of Emerging Markets, 18*(9), 2455–2476.
Marques-Mendes, A., & dos Santos, M. J. N. (2022). The strategic character of Corporate Social Responsibility (CSR): A discussion and evaluation of the strategic corporate responsibility concept. In *Management for sustainable development* (pp. 83–110). River Publishers.
Mckinsey Survey Report. (2009, February). Retrieved from https://www.mckinsey.com/business-functions/strategy-and-corporate-finance/our-insights/valuing-corporate-social-responsibility-mckinsey-global-survey-results
Mg, Y. L. (2024). *A study on the private sector towards Corporate Social Responsibility (CSR) concepts.* (Doctoral dissertation). MERAL Portal.
Oliveira, M., Proença, T., & Ferreira, M. R. (2022). Do corporate volunteering programs and perceptions of corporate morality impact perceived employer attractiveness? *Social Responsibility Journal, 18*(7), 1229–1250.
Ouakouak, M. L., & Ouedraogo, N. (2019). Fostering knowledge sharing and knowledge utilization: The impact of organizational commitment and trust. *Business Process Management Journal, 25*(4), 757–779.
Ouyang, Z., Zhang, Y., & Hu, X. (2021). Negative publicity and potential applicants' intention to apply amid a discrimination scandal: A moderated mediation model. *Personnel Review, 50*(1), 129–142.
Presley, A., Presley, T., & Blum, M. (2018). Sustainability and company attractiveness: A study of American college students entering the job market. *Sustainability Accounting, Management and Policy Journal, 9*(4), 470–489.
Rahman, S., Haski-Leventhal, D., & Pournader, M. (2016). The effect of employee CSR attitudes on job satisfaction and organizational commitment: Evidence from the Bangladeshi banking industry. *Social Responsibility Journal, 12*(2), 228–246.
Remišová, A., & Búciová, Z. (2012). Measuring corporate social responsibility towards employees. *Journal for East European Management Studies, 17*(3), 273–291.
Ronda, L., Abril, C., & Valor, C. (2020). Job choice decisions: Understanding the role of nonnegotiable attributes and trade-offs ineffective segmentation. *Management Decision, 59*(6), 1546–1561.
Salanţă, I. I., Petre, A., & Stegerean, R. (2019). Corporate social responsibility effect of employer branding. *Managerial Challenges of the Contemporary Society. Proceedings, 12*(1), 52–56.
Santhosh, M., & Baral, R. (2015). A conceptual framework for exploring the impacts of corporate social responsibility on employee attitudes and behaviour. *Journal of Human Values, 21*(2), 127–136.
Sharp, Z., & Zaidman, N. (2010). Strategization of CSR. *Journal of Business Ethics, 93*(1), 51–71.
Silva Junior, A. da, Martins-Silva, P. de O., Feu, K. S., Komino, A. C., Silva, V. C. da, & Vasconcelos, K. C. de A. (2020). Corporate social responsibility in the perspective of Brazilian management students: The inversion of the pyramid. *Social Responsibility Journal, 16*(1), 50–72.
Sohn, M., Sohn, W., Klaas-Wissing, T., & Hirsch, B. (2015). The influence of corporate social performance on employer attractiveness in the transport and logistics industry: Insights from German junior talent. *International Journal of Physical Distribution and Logistics Management, 45*(5), 486 505.
Sommer, L. P., Heidenreich, S., & Handrich, M. (2017). War for talents—How perceived organizational innovativeness affects employer attractiveness. *R and D Management, 47*(2), 299–310.

Song, B. L., Liew, C. Y., Tee, P. K., & Wong, L. C. (2024). Corporate social responsibility and job pursuit intention: The role of job seekers' perception on employer prosocial orientation, value congruence and employer attractiveness. *Social Responsibility Journal, 20*(9), 1809–1831.

Spence, M. (1973). Job market signaling. *The Quarterly Journal of Economics, 87*(3), 355–374.

Stefańska, M. (2018). The relationship between the perception of the company and employee attitude in the context of CSR—The results of the empirical study. *Prace Naukowe Uniwersytetu Ekonomicznego we Wrocławiu, 520*, 136–149.

Strobel, M., Tumasjan, A., & Welpe, I. (2010). Do business ethics pay off? The influence of ethical leadership on organizational attractiveness. *Journal of Psychology, 218*(4), 213–224.

Turban, D. B., & Greening, D. W. (1996). Corporate social performance and organizational attractiveness to prospective employees. *Academy of Management Journal, 40*(3), 658–672.

Turker, D. (2009). Measuring corporate social responsibility: A scale development study. *Journal of Business Ethics, 85*(4), 411–427.

Wang, R. T. (2013). Modeling corporate social performance and job pursuit intention: Mediating mechanisms of corporate reputation and job advancement prospects. *Journal of Business Ethics, 117*(3), 569–582.

Wood, D. J. (1991). Corporate social performance revisited. *Academy of Management Review, 16*(4), 691–718.

Yin, J., & Zhang, Y. (2012). Institutional dynamics and corporate social responsibility (CSR) in an emerging country context: Evidence from China. *Journal of Business Ethics, 111*(2), 301–316.

Zhang et al. (2019). Revealed that corporate reputations substantially influenced CSR and customer satisfaction relationships

Zweers, A. (2015). *CSR in recruitment advertising: Does it really pay to be green?*

Chapter 13
CSR Shaping the Future of Sustainable Governance

Anchal and Md Safiullah

Abstract The paper emphasizes how the integral role of CSR and Corporate governance helps in influencing sustainable business practices. The paper highlights how Corporate Sustainability Reporting and CSR enhanced gradually due to the demands of the stakeholders, emerging from the late 1980s focusing on environmental accountability to encompass ethical responsibilities, reflecting a global trend toward integrating sustainability into corporate strategies.

While determining the linkage between CSR and corporate governance, the researcher shows how embracing CSR initiatives can enhance a company's reputation, brand equity, and financial standing while also promoting social responsibility and environmental stewardship and the special role being performed by the board. The present article attempted a study where 250 listed companies were taken into consideration to determine the comprehensive view of corporate practices and sustainability reporting among leading firms in the country. It found that larger companies have more resources and are more likely to engage in formal sustainability reporting, which increased the likelihood of obtaining the data for analysis, which is crucial for assessing the quality and effectiveness of CSR initiatives and at the end it highlights the interconnectedness of corporate governance, CSR, and corporate democracy in shaping sustainable governance practices. This paper highlighted the importance of strategic imperative and moral obligation for businesses to embrace these values for long-term success and societal impact. It stresses the importance of integrating corporate governance, CSR, and corporate democracy to build resilient and responsible businesses in today's dynamic business landscape.

Keywords CSR · Corporate Democracy · Corporate Governance · Sustainable Governance

---

Anchal (✉) · M. Safiullah
Chanakya National Law University, Patna, India
e-mail: anchalpusph000@gmail.com

M. Safiullah
e-mail: safiullah@cnlu.ac.in

V. Kandpal et al. (eds.), *Developments in Corporate Governance*,
Approaches to Global Sustainability, Markets, and Governance,
https://doi.org/10.1007/978-981-96-6366-8_13

Abbreviations

BRSR	Business Responsibility and Sustainability Reporting
BRR	Business Responsibility Reporting
CG	Corporate Governance
CPI	Corruption Perceptions Index
Cr.	Crore (Indian monetary unit)
CSR	Corporate Social Responsibility
ESG	Environmental, Social, and Governance
FY	Financial Year
GRI	Global Reporting Initiative
IBC	Insolvency and Bankruptcy Code
KYC	Know Your Customer
MCA	Ministry of Corporate Affairs
NFCG	National Foundation for Corporate Governance
NSE	National Stock Exchange
OECD	Organisation for Economic Co-operation and Development
RBI	Reserve Bank of India
SARFAESI	Securitisation and Reconstruction of Financial Assets and Enforcement of Security Interest Act
SEBI	Securities and Exchange Board of India

13.1 Introduction

The Organisation for Economic Co-operation and Development defines corporate governance as a system in which the company's directors, managers, shareholders, and other interested parties are each given certain roles and duties (Malla, 2010). The corporate governance oversees the company's goals and management's performance. Included in this framework are the owners' strategic goals, the employees' interests, the local area and environment, excellent connections with customers and suppliers, and compliance with regulatory authorities' orders. How well a corporation practices corporate democracy is a key indicator of its corporate governance. When all members take part in the company's decision-making process, it's called corporate democracy. A strong corporate democracy cannot be led by shareholders with short-term investments in the company (Mitra, 2014).

Corporate social responsibility, on the other hand, refers to the altruism that businesses engage in for the benefit of their stakeholders. It is the firm's dedication to meeting social, environmental, and economic demands. A company's credit rating and reputation in the business world can both benefit from CSR. Sustainable development, social performance, and corporate conscience are the three main components of the term Corporate Social Responsibility (Chaudhary, 2016).

Since the end of the 1990s, corporate sustainability reporting methods have gained importance among significant companies worldwide. It is a subject that is developing increasingly and is crucial to the commercial and academic worlds (Amin & Chaudhary, 2016). It has now been considered to be part of CSR as well. Businesses nowadays are attempting to integrate sustainability initiatives into their business practices. This is because socially responsible business practices have become more popular due to growing concerns about global warming, climate change, environmental degradation, resource depletion, and human rights (Sheldon & Park, 2010).

In the end, businesses must deal with each of these problems in order to restore the earth's resilience and equilibrium and to satisfy the needs of both internal and external stakeholders (such as suppliers, capital providers, and employees) as a part of CSR activities (Adams & Frost, 2008). Companies are willingly releasing their sustainability reports to show that (Dyllick & Hockerts, 2002; Ebinger et al., 2006). This trend has increased as a result of customer education and awareness campaigns, which put pressure on companies to report on their sustainable operations. Furthermore, to measure how well the economy, society, and environment are doing in critical areas using non-financial information that supports a broader picture of organization performance, organizations need to be more transparent and accountable to stock exchanges, markets, the government, society, and investors (Roca & Searcy, 2012).

As a result, senior management is compelled to enhance both the company's reputation and sustainability policies (Karabell & Cramer, 2010). Consequently, managers are expected to exert a greater degree of control on social and environmental performance criteria in addition to the profitability measurements that they now oversee. Globally, there has been an increasing recognition of the need to prioritize sustainability reporting (Michelon et al., 2015). As a result, businesses all around the world began to disclose information about sustainability. The number and quality of sustainability reporting were positively correlated with the implementation of the GRI guidelines (Kumar & Ruhela, 2021), and this has been accepted as the standard reporting process worldwide, including in India.

13.2 Literature Review

An empirical study conducted by Michelon and the team assesses the quality of CSR information through three complementary dimensions. Firstly, evidence indicates that many CSR reporting practices often prioritize symbolism over substance, suggesting that they may lack a genuine commitment to social responsibility. Secondly, disclosures presented in stand-alone reports tend to be more diluted, lacking the depth found in integrated reports that combine financial and non-financial information. Lastly, it appears that compliance with the Global Reporting Initiative (GRI) and obtaining external assurance generally do not correlate with enhanced quality in CSR disclosures (Michelon et al., 2015).

Companies engaging in CSR initiatives, such as providing medical care and environmental projects, enhance their brand image and foster customer loyalty. Research indicates that ethical practices significantly influence consumer decisions, making CSR essential for differentiation in a competitive market. Additionally, CSR integration into strategic decision-making offers companies global outreach opportunities and potential tax benefits, reinforcing the importance of socially responsible practices in building sustainable brands within the tourism industry (Safiullah et al., 2024a).

However, Govindarajan (2024) examines corporate responsibility trends between 2019 and 2023. It focuses on key areas like education, health, and rural development. The data shows a significant growth in CSR contributions: education funding rose from ₹6,111.7 crores in 2019 to ₹10,085.4 crores in 2023, while healthcare contributions increased from ₹3,617 crores to ₹6,830 crores. The paper also highlights the top CSR-spending companies and tracks year-on-year growth across various sectors, especially emphasizing the impact on rural development initiatives.

13.3 Origin of Corporate Sustainability Reporting and CSR

Modern organizations have shown a readiness to provide data on their social, economic, and environmental performance, indicating that they understand the need for sustainability. Moreover, the demands for accountability from stakeholders have compelled businesses worldwide to acknowledge the importance of better governance (Dodds & Kuehnel, 2010).

Companies are under growing pressure to disclose their corporate behavior in a more open manner (Kolk, 2006; Pollach et al., 2012).

According to studies, civil society organizations frequently saw themselves as "watchdogs" on corporate behavior, and consumers and the media express demands on behalf of these organizations. This holds true for every nation on the planet. Due to their negative effects on the environment and society, the chemical industry released the first sustainability reports in the late 1980s. Additionally, this approach was implemented by the tobacco industry to reveal global sustainability initiatives. It began by taking into account the serious environmental challenges that propelled the US sustainability reporting movement. Since then, a lot of organizations have started to share details about their sustainability, social, and environmental policies. Since the release of independent environmental studies in 1989, their significance has increased dramatically. When the concept gained popularity in the early 1990s, sustainability was mostly seen as a business term. As sustainability was increasingly viewed from an ethical and moral perspective in the corporate community, the concept gradually took on a new form that gave rise to Corporate Social Responsibility (Boudreau & Ramstad, 2005). Something went wrong with the movement that began to speak up in favor of environmental protection, and businesses and the government are now engaged in a conflict. Their main contention is that enterprises must be forced to adopt more environment-friendly practices and that consumer awareness and effective education programs are necessary to make voluntary action insufficient for environmental protection. Although laws and education are important, they might not be sufficient to address the issue entirely or promptly (Nidumolu et al., 2013). Ultimately, nevertheless, this sustainability strategy examined the right steps to prevent environmental calamities.

13.3.1 Major Changes in CSR and Corporate Governance in India

Company governance in India was in a bad situation after independence. There were no appropriate provisions in the Companies Act of 1956 that encouraged good corporate governance practices at the time. In India, there was no predetermined system for overseeing corporations. In many cases, the individuals exercising the Board's powers were close relatives or friends. It was common practice at the time to repress minority shareholders. Corporate reporting was of low quality, and the system for overseeing companies was equally dismal, at least up until the 1990s (Bhat, 2007) (Fig. 13.1).

Clause 49 of the Listing Agreement (which applies to companies listed on stock exchanges) requires compliance reports on corporate governance. After the U.S. enacted the Sarbanes-Oxley Act and governance issues arose, therefore Naresh Chandra Committee (Khurshid & Bandyopadhyay, 2009) in 2002 to recommend

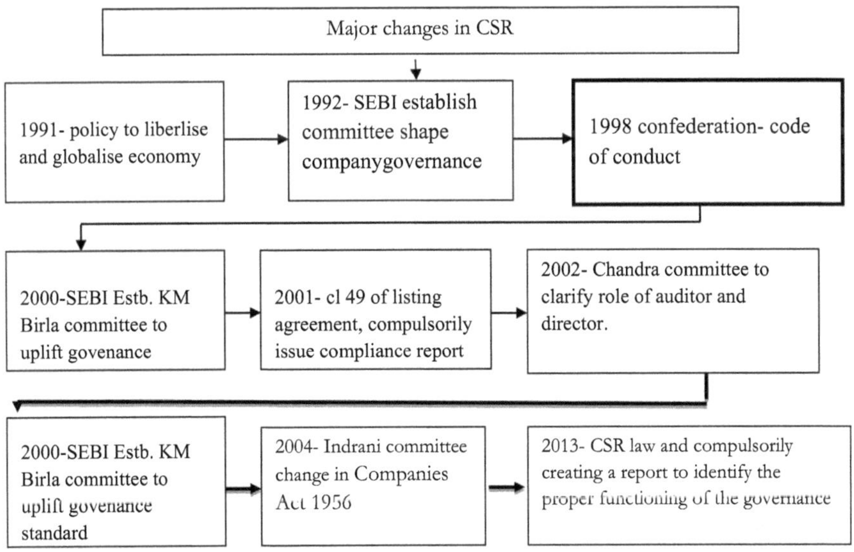

Fig. 13.1 Metamorphosis of "Companies Act 2013." *Source* Safiullah et al. (2024a)

mandatory roles for auditors and independent directors was formed. SEBI reviewed Clause 49 and, under the guidance of N. R. Narayana Murthy (Murthy, 2003), and Naresh Chandra committee released a revised Clause 49. However, the actual concept of CSR evolved in 2013 when the Companies Act 2013 came into the picture.

13.4 The Common Linkage Between Corporate Governance and Corporate Social Responsibility

"Corporate Social Responsibility" and "Corporate Governance" are complementary concepts that must be present in any discussion of effective business governance. In India, companies that have developed sustainable development models for society also receive CSR awards and best Corporate Governance awards. According to research, financial markets always reward companies with strong governance and punish those with weak governance. In both areas, firms are held to the greatest standards of accountability to their stakeholders. To boost stakeholder and investor confidence, both fields encourage openness, accountability, and disclosure in business operations (Berghe & Loche, 2011).

It also helps corporations reap long-term benefits and keep their businesses running smoothly; SEBI has recently drafted regulations requiring publicly traded companies to disclose all material matters, including financial reports, performance, ownership details, and governance frameworks. CSR shows the company's conscience and its dedication to stakeholders over the long term, while Corporate Governance assists the board in accepting ethical considerations while making decisions on shareholder interests. The two pillars of CSR, Corporate Governance and Corporate Social Responsibility, are now working together to end the exploitation of employees, communities, and resources by sharing information with accountable stakeholders (Gill, 2008).

CSR initiatives are the product of solid business governance. John Hancock identifies four cornerstones of corporate social responsibility (CSR) in his book "Investing in Corporate Social Responsibility": strategic governance, human capital, stakeholder capital, and the environment. The business should uphold these principles. Additionally, it aids organizations in increasing their future value. Without these tenets of good corporate governance, companies would never be able to achieve their CSR goals. Both Corporate Governance and CSR are essential for a company's success, which is another shared characteristic of the two fields. Companies should make sure they are equally a part of how things work. Companies with strong corporate governance safeguard their shareholders, while those with a focus on social responsibility safeguard the company's stakeholders. Many Indian businesses have included a Corporate Social Responsibility Committee into their organizational structure since it helps them gain value for their organizations. Their Board report still includes Corporate Governance and CSR. Here are some key instances where the two fields collaborate:

a. **Transparency**: In both fields, transparency plays a key role. Companies are required to provide the "full details of CSR policy" in their financial statement before approval and signing by the Board of Directors to encourage transparency in corporate activities. Companies' average and net profits, prescribed CSR funds, CSR expenditures, CSR funds spent during the immediate financial year, an explanation for not spending the "unspent amount," and a declaration from the CSR Committee that the monitoring and implementation of CSR policy are according to the company's strategy should all be included in the transparent Board report. The Board should make sure that all the information is available to the public to promote openness in CSR operations. Now that it has signed the UN Global Compact, India is a major player in it. The promotion of openness is a goal of the United Nations Global Compact. The United Nations Global Compact and its 10 principles are anticipated to have the backing of the participants. Despite India's new membership in the United Nations Global Compact, the country should work to increase openness in other areas as well (Farrar & Hanrahan, 2016).

   India came up at number 93 out of 180 nations on Transparency International's most recent Corruption Perceptions Index (CPI) 2023 (Transparency International, 2024). Our laws and regulatory organizations are not being adequately enforced, as shown by these statistics.
b. **Accountability**: The Board is perpetually responsible for CSR accountability. The Board will make sure that the money is being used for what it approves of. The company's Chief Financial Officer (CFO) or someone with decision-making authority over the company's finances will also verify it. Since this is an ongoing project to which the firm has contributed, it is the responsibility of the company's board to carry it out according to the agreed-upon schedule. The firms have the option of carrying out CSR initiatives in-house or entrusting them to third parties. The law has also made it possible to assist Sec 8 (Rajeev & Kalagnanam, 2015) corporations with their corporate social responsibility initiatives, ensuring that Section 8 firms are registered if they have entrusted them with CSR initiatives. The Board must also look at these businesses. Its performance record in handling Schedule VII activities over the past three years should be favorable. Nowadays, not only may one company work jointly on a project, but two or more can do so as well. On top of that, businesses can team up with global organizations to make their ambitions a reality. To ensure a smooth implementation, businesses must adhere strictly to the terms outlined in the MOU they signed with the implementing agency. Prompt notification of any modification to the agreement is required. The parties responsible for overseeing the CSR money should designate a point of contact.
c. **Disclosure**: Companies' motivating motivations can be altered through disclosure, which also serves to enlighten stakeholders. As it relates to CSR, the disclosure principles are crucial. Businesses must be transparent about the trusts and registered societies that they work with for their corporate social responsibility initiatives. In addition, the company's members are entitled to view the CSR fund utilization registry at any time. The company's register is open for inspection by

any member, who can also make copies and extracts from it. Each and every annual general meeting has to have this register inspected.

13.5 Role of the Board in Implementing Corporate Social Responsibility and Corporate Governance Practices

The Board is the highest decision-making authority. It shows the public how the company sees itself. In avoiding problems and enhancing the standing of the business they represent, it is crucial. When necessary, the Board must advice and counsels the company's senior management. In every company, the board of directors is responsible for keeping an eye on everything. The organization's core ideas and ideals are upheld. It is comprised of a number of Directors who are typically chosen by the company's shareholders. In most cases, the directors are chosen by the shareholders through a system of annual rotation, with a portion of the directors departing from office. By approving the resolution, shareholders have the power to remove the members of the Board. The company's CSR programs are the responsibility of the board of directors. It needs to make sure the company's CSR programs are getting enough attention. Every aspect of CSR is overseen by the Board. The CSR policy can be better planned, implemented, and designed with its support. Board duties include establishing the CSR Committee, writing the CSR policy, staffing CSR programs with qualified individuals, conducting impact assessments on a regular basis, and giving final approval to the CSR report (Kaushik, 2017).

Companies are required to form a "CSR Committee" after they reach a certain threshold as outlined in Section 135. If the sum is below fifty lakh rupees, however, the necessity to constitute a CSR committee is reduced. The Board decides who will serve on this committee. To ensure its smooth operation, the Board of Directors selects at least two directors, one of whom must be an independent director (Rajeev & Kalagnanam, 2015).

Some businesses are exempt from the need for a minimum of three directors and are instead limited to appointing two. Because it is the CSR Committee's responsibility to establish the firm's CSR policy, determine the activities for which the company must allocate its contribution, determine the total amount to be spent, and oversee the policy's development, this committee plays a crucial role. Following the Board's establishment of the CSR Committee, it is the responsibility of the Committee to draft the CSR Policy. To make strategic use of the donations, the company follows the CSR policy. Following Board approval and consideration of CSR Committee recommendations, corporations are required to submit a legally binding CSR policy. An effective policy will lay out the rules for choosing a project, carrying it out, and keeping tabs on corporate social responsibility initiatives. It is always up to the Board to implement CSR. Following the CSR Committee's approval, the Board must make sure that the designated beneficiaries get the appropriate amount of CSR monies. Since this is an ongoing project to which the firm has contributed, it is the responsibility of the company's board to carry it out according to the agreed-upon schedule. Firms have the option of carrying out CSR initiatives in-house or entrusting them to third parties. To foster openness in CSR practices, the Board needs to make sure that all the information about the Committee's establishment, CSR policy, and CSR projects is available to the public. An essential tenet of Corporate Governance is the Board's shared responsibility. Each director has an obligation to the company's affairs and to cooperate with their fellow directors on the board. Directors, in their official position, are required to carry out the obligations outlined in the new Companies Act 2013. The Articles of Association bind them. They ought to behave honestly and for the benefit of the business. Every director has an overarching responsibility to defend the company's workers, shareholders, community, and the environment. In carrying out his responsibilities, he must do it independently and with the utmost care. If his interests are being cast into doubt, he should stay out of it. Neither he nor anybody connected to him may get an unfair advantage (Iyer, 2011). The For directors who fail to uphold these responsibilities, the new law establishes fines starting at 1 lakh rupees and going up to 5 lakhs rupees (Gugnani, 2021).

13.6 Corporate Sustainability Reporting Ensuring Good Governance in India

Corporate reporting has grown in importance among Indian business people and scholars throughout the past ten years which ensures the good governance of the companies. Following the new economic reform in 1991, businesses were compelled to adopt a new paradigm and create long-term business practices that took into account the welfare of their employees, the environment, and future generations. Companies are also required to increase market share for present stakeholders, finance innovation, and boost profitability. The companies are embracing these strategies, and a huge number of firms are following suit. Companies are including all these activities as a part of CSR as well. Following SEBI's mandate that corporations publish their economic, social, and environmental actions in conjunction with their annual reports, reporting has become an obligatory field for all companies ensuring that the governance of the companies is being maintained with transparency and accountability. From FY 2022–2023 onward, corporations are expected to submit the "Business Responsibility and Sustainability Reporting (BRSR)" based on their market capitalization. Countries all over the world have recommended some adjustments to report their recent operations for a variety of reasons. In the context of India, it is also true. Good governance not only includes the proper functioning of the companies internally, rather it includes the external functioning of the companies in the societies as well. The proper functioning in society ensures that companies are protecting the interest of the societies either in the form of CSR or in the form of ensuring the environmental, social, or economic interest.

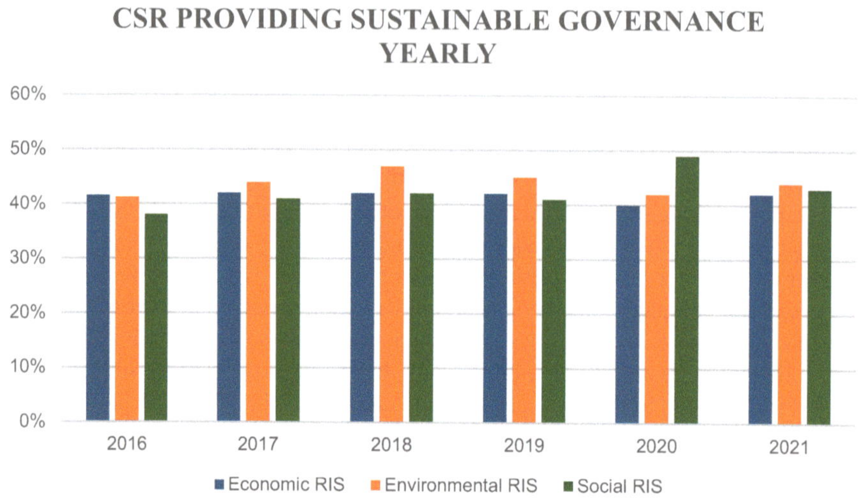

Fig. 13.2 Metamorphosis of "Companies Act 2013." *Source* Author's compilation

13.7 Reporting of CSR Proving Sustainable Governance

The study was conducted by selecting the top 250 NSE-listed companies in India, which followed the GRI—G4 standard, based on market capitalization from 2016 to 2020. From 2020 to 2021, the study was conducted by the researcher only (Fig. 13.2).

13.8 Analysis

The pie chart has been highlighted on the basis of different years on different dimensions consisting; economic reporting sustainability, environmental sustainability reporting, and social sustainability reporting which gave rise or fall in India.

a. **In an economic Dimension**—The economic dimension followed next (2016–2019), with reports showing a decrease in 2020 and then an increase in 2021. The reporting on the economic dimension consistently increased in 2016–2018 with 41.55%, 42.40%, and 42.85% respectively.
b. **Environmental disclosure** of sustainability reporting increased from 41.21%, 44.87%, and 47.15% over a period of three years between 2016 and 18 and declined subsequently over the next two years. Allocate a portion of the pie chart for environmental reporting practices, showing a consistent focus across all years.

c. The **social sustainability disclosure** practices over the years showed a pattern where an increase from 38.83% in 2016 to 2018 (42.62%), which decreased over the years in 2019 and 2020. Dimension-wise, the environmental dimensions 34.87% followed by economic 32.97% and social 32.16% dimensions were reported by the companies.
d. **Sectoral Differences**: You can represent sectoral differences by using different colors or patterns for primary, secondary, and tertiary sectors, highlighting their varying levels of sustainability reporting.
e. **Stakeholder Expectations**: Consider adding a legend to the pie chart to explain the significance of stakeholder expectations in driving sustainability reporting practices.
f. **Legitimacy Theory**: Include a key or labels to indicate the influence of legitimacy theory on reporting practices, particularly regarding the age and experience of organizations.
g. **Regulatory Pressure**: Show the impact of regulatory pressure on reporting practices by emphasizing the need for transparency and compliance in sustainability reporting.

13.9 CSR by the Companies in Various Sectors Ensuring Corporate Governance from 2019 to 2023

The factors that influenced are control variables. Large firms have a higher number of stakeholders including customers, employees, etc., who can influence firms' CSR activities. Both concepts emphasize the importance of stakeholders. Effective corporate governance requires considering the interests of all stakeholders in decision-making processes. Similarly, CSR initiatives often focus on creating value for stakeholders by improving environmental practices or engaging in community projects. Good governance can help identify and manage risks related to social and environmental issues. By incorporating CSR into governance structures, companies can proactively address these risks, leading to sustainable business practices. Good governance helps in identifying the effective governance, boost individual interest, and performance appraisal helps employee to work in healthy environment. Healthy environment enhances the efficiency of the work; therefore, the increase will certainly be there in the health of the people to work positively. The CSR in rural areas helps the working efficiency of the company as the reach of the company will be expanded and rural people will come to know about the specific institutions. Along with the development of the rural people, the companies' expansion will be there. The factor to reach to the rural areas is to expand the business and enhance the reach of the business. Expansion will certainly help in ensuring the corporate governance. CSR in corporate governance helps in enhancing the livelihood, as the positive environment enhances the working efficiency with the later impact of performance appraisal (Fig. 13.3).

Yrs./Sector	HEALTH SECTOR **(Rs. in Cr.)**	CSR on Rural devt. **(Rs. in Cr.)**	CSR on Livelihood **(Rs. in Cr.)**
2019	3617.2 Cr	2434.2 Cr	908 Cr
2020	4905.7 Cr.	2301 Cr.	1077.7 Cr.
2021	7325.8 Cr	1850.7 Cr	938.9 Cr
2022	7806.3 Cr	1832.8 Cr	854.5 Cr
2023	6830.4Cr	2005.4Cr	1654.4Cr

Fig. 13.3 Metamorphosis of "Companies Act 2013." *Source* Business Standard (9 September 2024). Safiullah et al. (2024b)

13.10 Impact of SEBI'S Business Responsibility and Sustainability Reporting (BRSR) Framework on Corporate Reporting Trends in India

The Securities and Exchange Board of India (SEBI) introduced the Business Responsibility and Sustainability Reporting (BRSR) framework to ensure Indian companies disclose measurable data on sustainability aspects, including human rights and environmental protection. India's ESG regulatory framework began with the **Business Responsibility Reporting (BRR)** guidelines in 2009. The updated **BRSR**, effective since 2023, aligns more closely with international reporting standards such as the **Global Reporting Initiative (GRI)** and the **Task Force on Climate-related Financial Disclosures (TCFD)**. It mandates the top 1,000 listed companies in India to respond to **140 indicators**, comprising **98 essential (mandatory) indicators** and **42 leadership (voluntary) indicators**.

Impact of the Sebi's BRSR are:-

Alignment with Global Standards: Companies are encouraged to incorporate elements from internationally recognized frameworks like the Global Reporting Initiative (GRI) and the Sustainability Accounting Standards Board (SASB) to ensure consistency and comparability.

Dual-Level Reporting: The framework requires reporting on both **essential** and **leadership** indicators. Essential indicators cover fundamental ESG aspects, while leadership indicators focus on advanced sustainability practices.

Data-Driven Disclosures: Companies must provide measurable data on environmental factors such as energy consumption, greenhouse gas emissions, water usage, and waste management.

Social Impact Considerations: The guidelines mandate disclosures on employee welfare, diversity and inclusion, human rights, and community initiatives.
Industry-Specific Flexibility: Recognizing sectoral differences, companies can justify the non-applicability of certain disclosures based on industry relevance.
Clarity and Transparency: Reports should be clear, precise, and well-supported by relevant documentation.

13.11 Conclusion

The companies are acting as responsible corporate citizens, and they cannot ignore the need for Corporate Governance or Corporate Social Responsibility. Those companies which will opt for the Good Corporate Governance model will ultimately enhance Corporate Social Responsibility. CSR has to be a part of ethical corporate governance. The corporate governance structure must interlink its activities with moral, ethical, social, and environmental issues. Without corporate governance and corporate social responsibility, no business enterprise can survive in the market world. In executing the Corporate Governance practices the companies can also take the help of trust like NFCG (National Foundation for Corporate Governance) which was set up by the Ministry of Corporate Affairs in the year 2003 to find the emerging areas related to corporate governance.

The growth of CSR depends upon the strategies opted by the CSR committee in executing the CSR policy. The working of the CSR Committee should be transparent. To make the CSR committee more effective the Board must try to rotate the directors of the committee on a rotational basis. By applying this method there will be ample chances that the CSR Committee will stick to its operations. This will also help the company to have a regular stream of appropriate projects rather than unplanned projects. The other mode through which CSR activities can be implemented in the company is by putting CSR into the governance of the company and information and management system. By putting CSR into information and management systems the companies will work more effectively. It will try to compete with the social and environmental impacts associated with the company.

No CSR or Corporate Governance can work without Independent Directors. The role of the "Independent Director" plays an important part in implementing the Corporate Governance norms of the Company. They are the ones who do not have any personal interest in the company. He represents the interest of all the investors and small shareholders. Similarly, he is also a part of the CSR Committee. He is the better person who can guide and suggest to the Board to take better decisions in the interest of stakeholders. On the other hand, the companies can promote workplace democracy where the employees can be part of decision making, especially concerning the governance structure.

Corporate Social Responsibility is an antecedent of Corporate Governance. Both can be achieved in the future years by the companies. Companies have to come forward to take these social initiatives. The companies must implement these objectives not on paper but in spirit, it is only then the purpose of corporate governance and corporate social responsibility will be achieved.

13.12 Recommendations

- The fact that companies disclosing methods to enhance corporate sustainability is encouraging, and it is advised that they are being supported in turning these into best practices. The study's findings indicate a growing understanding that sustainability reporting needs more attention.
- The study's findings showed a strong correlation between corporate sustainability reporting practices in India and profitability. First, industries began to report on their environmental practices. Later, they expanded to include the economic and social spheres. Encouraging industries to report on their environmental practices is important for following best practices as well as for regulatory compliance.
- The study's findings show that there is no significant relationship between the firm's age and degree of disclosure (Aust - Before Ehnert et al., 2015). Therefore, whether the company is young or old does not affect the present reporting practice trend, which should be supported.
- In terms of dimensions, the corporations reported the environmental, social, and economic factors. This tendency could be explained by the fact that businesses disclosed environmental information from the start. As they developed, social and economic practices became more transparent as profitability increased, and in the early twenty-first century, strategic human resource management gained significance. The current pandemic has also left its ugly imprints on the disclosures, though they should gradually improve as companies reclaim their former splendor because engaging in CSR activities is of the utmost importance.
- About sustainability reporting in Indian businesses, certain organizations have begun disseminating their reports to the public via independent websites and yearly reports (Chatterjee & Mir, 2008), which ensures that CSR activities are being performed in correspondence with maintaining sustainable governance.
- Nonetheless, the majority of poor nations including India have reports with insufficient data. It has been noted that reporting varies in scope and content throughout businesses, industries, and nations. As a result, there is room for variation in the sustainability reporting practices of businesses. For example, businesses may choose to begin with the Business Responsibility Sustainability Reporting format in FY 22–23 before expanding to global standards such as the GRI–G4 standards.

13.13 Implication and Future Scope

The present study highlighted the importance of CSR and corporate governance in India. It will help research scholars and practitioners in understanding the concept of evolution, development, and act related to CSR in India. The study will help in theory development in the area of laws and CSR where companies and practitioners can get benefit by understanding CSR in India. The study in future will help to ensure transparency among the governance of the companies. From the paper, we can infer that the better reporting obtained from the data will help in assessing the quality and effectiveness of CSR initiatives and will enhance the companies' effectiveness. This is a viable place for future work to evaluate the impact and efficacy of these disclosures in bringing about targeted positive change in the CSR-related activities.

References

Acharya, M. (2024, July 19). *Corporate social Responsibility under Section 135 of Companies Act 2013.* Cleartax. https://cleartax.in/s/corporate-social-responsibility

Adams, C. A., & Frost, G. R. (2008). Integrating sustainability reporting into management practices. *Accounting Forum, 32*(4), 288–302. https://doi.org/10.1016/j.accfor.2008.05.002

Amin-Chaudhry, A. (2016). Corporate social responsibility—From a mere concept to an expected business practice. *Social Responsibility Journal, 12*(1), 190–207. https://doi.org/10.1108/srj-02-2015-0033

Aust - Before Ehnert, I., Parsa, S., Roper, I., Wagner, M., & Muller-Camen, M. (2015). Reporting on sustainability and HRM: A comparative study of sustainability reporting practices by the world's largest companies. *The International Journal of Human Resource Management, 27*, 1–21.

Berghe, L., & Loche, C. (2011). The link between corporate governance and corporate social responsibility in insurance. *Geneva Papers on Risk and Insurance—Issues and Practice, 30*(3), 427. See also, Principle 1 of NVGs on Social Environment & Economic Responsibilities of Business.

Bergman, M. M., Bergman, Z., Teschemacher, Y., Arora, B., Jyoti, D., & Sengupta, R. (2019). Corporate responsibility in India: Academic perspectives on the Companies Act 2013. *Sustainability, 11*(21), 5939.

Bhat, V. (2007). Corporate governance in India: Past, present and suggestions for the future. *Iowa Law Review, 92*, 1431–1437–1438.

Boudreau, J. W., & Ramstad, P. M. (2005). Talentship, talent segmentation, and sustainability: A new HR decision science paradigm for a new strategy definition. *Human Resource Management, 44*(2), 129–136. https://doi.org/10.1002/hrm.20054

Chatterjee, B., & Mir, M. (2008). The current status of environmental reporting by Indian companies. *Managerial Auditing Journal*, 154496402. https://doi.org/10.1108/02686900810882138

Chaudhary, T. (2016). Companies Act 2013: Redefining Corporate Social Responsibility. *ALJ, 24*(265), 267–268.

Dodds, R., & Kuehnel, J. (2010). CSR among Canadian mass tour operators: Good awareness but little action. *International Journal of Contemporary Hospitality Management, 22*(2), 221–244. https://doi.org/10.1108/09596111011018205

Donaldson, W. (n.d.). *Shareholders Democracy.*

Dyllick, T., & Hockerts, K. (2002). Beyond the business case for corporate sustainability. *Business Strategy and the Environment, 11*(2), 130–141. https://doi.org/10.1002/bse.323

Ebinger, F., Cahyandito, M. F., von Detten, R., & Schlüter, A. (2006). Just a paper Tiger? Exploration of sustainability reporting as a corporate communication instrument. In S. Schaltegger, M. Bennett, & R. Burritt (Eds.), *Sustainability accounting and reporting* (Vol. 21). Springer, Dordrecht. https://doi.org/10.1007/978-1-4020-4974-3_23

Elijido-Ten, E. (2011). The impact of sustainability and balanced scorecard disclosures on market performance: Evidence from Australia's top 100. *Journal of Applied Management Accounting Research, 9*, 59–73.

Farrar, J., & Hanrahan, P. (2016). *Corporate Governance* (p. 550). LexisNexis Butterworths.

Ghosh, S. (2020). Altruistic CSR. In *Encyclopedia of sustainable management* (pp. 1–4). Springer International Publishing. https://doi.org/10.1007/978-3-030-02006-4_598-1

Gill, A. (2008). Corporate governance as social responsibility: A research agenda. *BJIL, 26*, 452.

Govindarajan, M. (2024). *CSR contributions by companies during 2023.* Tax Management India.com https://www.taxmanagementindia.com/visitor/detail_article.asp?ArticleID=12853#:~:text=It%20is%20reported%20that%20the,also%20been%20in%20the%20increase

Gugnani, V. (2021). The Companies Amendment Act, 2020: An initiative towards ease of doing business. *International Journal of Law Management & Humanities, 4*(1), 404.

Hancock, J. (2005). *Investing in Corporate Social Responsibility* (1st ed., p. 49). Kogan Page.

https://nsearchives.nseindia.com/web/sites/default/files/inlinefiles/ESG_Analysis_on_100_Listed_Companies_2022.pdf

Jamali, D. (2008). *Corporate governance and corporate social responsibility synergies and interrelationships* (pp. 447–448). Blackwell Publishing Ltd.

Iyer, L. V. V. (2011). *Guide to company directors; Powers, rights, duties, liabilities and corporate social responsibility* (pp. 218–219). LexisNexis Butterworths.

Karabell, Z., & Cramer, A. (2010). *Sustainable excellence: The future of business in a fast-changing world.* Rodale Books. https://books.google.co.in/books?hl=en&lr=&id=yJj23l9RRhEC&oi=fnd&pg=PA1&dq=embracing+sustainability:+the+future+of+international+business+strategies&ots=Lx-5C4MXTq&sig=fHB0609_eQzQD0SvmMRJjC7LD0g&redir_esc=y#v=onepage&q=embracing%20sustainability%3A%20the%20future%20of%20international%20business%20strategies&f=false

Kaushik, K. V. (2017). *CSR in India: Steering business towards social change* (p. 87). Lexisnexis.

Khatri, I., & Kjærland, F. (2023). Sustainability reporting practices and environmental performance amongst nordic listed firms. *Journal of Cleaner Production, 418*, Article 138172. https://doi.org/10.1016/j.jclepro.2023.138172

Khurshid, S., & Bandyopadhyay, R. (2009). *Corporate Governance Voluntary Guidelines 2009.* Ministry of Corporate Affairs, Government of India. https://www.mca.gov.in/Ministry/latestnews/CG_Voluntary_Guidelines_2009_24dec2009.pdf

Kolk, A. (2006). Sustainability, accountability and corporate governance: Exploring multinationals' reporting practices. *Business Strategy and the Environment, 17*(1), 1–15. https://doi.org/10.1002/bse.511

Kumar, R., & Ruhela, P. (2021). Corporate Social Responsibility (CSR) in India—Issues and Challenges. *PalArch's Journal of Archaeology of Egypt/Egyptology, 18*(4), 285–294. https://archives.palarch.nl/index.php/jae/article/view/5740

Lawless, J. (2024). *What is the relationship between corporate governance and social responsibility?* Purple Griffon. https://purplegriffon.com/blog/relationship-between-corporate-governance-andsocialresponsibility#:~:text=Good%20governance%20practices%2C%20such%20as,initiatives%20are%20developed%20and%20managed

Malla, P. (2010). *Corporate Governance* (p. 17). Routledge Taylor & Francis Group.

Marchetti, G., &Ventoruzzo, M. (2001). *Corporate Law* (pp. 2803–2810). Elsevier eBooks. https://doi.org/10.1016/b0-08-043076-7/04293-5

Michelon, G., Pilonato, S., & Ricceri, F. (2015). CSR reporting practices and the quality of disclosure: An empirical analysis. *Critical Perspectives on Accounting, 33*, 59–78. https://doi.org/10.1016/j.cpa.2014.10.003

Ministry of Corporate Affairs. (n.d.). https://www.mca.gov.in/Ministry/reportonexpertcommitte/chapter1.html

Ministry of Corporate Affairs. (n.d.). *Penalties and Offences under Companies' Act, 2013*. https://www.mca.gov.in/Ministry/pdf/Penalties-and-Offences-under-Companies-Act-2013.pdf

Mitra, N. L. (2014). Corporate Governance: A Sojourn to find a yardstick. *JILI, 56*(437), 446.

Nidumolu, R., Prahalad, C. K., & Rangaswami, M. R. (2013). Why sustainability is now the key driver of innovation. *IEEE Engineering Management Review, 41*(2), 30–37. https://doi.org/10.1109/emr.2013.6601104

OECD. (2024). *Recommendation of the Council on Principles of Corporate Governance*. In OECD Legal Instruments (OECD/LEGAL/0413). https://legalinstruments.oecd.org/public/doc/322/322.en.pdf

OECD Principles of Corporate Governance. (2004). OECD Publishing. https://doi.org/10.1787/9789264015999-en

Pollach, I., Johansen, T. S., Nielsen, A. E., & Thomsen, C. (2012). The integration of CSR into corporate communication in large European companies. *Journal of Communication Management, 16*(2), 204–216. https://doi.org/10.1108/13632541211217605

Rajeev, P. N., & Kalagnanam, S. (2015). *Section 135 of Companies Act (2013): Implementation challenges faced by Companies and NGOs*.

Rebello, A. (2024, July 5). *BRSR reporting in India: Full form, applicability, format, guidelines, benefits. Cleartax*. https://cleartax.in/s/brsr-reporting

Roca, L. C., & Searcy, C. (2012). An analysis of indicators disclosed in corporate sustainability reports. *Journal of Cleaner Production, 20*(1), 103–118. https://doi.org/10.1016/j.jclepro.2011.08.002

Safiullah, M., Anchal, & Parveen, N. (2024a). Brand building through CSR initiatives during Hajj and Umrah: A study of tourism industry. In *Corporate social responsibility, corporate governance and business ethics in tourism management: A business strategy for sustainable organizational performance* (pp. 85–92). Emerald Publishing Limited.

Safiullah, M., Iqbal, M. I., & Parveen, N. (2024b). Challenges and opportunities within the evolving CSR landscape in India. *Technology-Driven Evolution of the Corporate Social Responsibility Ecosystem*, 46–60.

Sarbanes Oxley compliance association, 2002, The Sarbanes Oxley Act. (n.d.). https://sarbanes-oxley-act.com/

Securities and Exchange Board of India. (2003). *The report of Shri N. R. Narayana Murthy Committee on corporate governance for public comments*. SEBI. https://www.sebi.gov.in/reports/reports/mar-2003/the-report-of-shri-n-r-narayana-murthy-committee-on-corporate-governance-for-public-comments-_12986.html

Sheldon, P. J., & Park, S. Y. (2010). An exploratory study of corporate social responsibility in the U.S. travel industry. *Journal of Travel Research, 50*(4), 392–407. https://doi.org/10.1177/0047287510371230

Stakeholders Empowerment Services (SES). *ESG analysis of 100 listed companies—India INC's readiness for regulatory push* (p. 24).

The institute of company secretaries of India. (2015). *Corporate social responsibility an engine for inclusive growth*. https://www.icsi.edu/media/webmodules/companiesact2013/CSR%20Final%2002022015.pdf

Transparency International. (2024, September 12). *2023 Corruption Perceptions Index: Explore the results*. Transparency.org. https://www.transparency.org/en/cpi/2023

Chapter 14
Integrating Community Participation Within the Policies of Corporate Governance for Ensuring Sustainable Trade in Forest Resources in India

Jayanta Boruah and E. Prema

Abstract Forests play a crucial role in the ecological and economic stability of nations, yet they face challenges due to their status as common property resources both subtractable and non-excludable. The trade in forest resources involves multiple stakeholders, including local communities, necessitating regulation for collective welfare. Unfortunately, unsustainable practices have resulted in ecological crises such as climate change and biodiversity loss, along with economic repercussions like the vulnerability of forest-dwelling communities and supply chain disruptions. Globally, community participation in forest management has proven effective for sustainable forestry, provided communities are empowered through integrated corporate governance. This approach aims to ensure sustainable trade in forest resources while improving local livelihoods. The chapter critically examines forest management issues such as deforestation and illegal logging driven by unsustainable practices, advocating for community involvement as a solution. Corporate governance, evolving to encompass community engagement and environmental sustainability, forms the theoretical foundation for integrating local communities as stakeholders in forest resource management. Successful case studies from various countries, as documented in SOFO 2022 and GFRA 2020, highlight instances where community integration in corporate governance has promoted sustainable forestry. In conclusion, integrating local communities into forest resource trade is crucial for mitigating degradation and addressing livelihood needs, aligning with India's Sustainable Development Goals (SDGs). This necessitates policy adjustments to foster community participation within corporate governance frameworks governing forest resource trading.

J. Boruah (✉)
Department of Law, Central University of Karnataka, Gulbarga, India
e-mail: jayanta.boruah94@gmail.com

E. Prema
Vellore Institute of Technology, Chennai, India
e-mail: prema.e@vit.ac.in

V. Kandpal et al. (eds.), *Developments in Corporate Governance*,
Approaches to Global Sustainability, Markets, and Governance,
https://doi.org/10.1007/978-981-96-6366-8_14

Keywords Community participation · Corporate governance · Forest resources · Sustainable trade · Sustainable forestry

14.1 Introduction

Community participation with productive goals in the management of natural resources has become an important agenda for global discussion (Ostrom, 1990). Several instances of successful community-based management approaches have encouraged policymakers to advance the scope of this approach across local borders. Such advances have shaped the academic discussions to explore ideas for making businesses integrate DEI (Diversity, Equity, and Inclusion); ESG (Environmental, Social, and Governance), and the SDGs (Sustainable Development Goals) for productive outcomes and mutual benefits of all the stakeholders. (Toke, 2024). This approach is also warranted for achieving sustainable forestry where both indigenous forest-dwelling populations will benefit as well and the forest ecosystem will be protected. This will further help in mitigating climate change issues and addressing the historical injustice among the Indigenous communities (FAO, 2022).

India has a forest cover of approximately 22% of its total land area. This makes the Indian forest economy a multi-billion economy capable of contributing significantly to the nation's GDP. The forest economy is also significant because more than 200 million local forest dwellers are residing in or near the forest areas whose involvement in the management of the forests will increase both the productivity of the forests as well as the livelihood opportunities for such forest dwellers (Chhatere & Kattamuri, 2022). However, forests in India are mostly managed by the State. The local communities that have been residing in and around the forests were historically exploited through the 'People Free Zone' policies of the State and thereby they lack eminent control over the natural resources (Pratap, 2010). Further, the Common-law system in India favors *eminent domain* through which the State has assumed enormous control over the forests of India. Such practices have expelled the local forest dwellers from the forest lands where they were traditionally residing. Moreover, although the local communities were expelled from their forest lands, commercial exploitation of forests took place in a manner detrimental to the health of the forests (Jain & Das, 2019). After independence conservation of forests became the core of the forest policies, however, the rights of the forest-dwelling communities gained a very late recognition. It was only by the National Forest Policy, 1988 that community participation in the management of forests through a sustainable commercial approach was acknowledged (Joshi et al., 2011). However, the policy failed to do complete justice and the same came into conflict with the provisions of the Forest Rights Act, 2006 which explicitly focuses on addressing the historical injustice faced by the local forest dwellers by ensuring community and individual ownership rights over the forest resources. This new legislation paves the way for an inclusive approach through which the community-based management of forest resources in India might achieve new heights. It further provides a scope for the forest-dwelling communities to establish themselves as small forest-based enterprises that could deliver several services for large-scale industries promoting a bottom-up approach and decentralized governance in the corporate world. However, the corporate mechanism in the management of forests involving the local communities as a stakeholder is critical owing to the insecurity of their tenures, and the nature of their property regime over the forest resources. The paper therefore will deal with the corporate aspects of community-based management of forest resources within the legal framework in India for analyzing the future opportunities of such management in achieving the goals of sustainable forestry and development.

The paper is expected to enhance the scope of CSR in integrating local communities for sustainable trade in forest resources at the global level where global companies can adopt such approaches for mitigating global issues of climate change adverse effects due to loss of forest cover. It also supports the notion of 'think globally act locally' whereby integration of community participation in corporate policies at the global level can make local people contribute towards the mitigation of global issues. Further, the paper examines the scenario concerning the Indian context, while similar approaches can be adopted by different countries within the domestic policy framework securing the rights of their respective local communities.

14.2 Review of Literature

Forests assume the attributes of common property resources (CPR) due to the characteristics of subtractability and non-excludability (Basu et al., 2017). And the debates over community management of common pool resources including forests gained popularity after the publication of *The Tragedy of the Commons* by Garett Hardin in 1968 where centralized regulations were preferred over community ownership of natural resources because community management of resources results in degradation (Hardin, 1968). Since then, community management of resources and their economic viability have been studied in different parts of the world. In India, N. S. Jodha highlighted the importance of community management of natural resources for the sustainable livelihood of the poor rural population through empirical data (Jodha, 1986). Similarly, Robert Wade proved that both privatization and centralized regulation are ineffective in the sustainable utilization of CPR and hence advocated for collective action (Wade, 1987). Elinor Ostrom further introduced and prioritized a new model for managing CPR via, durable cooperative institutions that are governed and organized by users of CPR themselves and this model was also introduced as an alternative to privatization and centralized regulation of CPR for protecting them (Ostrom, 2009). After considering the importance of CPRs including forests for the sustainable livelihood of the poor and dependent populations as well as the efficiency of local communities in the management of such resources, researches were done advocating the integration of community management in policies. For instance, Abhi A. Ene highlighted the requirements for legal measures to ensure efficient community forestry and established the role of CPR in achieving sustainable forestry (Abi,

2016). In addition to the above works the Global Forest Resource Assessment Report 2020 and the Global Forest Goals Report 2021 have depicted successful stories of community management of forests gaining economic as well as ecological advantages. Thus, it is evident that there are innumerable advantages of community participation in corporate governance. Naia Toke through his research has proved that such community involvements yield better resource management, ensure transparency, and develop a mutual trust between corporations and local communities that can ultimately result in sustainable practices since it develops a sense of responsibility and ownership among the community members (Toke, 2024). Community-managed forests in practice have shown better results in ecological conservation and economic development compared to sole management by corporations and governments in many case studies (Raj, 2023).

However, despite all these community involvements, there exist several challenges like lack of trust among the stakeholders, conflicts of power, and unavailability of adequate resources in the practical implementation of such an approach as highlighted by Naia Toke in his research. Along with those issues, in India studies prove that the situation is even more complicated due to political ignorance leading to inefficient legislations on this behalf (Dasgupta, 2023). Although the legal framework recognized the rights of local communities in the management of forests in India in the late 1980s (Kumari et al., 2019), the integration of such management in the corporate policies was never recognized. Hence, a gap exists to the recognition of the role of local communities in harmonizing economic benefits with ecological stability in the sustainable management of forests through corporate policies. This study is therefore expected to investigate the aspects of corporate governance integrating local community management through policies for sustainable trade in forestry.

14.3 Community-Based Corporate Approach for Sustainable Trade in Forest Resources

There is both demand and supply of forest resources but both these two have not fully reconciled to harness the big economic opportunity in the forest economy. Industry demands a formal, assured, and cost-effective source of raw materials at scale irrespective of whether it be for existing, or future sectors of energy or as alternatives to fossil fuels. For such reasons, forests are required to be nurtured and cared for without disrupting productivity. This is where the local communities play a major role (Chhatere & Kattamuri, 2022). The traditional knowledge of the local indigenous communities, their experience of being in harmonious relations with nature for centuries, their best practices of management, and the mutually dependable relationship between the forests and the forest-dwelling communities are the major factors that allow the traditional forest-dwelling communities to be at a better position as a stakeholder in the management of forests (Sick, 2008). Further forests managed sustainably by local communities play a vital role in carbon sequestration and climate change mitigation. Species like *Dalbergia sissoo* and *Tectona grandis*, known for their carbon sequestration capacity, thrive under community stewardship. By maintaining healthy forest ecosystems, local involvement in resource trade helps mitigate climate change impacts through carbon storage and ecosystem restoration efforts (Mondal et al., 2022).

Forests shall be managed considering them not only as ecological systems but also as biological systems that are under the influence of human beings who are again influenced by several aspects of the socio-economic environment. Therefore, the dynamics of forest management involve both direct and inverse relationship between ecological and social characteristics. Such arrangements demand for human adjustments with natural complexities within a particular geographical locality that is different from one locality to another. These diversified local arrangements increase the demand for corresponding diversified solutions required to counter complex natural phenomena involved in the management of forest resources. Given such circumstances, the Indigenous communities who developed harmonious relationships with nature for a long period become effective in managing the resources (Singhal & Rishi, 2012). Therefore, a single blueprint approach might not be adequate for managing such resources in different local situations and thus the role of local communities needs to be recognized in the policies for management of forest resources. This system of management is well contemplated through the mechanism of Community Forestry that allows a respectable share to the local communities in the decision-making process and also to retain the benefits making them a stakeholder in the forest governance (Abi, 2016). But such a system is different from the traditional community institutions managing the natural resources. Here legitimacy of the rights of the communities is attributed by external factors rather than internal legitimacy and sovereignty. Thus, now recognition of community rights over forest resources depends on the law and governance.

Empowering local communities through initiatives like Joint Forest Management (JFM) and High Conservation Value Forest (HCVF) safeguards natural habitats while supporting livelihoods. By combining indigenous knowledge with modern conservation practices, these approaches can promote sustainable forest use and enhance ecological resilience. Participation of local communities in forest management not only delivers economic and ecological benefits but also fosters social cohesion and community development, granting local residents decision-making power over forest resources. This empowerment reduces dependency on external resources, enhances community resilience, and promotes inclusive governance, allowing all community members to influence local policies directly (Chakraborty & Chanda, 2022).

To cite some successful case studies, the Maya Biosphere Reserve can be referred to where granting of concessions to the local communities along with 2 concessions to timber-based industries have resulted in significant progress in the management of the forests and the development of the communities including the rise of employment opportunities, social investments, capacity-building, and improved access to bank credits as a result of the concessions' increasing credibility (Stoian et al., 2018).

Even the Royal Society for the Conservation of Nature was given management authority over Jordan's forests after the government designated them as national reserves. With the help of the local communities, they managed livestock grazing in a balanced manner, planned capacity-building initiatives, increased tourism to create jobs, and obtained loans for community projects (RSCN, 2014). The establishment of Makuria National Park in Colombia resulted in disputes with the *Wayúu* community. They were given collective land tenure by the government in 1984, which settled their differences. This partnership demonstrated the advantages of integrating local communities in forest management by facilitating effective park protection, sustainable trade, and forestry. There are several similar stories where the involvement of local communities in the management of forest resources has led to both commercial as well as conservational benefits. It has also been acknowledged by the Global Forest Resource Assessment Report (GFRA) 2020 that the knowledge and values of the local communities associated with forest conservation contribute towards the protection of forests during the utilization of such forests.

Attempts are being made to involve local communities in India's corporate governance structure. The Access and Benefit Sharing mechanism is one such example where the Biological Diversity Act, of 2002 provides for the constitution of Biodiversity Management Committees (BMCs) at the local level with the member of the village Panchayat as the President of the Committee, one Forest Officer as the Secretary, and local villagers as members (Alaknanda, 2020). This committee is responsible for the sustainable trade of biological resources where local communities are provided a share in the benefits derived from such utilization. In addition, any access to such biological resources shall be made only with Prior Informed Consent (PIC) from such communities who are designated as owners of the resources. Further, the corporations who access such resources are responsible for sharing the profit in the form of benefits equitably among such communities and the corporations are also provided incentives for such access. Government intervention is ensured through the National Biodiversity Authority (NBA) at the national and State Biodiversity Boards (SBBs) at the state level for monitoring and regulating the Access and Benefit Sharing agreements between the corporations and the local communities (Pisupati, 2015). This model can even prove beneficial if implemented in the forestry sector. However, such a model suffers from its limited applicability and the temporary nature of its members. Another initiative Forest Certification involves two distinct categories-Forest Management Unit Certification (FMU) and Chain of Custody (CoC) ensures suppliers of forest produce from well-managed forests are rewarded against suppliers from illegal and unknown sources and also provides the buyers a kind of guarantee of the quality of the products certified. In India too, the Ministry of Environment, Forest and Climate Change (MoEF&CC) is emphasizing establishing a National Forest Certification Council with the support from the Centre for Sustainable Forest Management and Forest Certification (CSFM&FC) and the WWF (WWF-India, 2008). Such an initiative will develop a relationship of dependability between the local forest communities and the corporate stakeholders but this process has yet to develop to its full strength in the country.

Moreover, it is to be admitted that despite decades of calls for decentralization, over 85% of the world's forests remain state-owned, reflecting entrenched political and economic interests, especially in countries like India and Tanzania. International frameworks like the United Nations Framework Convention on Climate Change, 1992 (UNFCCC) have largely failed to support local land rights, maintaining state dominance. Although involving local forest dwellers in resource trade fosters grass-roots engagement and strategic partnerships, empowering communities in national policymaking and aligning with Sustainable Development Goals (SDGs), it has not received adequate policy recognition. Traditional practices like swidden farming are crucial for forest-dependent populations but often lack formal land rights, leading to marginalization. Therefore, integrating local communities into forest management is essential for economic development, environmental stewardship, and social justice through recognized land rights and sustainable resource practices (Keshwan, 2017).

14.4 Implication of Trade on the Forest Health and Life of Forest Dwellers in India

Forests provide a variety of ecosystem services that are vital for the growth and survival of humanity. Forests are necessary to mitigate climate change, to prevent soil degradation, for water conservation, to prevent natural disasters like floods, etc. (FSI, 2011) Besides being the essential element of nature, it supplies a large quantity of natural resources that directly as well as indirectly support the economy of a nation. India being a developing and agriculturalist nation, with approximately more than 200 million population of forest-dependent communities, is heavily dependent on its forest resources. The category-wise total forest cover in India has been provided in Table 14.1 where VDF implies Very Dense Forests; MDF implies Moderately Dense Forests; and OF implies Open Forests. This categorization was implemented by the Food and Agricultural Organization in its Global Forest Resources Assessment Reports (FAO, 2020) and the same has been applied in India while preparing its India State of Forest Reports. Table 14.1 shows the category-wise extent of forest cover in India.

For estimating the productivity of the forests, the assessment of the growing stocks becomes significant. Such an assessment is applied as a quantifiable parameter to assess the economic value of the trees inside forests. While assessment of the growing stocks outside the forests is essential to ascertain the productivity of the trees outside the forests that can be used for the nation's development. Figure 14.1 shows the State-wise volume of growing stocks in India in 2021.

Table 14.1 Category Wise Forest Cover in India in 2021 (FSI, 2021a)

Category	Forest Cover Area in sq. kms
VDF	99,779
MDF	3,06,890
OF	3,07,120
Total	7,13,789

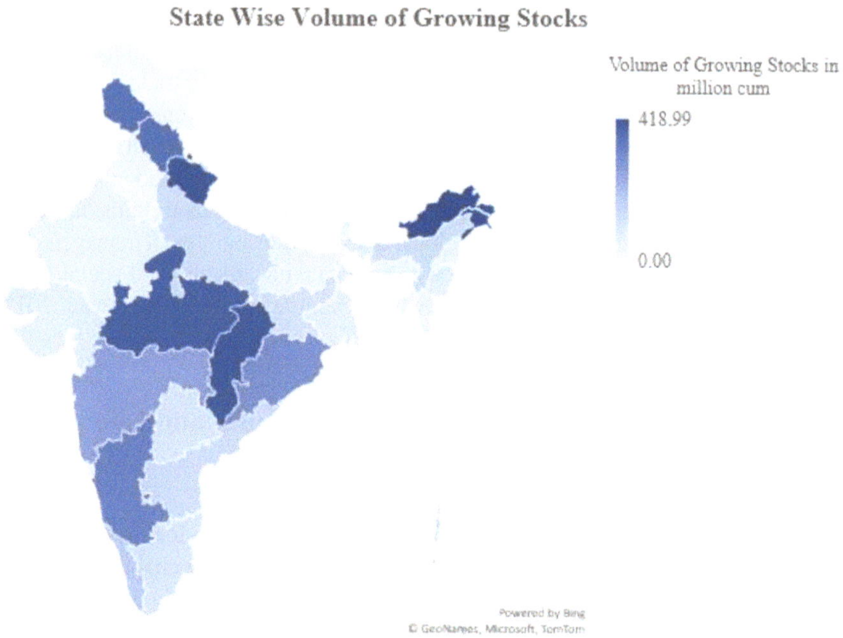

Fig. 14.1 State Wise Volume of Growing Stocks in India in 2021 (FSI, 2021b)

The total growing stock of wood in the country in 2021 is estimated to be 6,167.50 m. cum including 4,388.15 m. cum inside the forest areas and 1,779.35 outside the recorded forest areas. It has been reported that there is an increase of 4.26% of growing stock in the country compared to 2019. Figure 14.1 highlights that the highest growing stock per hectare among the states is in Kerala followed by Uttarakhand and among the UTs the highest per hectare growing stock is in Ladakh followed by Jammu & Kashmir (FSI, 2021b).

Although forests provide a variety of ecosystem services, some of them are directly influential on the economy of India. Such services are quantifiably measured in three categories—Timber Provisioning Services; Non-Timber Forest Products (NTFP) Provisioning Services; and Carbon Retention Services. The contributions of the forestry and logging sector to the economy of India have been derived from the National Accounts Statistics that calculate the Gross Value Added (GVA) from the forestry sector in two broad categories via—timber products comprising of Industrial Wood and NTFP comprising non-timber forest materials including bamboos, lac, sandalwood, etc. (EnviStats, 2022). Figure 14.2 demonstrates the State value of timber provisioning services in India in 2019–2020. Figure 14.3 provides the state-wise value of NTFP provisioning services in 2019–2020. NTFP generally includes all biological resources that are derived from the forests and are used for anthropogenic purposes. The value of carbon retention services is calculated based on the Social Cost of Carbon (SCC) approach. It represents the economic cost of climate damage or benefit due to the increase in the emission of CO_2 by one ton (Ricke et al., 2018) and this is often used as a carbon price estimate. By using the GDP deflator, India's social cost of one ton of CO_2 in 2019–2020 has been estimated as US$91 by EnviStats India 2022. Figure 14.4 provides the year-wise comparative analysis of the value of carbon retention services in India.

Figures 14.2 and 14.3 highlight that the timber provisioning services per hectare at current prices are maximum in Kerala followed by Goa and Punjab. On the other hand, the highest value to NTFP provisioning services is observed in Tripura, Kerala and Bihar (EnviStats, 2022).

Timber trade has always been the highest concern for production forestry since the colonial period. However, the local communities are rarely allowed a share in such trade and the State along with private entities are the biggest monopolies. This is due to the regulatory restrictions in the reserved and protected forest areas. Thus, the local forest dwellers are rarely allowed any share in the benefits of timber trade. On the other hand, approximately 275 million poor rural population amounting to 27% of the total population in India depends on NTFPs for subsistence. This dependency

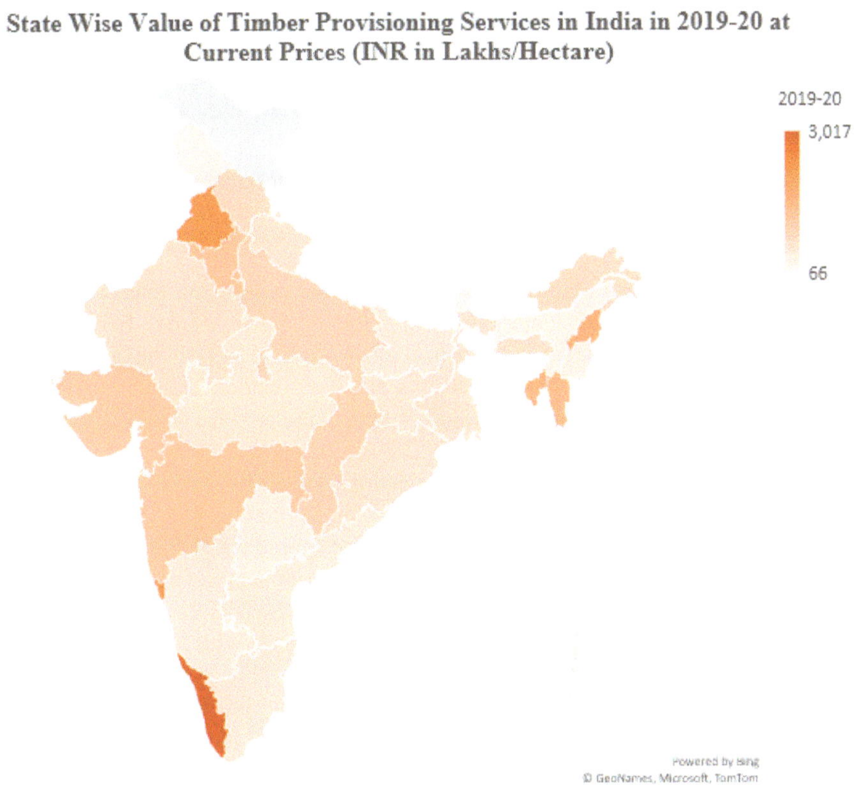

Fig. 14.2 State wise value of Timber Provisioning Services in India at current prices in 2019–2020 (Mational Accounts Division, 2024)

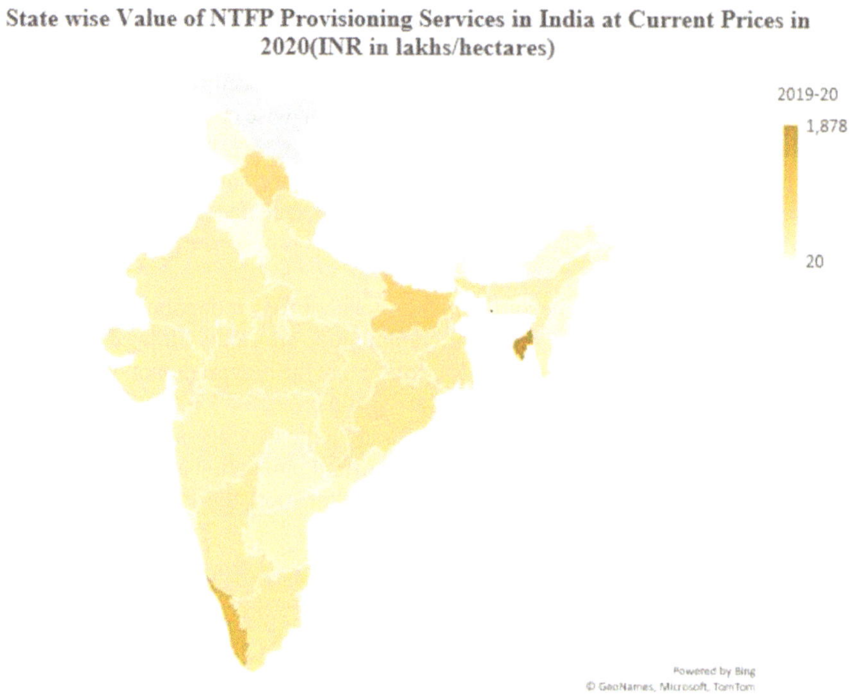

Fig. 14.3 State wise value of NTFP Provisioning Services in India at current prices in 2020 (Mational Accounts Division, 2024)

is however more intense among the 89 million tribal population that is the most disadvantaged section of the society residing in the forest fringe areas. Even 70% of the NTFP collection takes place in the tribal belt and around 55% of the employment in the forestry sector is generated from this NTFP collection (Pandey & Bisaria, 1996). NTFP collection by the tribal differs from state to state where in Manipur alone 90% of the population depends on NTFP as a major source and approximately 2,50,000 women are employed in the collection of NTFPs. In Bastar district of Chhattisgarh around 75% of the population supplement their food with NTFPs. The government of India reported that at least 35 million man-days of employment were generated in the trading of NTFPs and such trading was inclusive of the collection and reprocessing of economically viable NTFP species. These NTFPs are vital sources of livelihood for the tribal population especially for those who have limited non-agricultural income opportunities (MOEF, 2006).

Figure 14.4 highlights the Value of carbon retention at the current price in India during the year 2019–2020, which is INR 546.95 thousand crores, while it was only INR 438.79 thousand crores in 2017–2018. This shows that the value of carbon retention has increased which is due to the increase in the carbon stocks (EnviStats, 2022). Carbon retention services are highly significant for the local forest dwellers in the contemporary period since they can generate adequate incentives for the local

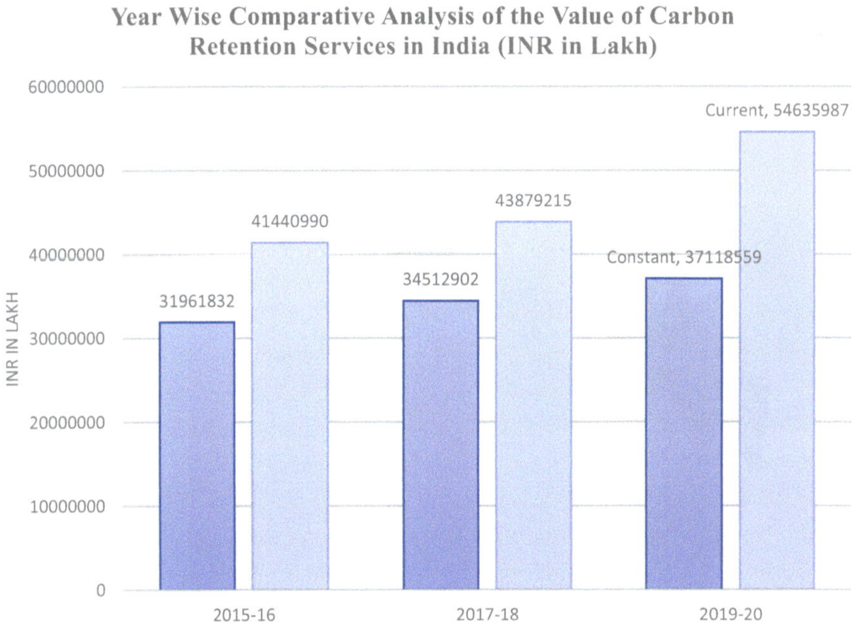

Fig. 14.4 Year wise comparative analysis of the value of carbon retention in India (EnviStats, 2022)

communities engaged in the conservation of natural forests if the benefits of such trading are shared with such communities (FAO, 2022). This would not only generate an alternative income for the local communities but would also encourage such communities to protect the forests rather than destroy them for survival.

From the above analysis, it becomes clear that these three forestry services have a major role to contribute for the country's GDP where it has been observed that in 2019–2020 timber provisioning services, NTFP provisioning services, and carbon retention services have contributed to 0.13%, 0.08%, and 3.75% of India's GDP respectively (EnviStats, 2022).

In addition to the above economic aspects of the forest resources, bamboo as a resource has been separately analyzed by the Forest Survey of India due to its immense significance for the 2 million local artisan community whose livelihood is dependent on the production and sale of bamboo. A majority of these communities are tribal and forest dweller communities. Bamboo is also essential because of its versatile nature for making a variety of products without much investment by small entrepreneurs (FSI, 2021c).

Figure 14.5 highlights that Madhya Pradesh in India has the highest bamboo-bearing area with 1.84 million hectares followed by Arunachal Pradesh with 1.57 million hectares. However, it has also been reported that the total bamboo-bearing area of the country in 2021 has decreased by 1.06 million hectares compared to 2019. Madhya Pradesh having the highest bamboo-bearing area has shown the highest decrease of around 2,473 sq. km area as reported in 2021.

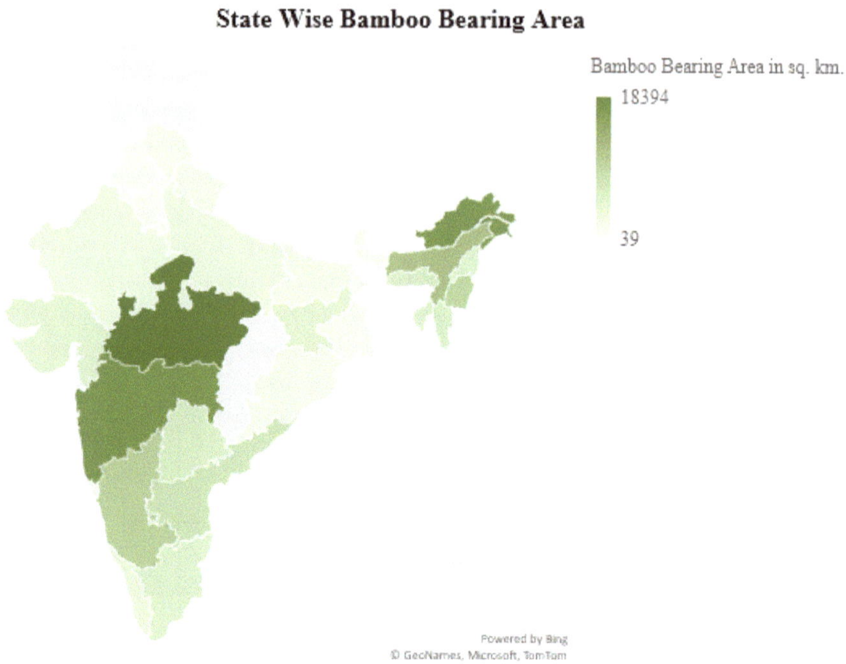

Fig. 14.5 State Wise Bamboo Bearing Area in India in 2021 (FSI, 2021c)

Figure 14.6 highlights that a maximum number of green culms is found in Arunachal Pradesh followed by Assam and Madhya Pradesh. The increase in the soundness of bamboo culms has been highest in Arunachal Pradesh followed by Assam and Orissa (FSI, 2021c).

Although the above stats show the economic potential of the forest resources in India, it is to be noted that around 40% of the forest cover is degraded and 70% of the forests have lost their natural regeneration potential (Patra et al., 2022). Since poverty and ecological degradation are reciprocal, in countries like India with approximately 21.9% of the total population being below the poverty line out of which 275 million people are dependent on forest for subsistence, degradation of forests impacts such populations while the poverty of such population results into the degradation of forests due to their dependency (Parogaien, 2022). Unsustainable trade practices have impacted negatively on forest health (Roy, 2020). For instance, although India has witnessed a slight increase in its forest cover in the last few years, in some states the forest cover rate has decreased. This is alleged to be mainly because of shifting cultivation and other unsustainable trade practices (Kumari et al., 2019). It was found that over 14,000 sq. km. of forest land had been destroyed for industrial purposes in India. Exporting primary forest products to developed nations has resulted in a grave decline in the forest health of the nation (ISBINSIGHT, 2018).

Table 14.2 highlights how forests in India have suffered deforestation due to the diversion of forest lands for non-forest purposes. This shows that the State is a major

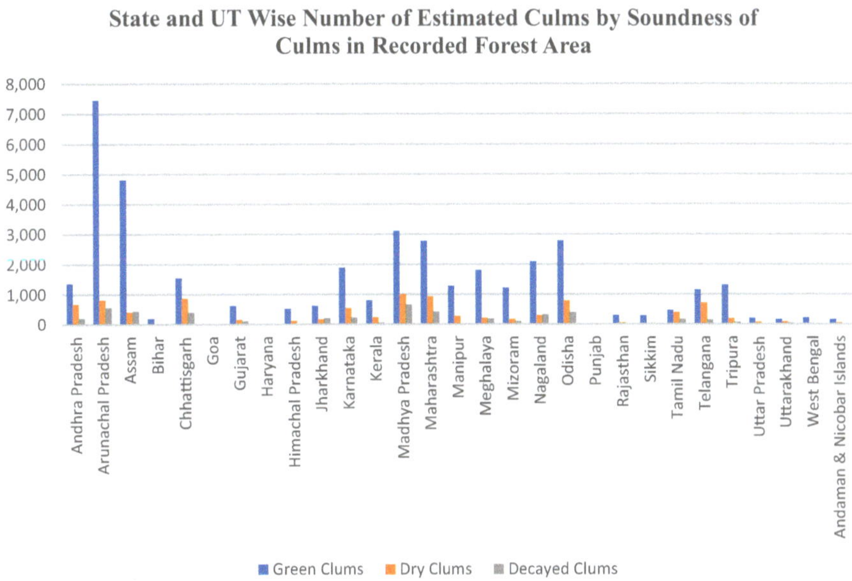

Fig. 14.6 State and UT Wise Number of Estimated Culms by Soundness of Culms in Recorded Forest Area (FSI, 2021a)

factor responsible for the destruction of forests and the blame is often placed on the forest dwellers. Until recently, plantation for increasing forest areas in India was massively done to increase revenue and production of timbers through various silviculture systems. Such plantations although based on sophisticated research across the country failed to recognize the unique geographical conditions of the different localities (Parogaien, 2022). As a result, such plantations failed to achieve the desired objective of forest restoration in the manner expected and the growing stocks decreased significantly.

The ISFR 2021 claimed an increase of 0.22% of the forest cover in India amounting to a net increase of 1,540 sq. km. in 2021 compared to its previous assessment in 2019. But the report failed to recognize that the quality of the forests has significantly deteriorated where 15,183 sq. km. of the natural forest land got deforested within these two years. This deforested area is nearly 10 times higher than the increased area of new forests as claimed by ISFR. Around 9.117 sq. km. of the degraded forests have turned into either scrubs or non-forest lands. From 2015 to 2021, ISFR reports provide that around 74,457 sq. km. of the forest cover has been degraded including 31,367 sq. km. of VDF and MDF out of which 10,979 sq. km. of such forest cover has been entirely cropped down. In the recent 2021 ISFR, the increase of the VDF is due to the conversion of MDF, OF, scrubs, and Non-Forest Lands into VDF. However, it has failed to recognize the loss of natural forests and demonstrated the increase of anthropogenic plantation forests as the increase in total forest cover (Rupavath, 2022).

Table 14.2 State wise diversion of forest lands for non-forest purposes till 2019 (Kukreti, 2020)

State	Number of projects	Total Forest Land approved for diversion (in ha)
Andhra Pradesh	3	37.82
Bihar	28	453.43
Chhattisgarh	1	207.99
Goa	1	0.93
Gujarat	99	114.01
Haryana	251	519.53
Himachal Pradesh	52	434.36
Jharkhand	11	869.99
Karnataka	11	162.61
Kerala	2	0.26
Madhya Pradesh	220	795.36
Maharashtra	2	151.81
Mizoram	1	23.69
Odisha	14	4514.00
Punjab	123	411.07
Rajasthan	27	370.34
Tamil Nadu	6	18.45
Telangana	11	2055.05
Tripura	2	1.80
Uttar Pradesh	1	63.27
Uttarakhand	64	159.74
West Bengal	2	102.33
Total	932	11,467.83

Table 14.3 highlights that the MDF has decreased while OF has increased significantly with a decrease in the non-forest land in 2021 compared to 2019. It shows the result of the increase of plantation forests with the downfall of MDF forests. While VDF has a minor increase due to the conversion of MDF, OF, and Scrubs into VDF.

Moreover, the authenticity of the Government reports is doubtful due to inconsistent definitions of forest, as a result of which any tea gardens or courtyard gardens can be treated as forests since forests in India are measured only through remote sensing satellites. Furthermore, the northeast of India which is rich in biodiversity and consists of a large number of tribal populations, has lost nearly 1,020 sq. km. of forests in 2021 compared to 2019 assessments (Agarwal, 2022). It is to be noted that the larger the green cover with older and healthier trees, higher the sequestration levels of carbon. Thus, deforestation of natural forests will lead to loss of ecosystem value since carbon restoration value will get decrease. Consequently, due to such deterioration of the forest health, India has suffered a loss of approximately 1.4% of GDP annually (Sethy, 2018).

Table 14.3 Forest cover change matrix between 2019 and 2021 in India (FSI, 2021a)

Class	VDF	MDF	OF	Scrub	Non-Forest	Total ISFR
VDF	97,770	382	348	28	150	99,278
MDF	1696	3,02,216	2,736	331	1,493	3,08,472
OF	245	2,939	2,94,200	1,471	5,624	3,04,499
Scrub	31	241	3,048	40,977	2,000	46,297
Non-Forest	37	312	6,788	3,712	25,17,874	25,28,923
Total ISFR 2021	99,779	3,06,890	3,07,120	40,539	25,27,141	32,87,469
Net Change	501	−1,582	2621	242	−1,782	

14.5 Status of the Rights of Local Forest Dwellers in India

'*Forests in India is a business, which is based more upon mathematics than upon botany*' was a statement given by Dr. Dietrich Brandis, Former Inspector General of Forests. This shows the colonial intention behind forest regulations during the imperial rule which was aimed at preserving forests for future exploitation expelling all the native tribal communities who had such resources. (Dasgupta, 2023). As a result, forests got deteriorated and the forest dwellers lost their attachments to nature causing ecological imbalance. Although management and protection of forests by the local communities has been prevalent in India since time immemorial, for instance, the *Bishnoi* Community of Rajasthan who were known for its protective nature towards their local forests and animals, it was only between 1970 and 1990 that community-based forest management emerged at the administrative level after independence (Kumari et al., 2019).

An analysis of forest laws in India reveals that the state perceives forest communities primarily from an economic aspect, as economic factors are dominant themes in these laws. Some laws recognize both cultural and economic aspects, while others fail to acknowledge any claims of forest communities over forests. Community-based organizations can be classified as traditional or modern, with traditional ones existing before the passage of laws, and modern ones being created by statutes for specific objectives. The legal documents are creating more modern community institutions rather than strengthening traditional ones. Legal documents recognize various aspects of these communities but often neglect intra-village forest use and changes in demographic profiles. Forest communities are defined as socio-economic groups organized through traditional or modern institutions, spatially living in and around forests, economically dependent on forests for livelihood needs, and conserving biodiversity and forest resources. However, this definition renders forest communities apolitical and static, used by the state for its economic interests while ignoring the traditional values associated with forests (Jain & Das, 2019).

So far ownership of forests in India is concerned, 13,081,000 hectares of forest area is under private ownership and 57,747,000 hectares of forest area is under public ownership till 2020. The public administration has control over 32,998,000 hectares of forest lands while the local and tribal communities hold rights over 24,759,000 hectares (FAO, 2020). This holding of the tribal communities is only because of the JFM system which is a collaboration between the State and local communities, granting limited control to tribes without absolute rights. This creates a contractual property regime where the State retains dominant authority. Consequently, State-controlled Forest lands become open access areas with local communities having insecure tenure, leading to inferior villages. Even though the land reforms in India permit for having common property resources in the villages, such resources have faced massive deterioration due to their improper management under the influence of rural rich (Ali, 2011), privatization, and decentralized patterns of the government (Singh, 2012).

State-sponsored mining activities in forest areas have led to the marginalization of the local communities and allowed the private sector to gain a monopoly. Although institutions like *Panchayati Raj* have been developed for decentralizing the control over natural resources, such institutions could not do justice to the marginalized non-tribals and the *Adivasi* communities since they are unable to contest elections due to lack of funds and they even could not contest without support from a central political party. The status of the Commissions framed for the welfare of such communities under Articles 338 and 338A of the Constitution of India is merely advisory with very minimum executive powers (Expert Group, 2008). As a result, the concerns of such sections of the population have become more vulnerable at the onset of globalization due to large-scale commodification and commercialization of land. The land market has opened for mining, farming, and industrialization, with laws favoring these uses. The Forest Commission and amended land policies facilitate privatization, often at the expense of forest dwellers. the greatest example is the definition of landless poor where the earlier meaning of landless meant persons having less than 2 decimals of land has been altered to mean no male member of a family should owe 2 decimals of land which has excluded a majority of landless people with no fault of their own. The State's policies prioritize commercial value over local community needs, allowing private sector encroachment on tribal and agricultural lands (Partners for Law in Development, 2002). The phrase—where there is bauxite there is spring|| is a better example in Orissa.

The State has thus committed a breach of the Public Trust by extracting resources more than was expected causing the local communities to protest for protecting and utilizing such resources, the *Chpko* Movement being one fine example (Petruzzello, 2024). The State's mismanagement of natural resources led to conflicts with local communities through a one-sided participatory approach and commercialization of commons. Weak State control resulted in community rejection of State claims, sparking political debates. A 2008 expert report cited failures in protections under Schedule V, Tribal Advisory Councils, and delayed legislation, devaluing tribal autonomy and leading to insurgencies like *Naxalites.* The Panchayat Extension to Scheduled Areas Act, 1996 (PESA) aimed for decentralized management but failed to grant real autonomy. States bypassed PESA, exploiting tribal lands, while bureaucratic-industrial nexus worsened issues. These problems significantly harmed the tribal economy and sustainable resource management (Expert Group, 2008).

The increase in industrial land by 24 million acres and forest land by 28 million acres in India since independence are surprising figures but this is mainly because of reclassification of the wastelands as private and protected areas. Such an increase denies the traditional forest dwellers access to their forest lands while collection of NTFPs by private entities benefits companies and increases state revenue. Again, forest dweller's rights are restricted in the protected areas but the same protected areas are opened for eco-tourism. And if the tribal people protest then such protests are contended through regulations and their livelihood concerns are subjected to a law-and-order situation. Even the Dhebar Commission reported that policies from 1884 to 1952 have only strengthened state control over forests by transforming tribal rights from privileges to concessions (Dhebar Commission, 1961).

Although the JFM and PESA both aim at empowering the tribal communities and thus in Scheduled Areas the JFM should have considered the provisions of PESA but both 2000 and 2002 guidelines of JFM failed to recognize PESA. Moreover, the state conservationist agencies argue that PESA should not be used to secure the land rights of the tribals in protected forests. Thus, policies like JFM have given a new form of state control over forests rather than balancing the powers or reducing conflicts between local interests and the state. In Andhra Pradesh, the State took control over 40,000 acres of forest land (Tewari, 2015). Another example where the villagers in *Sukhomajari*, Gujarat, formed an organization to conserve soil and forests, sharing resources equally. But after 20 years, the forest department broke its promise, and distributed the land to another village leading to inter-village conflicts (Partners for Law in Development, 2002).

Even though the Forest Rights Act, of 2006 was enacted to address the historical injustice with the traditional forest dwellers (Indian Parliament, 2006), it faced severe issues in the implementation. Creating two categories of beneficiaries, via- Scheduled Tribes and Other Traditional Forest Dwellers (OTFDs), has led to systematic discrimination of the OTFDs who are mostly denied their rights on the ground of lacking stringent residential evidence (Kumar, 2016). As a result, such OTFDs face exploitation at the hands of bureaucrats also leading to the policy of 'Divide and Rule' (Sharma, 2018). It has also failed to recognize the community rights at the pace expected, and the provisions of the 2006 Act came in conflict with the JFM Committees (Rao, 2022).

Further, the implementation of this Act of 2006 often conflicts with the Forest Conservation Act 1980 and Wildlife Protection Act 1972, which restrict activities in protected areas. Although the Act grants land rights to forest dwellers, it limits their ability to use land for economic support due to restrictions on alienation or mortgage. The 2007 guidelines for the Act aimed to include tribal women in decision-making, but many committees lack female representation. Additionally, the forest department's veto power and ineffective appeal mechanisms have led to the rejection of tribal claims and immediate eviction of rejected claimants.

14.6 Achieving Sustainable Development Goals Through an Inclusive Corporate Governance Approach in Forestry in India

India is a signatory to all the major International Conventions and Treaties regulating forests and forest resources. India has also committed to restore 21 mega hectares of degraded forests by 2030 under the Bonn Challenge (IUCN, 2018). The forestry sector in India has also been acknowledged as an integral component in the Nationally Determined Contributions (NDC) and several initiatives like the Green India Mission and Compensatory Afforestation (Press Information Bureau, 2019). All these initiatives are also expected to simultaneously generate an alternative income for the forest-dependent communities (Patra et al., 2022). India's roadmap towards sustainable forestry is drawn based on such commitments where the participation of local forest-dependent communities in the management and distribution of resources assumes significant importance. Such communities shall not only be protected against historical injustice by addressing their core issues rather they shall be provided with positive incentives through a definitive value-chain approach for their contribution towards sustainable forestry. In 2019 due to the regulatory requirement of spending 2% of the net profits of the Companies as a contribution towards Corporate Social Responsibility (CSR) resulted in INR 24,689 crores. A considerable amount of this fund can be contributed towards the empowerment of the communities for the restoration of forests and supplying sustainably generated forest resources (Parogaien, 2022). Further by enhancing the manufacturing capacities and export of finished goods, the export of primary forest-based products will be reduced and subsequently, the increasing pressure on the forest cover will be reduced along with allowing the surplus benefits from the export of manufactured goods to reach the real beneficiaries (ISBINSIGHT, 2018). And for increasing the volume of such manufactured goods, the skills of the forest-dependent communities can be enhanced by encouraging their involvement through market incentives.

NTFPSs are also acquiring higher commercial value in trade and industries on account of an upsurge in demand for natural resources. The trade in NTFP can be considered an incentive for forest conservation since it generates an income for forest-dependent communities. Furthermore, due to the rise of globalization, trade in NTFPs has risen to new heights where unseen hands are contributing to the price determination for those of NTFP collectors to the food and drug companies sitting at distant places from across the world (Pandey et al., 2016). Engaging local communities in NTFP trade not only sustains their income but also promotes forest conservation by creating economic incentives for sustainable harvesting. For instance, communities in Bankura, West Bengal, generate income from Sal leaf plates and *Belmala* beads. Women play a vital role in this sector, although decision-making power varies. This will ensure gender equality and eliminate various forms of discrimination within the communities Despite the economic and social potential, challenges such as reliance on middlemen, lack of modern techniques (Nandi & Sarkar, 2020), narrow market access, tenure insecurity, and lack of processing skills (MOEF, 2006) restricts the benefits of increasing trade in NTFPs. Such hindrances can be removed through efficient regulations whereby the local communities are provided with tenure securities and ownership rights over their land and resources. Although the Forest Rights Act 2006 promises to provide such rights to the traditional forest dwellers, the Act itself has contributed towards systematic discrimination among the forest dwellers communities.

Right economic incentives at the community level will ensure sustainable forestry by local communities and will also deliver better value to both the community and the industry. The Forest Rights Act 2006 in India is expected to provide an institutional framework where through secured community rights and tenure, such communities can manage themselves into medium-scale formal enterprises. If such an institution is framed then it is expected to increase visibility and traceability in the supply chain along with creating a guaranteed supply source to the industry. The accountability for sustainable extraction and management can be easily fixed with the local communities through encompassing legal authority at the local level and the formal structure of enterprises (Chhatere & Kattamuri, 2022). However, the Act has suffered from serious defects at the implementation level. Thus, a shift in the ideology of the policymakers is required from People Free Zone Policies to a Participatory Management Process to recognize the role of local forest dwellers.

For this Participatory Forest Management (PFM) approach, an initiative can be adopted where local communities are provided with the rights of harvesting and collecting NTFPs in exchange for their assurance of protecting and conserving the forests for 5–10 years. It is observed that local communities in India under the JFM system have contributed to an increase from 22,017,583 hectares to 2,144,000 hectares till March 2006 which was similar to the informal contract for PFM. However, still in India, the local communities have no legal authority while in countries like Nepal, the Community Forest User Group (CFUG) is registered under the Forest Act,1983 (Kumari et al., 2019). Private-community partnerships can further benefit the process of sustainable trade through community involvement. Such initiatives have even been started in India and are showing positive results. For instance, BILT Graphic Paper Products Limited (BGPPL) is sourcing bamboo from forests that are owned and managed by communities in Maharashtra at scale. Similarly, AAK is initiating to sale of Sal seeds harvested by women-led communities in Jharkhand. A combination of industrial and emerging technologies will enable value creation at scale, with additional elements of improved visibility and traceability (Chhatere & Kattamuri, 2022).

Even for recording precise data on forest inventory and for predicting yields for business planning, data provided through satellite imageries can be supplemented by ground-level data provided by the local communities. This same technology can serve the purpose of sustainable forestry since any change in the tree cover or any of the basic parameters like carbon sinks at the landscape level will get reflected in real-time which will benefit all the stakeholders including local communities, industrialists, and regulatory agents for ensuring sustainable trade throughout the supply chain. Involving local communities in the trading of forest resources is a triple-win approach where quality products will be produced, and monitored with precise data through the involvement of modern technologies, industries will be ensured a guaranteed quality supply of raw materials, and the consumers will have a reliable source to verify products manufactured and sold through such approaches.

Thus, involving forest-dependent local communities in the corporate governance over forest resources will ensure DEI (Diversity, Equity, and Inclusion) and ESG (Environment, Social, and Governance) that will further ensure the achievement of the SDGs (Toke, 2024). DEI stands for Diversity, Equity, and Inclusion, where allowing the local communities to be a stakeholder in the management of forest trade will enable diversity in the problem-solving approach that suits best the different geographical issues related to forest management, equal distribution of profits among the diversified groups of local communities will ensure equal protection of their rights, and their involvement in the decision-making process will bring an inclusive approach towards sustainable forestry by making them equally responsible. The partnership between the local communities and the corporation will enable the corporates to contribute towards ecological development as a part of Corporate Social Responsibility through the empowerment of such communities. In this case, the empowerment of the local communities by the corporations will ensure greater security for the rights of such communities providing social benefits. The governance can monitor and regulate the functioning of such a model through efficient policies providing security of tenure and ownership rights. Such an approach thus will serve the purpose of achieving the following SDGs: SDG 1: No Poverty; SDG 5: Gender Equality; SDG8: Decent Work and Economic Growth; SDG 10: Reduced Inequalities; SDG 12: Responsible Consumption and Production; SDG 13: Climate Action; and SDG 15: Life on Land. Figure 14.7 illustrates the idea of such an inclusive model.

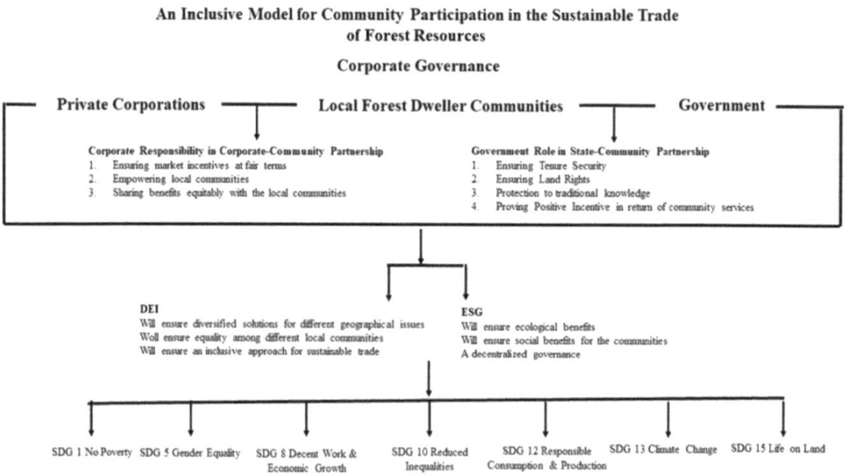

Fig. 14.7 An inclusive model for community participation in the sustainable trade of forest resources (Self-authored)

14.7 Conclusion

The paper has highlighted the importance of community participation in the management and the trade of forest resources along with the implications of such an inclusive approach on the forest health and the life of forest dwellers. It has also analyzed the status of the rights of forest dwellers under the legal framework in India. Finally, the paper has attempted to forecast the possibility of an inclusive approach involving local communities in the corporate governance for sustainable trade in forest resources. It has been observed that such an approach will lead to a win-win situation for all the stakeholders in the management of forests and will move a step towards sustainable forestry along the lines of the SDGs. However, the forest policies are required to address the issues of inconsistencies in the forest definitions and policies, discrimination between different forest dwellers communities, and disparities between the private corporations and the local communities. Moreover, security of land tenures, and ownership rights through efficient implementation of the Forest Rights Act 2006, facilitation for positive market-based incentives for the forest-based products produced or sold by local communities, and greater autonomy through adequate representation of the traditional knowledge of the local forest dwellers shall be provided.

References

Abi, E. A. (2016, March 21). Binary logistic models for the evaluation of Community Forestry as a tool for effective forest law in Cross River State Nigeria. *Arboricultural Journal, 37*(4), 243–251. Retrieved June 22, 2024, from https://doi.org/10.1080/03071375.2015.1136147

Agarwal, M. (2022, January 24). *The state of India's forests: Losing forests, gaining plantations.* Retrieved June 12, 2024, from Mongabay: https://india.mongabay.com/2022/01/the-state-of-indias-forests-losing-forests-gaining-plantations/

Alaknanda. (2020). A critical analysis of the implementation of access and benifit sharing in India. *E-Journal of Academic Innovation and Research inIntellectual Property Assets (E-JAIRIPA), 1*(1), 29–46. Retrieved October 24, 2024, from https://cnlu.ac.in/wp-content/uploads/2022/08/3-Alaknanda.pdf

Ali, J. (2011, June 8). Commentary: Livestock, common property resources and rural smallholders in India. *International Journal of Agricultural Sustainability, 4*, 265–268. https://doi.org/10.1080/14735903.2007.9684826

Basu, S., Jongerden, J., & Ruivenkamp, G. (2017). Development of the drought tolerant variety Sahbhagi Dhan: Exploring the concepts commons and community building. *International Journal of Commons, 11*(1), 144–170. Retrieved May 2021, 21, from https://www.readcube.com/articles/10.18352/ijc.673

Chakraborty, S., & Chanda, A. (2022). Assessment of forest cover change, community responses, and conservation strategy: Evidence from North Sikkim District, India. In M. Sahana, G. Areendran, & K. Raj (Eds.), *Conservation, management and monitoring of forest resources in India* (p. 221). Springer.

Chhatere, A., & Kattamuri, S. (2022, November 15). *Forests of the futur: A national economic opportunity.* Retrieved January 12, 2024, from Forbesindia.com: https://www.forbesindia.com/article/isbinsight/forests-of-the-future-a-national-economic-opportunity/81239/1

Dasgupta, S. (2023). Timber exploitation: The beginning of early colonial onslaught on Indian forests in the background of industrial revolution in the 19th century. *International Journal of Applied Research, 8*(1), 99–106. https://www.allresearchjournal.com/archives/2022/vol8issue2/PartB/7-12-141-605.pdf

Dhebar Commission. (1961). *Report of the scheduled areas and scheduled tribes commission.* Government of India. Retrieved January 12, 2014, from https://tribal.gov.in/downloads/other-important-reports/Dhebar-Commission-Report-02.pdf

EnviStats. (2022). *Forest Ecosystem, Vol-11 Environment Accpunts.* EnviStats India.

Expert Group. (2008). *Development challenges in extremist affected areas.* 2008: Government of India. Retrieved June 11, 2024, from https://tribal.nic.in/downloads/Statistics/OtherReport/DevelopmentChallengesinExtremistAffectedAreas.pdf

FAO. (2020). *Global Forest Resource Assessment Main Report 2020.* Food and Agriculture Organization, United Nations.

FAO. (2022). *The State of World's Forest 2022, Forest pathways for green recovery and building inclusive, resilient and sustainable economies.* Food and Agriculture Organization, United Natios.

FSI. (2011). *India State of Forest Report 2011.* Forest Survey of India.

FSI. (2021a). *02, Forest Cover, India State of Forest Report, 2021.* Forest Survey of India.

FSI. (2021b). *07, Growing Stock, India State of Forest Report 2021.* Forest Survey of India.

FSI. (2021c). *08, Bamboo Resources of the Country, India State of Forests Report, 2021.* Forest Survey of India.

Hardin, G. (1968, December 13). The tragedy of the commons. *Science, 162*(3859), 1243–1248. Retrieved December 12, 2019, from https://pages.mtu.edu/~asmayer/rural_sustain/governance/Hardin%201968.pdf

Indian Parliament. (2006). *The Scheduled Tribe and Other Traditional Forest Dwellers (Recognition of Forest Rights) Act, 2006, Act No. 2 of 2007.* Government of India.

ISBINSIGHT. (2018, August 22). *Can trade save forests?* Retrieved June 22, 2024, from Forbes India: https://www.forbesindia.com/article/isbinsight/can-trade-save-forests/51071/1

IUCN. (2018, August 14). *India first of all Bonn Challenge countries to develop progress report.* Retrieved June 15, 2024, from iucn.org: https://www.iucn.org/news/india/201808/india-first-all-bonn-challenge-countries-develop-progress-report#:~:text=At%20the%20UN%20Framework%20Convention,8%20million%20

Jain, A., & Das, S. (2019). Crisis of the forest community in postcolonial Indian forest policies and laws. *Environmental Sociology, 5*(1), 23–32. https://doi.org/10.1080/23251042.2018.1492663

Jodha, N. S. (1986). Common property resources and rural poor in dry regions of India. *Economic and Political Weekly, 24*(27), 1169–1181. Retrieved May 21, 2021, from https://www.jstor.org/stable/4375858

Joshi, A. K., Pant, P., Kumar, P., Giriraj, A., & Joshi, P. K. (2011). National forest policy in India: Critique of targets and implementation. *Small-Scale Forestry*, 83–96. https://doi.org/10.1007/s11842-010-9133-z

Keshwan, P. (2017). *Democracy in the woods: Environmental conservation and social justice in India, Tanzania, and Mexico.* Oxford Academic. Retrieved July 12, 2024, from https://doi.org/10.1093/acprof:oso/9780190637385.001.0001

Kukreti, I. (2020, January 9). *India continues to lose forest land to non-forestry projects.* Retrieved May 21, 2024, from DownToEarth: https://www.downtoearth.org.in/forests/india-continues-to-lose-forest-land-to-non-forestry-projects-71633

Kumari, R., Banerjee, A., Kumar, R., Kumar, A., Purabi, S., & Mohammed, K. L. (2019). Deforestation in India: Consequences and sustainable solutions. In M. N. Suratman, A. L. Zulkiflee, D. O. Gabriel, B. Nathaniel, S. Yosio, & A. C. Carlos (Eds.), *Forest degradation around the world* (IntechOpen, Trans.). https://doi.org/10.5772/intechopen.77433

Kumar, V. (2016). A critique of the scheduled tribes and other traditional forest dwellers (recognition of forest rights) act 2006. *Environmental Law, Research and Advocacy, 1*, 1–5.

Mational Accounts Division. (2024). *Statistical Forestry on Value of Output from Agriculture and Allied Sectors (2011–12 to 2022–23)*. National Accounts Division, National Statistical Office, Ministry of Statistics and Programme Implementation, Government of India. https://mospi.gov.in/sites/default/files/publication_reports/Brochure_2024.pdf

MOEF. (2006). *Annual Report 2005–2006*. Ministry of Environment, Forests and Climate Change. New Delhi: Government of India.

Mondal, S., Chattopadhyay, P., Banerjee, A., & Palot, D. (2022). Floristic composition and inventorization of forest resources in some selected forest areas of Paschim Bardhaman District, West Bengal, India. In M. Sahana, G. Areendran, & K. Raj (Eds.), *Conservation, management and monitoring of forest resources in India* (p. 191). Springer.

Nandi, D., & Sarkar, S. (2020). Non-Timber Forest Products based household industries and rural economy—A case study of Jaypur Block in Bankura District, West Bengal (India). In P. K. Shit, H. R. Pourghasemi, P. Das, & G. S. Bhunia (Eds.), *Spatial modeling in forest resources management: Rural livelihood and sustainable development* (p. 505). Springer.

Ostrom, E. (1990). *Governing the commons, the evolution of institutions for collective actions*. Cambridge University Press.

Ostrom, E. (2009). *Governing the commons: The evolution of institutions for collective action (Political Economy of Institutions and Decisions)*.

Pandey, A., & Bisaria, A. (1996). Rational utilization of important medicinal plants: A tool for conservation. *The Indian Forester, 124*(4). https://doi.org/10.36808/if%2F1998%2Fv124i4%2F4977

Pandey, A., Tripathy, V., & Kumar, A. (2016). Non Timber Forest Products (NTFPs) for sustained livelihood: Challenges and strategies. *Research Journal of Forestry, 10*(1), 1–7. https://scialert.net/abstract/?doi=rjf.2016.1.7

Parogaien, M. C. (2022, February 02). *[Commentary] Forest restoration: challenges and opportunities for India*. Retrieved June 22, 2024, from Mongabay: https://india.mongabay.com/2022/02/commentary-forest-restoration-challenges-and-opportunities-for-india/

Partners for Law in Development. (2002). Forests. In *4th annual workshop on common property resources and the law* (pp. 6–13). Partners for Law in Development. Retrieved June 11, 2024, from https://feministlawarchives.pldindia.org/wp-content/uploads/Annual_Workshop_on_Common_Property_Resources_and_the_Law.pdf

Patra, S., Kumar, A., & Saikia, P. (2022). Deforestation and forests degradation impacts on livelihood security and climate change: Indian initiatives towards its mitigation. In V. Singh, S. Yadav, K. Yadav, & R. Yadava (Eds.), *Environmental degradation: Challenges and strategies for mitigation* (Vol. 10A). Water Science and Technology Library. https://doi.org/10.1007/978-3-030-95542-7_18

Petruzzello, M. (2024, June 17). Chipko movement. *Britannica Encylopedia*. Retrieved July 12, 2024, from https://www.britannica.com/topic/Chipko-movement

Pisupati, B. (2015). Access and benefit sharing: Issues and experiences from India. *Jindal Global Law Review, 6*(1), 31–38. https://doi.org/10.1007/s41020-015-0007-2

Pratap, D. (2010, November 29). Community participation and forest policies in India: An overview. *Social Change, 40*(3), 235–256. https://doi.org/10.1177/004908571004000301

Press Information Bureau. (2019, December 13). *National Afforestation Programme*. Retrieved June 14, 2024, from plb.hov.in: https://www.pib.gov.in/Pressreleaseshare.aspx?PRID=1596332#:~:text=National%20Afforestation%20Programme,&%20Management%20Scheme%20(FFPM)

Raj, A. (2023). Community participation: A windfall for corporate social responsibility implementation in developing countries. In P. Mishra, A. Sharma, S. Khanra, S. Kundu, & S. Mishra (Eds.), *Digital economy post Covid-19 era. INDAM 2023*. Springer. https://doi.org/10.1007/978-981-99-0197-5_33

Rao, V. G. (2022, September 22). *Forests Rights Act 2006: A status update*. Retrieved January 12, 2023, from cdn.cseindia.org: https://cdn.cseindia.org/attachments/0.95350600_1663835835_fra-status-ygirirao.pdf

Ricke, K., Drouet, L., Caldeira, K., & Tavoni, M. (2018). Country-level social cost of carbon. *Nature Climate Change, 8*(10), 895–900. https://doi.org/10.1038/s41558-018-0282-y

Roy, A. (2020, March 24). *Is government telling a distorted story of India's forests? Here's what it must do.* Retrieved June 11, 2024, from Times of India: https://timesofindia.indiatimes.com/blogs/toi-edit-page/is-the-government-telling-a-distorted-story-of-the-state-of-indias-forests-heres-what-it-must-do/

RSCN. (2014, September). *The Full Story.* Retrieved January 10, 2024, from Jordan Biodiversity: http://www.jordanbiodiversity.com/content/full-story

Rupavath, P. (2022, January 27). *[Analysis] Indian forests around the size of Nagaland, thinned down in two years.* Retrieved June 11, 2024, from Mongabay: https://india.mongabay.com/2022/01/analysis-indian-forests-around-the-size-of-nagaland-thinned-down-in-two-years/

Sethy, P. (2018, April 19). *Study on Economics of Desertification, Land Degradation and Drought (DLDD) in India.* Retrieved June 10, 2024, from The Energy and Resources Institute: https://www.teriin.org/project/study-economics-desertification-land-degradation-and-drought-dldd-india?utm_source=newsletter&utm_

Sharma, A. R. (2018, July 2). *Why India's Forest Rights Act is discriminatory against non-tribals.* Retrieved January 10, 2023, from Business Standard: https://www.business-standard.com/article/economy-policy/why-india-s-forest-rights-act-is-discriminatory-against-non-tribals-118070200116_1.html

Sick, D. (2008). Social contract and institutional chane in common-pool-resource management. *Society and Natural Resources, 21*(2), 94–105.

Singhal, R., & Rishi, P. (2012). Conflicts in institutional development: Institutions for joint forest management. *Forest, Trees and Livelihood*, 261–270. Retrieved January 21, 2021.

Singh, S. (2012, April 4). Communities in transition: Implications for common property resource management. *Forests, Trees and Livelihood, 16*(4), 311–328. https://doi.org/10.1080/14728028.2006.9752571

Stoian, D., Rodas, A., Butler, M., Monterroso, I., & Hodgdon, B. (2018). *Forest concessions in Petén, Guatemala: A systematic analysis of the socioeconomic performance of community enterprises in the Maya Biosphere Reserve.* Center for International Forestry Research (CIFOR). https://www.cifor-icraf.org/knowledge/publication/7163/

Tewari, V. (2015, April 16). Forest inventory, assessment, and monitoring, and long-term forest observational studies, with special reference to India. *Forest Science and Technology, 12*(1), 24–32. https://doi.org/10.1080/21580103.2015.1018962

Toke, N. (2024, June 4). *Triple the impact—Integrating DEI, ESG, and UN SDGs for maximum corporate responsibility and business opportunity.* Retrieved June 10, 2024, from Diversity for Social Impact: https://diversity.social/dei-esg-un-sdg-impact-strategy/

Wade, R. (1987). *The management of common property resources: Collective action as an alternative to privatization or state regulation (English).* Agricultural Research Unit. World Bank Group. Retrieved May 21, 2021, from https://documents.worldbank.org/en/publication/documents-reports/documentdetail/350041467997569278/the-management-of-common-property-resources-collective-action-as-an-alternative-to-privatization-or-state-regulation

WWF-India. (2008, April 29). *Consultative Meeting on Forest Certification in India.* Retrieved October 22, 2024, from wwf.india.org: http://awsassets.wwfindia.org/downloads/forest_certification_meeting__background_1.pdf

Chapter 15
CSR Towards Building Sustainable Cities and Communities Through Sustainable Governance

Rashmeet Kaur Bhupinder Singh Saluja and Priyanka Srivastava

Abstract This chapter explores the relationship between Corporate Social Responsibility (CSR), Sustainable Development Goals (SDG), and sustainable governance in the real estate sector. The real estate sector is expected to provide strong, sustainable, and inclusive human settlements as India's demand for urban spaces is increasing rapidly. According to SDG Report 2024, India's contribution towards building sustainable cities and communities has shown an overall downward trend. The percentage of urban residents living in slums has remained constant, while fewer people in metropolitan areas get clean water. As the Indian population continues to urbanize, the business sector may encourage sustainable urban development by utilizing CSR funds and sustainable governance. This chapter begins with mapping the relationship between the CSR targets mentioned in Schedule VII of the Companies Act, 2013 with SDGs, and discusses the opportunities and challenges for developing sustainable cities and communities. The chapter further includes the case studies of sustainable cities and communities built in India through CSR and the significant role of sustainable governance in providing a decent quality of life to the citizens. The chapter concludes with the author's insight regarding the present scenario of sustainable cities and communities in India, thus giving a clear understanding to the readers.

Keywords Corporate Social Responsibility · Sustainable governance · Sustainable Development Goals (SDG) · Real estate sector · Sustainable cities and communities

R. K. B. S. Saluja (✉)
School of Commerce, Manav Rachna International Institute of Research and Studies, Faridabad, Haryana, India
e-mail: rashmeetsaluja88@gmail.com

P. Srivastava
School of Leadership and Management, Manav Rachna International Institute of Research and Studies, Faridabad, Haryana, India
e-mail: priyankasrivastava.slm@mriu.edu.in

V. Kandpal et al. (eds.), *Developments in Corporate Governance*,
Approaches to Global Sustainability, Markets, and Governance,
https://doi.org/10.1007/978-981-96-6366-8_15

15.1 Introduction

15.1.1 The Emergence of CSR

The concept of Corporate Social Responsibility (CSR) practice was propounded by various British industrial philanthropists in the nineteenth century. They promoted the idea of "enlightened self-interest" (Carroll, 2015). The concept was unknown for the last few decades until Bowen first used the term in his famous book "Social Responsibilities of Businessman," published in the year 1953. CSR is defined as the ethical behavior of a firm towards society and other stakeholders with a legitimate interest in the business (WBCSD, 1999, p. 6). In India, CSR emerged from being a mere philanthropic or charitable activity to being the central focus of the socially conscious corporate movement. CSR has been linked to corporate philanthropy and the Gandhian trusteeship phase, when firms contributed to social development and supported government efforts for community development (Sundar, 2000). The nation observed a huge transition in its CSR activities from philanthropy-based philosophy to the awareness and growth of educational institutions for imparting education to all. In 1991, the implementation of Liberalisation, Privatisation, and Globalisation (LPG) in India encouraged public sector undertakings to contribute towards the economy's growth by including CSR as a part of their sustainable business plan.

Since then, numerous examples have been of corporations diverging their CSR funds for sustainable practices. One such example has been put forth by Ambuja Cements. According to the data in CSRBOX, the corporation, through its CSR arm, mobilized sources and worked with local authorities to implement a comprehensive water conservation plan at Magarway hamlet. The project entailed carefully excavating and enlarging the pond's potential, guaranteeing that it could save a significantly greater amount of water. Throughout the operation, great effort was made to minimize environmental impact and adhere to sustainable methods. Hyundai Motor India Foundation (HMIF), a leading automobile company has started a healthcare initiative called Sparsh Sanjeevani offering complete healthcare services at an affordable cost. Tech Mahindra Foundation has started a program SMART with 50% of its beneficiaries as women. This program aims to empower women who are school dropouts, women with disabilities, and those who cannot complete their education. Thus, corporations are engaging in various CSR initiatives leading to sustainable development of the nation. Graph 15.1 depicts the amount spent by Indian corporates in multiple sectors in the financial year 2021–2022.

15.1.2 History of Sustainable Development Goals

The UN Sustainable Development Organization announced the adoption of seventeen sustainable development goals on September 25, 2015, encompassing 169 targets and aiming to develop many more to be achieved by 2030. The objectives demand

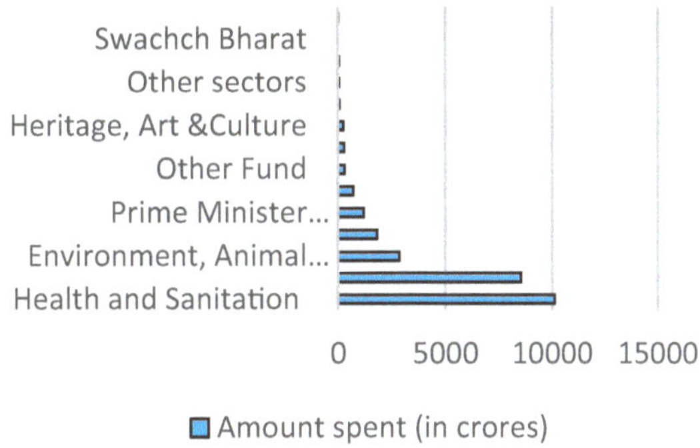

Graph 15.1 CSR spent across sectors. *Source* Ministry of Corporate Affairs (2021–2022)

a sustainably developing agenda through cooperative efforts by governments, businesses, and civil society organizations. These goals address global environmental, social, and economic issues. According to the UN Department of Economic and Social Affairs, the history of Sustainable Development, beginning with the UN Conference on the Human Environment, held in 1972, to the Paris Agreement in 2015, is discussed below.

- The UN Conference on the Human Environment, held in 1972, marked the beginning of sustainable development. The Conference addressed environmental concerns and began a discussion between industrialized and developing countries about economic progress, pollution, and human well-being.
- The Brundtland Report was designed to debate ways for achieving sustainable development by 2000.
- At the Rio+5 Conference in 1997, a five-year evaluation of the Earth Summit's progress was undertaken.
- In September 2000, leaders from 189 nations signed the Millennium Declaration, which set eight Millennium Development Goals to be achieved by 2015. The goals were designed to consider economic, environmental, and social factors, but unequal growth made them unattainable.
- The post-2015 discussions aimed to establish an egalitarian framework that balances all three dimensions equally, ensuring no one falls behind. The World Summit emphasized the importance of collaborative responsibility in advancing and strengthening sustainable development's three pillars: economic, environmental, and social. The Summit recognized the need to build a viable and realistic plan for poverty eradication and human development.
- The Rio +20 Summit's "Future We Want" paper underlined the necessity of using the Millennium Development Goals as a foundation to build on the SDGs. This document highlights the necessity of using the Millennium Development Goals to build on the SDGs.
- A few weeks after the UN Summit in Paris, another worldwide agreement was struck to combat global warming and human-caused climate change. The "Paris Climate Agreement" was inextricably linked with the 2030 Agenda. These two fundamental agreements, taken together, formed an overarching framework.

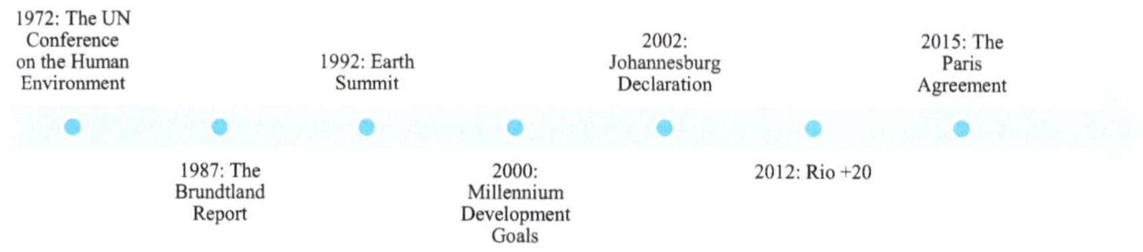

Fig. 15.1 History of Sustainable Development. *Source* Created by authors using data from the UN, Department of Economic and Social Affairs https://sdgs.un.org/goals

The history or stages of sustainable development is depicted in the Fig. 15.1.

15.1.3 CSR and SDG

The companies must prioritize their SDGs and set objectives that CSR projects could achieve in line with the 169 targets. The convergence of SDGs and CSR paints a broader landscape for the company and allows exchanging views with the global community through the same channel. On channeling their efforts through CSR programs that are estimated through the lens of SDGs, the businesses are most likely to have an eminent position in the conversation of sustainable development (KPMG, 2018).

The overlap between SDGs and CSR activities has become the global language of sustainable development. As the sustainable development goals and the targets mentioned in Schedule VII of the Companies Act 2013 coincide, it provides a great deal of flexibility for enterprises to integrate their SDGs with the CSR programs and develop a comprehensive SDG target-oriented CSR approach in their policy. The CSR activities and the related SDGs are discussed in Table 15.1.

15.1.4 Sustainable Cities and Communities

The 2030 Agenda for Sustainable Development, adopted by all UN Member States in 2015, reflects a common aim of prosperity and peace. It understands that getting rid of poverty and other deprivations must be linked with programs to promote education and health and fight climate change. In 2015, the UN General Assembly set SDG 11, themed "Sustainable Cities and Communities" and its official mission is to "make cities inclusive, safe, resilient, and sustainable." This goal deals with green and inclusive urbanization, sustainable transport, and affordable housing. The proposed indicators for measuring SDG 11 are discussed in Table 15.2.

Table 15.1 CSR activities and mapped SDGs

Activity	CSR activities	Mapped SDGs
Activity 1	Removal of hunger, poverty, and malnutrition	SDG 1: No poverty SDG 2: Zero hunger SDG 3: Good health and well-being SDG 6: Clean water and sanitation
Activity 2	Promoting education	SDG 1: No poverty SDG 2: Zero hunger SDG 4: Quality Education SDG 8: Decent work and economic growth
Activity 3	Promoting gender equality, empowering women	SDG 1: No poverty SDG 5: Gender equality SDG10: Reduced inequalities
Activity 4	Ensuring sustainability of the environment	SDG 6: Clean water and sanitation SDG 7: Affordable and clean energy SDG 9: Industry, innovation and infrastructure SDG 11: Sustainable cities and communities SDG 13: Climate Action SDG 14: Life below water SDG 15: Life on land
Activity 5	Preservation of national heritage, art, and culture	SDG 9: Industry, innovation and infrastructure SDG 11: Sustainable cities and communities
Activity 6	Measures for the welfare of military veterans, war widows, and their dependents	SDG 3: Good health and well-being SDG 4: Quality Education SDG 8: Decent work and economic growth
Activity 7	Training to promote rural sports, nationally recognized sports, paralympic sports, and Olympic sports	SDG 8: Decent work and economic growth SDG10: Reduced inequalities

(continued)

Table 15.1 (continued)

Activity	CSR activities	Mapped SDGs
Activity 8	Contribution to the Prime Minister's National Relief fund	SDG 1: No poverty SDG 2: Zero hunger SDG 3: Good health and well-being SDG 4: Quality Education SDG 6: Clean water and sanitation SDG 9: Industry, innovation and infrastructure SDG 15: Life on land
Activity 9	Contribution to incubators for research and development initiatives in science, technology, engineering, and medicine	SDG 9: Industry, innovation and infrastructure SDG 12: Responsible Consumption and Production SDG 13: Climate Action
Activity 10	Rural development projects	SDG 1: No poverty SDG 2: Zero hunger SDG 3: Good health and well-being SDG 4: Quality Education SDG 9: Industry, innovation and infrastructure
Activity 11	Slum area development	SDG 1: No poverty SDG 2: Zero hunger SDG 3: Good health and well-being SDG 4: Quality Education SDG 6: Clean water and sanitation SDG 7: Affordable and clean energy SDG 9: Industry, innovation and infrastructure
Activity 12	Disaster management, including relief, Rehabilitation, and reconstruction activities	—

Source KPMG (2017)

Table 15.2 Proposed indicators for measuring SDG 11

The National Indicator Framework, 2022	
Targets	Indicators
11.1 Ensure that everyone has access to essential housing amenities	11.1.1. Proportion of urban population residing in slums, informal settlements, or substandard housing
11.2. Access to affordable and safe transport systems for all, including those in vulnerable situations by 2030	11.2.1. People who can use public transportation irrespective of age, gender, and disabilities
11.3. Improve sustainable and inclusive urbanization and create dimensions for integrated and sustainable human settlement planning by 2030	11.3.1. Ratio between the growth rate of population and land usage 11.3.2. The percentage of cities that have a framework for involving civil society directly in planning and running the city
11.4. Take care of and protect the world's cultural and natural heritage	11.4.1. Type of private funding, total expenditure for preservation and protection of heritage, and level of government
11.5. By 2030, decrease the number of fatalities relative to the global GDP	11.5.1. Death rate per 100,000 people 11.5.2. Loss of infrastructure, and essential services and loss to global GDP due to disasters
11.6. Improve waste management and air quality to reduce cities' negative influence on the environment	11.6.1. Cities create a proportion of urban solid waste that is collected and discharged properly 11.6.2. Annual average of tiny particulate particles
11.7. Provision of open spaces that are safe, accessible, and inclusive by 2030	11.7.1. Share of built-up land for public usage 11.7.2. Offense of any kind and location of occurrence, in the last year
Means of Implementation of SDG 11	
11. a. To promote strong links between urban, peri-urban, and rural regions, national and regional development planning should be strengthened	
11. b. There is a need for integrated policies and strategies for inclusiveness, optimum use of resources, and mitigation and adaptation of climate change by 2020	
11. c. Develop strong buildings through assistance	

Source Ministry of Statistics and Programme Implementation (Government of India). 2022. SDG National Indicator Framework Baseline Report, 2022–23. New Delhi

15.1.5 Sustainable Governance

Governance is a joint duty of government, business, and civil society representatives addressing societal issues (Driessen & Glasbergen, 2002). There are four perspectives of governance on sustainable development; ecological sustainability perspective (focus on the environment), well-being perspective (quality of life), hierarchical (top-down approach with government as the lead player), and deliberative governance (bottom-up approach and equal relation with all the parties involved) (van Zeijl-Rozema et al., 2008). To solve governance issues, eleven principles are

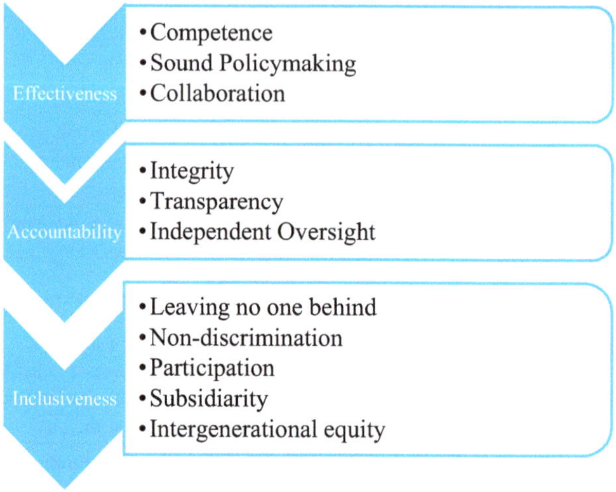

Fig. 15.2 Values of Effective Governance for Sustainable Development. *Source* Created by authors using data from the Committee of Experts on Public Administration (UN, Department of Economic and Social Affairs). https://publicadministration.un.org/en/cepa

shown in Fig. 15.2 focusing on three main aspects; effectiveness, accountability, and inclusiveness.

This chapter aims to understand the impact of CSR on creating communities and cities that are sustainable (SDG 11) through sustainable governance as minimal studies highlight the relationship between these variables. As India accounts for almost 17.96% of the global population, and the second-largest employer in the nation is the real estate industry, this becomes an important case study to analyze the efforts to achieve SDG 11. The involvement of this industry is critical because, despite efforts to enhance CSR implementation, the question of sustainable CSR implementation has not been fully addressed. The sustainable effort for increasing green buildings is caught in the vicious circle of blame and it needs to be broken as this mental atmosphere can create an infinite circle (Andelin et al., 2015). However, the absence of thorough research on SDG 11's implementation and localization in India (Kandpal & Okitasari, 2023) may put sustainability on the real estate industry's back burner. By illustrating the case studies of how CSR money may be utilized in generating sustainable and inclusive urbanization through sustainable governance, this study aims to close this research gap.

The relationship between CSR and SDG 11 achievement must be investigated, particularly in developing countries, where the proportion of slum dwellers is increasing, the environmental footprint is increasing, there is a lack of coordination between different government hierarchical structures, and even access to basic infrastructure is limited. Thus, there is a need to integrate sustainable development into the CSR practices of the company, as it can have a long and sustainable impact on the built environment.

15.2 Research Objectives

1. To identify the link between CSR targets in Schedule VII of the Companies Act, 2013, and the SDGs.
2. To discuss the opportunities and difficulties in creating sustainable cities and communities in India.
3. To discuss the case studies of sustainable cities and communities established in India using CSR and sustainable governance.

The rest of the chapter is laid out as follows: Section 15.3 addresses the literature review that emphasizes the importance of CSR in making communities and cities sustainable through sustainable governance. Section 15.4 describes the research methodology wherein the case study method has been adopted. Section 15.5 discusses the opportunities and challenges in developing sustainable cities and communities in India, followed by the case study, implications, and future work.

15.3 Literature Review

Ebekozien et al. (2024) investigated how SDG 11 may be achieved and how homeownership among Nigerian junior employees in the private sector can be enhanced using the extended CSR (ECSR) framework. The researchers gathered data from knowledgeable individuals in government housing like staff from ministries, departments, agencies, mortgage institutions, housing developers, private organizations, and junior staff members. The study covered two major commercial cities: Lagos and Federal Capital Territory. The study identified 18 sub-factors to enhance house ownership among junior personnel in the private sector. These included integrating ECSR into low-income earners' homeownership, financing for junior staffers, incentives, stakeholder participation, government policy, legal framework, and construction costs. The study found that implementing these variables and sub-variables can improve Low-Cost Housing (LCH) and align with SDG 11. Thus, Nigerian housing officials and developers can use this approach to encourage private sector investment in housing.

Emeka-Okoli et al., (2024) explored how corporate governance frameworks impact CSR efforts, emphasizing the significance of board monitoring and stakeholder participation in promoting sustainable practices. Case studies demonstrate effective CSR implementation, creating shared value for stakeholders while achieving positive environmental and social consequences. Companies are incorporating Corporate Social Responsibility (CSR) into their business strategy to ensure long-term success and profitability. Companies are integrating CSR into their risk management procedures to identify and manage ESG risks that may affect operations, reputation, and financial performance. This encompasses hazards associated with climate change, community relations, and regulatory compliance.

Kandpal and Okitasari (2023) examined the localization of SDG 11 in India by evaluating national-level developments and analyzing sub-national reactions. This study uses a descriptive and exploratory method, utilizing a qualitative review of government documents and reports. MoSPI's National Indicator Framework identifies 15 national-level indicators to track progress against SDG 11 targets. Localized indicators cover 11 of the 14 SDG 11 indicators. Punjab and Chandigarh have the highest overall ratings. Goa, Gujarat, and Maharashtra are also among the top performers. West Bengal, Arunachal Pradesh, and Nagaland have the lowest scores in India's north and northeast. The study found that the SDGs' political impact on sub-national governance in India has mostly been discursive. State governments also differ in their approach to localizing SDG 11 and responding to national changes while monitoring and adhering to national requirements.

Khan and Ghouri (2022) studied the link between CSR, sustainability governance, and sustainable performance in the tourism sector. The study employed a content analysis technique. CSR was divided into two categories internal (employees) and external (society). It was found that CSR plays a critical role in reaching sustainable objectives. Organizations with a higher level of CSR have a higher sustainable performance. The sustainable performance of tourist organizations, sustainable governance, and Corporate Social Responsibility (CSR) are positively correlated, with sustainable governance acting as a mediator in this connection. Thus, sustainability governance incorporates CSR practices for admittance to reduce vulnerabilities and increase organizations' sustainable performance.

Wu and Jin (2022) looked at the connection between corporate social responsibility and sustainability. The study also looked at how internal governance affects relationships. Fixed-effects regression research on Chinese A-share listed companies from 2015–2019 was conducted. The result of the study indicates that the sample companies had poor overall sustainability, with considerable differences among them. CSR showed a significant positive correlation with corporate sustainable development, indicating that fulfilling CSR can help listed companies improve their sustainability. Accounting information quality impacts firms' ability moderately to fulfill social duties and achieve sustainable development goals. Thus, enterprises must prioritize social responsibility, business improvement, and sustainable development initiatives and the government departments should guide them in performing these social duties.

Hiep et al. (2021) conducted a systematic literature analysis of CSR and sustainability concerns in developing countries, with a particular emphasis on the real estate business between 1987 and 2019. Based on the literature research, a CSR-focused sustainable business model (contractual, ethical, legal, social, and economic responsibility, as well as stakeholders) was proposed and tested in the real estate industry of Vietnam. This industry has primarily focused on the high-end market, even though the bulk of the population earns low to moderate wages. This trend leaves average people homeless. Property developers concentrate on the high-end and luxury segments of the housing market due to higher profit margins. Furthermore, most real estate developers are accused of having a negative environmental impact, owing to the unclear legality of real estate items and initiatives that defy urban planning strategy and architectural landscape. Even as sales have increased, Vietnamese developers have failed to improve their social performance and environmental sustainability. Therefore, in order to achieve sustainable development in the real estate sector, sustainable leadership and the integration of corporate culture into a sustainable business model are necessary.

Lu et al. (2021) addressed CSR evaluation concerns and proposed a model that incorporates SDGs. The authors found that the selective application of SDGs led to concealing a company's harmful acts or misleading its customers by falsely emphasizing environmental activities and products. Many CSR assessment methodologies fail to evaluate a company's impact on SDGs, making it challenging to discern its true social responsibility. Furthermore, because different industries or corporations placed differing emphasis on CSR measures, only aggregate comparisons of enterprises were available. Thus, effective communication of the impact of sustainable development is crucial for stakeholders, in addition to improving implementation.

Aksoy et al. (2020) investigated the motivations that led to Corporate Sustainability Performance (CSP) level among Turkish enterprises included in the Borsa Istanbul Sustainability Index. This study examined the relationship between CSP, the structure of ownership, and board characteristics. The structure of ownership included foreign, family, institutional, and public ownership, while board qualities included the size of the board, female representation, independent board membership, and CEO duality. The logit and probit models showed a positive correlation between CSP and both institutional and foreign ownership, supporting stakeholder theory, whereas no link between CSP and family or public ownership was found. There was a positive correlation between board size and the share of independent members and CSP in Turkish firms.

Ionaşcu et al. (2020) examined how the real estate industry uses the SDGs to justify its position in society while meeting the needs of the stakeholders. There has been a continuous increase in integrated reporting among EU real estate entities. In 54% of cases, the presentation of the SDGs is unrelated to the remainder of the report's content. The real estate entities prioritized only eight SDGs, with SDGs 11, 13, and 7 being the most important. Real estate businesses must include all SDGs in their business strategies to have complete growth. There arises a need for open communication between all the stakeholders at different levels of the construction of a building to achieve sustainable outcomes.

Bae et al. (2018) investigated the impact of corporate governance and provided insight into the sustainability disclosure policies of South Asian nations. The paper examined both agency and signaling theories to understand the motivations and drivers for sustainability reporting. The data was compiled from annual sustainability reports from the GRI database focusing on 88 listed enterprises in India, Pakistan, and Bangladesh from 2009 to 2016. OLS regression analysis was employed to investigate the correlation between corporate governance and total sustainability disclosure (TSD). The findings indicate that foreign shareholding has a favorable relationship with TSD. Institutional shareholders have a positive and important association with TSD due to their focus on wealth maximization. TSD and director shareholding had a negative relationship as the directors' shareholders did not want the management to disclose information about the triple bottom line dimensions. The results show that independent directors reduce agency conflict and have a significant influence on the three sustainability criteria.

Huang et al. (2017) aimed to assess CSR implementation in the construction industry and identify achievable benchmarks for interested organizations. Secondary data from CSR reports of nine construction businesses listed in Taiwan's stock and OTC markets were analyzed and compared. An expert questionnaire survey was performed to gather thoughts on CSR and its implementation levels within their firms. The questionnaire survey aimed to assess the CSR implementation levels of large Taiwanese construction enterprises. Among the 69 companies, nine have issued CSR reports. The majority of the nine organizations' CSR reports included information about their CSR committees, risk management committees, and stakeholders. The construction companies have transparent corporate governance practices, including internal management rules, procedures, company conditions, and governance. Both contractors and real estate developers prioritize environmental sustainability and prioritize public welfare and charitable efforts to support the needy.

To transform the blame cycle for sustainable buildings into virtuous loops of adaptation, Andelin et al. (2015) explored the drivers for investors and tenants in sustainable buildings and their corporate responsibility (CR) actions. To comprehend the CR and sustainability drivers of renters and investors, the study used qualitative analysis. Tenant demand was regarded by the respondents as the second most significant motivator, while the property-level variables, including greater market value, lower costs, and marketing/branding potential, were the key CR drivers. According to the report, tenants and investors have adopted CR initiatives, realizing the significance of sustainability for the triple bottom line. The survey found that investors prioritized securing rental income and maximizing investment returns. Sustainable buildings were preferred by investors due to potential benefits such as cheaper property expenses, improved rental income, and marketing opportunities, while users sought premises that complement their business operations, reducing expenses and enhancing efficiency.

Warren-Myers (2012) examined how commercial real estate owners and managers approach sustainability. The survey method was used to acquire qualitative data on owners' behavioral views and was analyzed using content analysis, followed by coding and quantifying the responses. The sample consisted of 30 real estate trusts, funds, and corporations, with 16 from Australia and 14 from New Zealand. The analysis focused on three major themes: when sustainability becomes crucial, making people aware of sustainable commercial real estate and emphasizing the significance of adopting it. The study found a significant level of difference in perceiving sustainability between Australia and New Zealand. The analysis revealed that owners emphasized operating expenses and energy consumption as part of sustainability. The results showed a significant emphasis on operating expenditure (49%), followed by 18% for indoor environment quality and energy use. According to the findings, respondents in Australia perceive sustainability in commercial real estate as a cost-cutting measure, while New Zealand looked at it as a marketing gimmick to attract and keep tenants. Thus, the inability to assess the monetary worth of sustainability efforts limits their capacity to justify their implementation.

Sardinha et al. (2011) used data from websites, annual, environmental, and sustainability reports, and customer service departments to conduct a comparative CSR benchmarking on 23 real estate companies with shopping centers in Europe, China, Australia, and the United States between 2004 and 2010. The study examined real estate corporations that invest in, create, and maintain shopping centers. According to the survey, there are now fifteen corporations instead of only ten that produce annual sustainability reports. In 2010, 5 of 15 companies had CSR or sustainability reports meeting GRI requirements. The nine companies that dominated the UK real estate market in 2004 still have the most information available on their websites and annual reports as of 2010. Thus, the literature and benchmarking studies in the 2004 examination revealed significant possibilities for developing greener transportation solutions for large shopping complexes in town centers. However, 2010 studies showed no signs of evolution in this regard.

Eichholtz et al. (2009) identified the firms and industries that use green office buildings. The statistics were gathered from 11,000 occupants in 1100 green office buildings and 3900 non-green office buildings. The study's findings suggested that people preferred green office buildings over non-green ones due to their quality. Legal and financial organizations frequently lease green office space, whereas tertiary sector firms recognize the productivity benefits of green buildings. Thus, the incremental cost of occupancy or refurbishment can be recouped through lower risk premiums, energy savings, and higher net rents.

15.4 Research Methodology

The chapter employs the case study method to analyze how CSR practices can help in attaining the goal of sustainable cities and communities through sustainable governance. The research was carried out by performing a case study analysis of various companies' CSR practices, including Tata, Hyundai, Infosys, Hindustan Zinc, DHL, and DLF Limited which are aligned with the SDG 11 target and have a strong sustainable governance.

In their paper "Designing Case Studies from secondary sources—a conceptual framework," Reddy and Agrawal (2012) emphasized the importance of case study design, creating a strong subject foundation through the development of writing style, creating concepts, and contributing to literature. Case studies were classified into three types: Type I, Type II, and Field Cases. A Type I case study deals with the primary source of data and is written through direct interviews with respondents. Field Cases are case studies designed using field research methods. A Type II Case Study is created using secondary information that is easily available. This study is a Type II case study as it is designed using secondary data sources, including annual reports, ESG reports, CSR reports, and websites, which were used to create a conceptual framework for the study. A Type II Case Study was undertaken to gain a deeper grasp of the topics using secondary data sources. In this chapter, only six companies were chosen as they are undertaking SDG 11-aligned CSR activities through sustainable governance in India.

15.5 Opportunities for Developing Sustainable Cities and Communities in India

a. Smart Cities Mission (SCM)

To deal with rapid urbanization and ensure a better standard of living, SCM was launched on 25th June 2015. IBM initiated this mission as the "Smarter Planet Initiative" in 2008. The mission's primary goal was to support communities through an equal partnership of the State and Central Governments. It aimed to develop 100 cities from across 21 sectors, by 2020. The shortlisted cities aimed to transform themselves into self-sustainable urban settlements. It had two main components; area-based development and pan-city solutions.

b. Pradhan Mantri Awas Yojna (PMAY) and Swachh Bharat Mission

This mission was designed to provide concrete houses to the poor from urban areas and migrants. Prime Minister Mr. Narendra Modi launched the PMAY on 25th June 2015. Around 83.67 lakh houses are completed and 1.14 crore houses are grounded for construction. In terms of finance, Rs. 1,63,926 crores have been released for the implementation of this scheme, and additional houses using modern and innovative technology, reflecting sustainable practices in the construction sector will be constructed soon (MoHUA, Government of India). This scheme has created a robust and positive impact on the urban housing development in India. Graph 15.2 shares the data about the houses under PMAY.

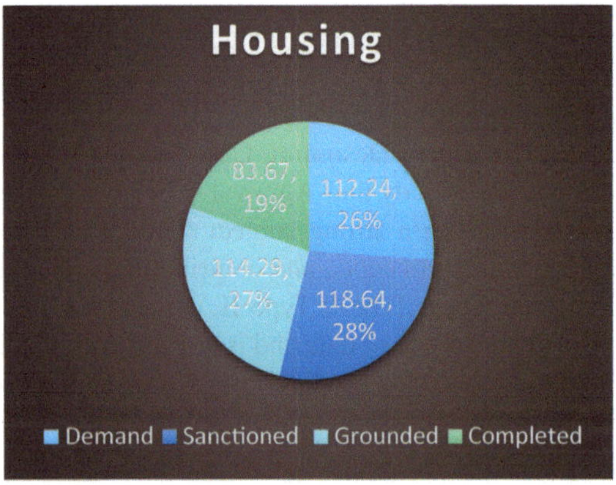

Graph 15.2 Houses under PMAY. *Source* Ministry of Housing and Urban Affairs. https://pmaymis.gov.in/

Another flagship program, Swachh Bharat Abhiyan was an initiative to make India Open Defecation Free (ODF). By designating more than 6 lakh villages as ODF by October 2019, a noteworthy accomplishment was made in honor of Mahatma Gandhi on the 150th anniversary of his birth. This accomplishment also matched SDG Target 6.2, which effectively put an end to open defecation across the country. Once the ODF status was attained, SBM-G Phase II was initiated to attain Sampoorn Swachhata, which is maintaining the ODF status, handle solid and liquid waste by 2024–2025, and converting every village from the ODF to the ODF Plus Model.

c. Affordable and Sustainable Transport System Through SATAT

This scheme was launched by the Government of India in October 2018. The mission aimed to support vehicle users, farmers, and entrepreneurs by discussing alternatives that help reduce emissions from vehicles and pollution caused by burning agricultural or organic waste.

15.6 Challenges in Developing Sustainable Cities and Communities in India

a. Rise in the Percentage of Slum Dwellers

Growing population and poverty are the prime reasons for an increase in the population dwelling in slums. Unequal regional development causes people from rural areas to migrate to cities, causing an increase in the urban population density. This leads to a shortage of land, and further, an increase in material and labor costs makes it difficult for the developers to deliver affordable houses. There needs to be proper planned city management and equal efforts from the urban local bodies are required to reduce the creation of slum areas. According to the SDG 11 Synthesis report by UN-Habitat, more people are living in slums around the world that can be addressed. In contrast to the increasing pace of growth in the overall number of slum residents, there was a slowdown in the percentage of the population living in slums between 2014 and 2020. According to the SDG Report 2024, 49% of Indians who live in cities live in slums; this percentage has not changed since 2020 and shows no indications of improvement.

b. Air Quality and Water Management

India secured the 177th position for air quality, barely above Bangladesh, Nepal, and Pakistan (EPI Report, 2024). The yearly average of PM 2.5 (ug/m^3) is 50 μg/per cubic meter since 2022 (SDG Report, 2024). This exceeds the WHO limit of 10 μg per cubic meter, leading to adverse health effects. The percentage of individuals having access to quality piped water in urban areas has also decreased.

c. Localizing of SDG

The widening gap between the top-down authorities and bottom-up environment, and the large number of stakeholders have made it critical to localize the SDGs. City governments should make plans ahead of time and adhere to critical procedures, such as creating indicators, prioritizing activities, defining a vision, forming alliances and collaborations, and investing in training and capacity-building. Preliminary analysis, however, suggests that a significant barrier has been the implementation and oversight of SDG 11 at the lowest spatial unit—municipal wards (Kundu & Pandey, 2021).

d. Lack of Coordinated and Sustainable Governance Structure

The hierarchical structure of the government in India has too many levels resulting in delays in making decisions and accountability issues. A lack of coordination and integration may confuse. Urban Local agencies (ULBs) have less authority as parastatal agencies like development authorities, housing boards, water supply, sewage boards, etc. operate in parallel (Kundu & Pandey, 2021). A strong partnership between the authorities from the concerned departments is a must to have a strong and sustainable governance structure and hence the achievement of SDG 11.

e. Lack of Data and Funds

As per the article, "The Hopes and Holes in India's Schemes to Meet the Sustainable Development Goals," the schemes launched under SDG 11 had not only implementation inefficiencies but were also based on incomplete data. The slum population is estimated from the 2011 Census. The 2012 National Sample Survey Report served as the basis for AMRUT's estimation of the water facilities required in slum areas. The planning procedures not only use old data but also have exclusionary practices. Funding for basic facilities for slums was not specifically allocated under AMRUT, and unauthorized colonies were not taken into consideration when calculating the requirement for house toilets under SBM–U.

f. Monitoring and Review

There should be more stringent regulations to monitor the SDGs on a timely basis. Governments at the national as well as local levels must create appropriate frameworks to review the progress of the nation on the various targets and aforesaid indicators. Regularly collecting and updating data is essential for the achievement of the SDG.

15.7 Case Studies of Sustainable Cities and Communities Built in India Through CSR and Sustainable Governance

The following information has been adopted from the Integrated Annual and ESG Reports of the respective companies, stating their contribution towards CSR activities in building sustainable cities and communities by implementing sustainable governance. These companies have portrayed the principles of sustainable governance like transparency, collaboration, participation, non-discrimination, sound policymaking, and integrity.

A. *Case Study 1: Tata Trust working on the upgradation of slums by providing essential housing amenities for all (Target 11.1)*

Employees of the Tata Trust, such as Jaga fellows, and representatives of the Dhenkanal Municipality helped the slum dwellers by ensuring the safety of their land tenure and supplying the bare necessities. They achieved this by facilitating tenure security, and financial availability, and offering affordable housing design and technology models. The Trust assisted with the digitization of data, provided training to local non-governmental organizations (NGOs) on how to conduct household surveys, aided Municipalities in managing data and interpreting Act regulations, and developed skills for efficient implementation. The Trusts used drones as a cutting-edge technological tool to map slums and gather data, which was the first instance of geo-fencing slum boundaries and creating ortho maps using drones. Drone surveys have been used to map over 1,800 slums with over 125,000 households by actively involving local populations. To ensure fair community involvement and promote collective decision-making, the Trusts also enabled the establishment of the Slum Dwellers Association (SDA). More than 51,000 families received land rights certificates, and about 1,100 government employees and other interested parties received training on land rights and the elements of a liveable ecosystem. Hence from this case study, it is observed that Tata Trust is working towards providing better living conditions to slum dwellers through safe and affordable houses. Though the Trust is at its infancy stage of meeting the objectives in the aforementioned target, it has set an example for other organizations to work diligently towards achieving a strong and sustainable ecosystem. The organizations need to expand on effective community-based upgrading initiatives and improve the lives of the most disadvantaged and vulnerable urban dwellers by assisting national and local authorities that are ready to create upgrading programs for the entire city and country.

B. *Case Study 2: Hyundai Motor India Foundation and DHL providing safe and affordable transport to all (Target 11.2)*

As a part of the CSR initiative, Hyundai Motor India Foundation (HMIF) has initiated a "Road-Safety" campaign called the "Dhyan Do." This initiative dealt with identifying and spreading awareness about the black spots on the roads of the districts of Gurugram and Faridabad. The initiative focused on ensuring a safe and sustainable transport system on the roads. The drivers were given free health check-ups, including eye tests, blood tests, and sugar tests along with free first aid kits. The initiative was not only limited to public transport but included people driving personal vehicles too.

By 2030, DHL, a prominent logistics firm operating in over 220 countries, plans to achieve sustainability. To make logistics more environmentally friendly and sustainable, the massive logistics company is stepping up its efforts. By 2030, the business wants more than 1000 of its cars in the Asia-Pacific area representing 30% of its fleet to be electric. By purchasing its first electric freight jet and investing in sustainable aviation fuel, the company hopes to lessen its carbon footprint. By indulging in these sustainable practices and providing sustainable transport, there will be a reduction of CO_2 emissions to below 29 million by 2030.

According to a report by UITP, 1.2 million people die, and up to 50 million are injured in traffic accidents each year. Hyundai Motor has started a great initiative in this area making it possible for people from a variety of socioeconomic backgrounds to travel with ease and little risk to their safety. This will eventually promote economic growth, improve quality of life, lower social inequality, and contribute to a sustainable environment. On the other hand, DHL is promoting the use of e-vehicles for building a resilient and sustainable transport system. As India's EV market is on a rapid growth trajectory, the Government is also supporting its use by providing tax subsidies and incentives. Electric vehicles can not only reduce air pollution, and reduce reliance on fossil fuels, but also support achieving another sustainable development goal i.e. affordable and clean energy (SDG 7). Other companies, as a part of their CSR activities, can join hands with the local and national authorities and create a strong awareness among vehicle users.

C. *Case Study 3: Hindustan Zinc Ltd. provides clean and safe water (Target 11.5.)*

In the state of Rajasthan, which lacks enough groundwater, Hindustan Zinc Ltd. has made a sustainable step by providing clean drinking water, relieving women of the arduous task of walking long distances to fetch water. About 63,000 people have been impacted by this project, which is being carried out in partnership with Neer Amrit LLP. It aims to give the community access to and availability of clean drinking water by installing RO/ATMs or providing water tankers that cover five districts and six locations in Rajasthan. The major activities include supplying water through tankers in 25 villages; guaranteeing safe and doorstep water facilities; providing 3,726 KL of safe water to the community through ROs and ATMs; and contributing ₹3.87 lakhs to Panchayats towards the sustainability of the RO/ATM project.

Hindustan Zinc Ltd. has set an example of providing clean and safe water, even in the state of Rajasthan where water scarcity is a major problem. Through initiatives like drilling wells, constructing water treatment plants, or encouraging water conservation practices, they are actively funding projects that give communities access to it, especially in places where clean drinking water is scarce. This promotes environmental responsibility while also improving the social well-being of the local population. The rejuvenation of the Yamuna River is another magnificent example set by the local and municipal authorities in Delhi.

D. *Case Study 4: Infosys Foundation working towards waste management (Target 11.6.)*

Using about 300 crores of its CSR funding, the Infosys Foundation embarked on the project of restoring Hebbal Lake, which is situated on the outskirts of Mysore, in collaboration with the Mysore District Administration and other stakeholders. The expanding residential neighborhoods damaged the lake, which was originally used as cattle grazing area and a bird breeding habitat. Sewage water would run into the lake because there were no water management measures in place. The removal of encroachments that obstructed the lake and the installation of natural drains to guarantee a smooth water intake and prevent undesired litter were recognized as the primary initiatives for the lake's revitalization by the Infosys Foundation, the charitable arm of Infosys Ltd. To secure the long-term future of the lake, a sewage treatment plant was installed that can treat around eight million liters of water per day. A fence has been built alongside the lake, jogging tracks have been built and landscaping has been done to beautify the lake.

Infosys Foundation is playing a pivotal role in improving cities and achieving important environmental objectives. It is setting an example for other organizations to diverge their CSR funds for long-term garbage management and have strong and proactive community involvement to reduce landfill usage. Waste management has an array of benefits like reducing pollution, protecting the environment, conserving resources, and providing a healthy living to the citizens. This can help in attaining other sustainable goals like climate action (SDG 13) and good health and well-being (SDG 3).

E. *Case Study 5: DLF Limited administering open and green spaces in Gurugram, Hyderabad, and Delhi (Target 11.7.)*

DLF has initiated plantation drives in various parts of Gurugram, Hyderabad, and Delhi. In New Delhi, two public parks totaling about eight acres of land have been built. Eight dustbins have been placed in these parks to guarantee hygienic conditions and cleanliness. DLF Limited's business premises are another illustration of environmental sustainability. The 26 acres of decarbonized and sustainable office space in DLF Cybercity in Hyderabad is an example of expanding the reforestation effort into public areas throughout the Hyderabad belt. The project adopted the mitigation hierarchy in addition to preserving the biodiversity in their workplace space. In collaboration with the Greater Hyderabad Municipal Corporation, this plantation initiative has also been expanded throughout Nawabi city. The US Green Building Council (USGBC) awarded it the LEED Zero Waste Certification, making it the nation's first workplace to do so.

DLF plays a crucial role in positively affecting the mental health of the people as these open spaces promote sociability, community cohesiveness, relationship development, and stress-relieving relaxation. Additionally, the areas promote walking, jogging, and bicycling, all of which reduce stress and depression. Such initiatives can help build strong relationships with the stakeholders and build a strong positive image in the minds of the consumers. Thus, strong and sustainable societies can be built through such measures which is possible through an effective governance system.

15.8 Present Scenario of Sustainable Cities and Communities in India

Rapid industrialization and massive urbanization have resulted in growing pressure on the housing industry, especially in metropolitan areas. The total world urban population is projected to increase by 13% by 2050. 90% of this increase will be seen in the continents of Asia and Africa (United Nations, 2018). According to a report by RICS and Knight Frank consultant (2019), 33% of the Indian population is living in urban India presently, which is anticipated to reach around 40% by the year 2030, creating an additional demand of 25 million mid-segment and affordable housing units. According to an article published in Business World, nearly 40% of the Indian population will not have drinking water by 2030 as the country will be unable to fulfill the huge demand gap in drinking water supply. The energy demand is also going to reach 1500 million tons by 2030. This calls for development using the available replenishable resources.

Thus, the current situation stimulates the need for more buildings to accommodate the increased population with the facility of reduced environmental impact (Owensby-Conte & Yepes, 2012). Based on a five-year review, there has been a significant shift in the availability of green buildings. Specifically, the supply of certified buildings has increased by 37%, adding 78 million square feet of certified stock. This shows that the Real Estate sector understands its responsibility towards the nation's Sustainable Development (CBRE, 2022).

As per an article published in the *Economic Times* in June 2024, the Indian smart home market will expand by 9.14% within a span of four years from 2024 to 2028. All PSU and central government buildings must receive at least a 3-star GRIHA certification, according to government regulations. The developers are also playing a major role in designing a sustainable future for the country. Sustainable and eco-friendly ideas like using recycled or locally sourced raw materials, HVAC systems, rainwater harvesting systems, solar panels, and wastewater treatment plants have significantly reduced the carbon footprint of construction projects. A major part of this change is played by builders and developers, who use cutting-edge techniques and materials to create a more sustainable future.

15.9 Conclusion

Urbanization has emerged as a major worldwide issue and can take the lead in mitigating climate change due to its substantial contribution to global emissions and energy consumption. It has thus become essential to alter the way cities and human settlements are being managed to avert the catastrophic outcomes that lie ahead. The goal of building sustainable cities and communities requires a sustainable urban and local component based on multilevel governance. Cities, municipal governments, and urban stakeholders can drive dramatic shifts in the current situation, leading to more sustainable development outcomes. On the other hand, well-planned migration and a strongly resourced ecosystem of the cities can create positive growth leading to sustainable development. Thus, by incorporating CSR funds and strong governance in the planning and execution of achieving SDG 11, positive growth can be expected.

As community development is an important agenda, there lies a need to develop the assets that surround the community. This chapter gave a glimpse of the strong partnership between the corporate sector and the government in building a sustainable nation. Strong partnerships, multistakeholder consultations, and effective public participation can help in achieving the Sustainable Development Agenda 2030.

15.10 Research Implications

The study is an example of how companies can help in creating sustainable cities and communities through sustainable governance. Through the case study analysis, it was observed that the business community along with governments, corporations, and civil society are working together to develop solutions for a shared sustainable development future. The aforementioned targets laid down in SDG 11 like affordable housing initiatives (PMAY) are advancing into their subsequent phase, thus emphasizing sustainability and scalability. Sustainable and scalable measures are being undertaken in the area of waste management, expanding them beyond pilot projects, and replicating them in other communities dealing with comparable issues or situations that have enormous potential. The transportation sector, on the other hand, is aiming to deliver not only lower climate emissions but also more resilient economies. These initiatives have emphasized the need for more robust solutions that can simultaneously cut costs for people while aiding the transition away from unsustainable practices.

One of the first solutions would be to reduce and eventually eradicate slums. States have designated real estate regulators, but they seldom address the issue of affordable housing. In a nation the size of India, the national concept of free housing has long been acknowledged as being flawed. Planning affordable dwellings on public property and providing soft long-term debt would be a better course of action. This requires concerted efforts on behalf of State and Central Governments in the form of regulatory laws and policies. States should implement the housing policies recommended by the Government of India in the rental and affordable housing sectors, making it attractive for private companies to enter these markets. The Municipal bodies need to work on waste management solutions and provide a healthy living to the citizens.

The research findings and the challenges mentioned can help policymakers create a CSR framework tailored to achieving SDG 11. Thus, evaluating the implementation of CSR initiatives provides leaders with a comprehensive knowledge of their organization's CSR performance. This insight is especially important when developing future CSR strategies. This research will assist construction companies in implementing successful CSR practices and adopting a circular approach that prioritizes efficient and sustainable use, reuse, and recycling of natural resources. To ensure sustainability in the sector, developers should prioritize low-carbon alternatives, green technologies, and skill development. Lastly, the SDGs, the plan for a more resilient and affluent world and the path out of the current global problems, will remain elusive without significant investment and increased action.

15.11 Future Research Directions

Future studies are recommended to investigate the CSR practices of different companies and identify the SDG targets that are met. Researchers can also investigate CSR performance individually in the public and private sectors. A framework can be designed and tested empirically showing the relationship between CSR and SDG 11 through sustainable governance. The study does not go deeply into the specifics of the assessment techniques. Future publications on CSR can focus on elaborating and presenting the findings of employing such techniques for assessment.

References

Aksoy, M., Yilmaz, M. K., Tatoglu, E., & Basar, M. (2020). Antecedents of corporate sustainability performance in Turkey: The effects of ownership structure and board attributes on non-financial companies. *Journal of Cleaner Production, 276*, Article 124284.

Andelin, M., Sarasoja, A. L., Ventovuori, T., & Junnila, S. (2015). Breaking the circle of blame for sustainable buildings—Evidence from Nordic countries. *Journal of Corporate Real Estate, 17*(1), 26–45.

Bae, S. M., Masud, M. A. K., & Kim, J. D. (2018). A cross-country investigation of corporate governance and corporate sustainability disclosure: A signaling theory perspective. *Sustainability, 10*(8), 2611.

Carroll, A. B. (2015). Corporate social responsibility: The centerpiece of competing and complementary frameworks. *Organizational dynamics, 44*(2), 87–96.

CBRE Group, Inc. (2022, May 26). CBRE releases 15th annual corporate responsibility report. https://www.cbre.com/press-releases/cbre-releases-15th-annual-corporate-responsibility-report

Driessen, P. P., & Glasbergen, P. (2002). New directions in environmental politics: Concluding remarks. In *Greening society: The paradigm shift in Dutch environmental politics* (pp. 245–262). Springer Netherlands.

Ebekozien, A., Ahmed, M. A. H., Aigbavboa, C., Samsurijan, M. S., Awo-Osagie, A. I., & Al-Hasan, A. Z. (2024). Expanded corporate social responsibility framework: Private sector's role in improving junior staffers' homeownership to achieving Sustainable Development Goal 11. *Journal of Infrastructure, Policy and Development, 8*(8), 4589.

Eichholtz, P., Kok, N., & Quigley, J. M. (2009, August). Why companies rent Green: CSR and the role of real estate. In *Academy of Management Proceedings* (Vol. 2009, No. 1, pp. 1–6). Academy of Management.

Emeka-Okoli, S., Nwankwo, T. C., Otonnah, C. A., & Nwankwo, E. E. (2024). Corporate governance and CSR for sustainability in oil and gas: Trends, challenges, and best practices: A review. *World Journal of Advanced Research and Reviews, 21*(3), 078–090

Environmental Performance Index Report. (2024). https://epi.yale.edu/country/2024/IND

Hiep, P. M., Tien, N. H., Dana, L. P., Kuc, B. R., Van Tien, N., & Ha, V. X. (2021). Enhancing social responsibility and sustainability in real estate industry. *Turkish Journal of Computer and Mathematics Education (TURCOMAT), 12*(14), 4999–5013.

H. N. D. (2024, July 3). Smart Cities Mission extended till March 2025 allowing cities to complete ongoing projects. *Hindustan Times.*

Huang, C. F., Lu, W. H., Lin, T. T., & Wu, E. J. (2017). The current conditions of CSR implementation in construction industry: A lesson from Taiwan. *Applied Ecology & Environmental Research, 15*(2), 67–80.

Institute of Social Sciences. (2019). *Draft State of the Cities Report (restricted circulation).* New Delhi.

Ionaşcu, E., Mironiuc, M., Anghel, I., & Huian, M. C. (2020). The involvement of real estate companies in sustainable development—An analysis from the SDGs reporting perspective. *Sustainability, 12*(3), 798.

Kandpal, R., & Okitasari, M. (2023). Governance transformation towards localisation of sustainable development goal 11 in India. *World Development Sustainability, 2*, Article 100069.

Khan, M., & Ghouri, A. M. (2022). Corporate social responsibility, sustainability governance and sustainable performance: A preliminary insight. *Asian Academy of Management Journal, 27*(1), 1–28.

KPMG. (2017). *Sustainable Development Goals (SDGs): Leveraging CSR to achieve SDGs.* https://assets.kpmg.com/content/dam/kpmg/in/pdf/2017/12/SDG_New_Final_Web.pdf. Accessed 3 June 2024.

KPMG. (2018). *Bridging the gap: CSR to SDGs.* https://assets.kpmg.com/content/dam/kpmg/in/pdf/2018/12/Bridging-the-gap-CSR-to-SDGs.pdf. Accessed 2 June 2024.

Kundu, D., & Pandey, A. (2021). Sustainable urbanisation in India and Delhi: Challenges and way forward. In *23rd ASEF Summer University ASEF Education Department.* Asia Europe Foundation.

Lu, J., Liang, M., Zhang, C., Rong, D., Guan, H., Mazeikaite, K., & Streimikis, J. (2021). Assessment of corporate social responsibility by addressing sustainable development goals. *Corporate Social Responsibility and Environmental Management, 28*(2), 686–703.

Ministry of Corporate Affairs. (2021). *CSR Data & Summary.* https://r.search.yahoo.com/_ylt=AwrKCaVsyJNmQvA1m8a7HAx.;_ylu=Y29sbwNzZzMEcG9zAzEEdnRpZAMEc2VjA3Ny/RV=2/RE=1720989932/RO=10/RU=https%3a%2f%2fwww.mca.gov.in%2fcontent%2fmca%2fglobal%2fen%2fdata-and-reports%2fcompany-statistics%2fcsr-data-summary.html/RK=2/RS=UP.NR0F3lN_eqLk5sSN9mIIJNuM-. Accessed 1 July 2024.

Ministry of Statistics and Programme Implementation. (2022). *Sustainable Development Goals National Indicator Framework, 2022.* https://mospi.gov.in/sites/default/files/publication_reports/Sustainable%20Development%20Goals%20National%20Indicator%20Framework,%202022165649656l237.pdf

NGOBOX. (2018). *CSR project by: Tech Mahindra Ltd.* [online]. https://csrbox.org/India_CSR_Project_Tech-Mahindra-Ltd--Programme-SMART,-SMART+-and-SMART-T-Delhi_10914
NGOBOX. (2021). *Health equity: Hyundai's Sparsh Sanjeevani CSR initiative touches 1.5 million lives in rural India* [online]. https://csrbox.org/Impact/description/Impact-stories_full_Health-Equity--Hyundai%27s-Sparsh-Sanjeevani-CSR-Initiative-Touches-1.5-Million-Lives-in-Rural-India_679
NGOBOX. (2024). *Sustainability in Magarway village* [online]. https://csrbox.org/Impact-stories_full_Ambuja-Cements%27-CSR-Efforts-Transform-Water-Scarcity-into-Sustainability-in-Magarway-Village_684
Owensby-Conte, D., & Yepes, V. (2012). Green buildings: Analysis of the state of Knowledge. *International Journal of Construction Engineering and Management, 1*(3), 27–32.
Reddy, K. S., & Agrawal, R. (2012). *Designing case studies from secondary sources—A conceptual framework.*
Sardinha, I. D., Reijnders, L., & Antunes, P. (2011). Using corporate social responsibility benchmarking framework to identify and assess corporate social responsibility trends of real estate companies owning and developing shopping centres. *Journal of Cleaner Production, 19*(13), 1486–1493.
Sundar, P. (2000). *Beyond business: From merchant charity to corporate citizenship.* Tata Mcgraw Hills.
Sustainable Development Report. (2024). *The SDGs and the UN Summit of the future.* https://dashboards.sdgindex.org/
UN-Habitat. (2018). *Tracking progress towards inclusive, safe, resilient and sustainable cities and human settlements.* https://unhabitat.org/sdg-11-synthesis-report
United Nations Department of Economic and Social Affairs. (2015). *History of sustainable development.* https://sdgs.un.org/goals. Accessed July 2024.
United Nations Department of Economic and Social Affairs. (2018). Population division. *World urbanization prospects: The 2018 revision.* United Nations. https://population.un.org/wup/
United Nations Department of Economic and Social Affairs. (2019). *Principles of effective governance for sustainable development.* https://publicadministration.un.org/en/cepa
van Zeijl-Rozema, A., Cörvers, R., Kemp, R., & Martens, P. (2008). Governance for sustainable development: A framework. *Sustainable Development, 16*(6), 410–421.
Warren-Myers, G. (2012). Sustainable management of real estate: Is it really sustainability? *Journal of Sustainable Real Estate, 4*(1), 177–197.
World Business Council for Sustainable Development (WBCSD). (1999). *Corporate social responsibility: Meeting changing expectations.* http://www.wbcsd.org/pages/edocument/edocumentdetails.aspx?id=82. Accessed 16 May 2016.
World Green Business Council (WGBC). (2013). *The business case for Green buildings: A review of the costs and benefits for developers, investors and occupiers.* WGBC. www.worldgbc.Org
Wu, L., & Jin, S. (2022). Corporate social responsibility and sustainability: From a corporate governance perspective. *Sustainability, 14*(22), 15457.

Chapter 16
Reimagining Shareholder Engagement Through AI and Technology in India

Kirthana Singh Khurana

Abstract The chapter shall explore the potential of technological advancements, especially artificial intelligence (AI), and their impact on shareholder empowerment in India, which will lead to enhanced corporate transparency, disclosure mechanisms, effective communication, and data-driven decision-making. The chapter's main argument is that ongoing innovation and technological investment are crucial for promoting more equitable and participatory conditions in Indian corporate governance. In the above context, the chapter will propose several suggestions to improve technology adoption in corporate governance. A strong pitch will be made to create a regulatory sandbox for testing artificial intelligence in governance, making disclosure requirements stricter for transparency, and training shareholders and executives using learning schemes to improve their digital literacy. An attempt would also be made to factor ethical standards to address issues like privacy of information as well as algorithmic bias; the requirement of international collaboration to incorporate the best practices worldwide; the imperative of the involvement of stakeholders in AI governance processes; and the revision of the legal frameworks to keep pace with advances in technology. These steps would ensure that technology enables effective governance while ensuring adherence to ethics and promoting fairness.

Keywords Corporate governance · Shareholder engagement · Corporate law · Technology · Artificial intelligence

K. S. Khurana (✉)
Jindal Global Law School, O. P. Jindal Global University, Sonipat, India
e-mail: kirthana@live.in; kskhurana@jgu.edu.in

Peter A. Allard School of Law, University of British Columbia, Vancouver, BC, Canada

V. Kandpal et al. (eds.), *Developments in Corporate Governance*,
Approaches to Global Sustainability, Markets, and Governance,
https://doi.org/10.1007/978-981-96-6366-8_16

16.1 Introduction

All foundational technologies emerge like earthquakes. They shift the tectonic plates below industries at large, reshape them, and herald the arrival of a new world order. So, while the first three industrial revolutions produced agriculture, physical factory goods, and information, the ushering of the artificial intelligence ("**AI**") era, touted as the fourth industrial revolution (World Economic Forum, 2020), is another earth-shaking moment and is going to revolutionize the very nature of intelligence itself.

The arrival of AI as a transformative force in business has challenged traditional paradigms, leading to reshaping the decision-making process across industries. As modern firms rush to unlock greater efficiency and effectiveness by integrating AI into various aspects of creating value for their stakeholders, corporate boards have assumed a crucial role in influencing and guiding this change. Germinating from a curious Alan Turing's remark in 1950—"Can machines think" and initially considered a concept reserved for science fiction, AI is rapidly establishing itself as a trailblazing scientific idea of the twenty-first century.

The advent of technology has redrawn the business landscape with its all-encompassing impact and has brought forth both new opportunities and challenges. No wonder, businesses across industries increasingly leverage technology to propel competence, stakeholder outreach, productivity, cost cut-outs, and innovation. For instance, in its quest to improve agility, "Domino's Anywhere" redefined how the company's clients could order their favorite pizza via dozens of channels (Kelso, 2018). For Nike, the leading athletic footwear and apparel company, roping in Nike Run Club and Nike Training Club applications to integrate business technology into building personalized interface with their customers turned out to be a transformational step (Dominguez, 2023). Moreover, as Eric Pearson so aptly remarked, "It is no longer the big beating the small, but the fast beating the slow" (Kintone, 2019); we live in the age of blistering pace of innovation. It is no surprise that 60 percent of Apple's revenue comes from products that did not exist four years ago (Shvartsman, 2024).

With technology having taken center stage in today's world of digitization, its imprints are palpable in all aspects of the functioning of modern-day businesses (Singh, 2024). As AI's influence expands, corporate governance structures must also evolve to integrate its capabilities in decision-making and operational efficiency (PricewaterhouseCoopers, 2023). Integrating technology and AI in corporate governance has proven to have many significant benefits, including improved decision-making, operational efficiency, and risk management (Allen & Brooks, 2024). One area where technology has particularly made a pathbreaking impact is shareholder engagement. Virtual Annual General Meetings ("**AGMs**"), Electronic Voting, Live Streaming, and Question-and-Answer Sessions have changed the very paradigm of how companies interact with their shareholders (Chia & Lee, 2020). Thanks to technology, multiple engagement channels are available for communication, enhanced transparency, and increased participation. The information on the company websites is wholesome and updated. These changes have made AGMs more accessible to shareholders and company boards more accountable and have extended leverage to companies to make the relationship and trust between the board and the shareholders more productive and fruitful (Denis & Blume, 2021).

The chapter examines how integrating technology and AI into corporate governance can foster shareholder engagement, particularly in countries like India, where concentrated shareholding abounds across the business ecosystem. In recent times, the Indian corporate space has witnessed unprecedented shareholder activism, and a fresh debate has erupted on issues like excessive executive remunerations, the role of the independent directors and the auditors, related party transactions, and minority shareholders not having an adequate say in such matters.

The chapter shall explore the potential of technological advancements and their impact on shareholder empowerment in India through enhanced corporate transparency, disclosure mechanisms, effective communication, and impactful decision-making that is data-driven. The main argument of the chapter is that ongoing innovation and technological investment are crucial for promoting more equitable and participatory conditions in Indian corporate governance. Given that some studies suggest that ownership concentration, on average, negatively affects technological innovations, the chapter will identify the sticking issues, analyze them, and delve into their likely solutions. In the above context, the chapter will propose several suggestions to improve the adoption of technology in corporate governance. A strong pitch will be made to create a regulatory sandbox for testing artificial intelligence in governance, making disclosure requirements stricter for transparency, and training shareholders and managers using learning schemes to improve their digital literacy. An attempt would also be made to factor ethical standards to address issues like privacy of information as well as algorithmic bias; the requirement of international collaboration to incorporate the best practices worldwide; the imperative of the involvement of stakeholders in AI governance processes; and of the revision of the legal frameworks to keep pace with advances in technology. These steps would ensure that technology enables effective governance while ensuring adherence to ethics and promoting fairness.

16.2 Shareholder Engagement and Its Impediments in the Indian Context

Shareholder democracy is implemented by empowering shareholders to vote. Shareholders in India are entitled to both (a) individual rights and (b) corporate rights. Shareholders may enforce their individual rights as per their own will, while corporate rights may be exercised by the collective involvement of shareholders (Suresh Chandra Marwaha v. Lauls (P.) Ltd., 1978).

16.2.1 *Individual Rights*

Some of the individual rights available to shareholders are as follows:

16.2.1.1 Right to Receive Information

This includes a right to obtain copies of the audited financial statement, including the balance sheet and profit and loss account, report of the cost auditor upon a direction given by the government, copies of the contracts regarding the appointment of managing director or manager of the company and disclosure of interest by directors, and copies of the notices of general meetings of the company and documents annexed with such notice.

16.2.1.2 Inspection Rights

These include the right to inspect statutory registers such as the register of charges, register of members and debenture holders, shareholders' minutes book, and register of directors.

16.2.1.3 Right to Vote, Attend General Meetings, and Other Allied Rights

Shareholders enjoy a right to receive notice of general meetings, participate in such meetings, and vote at them personally or by proxy. Equity shareholders have voting rights in proportion to their paid-up shares; they also have a right to receive share certificates and dividends as and when declared. They are entitled to rights shares and share in the company's surplus assets if it winds up.

16.2.1.4 Right to Transfer Shares

The Act and thc company's articles of association allow shareholders to transfer their shares. However, they may contain certain conditions that must be met by the shareholders before the transfer takes place.

16.2.2 Collective or Corporate Membership Rights of Shareholders:

Certain rights can be exercised collectively by a company's members via a democratic process. Each member has agreed that a majority vote governs these rights at the general meetings. These rights are also referred to as corporate membership rights. Some of the corporate membership rights enjoyed by shareholders are as follows:

16.2.2.1 Decision Rights

Although the company's board of directors is primarily responsible for decision-making, the Act allows for restrictions to be imposed on the board's powers and permits the shareholders to decide in case of specific vital matters related to the company. Their approval can be attained by passing an ordinary or special resolution per the Act. These include amending the company's charter documents—the memorandum and articles of association, appointment and removal of directors, obtaining loans, and sale of undertakings exceeding certain thresholds. The company's statutory auditors are also required to be appointed by the shareholders in the AGM among those recommended by the board or the company's audit committee. Similarly, the shareholders' approval by an ordinary resolution is required when the company desires to enter related party transactions exceeding certain financial limits. Such provisions ensure the participation of shareholders in matters that are significant to the company.

16.2.2.2 Right to File a Case for Oppression and Mismanagement

In case of oppression and mismanagement, §244 of the Act confers a right to not less than one hundred members of a company or not less than one-tenth of the total number of members, whichever is less, or members holding not less than one-tenth of the issued share capital of the company to apply to the National Company Law Tribunal ("**NCLT**") under §241 of the Act for claiming relief. In case the rights of any of the members are infringed, or the conduct of the management is prejudicial to the interest of the company or the members themselves, a class action may be instituted by the members against the company meeting certain thresholds against the company, its directors, and third-party advisors.

16.2.2.3 Right to Requisition an Extraordinary General Meeting

§100 of the Act confers a right on members holding not less than 1/10th of the company's paid-up share capital to requisition an extraordinary general meeting of the company.

16.2.2.4 Right to Make an Application to the Central Government

Members also have a right which may be collectively exercised in respect of making an application to the Central Government for carrying out an investigation into the company's affairs and for the appointment of government-nominated directors. On receiving such a request, the government can order the Serious Fraud Investigation Office to intervene.

16.2.2.5 Right to File Complaints

In order to address security holders' concerns, every listed company that has over 1,000 shareholders, debenture holders, deposit holders, and any other security holders during the course of a financial year must establish a Stakeholders Relationship Committee. Listed companies are also required to register themselves on the SCORES portal, which comes under the direct supervision of SEBI, the market regulator. Investors can use this platform to file complaints and track the progress of the complaint. Apart from the abovementioned rights, an individual shareholder, regardless of their stake in the company, also has the right to challenge corporate decisions detrimental to the company's interests by instituting a derivative suit. The Act does not cover such cases; however, the shareholder may seek redressal in the civil court and follow the procedure set out in the Code of Civil Procedure, 1908. In such cases, the "clean hands" doctrine will be applicable.

One of the fundamental principles of corporate law is the participation of both shareholders and directors in the governance of a corporation (Business Roundtable, 2016). As shareholders invest their money in a corporation, they delegate some of their decision-making powers to directors, who are responsible for managing the corporation. Here lies the problem of information asymmetry, as directors are left on their own to decide the company's destiny (Hansmann & Kraakman, 2004).

The law requires directors and other corporate executives to disclose crucial corporate information regularly—either through quarterly and annual reports or through notices of general meetings. In India, the number of disclosures that corporations are expected to make is one of the most stringent in the world. However, unsophisticated shareholders find comprehending and utilizing the information purposefully tricky (Macbride, 2011).

16.2.3 Shareholder Engagement Issues in India

The Indian corporate sphere is plagued with the issue of concentrated shareholding, where certain groups of majority shareholders exert significant control over corporate decision-making, which can lead to the possible marginalization of minority shareholders and harm corporate governance (Grover, 2020) Furthermore, management can exert influence over retail investors through strategic disclosure methods.

Misclassifying information to distract shareholders' attention and, as a result, produce market reactions that are not favorable is an example of this. The use of strategic disclosure strategies can affect the decisions that shareholders make, which further emphasizes the necessity of regulatory measures to prevent investors from being misled by firms.

Several Key Factors Compound These Issues:

16.2.3.1 Information Asymmetry

Information asymmetry is at the core of the agency problem that exists between shareholders and directors of a company. It occurs because the directors and executives have more information about the company's performance, strategies, and risks than shareholders. This gap can lead to decisions that benefit the insiders at the expense of the shareholders. This constitutes one of the main issues of corporate governance that the law seeks to address by implementing strict disclosure requirements and transparent communication between management and shareholders (Armour et al., 2017).

16.2.3.2 Concentrated Shareholding

In India, concentrated shareholding is a common phenomenon, with most publicly listed companies having large blocks of shares held by a small number of shareholders, such as promoters or family members. As the directors are often compelled to toe the line of the controlling shareholders, they often make decisions under their influence, potentially sidelining the interests of minority shareholders. Thus, in such companies, minority shareholders are left at the mercy of the majority shareholders.

16.2.3.3 Shareholder Passivity

Across the world, especially in India, there is an evident lack of motivation from the shareholders' side to participate in the company's governance. In India, retail shareholders often remain passive for a multitude of reasons, such as lack of awareness, the complexity of information disseminated by the company, and the perception that their individual vote will not make any difference. In case of any issues with the company, shareholders exercise their exit right and usually prefer to sell their shareholding rather than exercising their voice and advocating against certain business practices and decisions of the company (Preetha, 2021).

16.2.3.4 Enforcement Constraints

Enforcement of shareholder rights and corporate governance regulations is another significant hurdle in India. While there are a plethora of laws and regulations in place to protect the interests and rights of shareholders and motivate them to engage in corporate decision-making, the actual enforcement of these regulations has been somewhat inconsistent and ambiguous, exacerbated by delays and bureaucratic hurdles (Khanna, 2017). Judicial proceedings in India tend to move languidly, and regulatory bodies face resource constraints. Resultantly, the resolution of shareholder grievances often gets delayed (Khanna & Varottil, 2012).

16.3 Leveraging Technology for Enhancing Shareholder Engagement

16.3.1 India's Path to Digital Transformation

In India, corporate governance has traditionally been heavily reliant on manual processes. Corporate meetings—both board and shareholder—were conducted physically, and these often came with their own logistical challenges. This led to slower decision-making processes due to the need for physical presence and an exchange of manual documents. Corporate records were maintained in physical ledgers and folders, and much of the reporting and compliance documentation was also prepared and audited manually. Communication between executives and stakeholders was primarily through written notices, reports, and face-to-face interactions.

Many challenges existed due to lack of technology, e.g., delays in decision-making and information dissemination, higher risk of anomalies and errors in corporate records and accounts, the limited ability of shareholders to track and analyze corporate disclosures properly, and challenges in ensuring transparency and accountability in corporate governance.

With the advent of rapid digitalization starting in the 1990s, companies began to adopt computing for regular business operations, and we saw the introduction of software applications for accounting and bookkeeping purposes. This transition from manual to digital record-keeping brought about considerable improvements in data maintenance and accessibility. The Economic Reforms in 1991 ushered in an era of increased foreign investment and competition, compelling companies to adopt technology in order to keep pace with their multinational competitors and the demands of international trade and business.

The Internet has also facilitated the quick sharing of information among various corporate stakeholders, enhancing coordination and collaboration. Companies have also started disseminating important information through websites, where they upload their financial reports, company policies, and other important financial and non-financial information.

By mandating a regime of electronic filing of disclosures and reports, the Securities and Exchange Board of India ("**SEBI**") has emerged as the harbinger of technological reforms in the Indian corporate landscape. Electronic voting has streamlined the voting procedure for shareholders. Before 2014, shareholders were required to attend the AGM to exercise their voting privileges physically. The necessity of travel frequently dissuaded small shareholders from participating in the AGMs, impacting their attendance at such meetings. Shareholders showed limited interest in the postal ballot, which was provided as an alternative because of its cumbersome nature (Kundnani, 2023).

Further, in response to the COVID-19 pandemic, Indian regulatory authorities permitted companies to conduct their AGMs and extraordinary general meetings virtually. This also resulted in an increase in shareholder attendance and participation. Reliance Industries organized its first virtual AGM in 2020, which received the participation of over 300,000 shareholders. It is a shining example of the scalability of digital solutions. Virtual meetings tend to be shorter and more productive and enable focused discussion.

Traditional AGMs involved long waiting queues, manual registration processes, and restricted opportunities for shareholders to express concerns or ask questions. Virtual AGMs overcame most of these issues by allowing shareholders to join from anywhere in the world and access company documents and reports in real time while the meeting is ongoing. They also have a chance to communicate their questions or comments through the live chat feature—which may also be addressed by the directors and officers during the meeting. This improves the overall experience for shareholders and ensures that they can participate in meetings and contribute to discussions on corporate governance issues. It also allows the management to address shareholder concerns promptly and gain their approval and trust.

It is now essential to examine how technology, especially AI, is reshaping the role of boards of directors and shareholders in effective corporate governance.

16.3.2 The Role of the Board of Directors

In the corporate world, AI is making significant strides and assisting directors in decision-making. Three different categories have been identified:

16.3.2.1 Assisted AI

This involves humans as the primary decision-makers, with AI carrying out specific actions as directed by humans. This tool can collect and process massive amounts of data on shareholder-related issues, such as emails, social media posts, and regulatory filings. Shareholder grievances can be classified and prioritized according to their severity and potential impact on the company's performance and reputation. It can also analyze shareholder communications to assess and understand dissatisfaction.

16.3.2.2 Augmented AI

This category emphasizes how humans and machines should cooperate in making decisions. In this sense, AI can either assist or occasionally surpass human intellect. This instrument allows the board to generate reports addressing shareholders' apprehensions and underlying causes of concern. It can study industry best practices and compare corporate governance policies among companies. It can also suggest possible fixes and plans of action to handle particular issues and assuage shareholder concerns.

16.3.2.3 Autonomous AI

This is the most advanced level of AI, where systems are run automatically with little to no human help. AI systems can make complicated choices and act without constant human guidance. Companies can use these tools to generate regular financial reports and disclosures for shareholders. It can also be used to set up systems that monitor shareholder and consumer sentiment changes and regulatory updates. Alerts can be generated for immediate attention in case of significant developments. It can also draft standard responses to common shareholder concerns, speeding up the resolution process and ensuring that communication dissemination across all channels remains uniform.

16.3.3 Tools to Reduce Information Asymmetry and Increase Transparency

AI offers innovative solutions to reduce the challenges of information asymmetry. For example, natural language processing (“**NLP**”) tools can be used to evaluate vast amounts of unstructured data circulated by the company through its communications, including emails, reports, and meeting minutes. Shareholders and stakeholders can use AI to gather specific information, which may often get overlooked in hordes of corporate disclosures. Text analysis can be undertaken to extract insights, thus making the information more digestible and accessible for shareholders.

Machine learning (“**ML**”) algorithms can swiftly handle vast volumes of corporate data and help shareholders identify trends, disparities, or outliers that may impact their interests. As these algorithms are used, they evolve and become more competent, thus improving their accuracy and relevance over time. These tools can be handy when identifying patterns in a company's financial data and operational metrics to evaluate corporate performance properly. They can also be used to identify anomalies and irregularities in corporate disclosures, thereby becoming aware of future investment risks.

Institutional investors and proxy advisory firms can also use data processing tools. A study proposed that small passive investment funds can use generative AI language models like ChatGPT to improve their proxy voting accuracy and personalization and serve their shareholders better. For instance, Broadridge Financial Solutions Inc. has launched a proxy policies and insights database in the US that provides proxy voting data and analytics (Broadridge, 2021).

While AI tools such as NLP and ML can reduce the problem of information asymmetry and provide more accessible insights for shareholders, they are not free from risks. AI's reliance on algorithms can bring about inherent biases in decision-making processes, potentially complicating transparency rather than enhancing it. As Binns pointed out (Binns, 2019, 2022), this tension exists between transparency and algorithmic accuracy governance challenges, particularly in accountability and bias mitigation. Thus, companies must implement robust algorithmic auditing processes and ensure continuous oversight by human decision-makers to identify and mitigate biases.

16.3.4 Automating Financial Reporting and Improving Shareholder Communication

Traditional financial reporting and communication methods—through financial statements like balance sheets, income statements, cash flow statements, annual reports, and other similar disclosures—fail to make complex financial information accessible to unsophisticated investors. The notices attached to shareholder meeting notices provide the necessary information to help them decide how to vote and have similar issues. AI has transformative potential to automate financial reporting and improve shareholder communication, thus supporting informed decision-making.

Technology can propel a seismic shift in how companies deal with and communicate with their shareholders. Chatbots and virtual assistants, which are primarily AI-driven, have been instrumental in enabling companies to interact with their shareholders in real time, swiftly address inquiries, and provide tailored services and information. E.g., Reliance launched a WhatsApp chatbot to help shareholders and general viewers get access to frequently asked questions, video explainers, and documents used in preparation for the Reliance Industries Limited AGM 2021 (HT Tech, 2022). Through proper harnessing of the transformative power of AI, companies can improve transparency, build trust and accountability, and refine decision-making processes, which can undoubtedly lead to sustainable value creation and shareholder and stakeholder loyalty.

16.3.5 Leveraging AI for Market Sentiment Analysis

Market sentiment analysis involves scrutinizing public opinion and sentiments expressed by shareholders, stakeholders, and other corporate actors about the company and its practices. By using natural legal processing algorithms, companies can gauge public perception through posts on social media, news articles, blogs, forums, and other similar platforms. AI can be used as a tool to detect patterns and trends and aid human judgment.

This data can spur improved access to information, which shareholders can use to better forecast the impact of their decisions. While understanding market sentiment is a critical component of whether shareholders should invest in a particular stock, it may also influence how they cast their vote on important corporate matters such as mergers and acquisitions and appointment and removal of directors, among other vital issues. On the other hand, companies can use the data to mitigate conflicts of interest with shareholders and stakeholders of the company. For example, Tesla has been actively employing AI-powered sentiment analysis to analyze market sentiment and adjust its business strategies accordingly. The real-time insights allow Tesla to address concerns promptly, such as addressing apprehensions about its cars or responding to customer feedback, in the process winning the trust of its investors (Jean-Marc Beaujour, 2024).

Companies may also use market sentiment analysis to monitor possible concerns of the minority shareholders, which are expressed on social media and other platforms. Typical issues may be discussed and resolved in shareholder meetings or through regular updates, ensuring minority shareholders feel respected and heard.

Some studies also suggest that AI can reduce short-termism in companies. Expanding the dataset and enabling shareholders, predominantly minority shareholders, to contribute data helps mitigate the issue of exit over voice (Lee & Underwood, 2022).

16.4 Policy Recommendations

The integration of technology with corporate governance presents both opportunities and challenges. India is on a significant transformation journey with its booming digital economy and dynamic corporate sector. To ensure that technology and AI adoption improve corporate governance while maintaining ethical standards and promoting fairness, the following policy recommendations are proposed:

16.4.1 Establishing a Regulatory Sandbox for AI in Corporate Governance

A regulatory sandbox is a framework that permits controlled, time-bound experimentation of innovations under the supervision of a regulatory authority. It is a limited yet effective environment where established entities and newcomers can experiment with designs on the periphery or outside the current regulations (Shashidhar, 2023).

The AI tools subjected to sandbox testing can be developed to deliver real-time data to shareholders regarding the company's operations, financial performance, and strategic business decisions. These systems can be engineered to deliver personalized updates to shareholders according to their preferences and financial strategies. Moreover, experimental AI tools in the sandbox may include capacities to detect and potentially alleviate biases in decision-making contexts that could advantage specific groups. This will ensure that minority shareholders are treated equitably and fairly in decision-making processes.

This exercise will enhance regulators' comprehension of the genuine challenges and prospects of utilizing AI for effective corporate governance. Increased capital investment in AI and technology by corporations will transform the regulatory sandbox into a platform for informed decision-making that integrates innovation with heightened shareholder engagement and safeguarding their rights. This will provide Indian companies the opportunity to align their AI solutions with global competitors and receive guidance from international partners.

16.4.2 Enhancing Disclosure Requirements for Greater Transparency and Shareholder Education

Transparency is a hallmark of good corporate governance. Stringent disclosure requirements can ensure that shareholders receive accurate and timely information. Regulations requiring real-time disclosure of sensitive corporate information and financial information through digital platforms should be implemented. Blockchain technology should be used to ensure the authenticity and security of the information disclosed. Regarding financial disclosures, integrating AI and blockchain technology offers unprecedented opportunities for enhancing corporate transparency and security (Saleh, 2024; Susskind & Susskind, 2022). In this way, shareholders can have real-time, tamper-proof information about the company, enhancing their trust in its operations. AI can be used to automate the process of monitoring and reporting on company data so that real-time updates can be provided. Regulators should develop customized reporting systems for AI-driven analysis and decision-making processes to ensure clarity and comparability. The uniform template will make it easier for shareholders and regulators to compare data across sectors. Training sessions should be held for shareholders and managers to enhance their understanding of the disclosure requirements and underlying technology. Digital literacy programs should be used to help stakeholders decipher and interpret complex financial and non-financial information.

16.4.3 Addressing Ethical Issues in AI and Governance

Integrating AI into corporate governance can enhance efficiency, transparency, and decision-making. Nonetheless, the question of ethics enters this domain, and managers of companies dealing with AI systems face the need to resolve these critical issues to guarantee the ethical operation of AI tools. While ethical frameworks can be implemented to guide the implementation of AI, without adequate governance and oversight—they may prove insufficient and ineffective. As Mittelstadt also argues, implementing principles alone will be insufficient to guarantee ethical AI (Mittelstadt, 2019). Thus, companies must adopt active governance strategies to ensure fairness and accountability in their AI operations. Concerning data protection, there is a need to simplify and make more stringent company governance practices to comply with the levels of confidentiality set by data protection laws. This might include analysis of present laws and the development of new laws to cover new AI-derived data handling practices and to guarantee that they reach a higher level to protect private information. An external body needs to monitor the audits to ensure that companies comply with data protection laws.

The Digital Personal Data Protection Act 2023 ("**DPDP Act**"), which is a meticulous blend of recognition of the rights of individuals and the need for data processing, is a timely and appropriate step and restricts the type of information that Indian companies can collect. Until 2023, India did not have a standalone law or framework to govern data protection. The Information Technology Act, 2000, along with the rules notified under it, served as the core around which India's data protection framework developed. The new law bears similarities to the European Union's General Data Protection Regulation in scope, though there are some concerns about gaps and volume of exemptions (India Digital Personal Data Protection Act (DPDP Act): An Overview, 2024). The law applies to personal data collected in digital or non-digital form and digitized thereafter. It does not apply to data processed in a household/personal context.

The DPDP Act is applicable to any person who possesses digital personal data other than for personal or domestic context if the processing of personal data takes place within Indian territory or the overseas processing of digital personal data which offers goods or services to individuals in India.

The provisions of the DPDP Act have significant implications for Indian businesses. It imposes various obligations on organizations that gather, use, store, or disclose personal data, including:

- Obtaining informed consent before collecting personal data from individuals;
- Implementing adequate security measures to protect personal data from unauthorized access, use, disclosure, modification, or destruction;

- Providing individuals with access to their personal data and the right to correct, erase, or restrict its processing;
- In the event of a data breach, notifying the relevant authorities.

Organizations are under an obligation to fulfill the following requirements:

- Transparency: This comprises clear communication with individuals for data collection, usage, and sharing practices.
- Prior consent: Organizations are obligated to obtain informed consent from individuals before collecting and using their personal data.
- Optimal data collection: Organizations should collect only the data necessary for the specific purpose.
- Data security: Organizations need to have robust security measures for the protection of personal data from unauthorized access, use, disclosure, modification, or destruction.
- Data breach notification: Organizations are duty bound to notify the Data Protection Authority and affected individuals of any data breaches without undue delay.
- Data governance: Organizations need to put in place an effective data governance framework to keep a vigil over the data privacy practices. (Navigating Data Privacy Compliance in India: A Comprehensive Guide to DPDPA 2023 Compliance Audits, 2023)

The entities must modify their business, operations, and strategies to comply with this new law, which has been recently promulgated. To ensure they fully comply with this new law, entities must rewire their business operations, processes, and strategies to comply with the DPDP Act, which was recently passed into law. In a broad sense, this will fall under the purview of good corporate governance as well, given that protecting the confidentiality of data and minimizing the risk of data breaches, both contribute to the development of substantial trust among the various stakeholders of an organization. As an essential component of good governance, it is necessary for boards of directors of corporations to take the initiative in gaining an understanding of the law and ensuring that the statutory and compliance requirements are uniformly communicated throughout the organization across all functions and at all levels. In addition, boards will be responsible for ensuring that the fundamental principles of data protection and the prevention of its misuse and unauthorized use become ingrained in the organization's culture and become a function that cannot be negotiated. As data fiduciaries, directors and executives of companies need to ensure that the ethos of data responsibility is assimilated under the aegis of good corporate governance and the trust of stakeholders, as well as enabling assimilation from the top down, improved oversight, and the development of trust.

The DPDP Act compels companies to provide clear and accessible information to data principals about how companies use their data. For shareholders, such a move ensures greater transparency and more accurate disclosures about data handling, enabling them to make more informed investment decisions. It also has implications for minority shareholders, as with stricter data protection measures and greater transparency, they are better equipped to be informed and evaluate the actions of the majority shareholders and the board. It also helps them mitigate the risks associated with concentrated shareholding and protect their interests.

In a bit delayed response to the 2023 Act, the Ministry of Electronics and Information Technology released the draft Digital Personal Data Protection Rules, 2025, on January 3, 2025, for public consultation. The Ministry has invited feedback from the public and has invited feedback from the public on the draft rules by February 18, 2025.

The draft rules provide the implementation timeline for the new data protection regime which is slated to begin with the establishment of the Data Protection Board of India as the enforcement authority under the DPDP Act. While the board would come into existence immediately upon the official notification of the rules, the remaining provisions will come into force on such dates as are specified in the final version. The draft rules represent an important landmark in India's privacy framework. While the alignment with global standards is a welcome step, some unique features like consent management framework, verifiable parental consent, and restrictions against international data transfer which may influence foreign relations, give the rules a standout profile. News reports suggest the government seeks to allow businesses about two-year timeframe for the implementation of the new law. This augurs well for the whole fraternity of businesses. As for the international businesses, the draft rules provide not only clarity on important aspects but also compliance challenges. So, businesses considering entry into the Indian market should gear up for these new compliance standards in one of the fast-emerging digital economies of the world. (Chakraborty & Sonkar, 2025).

16.4.3.1 Recommendations for an Effective Execution of DPDP Act, 2023

In order to adapt to the DPDP Act, different sectors shall need to respond to challenges related to their data management practices, customer demographics, and their geographical reach. Yet, the following steps can help businesses in successfully navigating the new regulatory environment:

- Comprehensive data audit: Organizations must understand what data they gather, where is the data stored, and who has access to the same. Effective mapping of data flows enables businesses to identify potential risks and address them deftly.
- Data Protection Officer ("**DPO**"): Appointing a DPO can go a long way in overseeing compliance initiatives, liaising with regulatory authorities, and handling requests from data subjects.
- Use of technology: In order to monitor data flows, detect anomalies, and prevent unauthorized access to data, businesses will need to use tools like unified endpoint management solutions, encryption technologies, and data-loss prevention systems. These technologies can help the companies automate compliance and arrest the risk of human error.

- A culture of data privacy: Fostering a culture of data privacy can turn out to be the most effective measure to ensure the success of the DPDP Act implementation. Companies will have to establish a robust data breach response plan, encompassing the procedure for containment, assessment notification, and the remedial action.

16.4.3.2 A Case for Sandboxing as the Regulatory Approach for Emerging Technologies, Including AI

India's surge as a leader in the technology domain, with a huge pool of skilled human resources, makes it a perfect market for futuristic laws and robust policies. With its vast population, rich and varied data sets, and focus on digital public infrastructure, India can become the hub for developing successful technology models for the world. This would call for implementation of innovative regulatory approaches such as sandboxing, to regulate new and emerging technology/concepts in order to promote responsible innovation. Introduced in the UK, in 2015 by the Financial Conduct Authority to allow fintech businesses to test innovative propositions, sandboxes have gained in currency across the world. While across the globe one finds countries adopting regulatory sandboxes for technologies ranging from Web 3 to AI, in India, regulatory sandboxes, as soft-law mechanisms, have primarily been implemented within the fintech sector only, courtesy the farsighted and proactive measures taken by the regulatory bodies like RBI and SEBI (Tanvi Sharma et al.).

The successful implementation of regulatory sandboxes in India faces a plethora of challenges like the need for considerable financial resources and the balancing of innovation with consumer protection. It would be imperative to conduct an empirical study of the existing regulatory sandboxes in India to gauge the hurdles encountered during their implementation. For the starters, the sandbox framework, set up by the Indian banking regulator RBI, has been revised based on the experience of the last more than four years. RBI has advised the regulated entities under its regulatory sandbox framework to comply with the provisions of DPDP Act, 2023. (RBI Asks Banks to Comply with DPDP Act Provisions under Updated Regulatory Sandbox, 2024).

16.4.4 International Collaborations

As companies become more globalized, integrating AI into corporate governance practices cannot be constrained within national borders. Indian corporations and regulatory bodies must collaborate with their international counterparts to exchange information about industry practices and concerns related to AI adoption. Fostering partnerships with global technology firms, academic institutions, think tanks, and international regulatory bodies can facilitate the exchange of ideas and foster innovation.

In the 2024 season of AGMs, generative AI emerged as a new focus of shareholders, with shareholders of many big tech companies calling for detailed reports by the companies about how the technology is being used to create and circulate misinformation. Nearly 20 shareholder proposals have been submitted since late 2023 and are mainly aimed at companies at the forefront of the AI revolution, including Alphabet, Amazon, Apple, Microsoft, and Meta Platforms. There is a growing concern about the ramifications of AI usage, which reflects shareholder sentiment (Walsh, 2024).

In recent times, lawmakers and regulatory bodies globally have recognized the impact of AI across different industries and the potential risks and implications thereof, and legal frameworks have started to implement regulations to control its application. The first comprehensive legislation on AI, the AI Act and the AI Liability Directive was recently adopted by the European Parliament. The United Kingdom's AI regulation policy paper (2024) and US President Biden's executive order tasked federal agencies with creating guidelines for using AI are other welcome developments on this front. It is expected that other countries will soon follow suit.

As more foreign investors and multinational companies expand their presence in India, there is a greater need for Indian companies and regulatory bodies to benchmark their AI solutions against global standards. To improve their credibility and reputation in the global market, Indian companies can learn from the successes and failures of their international equivalents. A forceful beginning has already been initiated in India in this direction. India's National AI Strategy (2018) envisaged the establishment of a National AI Marketplace. This platform encourages business entities, government agencies, startups, AI researchers, system integrators, and academic research institutions to forge meaningful engagements. Taking this campaign forward, the "Open Loop India" program, which was a collaborative effort between Meta, the originator of the global open loop initiative, ArtEZ University of the Arts, Netherlands, and The Dialogue, one of India's leading public-policy think tanks, to develop a shareholder engagement framework that operationalizes the principle of human-centered AI, has been a pioneering step (Meta, 2023). The program involved working with a group of 12 Indian startups belonging to the Healthtech, Edtech, Fintech, and Agritech sectors that use AI to test the prototype and provide feedback. The program was rolled out between January and August 2023. The main findings and recommendations of the program are as follows: developing guidance on stakeholder engagement for AI actors, promoting interoperability and synergies, catalyzing capacity building and knowledge sharing, and ensuring accountability of ecosystem enablers. The findings also touched upon the desirability of the framework to work well not only with startups but with micro, small and medium enterprises and large corporations. This initiative will go a long way in crafting a robust AI integration framework for the Indian corporate sector. India has already proved its status as a global groundbreaker in IT services. A proactive approach toward technology and AI governance standards can position it as a world leader in this space.

16.4.5 *Implications and Future Research*

The impact of AI on corporate decision-making effectiveness and complying with disclosure requirements cannot be overstated. In countries where concentrated shareholding patterns abound, technology has the potential to transform the way companies operate by permitting real-time disclosure and analysis of data and enhanced communication with shareholders. However, as discussed in the above sections, integrating AI into corporate governance can lead to many ethical and legal issues that need to be addressed at the macro and micro levels. Future research can focus on the intersection of AI, corporate governance, and ethical compliance, particularly in promoting shareholders' interests while balancing the need for innovation within the company.

Future researchers could also work on developing AI-driven tools that are legally compliant, help retail investors engage with corporate boards, and empower them to exercise their voice. A comparative analysis may also be undertaken to study the extent to which artificial intelligence has been and can be used to protect the rights of minority shareholders in emerging economies such as Brazil, South Africa, and India. Moreover, developing legal frameworks for regulating the use of AI in corporate governance is another area that future researchers may work on. While the DPDP Act in India proposes to cover certain aspects of data security, more specific regulations may be necessary to enhance ethical AI deployments at the corporate level. Broad contours of AI governance should be harmonized with corporate law in emerging economies to foster growth and innovation.

Finally, corporate culture transformations involving AI must also be considered from the perspective of long-term AI integration. As more employees and officers use AI tools to improve operational efficiency, the extent of shareholder engagement through these tools would depend mainly on the cultural ethos of the company to adapt to technological changes. In emerging markets, corporate cultures usually tend to be authoritarian regarding technological changes and policies. Future research may be undertaken to evaluate how directors and other officers may foster efficiency, agility, and ethical stewardship in an AI-driven environment.

16.5 Conclusion

Looking ahead, the rapid development of technology, and by extension AI, is bound to have its existence felt and appreciated in the corporate governance space. For the last few years, the integration of AI into corporate governance has ceased to be a question of if and is now a question of when and how. Instead of being merely a technological shift at the periphery, it has repurposed how the company boards deal with strategic planning, shareholder engagement, and decision-making. It is no longer a matter of human versus machine, but rather human and machine working in tandem to negotiate complex situations. Besides, the integration of AI in corporate governance has the potential to address some of the long-standing agency problems. To make full use of what technology has to offer, the board of directors will have to prioritize transparency, data privacy, and accountability and have proper systems in place to regulate the use of AI in company operations.

16.5.1 Act: What's the Rush?

In India, the promulgation of the DPDP Act is a step in the right direction. It has paved the way for strict data protection measures and privacy standards that will require companies to implement robust data governance systems. These regulatory interventions are necessary to foster a secure and transparent environment where AI can be used effectively for better corporate governance.

To be equipped for what lies ahead, shareholders must have better digital literacy. They can make the best use of the tools available only if they know them and their functions. Indian regulatory bodies will also have to play an active role toward this end. There is also an opportunity for further global collaboration in AI and corporate governance in the future. Indian companies can evaluate their AI initiatives against global standards and enter cross-border alliances.

Though technology has extended the shareholders the opportunity to participate more actively in expressing their opinions and concerns, enhanced shareholder engagement presents some challenges also. While the use of technology can create the risk of disinformation, the companies would also be required to do a tight rope walking in ensuring a trade-off between being transparent and safeguarding proprietary knowledge. Aside from the technical, legal, and ethical considerations, a smooth integration of AI for enhanced shareholder engagement would also call for a broader cultural shift within organizations. With AI getting increasingly intertwined with decision-making processes, the boards would be required to cultivate a culture that values agility, adaptability, and ethical stewardship, and promotes open communication, a learning environment, and diversity of thought.

Companies can create a more responsible environment by leveraging regulatory advancements, enhancing digital literacy, fostering international collaborations, and embracing innovative AI-driven solutions. As India seeks its rightful place in the global business fraternity, there is a heightened necessity for Indian firms to catch up with the rest of the world in securing an efficient integration of AI into corporate governance to foster improved shareholder engagement. It would also be interesting to see how this integration helps Indian companies with concentrated shareholding succeed in ensuring that technology promotes fairness and trust in corporate governance. So, with the necessary precautions and diligent oversight, these attempts would help maximize the potential of AI in augmenting shareholder engagement to enhance sustainability goals, strengthen relationships between managers and investors, and catapult the overall efficiency of the corporate ecosystem of India.

Bibliography

Suresh Chandra Marwaha v. Lauls (P.) Ltd. [1978] 48 Comp. Cas. 110 (Punj, & Har.)
The Companies Act, 2013, §136.
The Companies Act, 2013, §184.
The Companies Act, 2013, §§101–102.
The Companies Act, 2013, §87.
The Companies Act, 2013, §94.
The Companies Act, 2013, §119.
The Companies Act, 2013, §170.
The Companies Act, 2013, §101.
The Companies Act, 2013, §105.
The Companies Act, 2013, §47.
The Companies Act, 2013, §s46 and 56.
The Companies Act, 2013, §51.
The Companies Act, 2013, §62(1)(a).
The Companies Act, 2013, §44.
The Companies Act, 2013, §114.
The Companies Act, 2013, §s13 and 14.
The Companies Act, 2013, §152.
The Companies Act, 2013, §169.
The Companies Act, 2013, §180.
The Companies Act, 2013, §139.
The Companies Act, 2013, §188(1) Proviso.
The Companies (Meeting of Board and its Powers) Rules, 2014, Rule 15.
The Companies Act, 2013, §245.
The National Company Law Tribunal Rules, 2016, Rule 84.
The Companies Act, 2013, §242.
The Companies Act, 2013, §178(5).
Securities and Exchange Board of India, Circular No. CIR/OIAE/1/2014 (Issued on December 18, 2014)
Allen and Brooks. (2024). *The role of technology and AI in corporate governance*. https://www.allenandbrooks.com/2024/06/10/the-role-of-technology-and-ai-in-corporate-governance/#:~:text=The%20rapid%20development%20of%20technology,operational%20efficiency%2C%20and%20risk%20management
Armour, J., Hansmann, H., & Kraakman, R. (2017). Agency problems and legal strategies. In *The anatomy of corporate law: A comparative and functional approach*. Oxford University Press. https://doi.org/10.1093/acprof:oso/9780198739630.003.0002
Beaujour, J. M. (2024). *Analyzing Tesla customer reviews: Insights unlocked. medium*. https://jmlbeaujour.medium.com/analyzing-tesla-customer-reviews-insights-unlocked-1d4db6c1816d
Binns, R. (2019). *On the apparent conflict between individual and group fairness*. CoRR, abs/1912.06883. http://arxiv.org/abs/1912.06883
Binns, R. (2022). Human judgment in algorithmic loops: Individual justice and automated decision-making. *Regulation & Governance, 16*(1), 197–211. https://doi.org/10.1111/rego.12358
Broadridge. (2021). *New AI platform provides proxy voting data from Broadridge*. https://www.broadridge.com/ca/press-release/2021/new-ai-platform-provides-proxy-voting-data-from-broadridge
Business Roundtable. (2016). *Principles of corporate governance*. Harvard Law School Forum on Corporate Governance. https://corpgov.law.harvard.edu/2016/09/08/principles-of-corporate-governance/
Chakraborty, S., & Sonkar, S. (2025). *Decoding India's draft DPDPA rules for the world*. IAPP. https://iapp.org/news/a/decoding-india-s-draft-dpdpa-rules-for-the-world

Denis, E., & Blume, D. (2021). *Using digital technologies to strengthen shareholder participation.* OECD Going Digital Toolkit Notes, No. 9. OECD Publishing. https://doi.org/10.1787/0fe52016-en

Dominguez, L. (2023). *Nike grows mobile app digital capabilities with sports pilot program.* https://consumergoods.com/nike-grows-mobile-app-digital-capabilities-sports-pilot-program

Douglas, K. C., & Ann, S. L. (2020). Report of the 2020 Multi-Stakeholder Working Group on Practices for Virtual Shareholder Meetings. Rutgers Center for Corporate Law and Governance Council of Institutional Investors Society for Corporate Governance.

Grover, N. (2020). *Shareholder activism: Will the Indian landscape change?* Deccan Herald. https://www.deccanherald.com/opinion/shareholder-activism-will-the-indian-landscape-change-850413.html

Hansmann, H., & Kraakman, R. (2004). The end of history for corporate law. In J. N. Gordon & M. J. Roe (Eds.), *Convergence and persistence in corporate governance* (pp. 33–68). Cambridge University Press; Cambridge Core. https://doi.org/10.1017/CBO9780511665905.002

HT Tech. (2022). Reliance WhatsApp chatbot launched; Here is how to download and why you may need it. *Hindustan Times.* https://tech.hindustantimes.com/tech/news/reliance-whatsapp-chatbot-launched-jio-5g-phone-jiobook-laptop-likely-to-be-unveiled-here-is-how-to-get-it-and-why-you-may-need-it-71624423049078.html

India Digital Personal Data Protection Act (DPDP Act): An Overview, 2024. Usercentrics. https://usercentrics.com/knowledge-hub/india-digital-personal-data-protection-act-dpdpa/

Kelso, A. (x2018). How becoming 'a tech company that sells pizza' delivered huge for Domino's. *Forbes.* https://www.forbes.com/sites/aliciakelso/2018/04/30/delivery-digital-provide-dominos-with-game-changing-advantages/

Khanna, V. (2017). Enforcement of corporate and securities laws in India: The arrival of the class action? In R. H. Huang & N. C. Howson (Eds.), *Enforcement of corporate and securities law: China and the world* (pp. 333–358). Cambridge University Press; Cambridge Core. https://doi.org/10.1017/9781316691397.017

Khanna, V., & Varottil, U. (2012). The rarity of derivative actions in India: Reasons and consequences. In D. W. Puchniak, H. Baum, & M. Ewing-Chow (Eds.), *The derivative action in Asia: A comparative and functional approach* (pp. 369–397). Cambridge University Press; Cambridge Core. https://doi.org/10.1017/CBO9780511998027.011

Kintone. (2019). *11 digital transformation quotes to lead change & inspire action.* Medium. https://medium.com/digital-transformation-talk/11-digital-transformation-quotes-to-lead-change-inspire-action-a81a3aa79a45

Kundnani, Y. (2023). Revolutionizing shareholder rights: Game-changing potential of e-AGMs and e-voting. *Economic Times.* https://economictimes.indiatimes.com/markets/stocks/news/revolutionizing-shareholder-rights-game-changing-potential-of-e-agms-and-e-voting/articleshow/103596259.cms?from=mdr

Lee, J., & Underwood, P. (2022). AI in the boardroom: Let the law be in the driving seat. *International and Comparative Corporate Law Journal, 16.*

Macbride, E. (2011). *Why disclosures don't work.* Wealthfront. https://www.wealthfront.com/blog/call-bullshit-disclosures-dont-work/

Meta. (2023). *Forging partnerships to advance AI technologies in India.* https://about.fb.com/news/2023/07/meta-and-india-ai-to-foster-advancements-in-ai-technologies-in-india/

Mittelstadt, B. (2019). Principles alone cannot guarantee ethical AI. *Nature Machine Intelligence, 1*, 501–507.

Navigating Data Privacy Compliance in India: A Comprehensive Guide to DPDPA 2023 Compliance Audits, 2023. Secure Privacy. https://secureprivacy.ai/blog/india-dpdpa-2023-compliance-framework

PricewaterhouseCoopers. (2023). *Global artificial intelligence study: Sizing the prize.* https://www.pwc.com/gx/en/issues/analytics/assets/pwc-ai-analysis-sizing-the-prize-report.pdf

RBI Asks Banks to Comply with DPDP Act Provisions Under Updated Regulatory Sandbox, 2024. Money Control. https://www.moneycontrol.com/news/business/rbi-asks-banks-to-comply-with-dpdp-act-provisions-under-updated-regulatory-sandbox-12372531.html

Preetha, S. (2021). Emerging prospects of shareholder engagement in India. *Indian Journal of Corporate Governance, 14*(2), 226–247. https://doi.org/10.1177/09746862211045760

Saleh, A. M. S. (2024). Blockchain for secure and decentralized artificial intelligence in cybersecurity: A comprehensive review. *Blockchain: Research and Applications, 5*(3), 100193. https://doi.org/10.1016/j.bcra.2024.100193

Singh, P. (2024). *Corporate boards and decision making: Impact of Artificial Intelligence (AI).* Metalegal Advocates. https://www.metalegal.in/post/corporate-boards-and-decision-making-impact-of-artificial-intelligence

Sharma, T., Saxena, S., Gupta, D., & Mahan, K. Regulatory sandbox for responsible AI. Chase India. https://www.chase-india.com/media/slljmxav/chase-india_regulatory-sandbox-for-responsible-ai.pdf

Shashidhar, K. J. (2023). *Regulatory Sandboxes: Decoding India's attempt to Regulate Fintech Disruption.* Observer Research Foundation. https://www.orfonline.org/research/regulatory-sandboxes-decoding-indias-attempt-to-regulate-fintech-disruption-66427

Shvartsman, D. (2024). *Apple Inc.: Our modern-day symbol of knowledge, power, and wealth.* Investing.Com. https://www.investing.com/academy/statistics/apple-facts/

Susskind, R., & Susskind, D. (2022). *The future of the professions: How technology will transform the work of human experts.* Oxford University Press.

Walsh, M. (2024). *How activist investors are pressing big tech for transparency on AI*. Raconteur. https://www.raconteur.net/artificial-intelligence/investors-transparency-ai

World Economic Forum. (2020). *Transforming paradigms: A global AI in financial services survey.* World Economic Forum. https://www3.weforum.org/docs/WEF_AI_in_Financial_Services_Survey.pdf

Part III
Corporate Governance, Compensation, and Data Protection

Chapter 17
Share Warrant Issuance and Stock Price Behavior: Evidence from the Indian Stock Market

Ardra Mani and Jose Joy Thoppan

Abstract We have been witnessing a great deal of progress in the area of corporate governance for the past several decades. Existing literature on emerging markets extensively addresses the improvement of corporate governance focusing on ownership structures. Considering the Indian economy, where a majority of the businesses are either family-run or promoter-controlled, studies on the ownership structure still hold relevance. Given the context, we aim to examine the impact of ownership structure on the returns of companies issuing share warrants in the Indian stock market. The sample consists of 130 share warrant issuances of listed companies in the National Stock Exchange (NSE) and Bombay Stock Exchange (BSE) during 2019–24. Share warrants issued to promoters can positively influence investors' perceptions; however, this may also facilitate perfectly timed insider trading. The informational asymmetry between the majority and minority stockholders may pave the way for profiteering in due course. We employ event study methodology to measure the price action to the share warrant issuance. The study can provide insights into investors' perceptions regarding the issuance of warrants to both promoters and non-promoter categories. The findings of the study could facilitate regulators and firms to enhance corporate governance by better management of information flow. Further, the results could enable investors and proxy advisory firms to make informed decisions and cast their votes judiciously on resolutions related to issuing warrants.

Keywords Share warrants · Seasoned equity offerings · Promoter group · Information asymmetry · Type II agency conflict

A. Mani (✉)
APJ Abdul Kalam Technological University and Saintgits College of Engineering (Autonomous), Kottayam, India
e-mail: ardra.mani@saintgits.org

J. J. Thoppan
Department of Business Administration Saintgits College of Engineering (Autonomous), Kottayam, India
e-mail: jose.joy@saintgits.org

V. Kandpal et al. (eds.), *Developments in Corporate Governance*,
Approaches to Global Sustainability, Markets, and Governance,
https://doi.org/10.1007/978-981-96-6366-8_17

17.1 Introduction

Financial markets have facilitated corporations to raise funds for their financial needs through various financial instruments like shares, bonds, and debentures. The fundraising process in the stock market happens in the primary market, also known as the "new issues" market, where companies raise funds by issuing new securities often through Initial Public Offerings (IPOs). Corporations that have already raised money from IPO can issue additional shares to the investors through Seasoned Equity Offerings (SEOs). In India, SEOs are often called Follow-on Public Offerings (FPOs). This approach is a common method for developing a sustainable capital structure through equity offerings by maintaining a balance between equity and debt (Abdullah & Zaby, 2021). Among the various methods of fundraising available post-IPO, share warrants are the not-so-popular financial instruments. While numerous studies contribute to the literature on IPOs and FPOs within the Indian stock market, research on share warrants and their impact on security prices is seldom explored. One reason attributed to the lack of attention might be that such events do not generate substantial profits for the merchant bankers and typically do not attract much attention as share warrant issuance is considered a routine event (Chen & Wu, 2001).

A share warrant is a right issued by a corporation that allows the holder to purchase a specific number of new shares at a predetermined price within a designated period (Suchard, 2004). In the Indian context, warrants are also issued as standalone instruments without being contingent on the purchase of another instrument. Schultz (1993) significantly contributed to the literature on share warrants by popularizing the theoretical support for warrant financing followed by Chemmanur and Fulghieri (1997) using the signaling model. In his paper, Schultz (1993) describes share warrants as a financing technique conducted in several phases that can mitigate issues of agency cost and free cash flow (Jensen, 1986). Managers, who are responsible for making financing decisions, might prioritize their profits when acting as agents for the principals (shareholders). Financing might not always reflect the company's actual needs but could instead address managers' objectives. A notable advantage of warrants is that they typically generate funds only when they are genuinely required. If the funds are directed toward the company's growth, this will likely increase the security price, prompting investors to exercise their warrants. Conversely, if the funds are not effectively utilized, the security's price may remain stagnant, and the warrants may not be exercised (Byoun & Moore, 2003).

The purpose of the work is to analyze the effect of share warrant issuance on the underlying stock prices in the Indian stock market during the 2019–24 period. The motivations of the current study are twofold. First, the study significantly contributes to the thin literature on share warrant issuances in India as share warrant issuances and their impact on stock price behavior have not been extensively studied in the Indian context. Secondly, the study explores how the impact of share warrant issuance varies depending on whether the warrants are issued to promoters, non-promoters, or both. Our study aims to enhance investors' understanding of stock trading behavior concerning warrant announcements enabling them to make more informed investment decisions.

The remaining structure of the paper is as follows: The next section provides a comprehensive literature review on the theoretical background of the study followed by hypothesis development. The subsequent section is a detailed description of the methodology. This is followed by a section that presents the empirical results of the study. The paper concludes with a discussion and conclusion section which discusses the significant insights of the study.

17.2 Literature Review and Hypothesis Development

Companies typically seek additional financing when viable business opportunities are present in the market (Abdullah & Zaby, 2021). While new issue offerings, IPOs tend to result in positive market returns (Gupta & Suri, 2017), SEOs are often negatively viewed by investors leading to negative abnormal returns and underpricing following their issuance. The negative returns are attributed to adverse selection problems as Myers and Majluf (1984) reported. Information asymmetry between corporates and investors is another explanation by Myers and Majluf (1984). Kim and Purnanandam (2013) bring in a surprisingly unnoticed explanation for negative returns, which is the strength of issuers' governance. The governance concern is supported by Jung et al. (1996), who present an argument suggesting that investors react adversely due to concerns about potential misuse of the proceeds. Empirical evidence from their study indicates that negative reactions are less pronounced when the issuing firm possesses substantial growth opportunities. We adopt a similar theoretical premise, by focusing on how preferential warrant issuances to promoters, non-promoters, and both are perceived by the investors.

Share warrants are often perceived as a mechanism of insider trading, strategically executed by promoters which is timed perfectly (Malik, 2024). When warrants are issued to promoters, they typically acquire company securities at lower prices and realize substantial profits. This practice disadvantages the minority shareholders of the company. The words promoter and insider are used interchangeably by Sarkar and Sarkar (2008). In the governance context, the issue of warrants to the promoters gives rise to Type II agency conflict. Type II agency conflict occurs when a large shareholder exploits their controlling position within the company to extract personal benefits, thereby disadvantaging smaller shareholders (Villalonga & Amit, 2006). This is where the governance of firms plays a role in influencing the returns from SEO announcements. Denis and McConnell (2003) define corporate governance as a framework comprising both institutional and market-based mechanisms designed to incentivize self-interested controllers of the company (promoters) to make decisions that enhance the firm's value for all its owners, including minority shareholders. Conflicts of interest between large shareholders and minority shareholders (Djankov et al., 2008) are one of the key governance issues prevailing in an economy with concentrated ownership. Thus, the study attempts to examine how preferential warrants issued to promoters impact the prices of companies issuing share warrants.

17.2.1 Price Reaction to SEOs

Significant negative abnormal returns have been observed during the SEO announcements in developed nations (Mikkelson & Partch, 1986). While we find no existing studies offering evidence for market reaction to share warrant issuances in India, there is a scope for understanding how share warrants work in the Indian equity market. India, characterized as a developing economy with relatively weaker investor protection (La Porta et al., 1998) and the highest number of family-owned businesses, provides an ideal context for our study on the potential misuse of SEOs, particularly in terms of their diversion of funds for the benefit of company promoters. Suryawinata et al. (2020) reports negative abnormal returns during the event of share warrant issuance in a study conducted in Indonesia. Literature suggests that preferential issuances through equity and share warrants pose positive reactions among investors due to future growth anticipation (Balachandran et al., 2017). Similarly, Suchard (2004) reports positive announcement effects with standalone warrant issues in Australia. Several studies reveal the impact of issuing share warrant units along with common equity or rights. Companies claim no advantages by attaching these warrants to the issue (Chemmanur & Fulghieri, 1997). A study by Susanti and Ardiana (2014) documents no significant difference between the Average Abnormal Returns (AAR) in the stocks of companies issuing rights alone and those issuing rights with attached share warrants.

17.2.2 Information Asymmetry

The pecking order hypothesis (Myers & Majluf, 1984) postulates that often companies prefer financing using internal funds (retained earnings) over external financing based on the assumption that insiders in the company are better informed than other shareholders. Therefore, it is argued that in markets with low asymmetries, firms may prefer to use retained earnings or debt financing (Miller, 1958). Firms are reluctant to pursue equity financing when they are undervalued. Managers, acting as agents for the principals of the firm, are incentivized to opt for equity financing when the firm is overvalued in the market (Myers & Majluf, 1984). Share warrants, as seasoned equity offerings, can be used as instruments to enter the market when it is overpriced, allowing informed investors to generate undue profits from the price differential. Lee (1997a, 1997b) reveals in his study that firms with favorable growth opportunities are likely to choose seasoned equity offerings (SEOs) to finance their investments if management believes the shares are overpriced. Smith (1986) also confirms these findings. Ambarish et al. (1987) demonstrated that the impact of announcing new stock issuance varies depending on the nature of existing assets and investment opportunities. Specifically, the announcement effect is negative for firms where private information predominantly concerns assets in place, whereas it is positive for firms where insider information primarily pertains to investment opportunities. In our study, we primarily examine the announcement effect of share warrant issuance in the Indian equity markets. Inadequacy in proper institutional settings, concentrated family ownership, poor governance, and weak investor protection could build on to the information asymmetry issues in developing nations like India as compared to developed nations (Chakraborty, 2018).

Our literature contributes to the continuously growing literature on Seasoned Equity Offerings (SEO) or Follow-on Public Offerings (FPO) by providing insights leading to price action during their issuances. The primary objective of the study is to examine the stock price reaction to warrant issuances conducted during 2019–24. This study also conducts a preliminary analysis to investigate the presence of information leakage in the system before the announcement date. Though there are several studies relating to the impact of SEOs in the market, to the best of our knowledge, no published paper has examined the impact of share warrant issuances in the Indian context.

With these objectives in place, the study aims to test three primary hypotheses. The first two hypotheses focus on the abnormal returns before and after the announcement of share warrant issuances, which is our primary objective. According to Elangovan et al. (2022), in a world where all information is available, the Efficient Market Hypothesis (EMH) should hold, as current prices are expected to incorporate all available information, resulting in no abnormal returns in the market.

Based on the above assumption, we formulate the following hypotheses:

H_{01}: The abnormal returns before the announcement date of share warrant issuance are close to zero.

H_{02}: The abnormal returns after the announcement date of share warrant issuance are close to zero.

The third hypothesis of the study tests for any potential information leakage in the system leading to the announcement date. Information asymmetry occurs when information is available to some privileged investors before its official public announcement (Bergh et al., 2018). Continuing with H_{01}, we formulate the following hypothesis:

H_{03}: There is no presence of information asymmetry in the period leading up to the event of warrant issuance.

17.3 Methodology

To examine the impact of share warrant issuance on security prices, we utilize event study methodology. The model was popularized by Fama (1991) as an alternative to analyze the semi-strong form efficiency of markets.

Table 17.1 Number of warrant issuances and allottee sub-sample

Year	Number of warrant issuances	Share warrant allottees		
		Promoters (P)	Non-promoters (NP)	Promoters & non-promoters (P & NP)
2018–19	23	17	1	5
2019–20	13	11	0	2
2020–21	26	19	5	6
2021–22	44	22	6	13
2022–23	23	10	4	8
2023–24	1	0	1	0
Total	130	79	17	34

17.3.1 Data

The data relating to Indian companies listed on the Bombay Stock Exchange (BSE) and National Stock Exchange (NSE) issuing share warrants between March 2019 and March 2024 were considered for the study. A total of 224 companies issuing warrants during the period were collected for the study. However, the data for 130 companies with their announcement dates were accessible from the Prowessiq database. The companies were divided into three subgroups: warrants issued to promoters (P), non-promoters (NP), and both promoters and non-promoters (P & NP). Table 17.1 provides a summary of the list of warrant issuances considered in the study.

17.3.2 Methodology

17.3.2.1 Standard Event Study—Market Model

The study uses the Standard Event study methodology to evaluate the abnormal returns around the warrants issue announcement. One of the pioneering studies conducted using event study is by Dolley (1993) to examine the effects of stock splits on prices, inspiring subsequent researchers to utilize this method for analyzing price reactions to specific events. In the current study, we adopt the event study method, a technique popularized by seminal papers from Ray Ball and Philip Brown (1968) and Fama (1965), which remains widely employed in event studies today.

The event study is conducted through the following steps:

17.3.2.2 Definition of an Event—Share Warrant Issue

In this study, the event is identified as share warrant issue announcement. Also, the announcement day is set as t_0.

17.3.2.3 Estimation Period

In determining the expected return of a security, an estimation period is defined as a specific interval preceding the event. This interval, known as the silent period, is chosen to minimize the influence of announcement-related noise. To align with the nature of this period, the estimation period is 150 days within the window (-175, -26). Conversely, event windows represent the duration over which the impact is averaged and cumulated to discern anomalies induced by the event. To evaluate the event's impact, various shorter windows before and after the announcement date (MacKinlay, 1997) are considered.

17.3.2.4 Calculation of Expected Returns

Evaluating the event's impact necessitates determining the abnormal return. Abnormal returns are measured as the return of the security during the window subtracted by the normal returns of the firm over the same period. Market model is used to find the expected returns of the security. The return of security in the Market model is:

$$E(R)_{it} = \alpha_i + \beta_i R_{mt} + \varepsilon_{it} \tag{17.1}$$

where $E(R)_{it}$ represents the expected return on the security i at period t, R_{mt} denotes the return of the market portfolio (Nifty 100) for the same period, and ε_{it} is the disturbance term with a mean of zero.

17.3.2.5 Calculation of Abnormal Returns

In Market model, we measure the abnormal returns (AR) as the following:

$$AR_{it} = R_{it} - E(R)_{it} \tag{17.2}$$

The Average Abnormal Returns (AAR) for period t is as follows:

$$AAR_t = \frac{1}{N}\sum_{i=1}^{N} AR_{i,t} \tag{17.3}$$

The Cumulative Abnormal Returns (CAR) for the security *i* over the windows are as follows:

$$CAR_{i,(tx,ty)} = \sum_{t=tx}^{ty} AR_{i,t} \tag{17.4}$$

AR_{it} is the abnormal return for each security *i* on a single day within the event window. Abnormal returns are not very informative since the stock price movements depend upon various other information unrelated to the event. Averaging the abnormal returns would improve the outcome of the analysis by canceling out the effect of the information unrelated to the event. CAR is the time-series aggregation of abnormal returns.

17.3.2.6 Test for Significance

T-test is used as the parametric test as it provides evidence on whether the event's impact is reflected as significant abnormal returns.

The primary objective of using event studies in management and business is to evaluate stock price reactions to announcements of events. Share warrant issuances can affect the prices of the securities positively anticipating potential growth prospects. Conversely, it can have negative effects due to information asymmetry between insiders (promoters) and outsiders (Chen, 2015). In event studies, the actual daily stock returns around the announcement date or on the date of the event are compared with the expected returns. Practically, as in the paper, the methodology differentiates the returns that would have been registered in the absence of the event, i.e., warrant issuance and the returns that are observed. Event study methodology helps in assessing the possibility of information leakage to insiders before the announcement of warrant issuance. This privileged information with the insiders can lead to potential manipulation in the market, generating undue profits to the advantageous investors. The empirical results provide insights into stock price behavior in and around the announcement of share warrant issuances.

17.4 Empirical Results

17.4.1 Price Reaction to Share Warrant Issuances—Average Abnormal Returns

In this section, the empirical results of event study on warrant issue announcements of Indian companies. Table 17.2 reports the AARs for 21 days leading up to the announcement and following the share warrant issue announcement.

Table 17.2 Average abnormal returns of 21 days in and around warrant issue announcements

Day (t)	Average Abnormal Returns (AAR)				t-statistics			
	All	P	NP	P & NP	All	P	NP	P & NP
−10	0.00348	0.00506	−0.00979	0.00642	1.03543	1.07223	1.53714	0.27273
−9	0.00337	0.00458	−0.00269	0.00358	1.17009	1.28625	0.29991	0.54388
−8	−0.00399	−5.1E−07	0.004183	−0.01736	1.25764	0.00011	0.49277	0.00027**
−7	0.00971	0.00705	0.014931	0.013301	3.32151**	1.87501	2.33863**	0.04082**
−6	−0.00271	−0.00099	−0.00626	−0.00494	1.0194	0.27035	0.88087	0.27384
−5	0.00421	0.00785	0.01092	−0.0076	1.26214	1.73513	1.36202	0.19078
−4	0.00539	0.00936	−0.00428	0.00100	1.73092	2.69367*	0.4978	0.89530
−3	0.00958	0.01317	0.012878	−0.00039	2.74215**	3.06426**	1.41668	0.95935
−2	0.00652	0.00726	0.002766	0.00667	1.98504**	1.72598	0.29863	0.31610
−1	0.01474	0.01504	0.015972	0.01343	3.94104**	3.07317**	1.38211	0.05383
0	0.01041	0.00570	0.02297	0.01507	3.31968**	1.63727	2.71061**	0.05692
1	0.02423	0.02493	0.025725	0.02187	6.28964**	5.05604**	3.10651**	0.01429**
2	0.00809	0.00948	0.007668	0.00508	2.22357**	2.21157**	0.73174	0.54794
3	0.00945	0.00780	0.010042	0.01300	2.62256**	1.86820	0.9261	0.12812
4	0.00606	0.00731	0.010317	0.00104	1.70324	1.54264	1.10587	0.87552
5	0.01125	0.00544	0.013071	0.02383	3.10189	1.09035	1.73058	0.00053**
6	0.00216	0.00302	0.003493	−0.00049	0.65706	0.69187	0.46549	0.94146
7	−0.00067	−0.00671	0.016272	0.00488	0.21353	1.98805	1.91717	0.53095
8	0.00228	0.00024	0.012519	0.00192	0.76541	0.06249	1.67958	0.75179

(continued)

Table 17.2 (continued)

Day (t)	Average Abnormal Returns (AAR)				t-statistics			
	All	P	NP	P & NP	All	P	NP	P & NP
9	0.00013	0.00296	0.004772	−0.00877	0.04563	0.79532	0.54893	0.10698
10	0.00053	0.00083	0.008766	−0.00426	0.18190	0.21363	1.11487	0.44987

Number of warrant issuances = 130. Study Period: 2019–24. **, * denote significance at 5% and 10% levels respectively (two-tailed *t*-test)

When considering the sample in which preferential allotment is made to all categories of investors, the abnormal returns seem concentrated in and around the announcement day. In the (−3, + 3) event window, the firms report positive abnormal returns before the announcement, indicating the possibility of a leak of privileged information among the insiders. The observed positive abnormal returns may be attributable to promoters or other privileged insiders who reacted to the news prior to its official announcement. The positive abnormal returns in the post-announcement period suggest that investors, regardless of whether the warrants are preferentially allotted to promoters, non-promoters, or both, generally perceive the issuance of warrants as a positive signal indicating potential firm growth. Investors likely anticipate business opportunities that justify the capital raised. These findings are consistent with the study by Balachandran et al. (2017), which reports positive abnormal returns across all examined periods.

Warrants issued exclusively to the promoter group experience significantly positive price reactions both in the days leading up to the announcement and following it. This phenomenon can be attributed to investors perceiving growth opportunities within the business, which is justified through promoters' willingness to infuse capital into the firm. The issuance of warrants to promoters signals their confidence in the company's prospects through the infusion of additional capital. In the context of an economy where a majority of businesses are family-run and family-controlled, investors tend to believe that such capital infusions are grounded in substantial reasons. The presence of abnormal returns before the public announcement raises concerns regarding the potential involvement of promoters in insider trading, at the expense of minority shareholders. In the case of warrant issuances made to investor groups (exclusively to non-promoters/both promoters and non-promoters) except the promoter group, the abnormal returns align with the expectations. A significant spike in abnormal returns is observed immediately after the public announcement. However, there are no notable abnormal returns leading up to the event announcement, except on days -7 and -8. Thus, unlike in the case of warrant issuances to promoters, the possibility of information leakage prior to the announcement is not substantiated.

Figure 17.1 presents the Average Abnormal Returns (AARs) both leading up to and following the warrant issuance announcements. The chart provides a comprehensive overview of the impact of warrant issuances on stock prices. The average abnormal returns fluctuate around the pre-announcement period. Significant spikes are visible during the days between −9 and −7 indicating potential information leakage which could be attributed to the warrants issued to promoter groups. On the announcement day, the AAR reaches a notable peak. This indicates a strong positive market reaction to the warrant issuance announcement. Investors likely perceive the issuance as a positive signal, leading to a significant rise in abnormal returns. In the post-announcement period, the sustained positive AARs prove that the market continues to perceive warrants positively, but the effect diminishes over time as the market fully incorporates the new information.

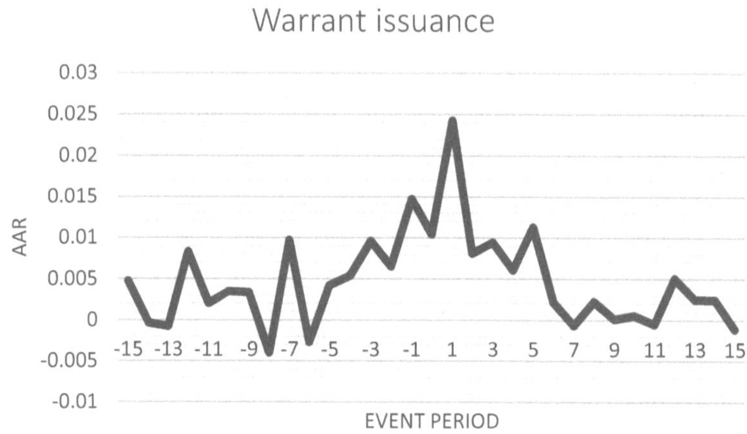

Fig. 17.1 AAR for the period −15 to + 15

17.4.2 Price Reaction to Share Warrant Issuances—Cumulative Average Returns (CAR)

To analyze the price reaction, we calculate the cumulative average returns over a 21-day window (−10, + 10). A short window of 3 days (−1, + 1) was also studied.

Table 17.3 shows the results of the event study methodology used for analyzing price reactions to warrant issuances in the Indian stock market. The mean abnormal returns of firms issuing warrants to all investor groups and exclusively to promoters are reported in the table. The mean abnormal returns are divided into three panels. Panel I is the CAR around the warrant issue announcements; Panel II is the CAR before the announcement; and Panel III is the CAR after the announcement of warrant issuance.

It is noteworthy that high positive statistically significant abnormal returns are observed around the warrant announcements, with significance at the 5% level. The period immediately preceding the announcement, specifically the window (-5, 0), documents the highest abnormal return of 5.08%. This substantial return is corroborated by the mean abnormal returns for warrants issued to promoters, which is similarly high at 5.83%. Insiders of the firms and information differentials could be responsible for the abnormal returns in the before-announcement period. Similarly, the windows (-4, 0), (-3, 0), (-2, 0), and (-1, 0) also document positive abnormal returns, which is corroborated by the returns observed for warrants issued to promoters. Similarly, warrants issued to investors other than promoters also exhibit significant abnormal returns in the pre-announcement period. The returns documented are as high as the other two groups. A plausible explanation is that the windows used for studying the pre-announcement period also encompass days following the announcement. In the case of promoters, the information is already disseminated within the system before the announcement, causing the effect to

Table 17.3 Event study results: cumulative abnormal returns of warrant issuances

	Warrant issuing firms		Warrant issued exclusively to promoters		Warrants issued to others	
Windows	Mean abnormal returns (%)	*t*-statistics	Mean abnormal returns (%)	*t*-statistics	Mean Abnormal returns (%)	*t*-statistics
Panel I: CAR around warrant issue announcement						
(−1, + 1)	+ 4.94	7.733***	+ 4.56	5.508***	+ 5.51	5.487***
(−2, + 2)	+ 6.40	7.122***	+ 6.24	5.366***	+ 6.64	4.653***
(−3, + 3)	+ 8.30	7.492***	+ 8.33	5.893***	+ 8.25	4.584***
(−4, + 4)	+ 9.45	7.146***	+ 10	6.013***	+ 8.85	3.925***
(−5, + 5)	+ 10.8	7.435***	+ 11.33	6.031***	+ 10.4	4.333***
Panel II: CAR pre-announcement date						
(−5, 0)	+ 5.08	5.201***	+ 5.83	4.701***	+ 3.92	2.466***
(−4, 0)	+ 4.66	5.219***	+ 5.05	4.407***	+ 4.06	2.827***
(−3, 0)	+ 4.12	5.159***	+ 4.11	4.062***	+ 4.13	3.153***
(−2, 0)	+ 3.16	4.934***	+ 2.80	3.399***	+ 3.73	3.630***
(−1, 0)	+ 2.51	4.820***	+ 2.07	3.157***	+ 3.19	3.737***
Panel III: CAR post-announcement date						
(0, 1)	+ 3.46	7.034**	+ 3.06	5.011**	+ 4.08	4.956**
(0, 2)	+ 4.27	6.123**	+ 4.02	4.857**	+ 4.68	3.759**
(0, 3)	+ 5.22	6.782**	+ 4.79	5.184**	+ 5.88	4.364**
(0, 4)	+ 5.82	6.310**	+ 5.52	4.964**	+ 6.29	3.899**
(0, 5)	+ 6.95	6.733**	+ 6.06	4.726**	+ 8.32	4.831**

Number of warrant issuances = 130. Study period—2019–24. **, * denote significance at 5% and 10% levels respectively (two-tailed t-test)

diminish more rapidly. Conversely, for firms issuing warrants to non-promoters, the announcement date accrues substantial abnormal returns, which take longer to dissipate post-announcement.

In the pre-announcement period, there is a similar pattern that is followed by firms issuing warrants to promoters and other investors. When examining the abnormal returns of warrant issuances to investors other than promoters, a notable difference is observed in the pre-announcement period. Warrants issued to promoters exhibit significant abnormal returns days before the announcement, suggesting potential front-running. In contrast, for other investors, the absence of insider trading could be the reason for much lower abnormal returns before the event date. Karpoff and Lee (1991) document that the insiders or promoters increase trading on their stocks months preceding any SEO announcements and decrease eventually. This trend needs to be further examined.

In the post-announcement period, abnormal returns for the promoter group declined, recording lower abnormal returns compared to the non-promoter group.

This can be explained by the fact that the news significantly impacts the stock prices of warrants issued to other investors because the information has just been released to the public. In contrast, for warrants issued to promoters, the information has already been integrated into the system for a considerable time, causing the returns to diminish more quickly than those for non-promoter investors.

17.4.3 Classification Matrix

The classification matrix allows us to understand how warrant issuances to different investor groups are reflected in stock prices. Figures 17.2 and 17.3 provide insights by indicating the number of significant abnormal returns over longer windows, specifically (-15, 0) and (0, + 15). We see that 17% of the stocks with warrant issues only to non-promoters and 15% of the stocks with warrant issues to both promoters and non-promoters exhibit abnormal returns. We observe that in the pre-event period, the abnormal returns are concentrated on issues involving the promoter group and in the post-event period, it is more equitably distributed indicating the presence of possible insider trading by the promoter group to earn undue profits.

In the above pre-event matrix, we can observe that 17% of stocks with warrants issued to the promoter group only show significant abnormal returns in the pre-event period, whereas none of the stocks with warrants issued to non-promoters only show abnormal returns. However, when we account for stock where issues are made to both groups combined around 5.88% show abnormal returns. Similarly, in the post-event matrix, we can observe that 27.7% of the stocks where warrants were issued to promoters alone show abnormal returns. The increase could be due to the higher levels of levels before the announcement contributing to sustained higher levels after the announcement.

	Sig. Abnormal returns	Insignificant Abnormal returns	
Promoters	14 (17.7%)	65	79
Non-promoters	0	17	17

	Sig. Abnormal returns	Insignificant Abnormal returns	
Promoters	14 (17.7%)	65	79
Others	3 (5.88%)	48	51

Fig. 17.2 Classification matrix for promoters & others for the window (−15,0)

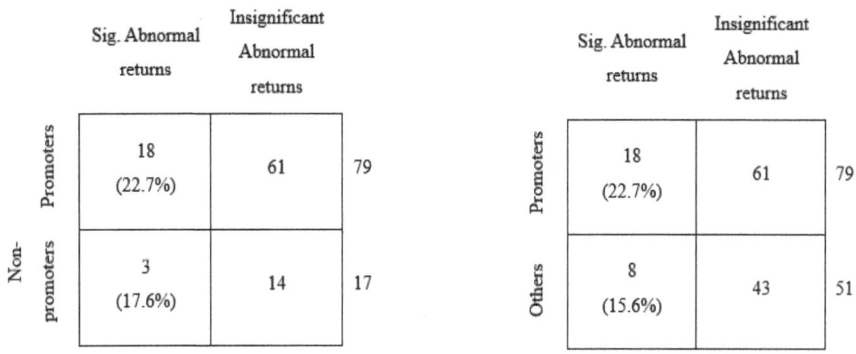

Fig. 17.3 Classification matrix for promoters & others for the window (0, 15)

Table 17.4 Summary of results from hypotheses testing

Hypothesis	Statement	Results
H_{01}	The abnormal returns before the announcement date of share warrant issuance are close to zero	Rejected
H_{02}	The abnormal returns after the announcement date of share warrant issuance are close to zero	Rejected
H_{03}	There is no presence of information asymmetry in the period leading up to the event of warrant issuance	Rejected

17.4.4 Results of Hypotheses Testing

See Table 17.4.

17.5 Discussion and Conclusion

The primary objective of the study is to examine the price reaction of share warrant issuance in the Indian stock market. The study employed event study methodology to measure the presence of abnormal returns around warrant issuance. In the Indian context, although several studies have examined the price reactions to seasoned equity offerings, the area of warrant issuance remains unexplored by scholars. Our study aims to fill this gap by analyzing the stock reactions to warrant issuances that occurred during 2019–24. This is possibly the first article to examine the price reactions to the event of warrant issuances in India. The study includes a total of 130 warrant issuances. Share warrants are often issued on a preferential basis to promoters, non-promoters, or both. In our study, we analyze the share warrants issued exclusively to promoters and others separately. Similar to other SEOs, share warrants are a popular method of equity financing among Indian companies. Specifically, SEOs are typically issued when management believes the firm is overvalued in the market. Myers and Majluf (1984) argues that the issuers of warrants often issue them when they perceive a mispricing in the market due to which SEOs are generally perceived negatively by the market due to the high information asymmetry they involve. Our study investigates the presence of information differentials by examining the returns prior to and post to the public announcement of warrant issuances. Family-owned and controlled businesses, potential insider trading among firms with large promoter holdings, and inadequate transparency in information exacerbate the information problems faced by Indian firms. Consequently, India provides a perfect environment to study the role of information asymmetry (Sony & Bhaduri, 2021).

The findings on the price reaction do not align with our expectations based on the existing SEO literature. The warrant issuances positively impacted stock prices, leading to positive abnormal returns around the event announcement. The outcome corroborates with Balachandran et al. (2017). The analysis reports positive abnormal returns in the shorter windows of (−1, + 1) and (−2, + 2), indicating that investors perceive the news of warrant issuances positively, anticipating favorable growth prospects for the firms. This reaction may also stem from investors' belief that promoters participating in the warrant issuance are confident about the company's future, as evidenced by their willingness to invest additional capital. Additionally, we observe significant positive abnormal returns prior to the announcement, indicating that information might have flowed into the market during the period. This may be due to the involvement of insiders with prior knowledge of the event, who might have engaged in excessive trading activities (Karpoff & Lee, 1991). This disadvantages minority investors, raising the issue of Type II agency conflict. This matter warrants further investigation, which falls outside the scope of our current research. The event study results indicate a differential impact of warrant issuances on the stock prices of firms based on the recipient group. Specifically, abnormal returns for stocks of firms that issued warrants to the promoter group exhibited significant increases before and around the announcement date, with the effect dissipating shortly after the announcement. Conversely, firms that issued warrants to other investor groups experienced less significant abnormal returns before the event but showed significant increases around and following the announcement while firms which issued warrants to non-promoters alone did not show any abnormal returns in the pre-announcement period. This pattern suggests that in the normal course the impact of the announcement takes time to be fully incorporated into the stock prices for warrants which is observed in the case of warrants issued to other investor groups, whereas if the recipient is the promoter group, the information appears to be reflected more rapidly into stock prices much before the announcement. The classification matrix confirms the existence of the pattern in the longer windows.

This study stands pioneer among the studies in warrant issuances as it provides a clear understanding of price action due to share warrants issued in India, especially in the case of firms that are managed and controlled by families. Theoretically, we provide evidence for Type II agency conflicts by exploring the issuance of share warrants, an under examined form of SEO.

The study has significant implications for regulators, proxy advisory groups, and investors. In an emerging economy like India with most family-controlled firms, establishing a constant supervision mechanism over the trading practices by the insiders could be an effective way to mitigate conflict between major and minor shareholders.

The regulator has an important role to play in ensuring a level playing field for all stakeholders. The Integrated Market Surveillance System (IMSS) of SEBI could be strengthened by configuring the surveillance models and AI-driven analytical engines to spot pricing anomalies and abnormal returns around the announcement of warrants. They can also then be triggered to check for any front-running, circular trading, related party transactions, or insider trading instances, which can then be investigated to ascertain the motives of the flagged transactions.

The proxy advisory groups may peruse the underlying information on warrant issuances to ascertain the true intent of the announcement by evaluating for any potential adverse consequences to any segment of the shareholders. Their recommendation is gaining significant traction among the investing community, and their role as vanguards to the efficient and frictionless functioning of the market is paramount. This does not preclude the investor from undertaking their independent research, and this paper provides the broad guidelines for an investor to carry out independent research to take informed decisions.

The findings suggest that regulatory bodies, functioning as market watchdogs, are critical in detecting and preventing irregular activities in the stock market. Strengthening compliance measures could help prevent participants from leveraging insider information for unfair profit generation thus enhancing governance measures. The study emphasizes the heightened need for awareness by investors regarding the risks of investing in firms affected by insider trading and inadequate information disclosure. The results could empower investors and proxy advisory firms to make well-informed decisions and cast their votes judiciously on resolutions about the issuance of warrants.

The study paves the way for further research in the area of share warrants. Despite the valuable insights provided, this study has certain limitations that open avenues for future research. Notably, the study does not differentiate between family and non-family firms issuing share warrants, focusing instead solely on promoter-held and non-promoter-held companies. Future research could usefully explore these distinctions to enhance the understanding.

Second, this study does not account for other governance factors that may influence stock price performance during share warrant issuance. Future research could focus on analyzing firm-level governance mechanisms to assess their impact on stock price behavior more comprehensively.

Third, the study does not account for the volatility during the issuance period. As noted by Brown et al. (1988), controlling for changes in variance is essential for a more accurate understanding of the abnormal returns observed. Thus, further research can accommodate volatility changes during share warrant issuance.

Funding Acknowledgements This research is supported by the Indian Council of Social Science Research (ICSSR), New Delhi. The corresponding author (Ardra Mani) is a recipient of the ICSSR Doctoral Fellowship [File no. ICSSR/RFD/2023-24/303].

Declaration and Conflicting Interests The authors have no conflicts of interest concerning the research, authorship, and/or publication of this article.

References

Abdullah, S. M., & Zaby, S. (2021). Seasoned equity offerings and differences in share-price impact by firm categories. *International Journal of Financial Studies, 9*(3), 36. https://doi.org/10.3390/ijfs9030036

Ambarish, R., John, K., & Williams, J. (1987). Efficient signaling with dividends and investments. *Journal of Finance, 42*(2), 321–343. https://doi.org/10.1111/j.1540-6261.1987.tb02570.x

Balachandran, B., Kanapathippillai, S., Krishnamurti, C., Theobald, M., & Velayutham, E. (2017). The issuance of warrants in rights offerings: Agency costs and signaling effects. *Australian Journal of Management, 42*(4), 608–636. https://doi.org/10.1177/0312896216682062

Ball, R., & Brown, P. (1968, Autumn). An empirical evaluation of accounting income figures. *Journal of Accounting Research*, 159–178.

Bergh, D. D., Ketchen, D. J., Orlandi, I., Heugens, P. P. M. A. R., & Boyd, B. K. (2018). Information Asymmetry in management Research: Past accomplishments and future opportunities. *Journal of Management, 45*(1), 122–158. https://doi.org/10.1177/0149206318798026

Brickley, J. A., Coles, J. L., & Terry, R. L. (1994). Outside directors and the adoption of poison pills. *Journal of Financial Economics, 35*(3), 371–390. https://doi.org/10.1016/0304-405x(94)90038-8

Brown, K. C., Harlow, W., & Tinic, S. M. (1988). Risk aversion, uncertain information, and market efficiency. *Journal of Financial Economics, 22*(2), 355–385. https://doi.org/10.1016/0304-405x(88)90075-x

Byoun, S., & Moore, W. T. (2003). Stock vs. stock-warrant units: Evidence from seasoned offerings. *Journal of Corporate Finance, 9*(5), 575–590. https://doi.org/10.1016/s0929-1199(02)00038-x

Chakraborty, S. (2018). Significance of BRICS: Regional powers, global governance, and the roadmap for multipolar world. *Emerging Economy Studies, 4*(2), 182–191. https://doi.org/10.1177/2394901518795070

Chemmanur, T. J., & Fulghieri, P. (1997). Why include warrants in new equity issues? A theory of unit IPOs. *Journal of Financial and Quantitative Analysis, 32*(1), 1. https://doi.org/10.2307/2331314

Chen, K., & Wu, L. (2001). Introduction and expiration effects of derivative equity warrants in Hong Kong. *International Review of Financial Analysis (Online)/International Review of Financial Analysis, 10*(1), 37–52. https://doi.org/10.1016/s1057-5219(00)00044-2

Chen, J. J. (2015). *A primer on corporate governance: China.* https://openlibrary.telkomuniversity.ac.id/home/catalog/id/159439/slug/a-primer-on-corporate-governance-china.html

Denis, D. K., & McConnell, J. J. (2003). International corporate governance. *Journal of Financial and Quantitative Analysis, 38*(1), 1. https://doi.org/10.2307/4126762

Djankov, S., La Porta, R., Lopez-De-Silanes, F., & Shleifer, A. (2008). The law and economics of self-dealing. *Journal of Financial Economics, 88*(3), 430–465. https://doi.org/10.1016/j.jfineco.2007.02.007

Dolley, J. C. (1993). Characteristics and procedure of common stock split-ups. *Harvard business review, 11*(3), 316–326. *100*(4), 859–887. https://doi.org/10.1086/261843

Elangovan, R., Irudayasamy, F. G., & Parayitam, S. (2022). Testing the market efficiency in Indian stock market: Evidence from Bombay Stock Exchange broad market indices. *Journal*

of Economics, Finance and Administrative Science, 27(54), 313–327. https://doi.org/10.1108/jefas-04-2021-0040
Fama, E. F. (1965). The behavior of stock-market prices. *The Journal of Business, 38*(1), 34–105.
Fama, E. F. (1991). Efficient capital markets: II. *The Journal of Finance, 46*(5), 1575–1617. https://doi.org/10.1111/j.1540-6261.1991.tb04636.x
Gupta, C., & Suri, S. (2017). Valuation and pricing of Indian IPOs. *Vision, 21*(4), 375–384. https://doi.org/10.1177/0972262917734706
Jensen, M. C. (1986). Agency costs of free cash flow, corporate finance, and takeovers. *American Economic Review, 76*(2), 323–329. https://econpapers.repec.org/article/aeaaecrev/v_3a76_3ay_3a1986_3ai_3a2_3ap_3a323-29.htm
Jung, K., Kim, Y., & Stulz, R. (1996). Timing, investment opportunities, managerial discretion, and the security issue decision. *Journal of Financial Economics, 42*(2), 159–185. https://doi.org/10.1016/0304-405x(96)00881-1
Karpoff, J. M., & Lee, D. (1991). Insider trading before new issue announcements. *Financial Management, 20*(1), 18. https://doi.org/10.2307/3666093
Kim, E. H., & Purnanandam, A. (2013). Seasoned equity offerings, corporate governance, and investments*. *European Finance Review, 18*(3), 1023–1057. https://doi.org/10.1093/rof/rft012
La Porta, R., Lopez-de-Silanes, F., Shleifer, A., et al. (1998). Law and finance. *Journal of Political Economy, 106*, 1113. https://doi.org/10.1086/250042
Lee, H. W. (1997a). Post offering earnings performance of firms that issue seasoned equity: The role of growth opportunities. *The Quarterly Review of Economics and Finance, 37*(1), 97–114. https://doi.org/10.1016/s1062-9769(97)90060-0
Lee, I. (1997b). Do firms knowingly sell overvalued equity? *The Journal of Finance, 52*(4), 1439–1466. https://doi.org/10.1111/j.1540-6261.1997.tb01116.x
MacKinlay, A. C. (1997). Event studies in economics and finance. *Journal of Economic Literature, 35*, 13–39.
Malik, V. (2024, May 5). *Stock warrants to promoters: How to analyse.* Dr. Vijay Malik. https://www.drvijaymalik.com/preferential-allotment-of-warrants-to-promoters/#when-share-warrants-are-allotted-above-market-price
Mikkelson, W. H., & Partch, M. (1986). Valuation effects of security offerings and the issuance process. *Journal of Financial Economics, 15*(1–2), 31–60. https://doi.org/10.1016/0304-405x(86)90049-8
Miller, M. H. (1958). The cost of capital, corporation finance and the theory of investment. *American Economic Review, 48*(3), 261–297. http://blog.bearing-consulting.com/wp-content/uploads/2012/09/The.Cost_.of_.Capital.Corporation.Finance.and_.the_.Theory.of_.Investment.pdf
Myers, S. C., & Majluf, N. S. (1984). Corporate financing and investment decisions when firms have information that investors do not have. *Journal of Financial Economics, 13*(2), 187–221. https://doi.org/10.1016/0304-405x(84)90023-0
Sarkar, J., & Sarkar, S. (2008). Debt and corporate governance in emerging economies evidence from India1. *Economics of Transition, 16*(2), 293–334. https://doi.org/10.1111/j.1468-0351.2008.00307.x
Schultz, P. (1993). Unit initial public offerings. *Journal of Financial Economics, 34*(2), 199–229. https://doi.org/10.1016/0304-405x(93)90018-7
Smith, C. W. (1986). Investment banking and the capital acquisition process. *Journal of Financial Economics, 15*(1–2), 3–29. https://doi.org/10.1016/0304-405x(86)90048-6
Sony, B., & Bhaduri, S. (2021). Information asymmetry and financing choice between debt, equity and dual issues by Indian firms. *International Review of Economics & Finance, 72*, 90–101. https://doi.org/10.1016/j.iref.2020.11.001
Suchard, J. (2004). The use of stand-alone warrants as unique capital-raising instruments. *Journal of Banking & Finance, 29*(5), 1095–1112. https://doi.org/10.1016/j.jbankfin.2004.05.032
Suryawinata, S., Tekadtuera, A., & Ervina, D. (2020). The Impact of Warrant Listing and Maturity on the Price of Shares Accompanied by Warrants in Indonesia. *Studi Akuntansi Dan Keuangan Indonesia, 3*(1), 74–99. https://doi.org/10.21632/saki.3.1.74-99

Susanti, P. A., & Ardiana, P. A. (2014). Reaksi Pasar Terhadap Pengumuman right issue Pada Perusahaan Warrant issuer Dan non warrant issuer. *E-Jurnal Akuntansi, 9*(1), 221–236. http://ojs.unud.ac.id/index.php/Akuntansi/article/view/8660/7648
Villalonga, B., & Amit, R. (2006). How do family ownership, control and management affect firm value? *Journal of Financial Economics, 80*(2), 385–417. https://doi.org/10.1016/j.jfineco.2004.12.005

Chapter 18
Executive Compensation: A Question of Sufficiency or Fairness?

Gobinda Gopal Pahari and Chandra Sekhar Mishra

Abstract Determining executive compensation is a crucial choice for every organization. Compensation is essential for attracting, retaining, and motivating an executive, and it influences the broader picture beyond the pay disparity. Moreover, executive compensation throughout the economy affects the public's image of capitalism. Executive compensation includes a whole set of monetary and non-monetary perks offered to top-level executive directors such as CEOs, CFOs, managing directors, and other executives in a firm. Agency conflict arises from the divergence of shareholders and managers in listed companies. Agency theory is critical to examining conflicts of interest between principals and agents. It is fundamental in elucidating executive compensation disparities and formulating strategies to mitigate conflicts of interest between shareholders and management. Other theories relating to executive compensation, such as tournament theory, optimal contracting theory, entrenchment effect, and alignment effect, are also essential. In India, the importance of executive compensation rose post-economic liberalization due to the increasing profitability margin of firms, increase in competition for managerial abilities, benchmarking Indian companies with global standards, etc. One unique characteristic exclusive to India (compared to developed nations) is the prevalence of "family-owned" firms. They essentially have "owner-managers," where business owners also serve as managers. Another uniqueness is that Indian family firms have a higher concentration of ownership, high growth opportunities, and an underdeveloped financial market. Overall, this study sheds light on over-compensation, pay disparity, family firms, and ownership concentration in Indian firms regarding executive compensation.

Keywords Executive compensation · Pay disparity · Legal aspects · Family firms · Corporate governance

G. G. Pahari (✉) · C. S. Mishra
VGSOM, IIT Kharagpur, Kharagpur, India
e-mail: Pahari.gobinda1@gmail.com

C. S. Mishra
e-mail: csmishra@vgsom.iitkgp.ac.in

V. Kandpal et al. (eds.), *Developments in Corporate Governance*,
Approaches to Global Sustainability, Markets, and Governance,
https://doi.org/10.1007/978-981-96-6366-8_18

18.1 Introduction

The board of directors is a committee set up to supervise the activities of the business. The board, particularly its executive directors, is responsible for making all significant decisions and formulating day-to-day strategies for a company. Determining executive compensation is a crucial choice for every organization. Compensation is essential for attracting, retaining, and motivating an executive, and it influences the broader picture beyond the pay disparity. Moreover, executive compensation throughout the economy affects the public's image of capitalism.

Consequently, executive compensation garners more scrutiny than almost any other corporate decision. Executive compensation[1] includes a whole set of monetary and non-monetary perks offered to top-level executive directors such as CEOs, CFOs, managing directors, and other executives in a firm. This typically encompasses salaries, bonuses linked to performance metrics, and stock options, with various benefits, such as healthcare and retirement benefits. Ownership and management are separated in a publicly listed firm, making the management substantially powerful. Agency issues arise between shareholders and managers because ownership and control are separated in publicly traded companies (Jensen & Meckling, 1976). Agency theory, which focuses on conflicts of interest between shareholders and managers, has been fundamental in explaining variations in executive compensation and in making strategies to mitigate or manage these conflicts (Gomez-Mejia et al., 1987; Tosi et al., 2000; van Essen et al., 2015). According to agency theory, CEOs may pursue their interests in publicly listed companies where ownership is dispersed, and professional managers are in control, diverging from the shareholders' goals. Effective corporate governance necessitates specific processes for determining executive remuneration and overseeing the performance of senior management. This encompasses the inclusion of "independent" directors on the board, separating duties between the chairman and the CEO, and forming remuneration committees. Albuquerque et al., (2024) assert that the mandatory disclosure of compensation peer groups enhances reciprocal benchmarking and results in reduced variability in CEO overall compensation levels. Albuquerque et al., (2024) indicate that CEO remuneration arrangements have increasingly grown intricate in recent years, partly owing to external influences, resulting in adverse effects on business performance.

The crucial question for shareholders pertains to the selection of management and the methods of incentivizing them to ensure their performance results in wealth maximization following consideration of the expenses associated with such processes. Executive compensation is a contentious issue, with proponents arguing it is necessary to attract and retain top talent, while critics suggest it increases income inequality and can incentivize unethical behavior. Consequently, regulators across the countries have introduced guidelines and regulations to ensure fairness and transparency in executive pay. These regulations often mandate public disclosure of executive compensation packages and may limit the types of compensation executives can

[1] Executive compensation, Executive remuneration, and Executive pay are alternatively used throughout the chapter.

receive. Executive compensation has brought significant attention in recent decades due to the substantial amount involved and potential conflicts of interest between executives and shareholders. A central issue in executive compensation is balancing the imperative of retaining top talent with aligning executives' long-term interests with those of the company and its shareholders. Advocates of high executive pay argue that it is essential to attract skilled executives and motivate them to make difficult decisions in the company's best interest. However, critics contend that excessive compensation can initiate a focus on short-term gains and unethical conduct. Corporate governance in India needs more transparency in deficient disclosure standards, concentrated ownership, and board rubber-stamping (Jaiswal & Bhattacharyya, 2016). All these variables diminish the efficacy of corporate governance systems in overseeing management and reconciling the interests of shareholders and managers. Agency problems are consequently more pronounced in countries such as India, necessitating an examination of the efficacy of corporate governance mechanisms in aligning managerial interests with those of shareholders.

Many executive compensation packages are negotiated privately, without disclosure to shareholders or the public, making assessing fairness challenging or preventing excessive pay challenging. Therefore, the Securities and Exchange Commission (SEC) in the United States requires that executive compensation data, such as wages, bonuses, and stock options, be publicly disclosed. Additionally, some companies have implemented "say-on-pay" policies, allowing shareholders to vote on executive compensation packages in the USA. Enhanced transparency regarding executive remuneration can foster consistency in compensation frameworks, enabling boards of directors to formulate compensation strategies by observing their counterparts (Cabezon, 2024). Edmans et al., (2023) provide survey evidence indicating that boards experience pressure to adhere to "best practices" contrary to their preferred compensation plans.

In India, executive compensation is a critical area of study due to the prevalence of family-owned businesses, pyramidal corporate structures, and owners holding key managerial positions. The compensation practices of Indian companies differ significantly from those in developed economies due to their unique structure. Indian companies typically emphasize cash components of compensation, whereas developed countries place greater emphasis on stock options and equity-based incentives. Many Indian firms increasingly offer generous compensation packages that include performance-based bonuses and stock options. Government regulations also play a significant role in shaping executive compensation in India. In recent years, the Indian government has implemented various rules to enhance transparency and accountability in executive pay. For instance, the Securities and Exchange Board of India (SEBI) requires publicly listed companies to disclose detailed executive pay information in their financial reports.

Additionally, the Companies Act 2013 introduced stringent rules regarding the appointment and remuneration of directors, including a limit on total compensation for non-executive directors. Despite these regulatory efforts, considerable debate remains surrounding executive compensation in India. However, some companies are adopting more transparent and equitable compensation practices. This includes linking executive pay to long-term performance metrics and seeking shareholder approval for executive compensation packages.

18.2 Background

18.2.1 Executive Compensation and Perspectives of Agency Problem

In executive compensation literature, addressing the underlying agency problem that impacts management decision-making is essential. In this paragraph, we discuss two prevalent theories on the effects of agency issues on executive compensation. The prevailing perspective among financial economists regards managers' compensation structure as a partial resolution to agency problem (Alolah, 2024). The "optimal contracting approach" perspective implies that boards create remuneration plans to motivate managers to maximize shareholder value effectively. An alternative viewpoint, the "managerial power approach," posits that executive remuneration addresses the agency problem and contributes to the problem itself. Several academics have observed that specific aspects of payment structures seem to indicate that managers are pursuing personal gain rather than providing compelling motivation (Bertrand & Mullainathan, 2000; Göx & Hemmer, 2020).

18.2.2 Importance of Cash Compensation in India

Stock options and restricted stock shares make up a significant part of CEO remuneration in the United States (Bryson et al., 2014; Conyon & He, 2016). Talavera et al., (2021) reported that from 2005 to 2015, more than 93% of US firms provided CEOs with equity stock options and restricted shares, and these stock-based forms accounted for 42.68% of total CEO compensation. In contrast, stock options are uncommon in India. Based on the findings of Chen et al., (2021), equity stock options have a negligible impact on the total pay for CEOs in India, accounting for less than 3.8% of the sample. Furthermore, the researchers could not assess the fluctuating value of unutilized equity options since there was inadequate data available on the cumulative number of stock options owned by CEOs in past years. As a result, their methodology diverges from conventional studies conducted in the US, which frequently use stock options as a substantial element of CEO compensation.

18.2.3 Executive Compensation in Public Sector Undertakings (PSUs) or Government Companies

Public sector enterprises (PSUs) in India are unique due to their ownership and management by the government. Several aspects set these organizations apart when it comes to compensating their managers. Calculating management remuneration in PSUs differs significantly from that in the private sector. The government determines the compensation in PSUs based on a "pay commission" model of managed pricing. This implies that these enterprises have limited or no independence in determining management remuneration. Furthermore, reports indicate that the overall compensation levels in public sector undertakings (PSUs) are considerably lower than those in the private sector. Additionally, PSUs often do not have substantial incentive pay schemes in place to acknowledge and recognize exceptional performance (Sridhar, 2021; Swami, 2005).

18.2.4 Trends of Executive Compensation in India

From this Fig. 18.1, it can be seen that the trends related to the components of executive compensation of the top 100 companies in India for the last 15 years. It can be seen from the figure that salary and bonus were the main contributors to the executive compensation. From the year 2015, we have seen a trend of stock options in these companies in India. It started at a good pace, but it lost its touch after the COVID-19. The perquisites seem to be growing at a stable pace. The other components, like contribution to provident fund, retirement benefits, and other remuneration did not contribute to the total compensation much. This shows how much cash compensation is important, considering executive compensation in a developing economy like India.

From Table 18.1, we can see salary %, bonus %, perquisites %, and stock options % as a part of total executive remuneration. The last column shows the growth rate year on year of the total executive compensation. This table shows that the salary as a % of total compensation has been increasing yearly. It was 29.87% in the year 2009, and in the year 2023, it has come to be 45.03% of the total compensation. Whereas the bonus/commission shows the trend to be the opposite, it was 60.41% in 2009, and in the year 2023, it became 33.69%. Overall, the cash compensation (salary + bonus) has declined very slowly from 90% in 2009 to 79% in 2023. Still, cash compensation remains an important part of pay, unlike in the USA and other developed economies, where stock options play a major role. Perquisites remained in a specific range of 6%–16% throughout the years. Stock options were majorly used from year 2015, instead of going in an uptrend, it has fallen year by year. In 2015, it was nearly 21%, but in 2023, it is currently at a mere 1.80%. This indicates that Indian executives still love the traditional form of payment, which is cash compensation (salary and bonus). In the last column, we see the growth of the executive compensation year by year, we see a variable trend in growth where in some years there was very high growth,

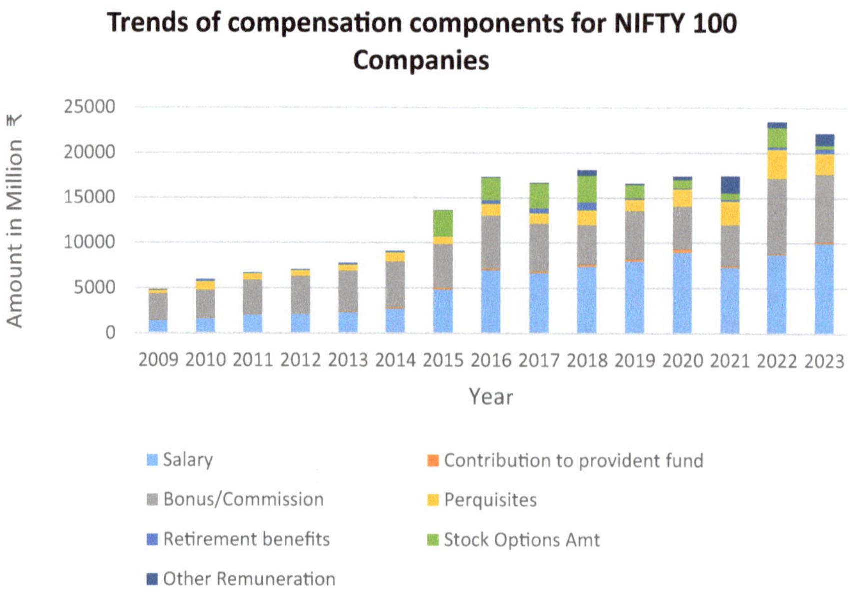

Fig. 18.1 Trends of compensation components for NIFTY 100 Companies. *Source* Author's calculation (data taken from Prowess IQ database)

but in some years the growth was also negative. Recently, after a sharp increase of 34.38% in 2022, again, it fell by 5.37% during the previous year.

18.3 Legal Aspects of Executive Compensation in India

Maximum limit of managerial remuneration payable by a public company (Section 197(1)): A public company cannot pay its managers, whole-time directors, managing directors, and other directors more than 11% of its net profits for any fiscal year, as computed according to Section 197 of the Companies Act of 2013. However, the compensation of directors is not deducted from the company's gross profit. A corporation can surpass the 11% limit by receiving the permission of its members during the annual general meeting and following the guidelines outlined in Schedule V of the Companies Act of 2013.

This restriction applies only to public limited companies; private limited companies are exempt from Section 197 and Schedule V, allowing them to pay any compensation.

Prescribed individual limit for each Director under Section 197(1) of the Companies Act, 2013:

Table 18.1 Trends of compensation components for NIFTY 100 Companies

Year	Salary in (million ₹)	Salary %	Bonus in (million ₹)	Bonus %	Perquisites in (million ₹)	Perquisites %	Retire benefit in (mn ₹)	Stock options amout in (million ₹)	Stock %	Total remuneration in (million ₹)	Growth of total comp%
2009	1,439	29.87	2,912	60.42	338	7.01	78		0.00	4,819	NIL
2010	1,715	28.69	3,049	51.02	960	16.05	210		0.00	5,977	24.04
2011	2,077	30.96	3,792	56.52	713	10.62	72		0.00	6,708	12.23
2012	2,188	31.06	4,060	57.61	670	9.50	73		0.00	7,047	5.04
2013	2,390	30.77	4,426	56.99	692	8.91	156		0.00	7,766	10.21
2014	2,769	30.36	5,062	55.51	1,013	11.11	142		0.00	9,119	17.42
2015	4,856	35.72	4,814	35.42	869	6.40	446	2,809	20.67	13,593	49.06
2016	7,019	40.53	5,806	33.52	1,367	7.89	375	2,527	14.59	17,319	27.41
2017	6,706	40.08	5,240	31.32	1,180	7.05	549	2,774	16.58	16,732	−3.39
2018	7,416	40.87	4,379	24.13	1,671	9.21	885	2,956	16.29	18,145	8.44
2019	8,011	48.14	5,320	31.97	1,311	7.87	154	1,467	8.82	16,642	−8.28
2020	8,994	51.56	4,768	27.34	1,932	11.08	128	872	5.00	17,441	4.80
2021	7,345	42.03	4,543	25.99	2,667	15.26	152	730	4.17	17,476	0.20
2022	8,772	37.35	8,316	35.41	3,210	13.67	275	2,118	9.02	23,486	34.39
2023	10,008	45.03	7,487	33.69	2,314	10.41	474	401	1.81	22,223	−5.38

Source Author's calculation (data taken from Prowess IQ database)

According to the second proviso, within the overall 11% limit specified under Section 197(1), individual directors can be paid remuneration up to the following limits:

- **Managing director, whole-time director, or manager**:

  Up to 5% of the net profits if only one such director exists.
  If there are many directors, their total pay should not surpass 10% of the net profit.

- **Directors who do not hold the positions of managing directors or whole-time directors:**

  If there is a managing director, whole-time director, or manager, they will receive up to 1% of the net profit as compensation.
  If there is an absence of a managing director, whole-time director, or manager, up to 3% of the net profit will be allocated.

Note Sitting fees are excluded from executive compensation.

18.4 Executive Compensation and Firm Performance

Conventional economic theory addresses the issue of managerial compensation by utilizing a "principal-agent" framework. Public companies maintain a distinct division between ownership and management, meaning that shareholders do not directly influence the company's resources. However, managers hold corporate control and possess the authority to utilize the company's resources. This gives rise to an issue of misalignment between the objectives of owners and managers, which has to be addressed to get an effective, wealth-maximizing resolution. Both theories of organization and finance entail that allocating power between company managers and owners substantially influences business performance. To fully comprehend managerial motivation, behavior, and remuneration, it is essential to consider a larger framework in which these factors operate. Shareholders are primarily concerned with selecting competent management and implementing effective incentives to achieve optimal performance that maximizes wealth while considering associated expenses.

A proficient labor market for managers, where workers possess the necessary skills, overcomes the problem of selecting managers. Rewarding good behavior as per performance has generally been a more complex problem. Creating pay packages linked to company performance or growth in shareholder wealth is a common way for economists to find incentives. This is usually done with bonuses, stock options, performance-based pay, etc. Nevertheless, these strategies must completely synchronize management and shareholders' goals. The process of ascertaining these solutions and the respective role played by each stakeholder remains complex. According to economic theory, the authority to select representatives and decide their payment should be held by the principals, namely the shareholders. Due to the widespread distribution of shareholders and their lack of majority ownership, it becomes expensive for them to participate actively, resulting in higher agency costs.

This condition complicates determining executive remuneration and raises concerns about shareholders losing their authority when not involved in decision-making. Companies have a structured system where the board of directors determines executive compensation and oversees performance evaluation. The board is responsible for safeguarding shareholder interests and prohibiting management from neglecting or disregarding these interests. This limits the authority and influence that managers possess over the financial assets of shareholders. So, it is common for high-ranking executives, including the CEO, to hold positions on the board, which might lead to collusion. This has the potential to undermine the process of incentivizing performance in the best interests of shareholders. Therefore, factors such as the composition of the board and the distribution of work among its members are crucial in guaranteeing effective oversight and alignment of objectives. The relationship between CEO compensation and firm performance has been a significant topic in corporate governance literature (Al Farooque et al., 2019; Frye, 2004; Holmström, 1979; Jensen & Murphy, 1990; Murphy, 1999). Research undertaken over the past three decades has consistently supported a correlation between CEO and top executive salary and business performance[2] (Grossman & Hart, 1983; Jensen & Murphy, 1990). Bhatia et al., (2024) find there is a positive and statistically significant correlation between executive compensation and corporate performance but this connection has deteriorated over time, underscoring the necessity of establishing robust governance regulations and enhancing transparency in the compensation determination process.

Pay-performance relationship is based on agency theory, which suggests that payment agreements should align the interests of managers and shareholders (Grossman & Hart, 1983; Holmström, 1979; Pollock et al., 2023). Strengthening the connection between executive compensation and performance can also help in the recruitment and retention of more effective managers. Introducing performance-based pay for top executives might reduce the negative effects of adverse selection. This type of compensation is incorporated into the executive selection process (Arya & Mittendorf, 2005; Darrough & Melumad, 1995). Bhuyan et al., (2020) study the relation between financial performance and CEO pay in the insurance sector and find a significant positive relationship after the crisis period.

Bebchuk and Fried (2003) argue that executives may place greater emphasis on their personal gain rather than on initiatives aimed at improving profitability. Murphy (1985) revealed a favorable association between the compensation of managers and corporate performance, particularly in regard to shareholders' earnings. Coughlan and Schmidt, (1985) found a link between the performance of share returns and changes in CEO salary and bonuses. Boschen and Smith, (1995) find a correlation between executive remuneration and with past and present performance of a

[2] Firm performance, Business performance, and Organizational performance are alternatively used in this chapter.

company. CEO remuneration was influenced significantly by both the performance of the current year and the performance of the preceding year (Core et al., 1999).

Different studies in India have investigated the correlation between the remuneration of top-level executives and with overall performance of these companies. Ramaswamy et al., (2000) found return on assets to be an important predictor, along with other factors. Ghosh, (2006) has demonstrated a direct relationship between the compensation of CEOs and the performance of business. This link is influenced by both current and prior performance since they both have a significant impact on executive pay. Raithatha and Komera, (2016) found accounting and market-based variables to have a substantial impact on executive compensation. Again, their sub-sampling analysis could not find any relation between CEO compensation and performance in small and group-affiliated firms. Parthasarathy et al., (2006) found that the compensation of CEOs was not affected by organizational performance, but rather by certain attributes of the firm, such as its size.

18.5 Executive Compensation and Persistence of Pay

We recognize that past executive remuneration and the firm's past and present performance influence current executive compensation. Remuneration committees sometimes utilize the prior year's remuneration as a benchmark for establishing remuneration for the present year (Bender, 2003; Bouteska & Mefteh-Wali, 2021). Thus, previous pay affects this year's compensation. Their estimate equation is dynamic, using previous pay as one of the independent variables. Main et al., (1996) examined the relation between the overall pay of the board and the performance of firms situated in the United Kingdom. In their model, they used previous compensation as a lagged dependent variable to account for the dynamic nature of compensation agreements, and they found a significant impact. Doucouliagos et al., (2012) performed a preliminary study on the variable nature of CEO compensation in large Australian companies. It has been shown that most effective CEO remuneration packages are not created suddenly, but rather, they develop and change gradually over a period of time.

18.6 Executive Compensation: Overcompensation and Under Compensation

There is a widespread belief that the remuneration received by executives is unreasonable and unfair. This topic became controversial in recent years, leading to discussions in scholarly papers about management, accounting, finance, and economics, the press, and media. The CEO usually has the highest salary and the most prominent public presence. Therefore, significant attention is being directed to CEO compensation. The issue of executive remuneration has generated much discussion and disagreement in firms and scholarly publications. As per arguments of equity and justice, executive pay has reached absurd levels, which asserts that the pay is inequitable compared to its employees or other CEOs. It has a lack of connection with performance. Recently, increments granted to CEOs have far exceeded those given to regular employees. Some contend that CEOs receive excessive compensation due to the principle of equity about the type of work they carry out. Establishing just remuneration in proportion to the labor performed is ideal, but achieving equitable salary levels across different job categories is complex.

For instance, what are the appropriate salary ranges for a seasoned nurse working in a trauma center compared to a novice worker on a garbage recycling truck? Nurse's remuneration should be increased based on the requisite education, specific expertise, degree of responsibility, and social significance of their profession. On the other hand, trash recycling workers may exert more physical effort, work in a more challenging workplace, and contribute substantially to maintaining a clean environment. While the fairness of wage disparities may be acknowledged, the precise magnitude of these disparities remains uncertain. For instance, what is a fair and justifiable wage gap between nurses and garbage recycling workers? What's correct 100%, 80%, 50%, 200%? Therefore, even in seemingly simple cases, determining work equity is complex, making it a problematic basis for discussing appropriate compensation levels across different types of work (Chaidir et al., 2024).

The increasing disparity in compensation between the CEO and the average employee cannot be considered conclusive proof of excessive remuneration for two main reasons. Initially, it is plausible that CEOs were previously receiving lower compensation compared to average employees, and substantial raises are intended to correct this disparity in salary. Furthermore, it is possible that a significant rise in CEO salaries is a result of evolving circumstances and growing complexities that CEOs now encounter, even if they were not previously receiving inadequate pay. Operating in a global, technologically advanced, and competitive environment may require more talent and place greater demands on successful CEOs than on average workers. Whereas, in family firms, founders or successors often significantly influence important decisions. Hambrick and Finkelstein, (1995) find CEO remuneration is linked to firm ownership. Cai et al., (2021) found that mixed ownership reform may enhance the internal governance mechanisms of firms, efficiently equilibrate managerial authority, and augment the sensitivity of executive remuneration to performance.

In firms with significant outside ownership, CEO compensation is decided as per demand and supply, contrasting to firms without a major outside shareholder, CEOs have more authority in setting their own compensation. CEO pay in family firms also varies depending on whether the CEO is a member of the family. CEOs from family are more likely to accept lesser salaries in exchange for job security and emotional investment (Gomez-Mejia et al., 2003). Therefore, CEOs from family-owned businesses may get lower salaries compared to CEOs who are not from family-owned businesses. Firms that have several family members in control may offer executive remuneration that is either greater or lower. The decision is contingent upon the preferences of the founder's family member and other family members, as well as the financial liquidity of the firm (Cheng et al., 2015).

In family firms, Michiels et al., (2013) find linking CEO pay to performance yields positive outcome. However, Gomez-Mejia et al., (2003) find no correlation between remuneration and performance in family enterprises where the CEO is from a family. Family business may differ from non-family business as founders invest for "amenity potential" or non-pecuniary benefits (Burkart et al., 2003). Cennamo et al., (2012) contend that family firms actively involve themselves with stakeholders in order to augment their socioemotional wealth (SEW). Therefore, compensation methods in family firms may be influenced by the desire for non-monetary advantages rather than financial results, resulting in a possible absence of a connection between pay and performance. The SEW in family firms significantly influences strategic decisions, including compensation policies. Cennamo et al., (2012) find family companies prioritize SEW over economic objectives, leading to proactive stakeholder engagement (PSE) activities. Consequently, family firms focused on SEW may avoid overpaying executives to prevent expropriating other stakeholders. Family firms often prioritize stakeholder interests over profits. Berrone et al., (2010) show that family firms are less likely to engage in environmentally harmful actions compared to non-family firms.

From Table 18.2, we can see a few examples of CEOs from Indian firms who are being compensated very heftily. The first company is Divi's Laboratories, where Dr. Murali K. Divi is the CEO from 2019 to 2023, he has charged ₹58.81 crores, ₹70.71 crores, ₹109.06 crores, ₹148.08 crores, ₹95.28 crores, respectively. Kalanithi Maran who was the CEO of Sun T V Network from 2009 to 2012 received ₹37 crores, ₹37 crores, ₹64 crores, and ₹57 crores, respectively. These amounts were the highest among Indian CEOs during that period, which raised eyebrows in the news media. He is currently the owner of the Sunrisers Hyderabad team. Harsh Pati Singhania from J K Paper Ltd and Salil Parekh from Infosys Ltd also received huge amounts of compensation from 2019 to 2023. Salil Parekh's compensation also included stock options in the year 2019 and 2022. For the calculation of total compensation, we have taken 25% of the stock options amount. Finkelstein and Boyd, (1998) and Gomez-Mejia et al., (2003) have employed a technique that values equity options at an exercise value of 25%, which was put forth by Lambert et al., (1993). In this table, we see that the CEOs of Sun TV Network and Divi's Laboratories were both founders of their respective companies and CEO duality was also present. CEO duality states that they were the chairman as well as MD/CEO of the company.

This Table 18.4 shows a few examples of underpaying or under compensation. CEO of Kothari Petrochemicals Ltd, Mr. B H Kothari, from 2009 to 2014, has taken ₹25 thousand, ₹25 thousand, ₹37.5 thousand, ₹41.5, and ₹42.5 thousand, respectively. CEO of M R O Tek Realty Ltd, Mr. Aniruddha Bhanupasad Mehta, received only ₹12 from 2020 to 2023. CEO of Thyrocare Technologies Ltd., Mr. A. Velumani, received ₹12 from 2018 to 2021; in 2022, he received ₹16. These amounts received by the CEOs are not so ordinary. It can also be seen from the table that all three CEOs are the family members of the owner's family and were also the chairmen of their companies. Unlike in Table 18.2, where we saw that the founders of the families

Table 18.2 Over compensation

Company name	Year	Director name	Duality	Founder	Salary (in million ₹)	Cash compensation (salary + bonus) (in million ₹)	Stock options (in million ₹)	Total compensation (in million ₹)
Divi'S Laboratories Ltd	2019	Murali K Divi (Dr.)	1	1	9	590		588
Divi'S Laboratories Ltd	2020	Murali K Divi (Dr.)	1	1	5	520		707
Divi'S Laboratories Ltd	2021	Murali K Divi (Dr.)	0	1	1	1		1091
Divi'S Laboratories Ltd	2022	Murali K Divi (Dr.)	0	1	0	1100		1481
Divi'S Laboratories Ltd	2023	Murali K Divi (Dr.)	0	1	0	700		953
Infosys Ltd	2019	Salil Parekh	0	0	170	170	76 m	189
Infosys Ltd	2020	Salil Parekh	0	0	57	170		343
Infosys Ltd	2021	Salil Parekh	0	0	57	180		497
Infosys Ltd	2022	Salil Parekh	0	0	180	180	520 m	318
Infosys Ltd	2023	Salil Parekh	0	0	67	250		565
J K Paper Ltd	2019	Harsh Pati Singhania	0	0	77	240		260
J K Paper Ltd	2020	Harsh Pati Singhania	0	0	77	80		95

(continued)

Table 18.2 (continued)

Company name	Year	Director name	Duality	Founder	Salary (in million ₹)	Cash compensation (salary + bonus) (in million ₹)	Stock options (in million ₹)	Total compensation (in million ₹)
J K Paper Ltd	2021	Harsh Pati Singhania	0	0	59	210		229
J K Paper Ltd	2022	Harsh Pati Singhania	0	0	65	240		281
J K Paper Ltd	2023	Harsh Pati Singhania	0	0	77	380		377
Sun T V Network Ltd	2009	Kalanithi Maran	1	1	110	110		370
Sun T V Network Ltd	2010	Kalanithi Maran	1	1	110	110		370
Sun T V Network Ltd	2011	Kalanithi Maran	1	1	110	640		640
Sun T V Network Ltd	2012	Kalanithi Maran	1	1	110	110		570

Source Prowess IQ database

received considerable amounts in compensation, we see in Table 18.3 that though the CEOs are family CEOs, they received a minimum amount in compensation. This shows that family CEOs do not always try to extract profits from the company; some CEOs try to reinvest their portion of profit in the firm for further growth.

18.7 Executive Compensation and Family Firms and Ownership Concentration

In developing nations, the issue of conflicting interests between the owner and the manager is less worrisome when firms are mostly held by a single family (Dharwadkar et al., 2000; Kohli & Gill, 2020). The reason for this is that family owners, who possess the majority of shares, are actively engaged in the management or may even be part of it, resulting in an inherent interest alignment among shareholders and managers. Nevertheless, family businesses encounter a distinct form of agency issue referred to as a principal-to-principal dilemma (Dharwadkar et al., 2000; Young et al., 2008). This problem arises when there is a conflict of interest among majority and minority stakeholders. Research has shown that when there is a lack of effective corporate governance systems, major shareholders with significant control over a company can exploit minority shareholders for their gain. This can have a detrimental effect on the overall value of the company in the market. Additionally, it can hinder the progress and growth of the economy as a whole.

Conflict among dominant stockholders and minority stockholders emerges in several ways: (1) Minority shareholders can spread their assets over other businesses, which reduces their risk exposure. Nevertheless, family owners, who own authority over the company, typically allocate a significant portion of their capital to these enterprises, resulting in a more prudent approach towards risk-taking. (2) Misappropriation of firm resources, wherein family executives may award themselves excessive rewards and privileges or engage in transactions with related parties. (3) Nepotism refers to the practice of family leaders appointing incompetent family relatives and partners to essential roles in a company, disregarding more competent individuals (Chen et al., 2021; Young et al., 2008).

Indian family firms are distinguished by their remarkably high average stock ownership and active involvement of the family in management, which stands at 51%. In comparison, Europe has an average of 38%, the U.S. has 18%, and Japan has 6%. The substantial amount of family ownership and participation in the companies, along with the lack of established capital markets, and legal and financial institutions leads to inadequate legal protections for external investors (Chen et al., 2021; Khanna & Palepu, 2000). As a result, external investors and creditors complain about founding families having a greater financial stake in the venture. In contrast to established countries, India offers greater growth potential and less competitive product marketplaces. Most family businesses and non-family businesses in India derive advantages from these significant growth prospects and limited competition

Table 18.3 Under compensation

Company Name	Year	Director name	Duality	Family—CEO	Founder	Salary (in ₹)	Cash compensation (salary + bonus) (in ₹)	Total compensation (in ₹)
Kothari Petrochemicals Ltd	2009	B H Kothari	1	1	0	0	0	25 k
Kothari Petrochemicals Ltd	2011	B H Kothari	1	1	0	0	0	25 k
Kothari Petrochemicals Ltd	2012	B H Kothari	1	1	0	0	0	38 k
Kothari Petrochemicals Ltd	2013	B H Kothari	1	1	0	0	0	42 k
Kothari Petrochemicals Ltd	2014	B H Kothari	1	1	0	0	0	43 k
M R O-Tek Realty Ltd	2020	Aniruddha Bhanupasad Mehta	1	1	0	12	12	12
M R O-Tek Realty Ltd	2021	Aniruddha Bhanupasad Mehta	1	1	0	12	12	12
M R O-Tek Realty Ltd	2022	Aniruddha Bhanupasad Mehta	1	1	0	12	12	12
M R O-Tek Realty Ltd	2023	Aniruddha Bhanupasad Mehta	1	1	0	12	12	12
Thyrocare Technologies Ltd	2018	A Velumani	1	1	1	12	12	12

(continued)

Table 18.3 (continued)

Company Name	Year	Director name	Duality	Family—CEO	Founder	Salary (in ₹)	Cash compensation (salary + bonus) (in ₹)	Total compensation (in ₹)
Thyrocare Technologies Ltd	2019	A Velumani	1	1	1	12	12	12
Thyrocare Technologies Ltd	2020	A Velumani	1	1	1	12	12	12
Thyrocare Technologies Ltd	2021	A Velumani	1	1	1	12	12	12
Thyrocare Technologies Ltd	2022	A Velumani	1	1	1	6	6	16

Source Prowess IQ database

Table 18.4 Pay disparity

Company name	Year	Director name	Duality	Family CEO	Founder	Pay Gap (Ratio)
Ahlada Engineers Ltd	2020	Ch Suresh Mohan Reddy	0	1	1	3152
Ahlada Engineers Ltd	2022	Ch Suresh Mohan Reddy	0	1	1	3506
Tech Mahindra Ltd	2017	C P Gurnani	0	0	0	3538
Tech Mahindra Ltd	2018	C P Gurnani	0	0	0	3151

Source Prowess IQ database, Annual reports (family and founder information hand-collected from annual reports)

in the product market. Family owners frequently allocate a substantial percentage of their capital to the company, which can lead to a lack of diversification and thus heighten their risk aversion (Shleifer & Vishny, 1997).

CEO compensation is commonly explained in the current literature using two primary theories: "optimum contract theory" (Core & Larcker, 2002; Gayle et al., 2022), and "theory of managerial power" (Bebchuk & Fried, 2003; Rehman et al., 2021). As per the theory of optimal contracting, CEOs' remuneration is determined through independent agreements among boards, which represent the interests of shareholders, with CEOs. It's anticipated that boards will offer CEOs effective incentives to mitigate agency issues arising from the division between ownership and control. Managerial power states, that CEOs have the ability to exercise power over board members and choose their own salary, with the sole limitation being a restraint on excessive behavior (Rehman et al., 2021). Therefore, CEO remuneration might contribute to the agency problem rather than resolving it. Shareholders have the ability to oversee the remuneration of CEOs, but they typically lack the motivation to do so unless they possess a significant number of shares (Shleifer & Vishny, 1997; Wang & Qiu, 2023).

Family ownership can impact CEO remuneration through two contrasting mechanisms: the entrenchment effect and the alignment effect. The entrenchment effect occurs due to disputes among dominant stakeholders, having the power to choose management and minority shareholders. Families who own a large portion of shares might unjustly take money from minority shareholders using several methods, such as high remuneration packages (Bertrand & Schoar, 2006; Chen et al., 2021). When member of family occupies the role of CEO, family could deliberately augment CEO's remuneration with the aim of favouring the family CEO, even if it results in a detriment to other shareholders. This statement aligns with the managerial power theory, which posits that CEOs with significant authority may select their own remuneration agreements.

Alignment effect, corresponds to "optimal contract theory" and contends that ownership concentration provides stronger reward for monitoring effective

(Shleifer & Vishny, 1997), which implies family firms would have a greater ability to supervise CEO salary and restrict its magnitude. De Massis et al., (2021) observe CEOs who are risk-averse in family enterprises, especially those who are related to owners, willing receive lesser salary for their enhanced job security. The presence of strong family bonds enhances the level of dedication to the organization and diminishes the appeal of external prospects, hence decreasing the likelihood of a family CEO departing to a different firm. Therefore, the alignment effect implies that family businesses, especially those with family CEOs, are less likely to participate in opportunistic behavior of offering excessive remuneration to their CEOs.

18.8 Executive Compensation and Pay Gap/Pay Disparity

Literature on executive remuneration predominantly concentrates on CEOs' remuneration, as CEOs bear the main accountability for company performance (Aggarwal & Samwick, 1999; Bebchuk & Fried, 2004). The majority of research focuses on total CEO pay without considering unique characteristics of a firm might impact the median compensation given to top-level executives in that company (Bebchuk & Fried, 2010; Eriksson, 1999). Edmans et al., (2017) found a sharp rise in executive compensation after the 1990s. CEOs are essential to a firm as they make all of its strategic and organizational decisions. Concerns regarding corporate governance, equity, and the long-term sustainability of the company have been raised in India by the growing disparity between CEO and executive/employee compensation.

Employee morale, business performance, and regulatory scrutiny are all significantly impacted by extreme pay disparities, especially CEO-to-executive pay ratio. Significant gaps may cause executives to feel unfairly treated and dissatisfied, which will diminish engagement and motivation. Operational efficiency may be impacted by decreasing productivity as a result of diminished engagement. Executive departures can result from large compensation disparities, particularly for senior executives who may feel underappreciated in comparison to the CEO. Excessive CEO compensation in comparison to other executives may indicate short-termism, rent extraction, and a weaker alignment with company performance. High pay inequalities are frequently scrutinized by institutional investors, particularly foreign portfolio investors (FPIs), which affects the valuation of the company. CEO pay gaps have been widening in Indian industries such as financial services and information technology. While some businesses use the global talent competition as an excuse for high CEO compensation, others encounter criticism from authorities and investors. Regulators have begun to place more emphasis on fair pay practices, and corporate governance standards in India are changing. According to SEBI's Disclosure Norms, to increase openness, listed companies are required to provide CEO compensation in relation to median employee compensation. In order to ensure alignment with performance, Section 197 of the Companies Act of 2013 requires shareholder approval for CEO compensation exceeding certain limitations. Since the government wants to reduce excessive CEO compensation, regulatory scrutiny is especially strong in banking and public sector companies (PSEs). A number of corporate governance issues are brought up by extreme pay disparities. CEOs may be able to influence pay choices due to inadequate board oversight, which might result in rent-seeking behavior. Although shareholder voting on CEO compensation is required by Indian laws, most resolutions pass with ease, making it primarily a symbolic process. Risks can be reduced by implementing a well-structured compensation ratio that is linked to company performance and ESG (Environmental, Social, Governance) standards.

In Table 18.4, we show the pay gap/pay disparity. The pay gap means it is the ratio of the executive's remuneration and the median employee's remuneration. In Table 18.4, we can see that the pay ratio for Suresh Mohan Reddy in 2020 was 3151.75, and in 2022 was 3505.84 for Ahlada engineers. Similarly, Tech Mahindra C P Gurnani had a pay ratio of 3537.72 and 3150.69 in 2017 and 2018. Pay disparity can be explained by two hypotheses, such as the hypothesis of "efficient contracting" and "managerial power."

18.8.1 The Efficient Contracting Hypothesis

18.8.1.1 Tournament Theory

The hypothesis of "efficient contracting" posits that discrepancies in CEOs' remuneration are influenced by the extent to which business actions correspond with the interests of shareholders and reduce agency problems. The considerable disparity in CEO income aligns with tournament theory, which draws a parallel between the CEO's role and its substantial remuneration to top reward in a competitive competition (Islam et al., 2022; Lazear & Rosen, 1981; Rosen, 1986a). This discrepancy reduces expenses associated with monitoring for shareholders and board, establishing systems that more effectively balance the interests of principal and agent, especially in companies where management autonomy might present a high risk (Lee et al., 2008). For example, there is a marathon race, and the winner only gets the prize. So, there will be many players competing to win the marathon and the analogy here is that the CEO is the only winner.

18.8.1.2 Managerial Ability

Alternatively, substantial discrepancy in CEO pay serves as compensation for remarkable contributions that CEOs make to firm's operations, which goes beyond contributions of other senior executives (Masulis & Zhang, 2013). According to Rosen, (1986b), even little variations in executive abilities can result in considerably amplified disparities in remuneration. As per Edmans et al., (2023) if a CEO is seen to be more competent than other senior executives, compensation committees have a valid reason to provide a larger pay differential in order to incentivize optimal performance.

18.8.2 The Managerial Power Hypothesis

18.8.2.1 Influencing Own Pay Design

The hypothesis of "Managerial power" posits that CEO remuneration exacerbates agency issues (Bebchuk & Fried, 2003; Göx & Hemmer, 2020). Income and pay disparity among CEOs are magnified as a result of their influence in determining pay rather than it being intended to facilitate effective contracting. The research elucidates many methods via which CEOs might exert influence over their remuneration. Although companies have an obligation to include non-executive directors in remuneration committees, the mandate restricts rather than eradicates CEOs' authority in determining compensation.

18.8.2.2 Larger Board Size and Board's Busyness

According to Core et al., (1999) and Nawaz, (2022), larger boards have lower cohesiveness, which enables CEOs to have greater influence and obtain better salaries. This problem is especially relevant to companies when independent directors serve on many boards, which can reduce their ability to effectively oversee a specific company (Bebchuk & Fried, 2004; Nawaz, 2022). In such situations, CEOs have greater authority in managing internal affairs, including their own remuneration. Moreover, when CEOs amass a more significant ownership stake in the company, they acquire increased authority and influence over the board (Core et al., 1999; Lambert et al., 1993; Murphy & Sandino, 2020).

18.8.2.3 CEO Duality

CEO duality is a condition in which a single person occupies the roles of both Chief Executive Officer (CEO)/managing director and Chairman of the board of a company (Al-Shaer et al., 2023; Bebchuk & Fried, 2004; Nawaz, 2022) Present data indicate that organizations in which the CEO also holds the chairman position of the board face significant challenges of power separation. This arrangement enables these CEOs to exercise a larger level of influence and secure big financial advantages.

18.9 Conclusion

In the Indian context, the importance of executive compensation rose post-economic liberalization for reasons such as increasing the profitability margin of firms, increasing competition for managerial abilities, benchmarking Indian companies with global standards, etc. One unique characteristic exclusive to India (compared to developed nations) is the prevalence of "family-owned" firms. They essentially have "owner-managers," where business owners also serve as managers. Another uniqueness is that Indian family firms have a higher concentration of ownership, high growth opportunities, and an underdeveloped financial market. One advantage of the concentration of ownership is that it incentivizes owners to supervise the firm and make necessary management changes. However, this concentrated ownership may result in entrenchment and extraction of private benefits, which leads to pay disparities and a lack of risk-taking by family owners.

Executive compensation has ignited a heated discussion in India, with both media and the general public expressing concern over excessive pay packages. This implies that the pay structure in India should be subject to closer scrutiny. Although there are SEBI regulations in place, there are some changes or clarity. In this paper, we have studied the issues relating to over-compensation, pay disparity, family firms, and ownership concentration in Indian firms about executive compensation. Analyzing executive compensation in India can provide valuable insights into the economic, cultural, and social dynamics shaping the country's business landscape. By understanding the factors that influence executive compensation in India, researchers can help promote more responsible and effective corporate governance, which can significantly benefit companies, shareholders, and society.

18.10 Implications and Scope of the Study

This paper investigates the pay-for-performance relationship, family firms, and pay disparity in executive compensation. We have explored issues related to stock options and the importance of cash compensation in India. We have also investigated what we consider a fundamental confusion in the literature—over-compensation and under compensation. We have also discussed how managerial power and ownership influence executive compensation. We have focused on theoretical specifications such as CEO duality, board size, managerial ability, tournament theory, family CEO, firm performance, and their influence on executive compensation.

Executive remuneration is essential for recruiting global talent in India. As the nation ascends as a prominent entity in the global economy, attractive compensation packages are vital for recruiting and retaining executives with international knowledge and skills. Executive remuneration is progressively associated with firm performance indicators, promoting a culture of accountability. Rigorous corporate governance regulations regulate executive compensation in India. The Companies Act of 2013, SEBI rules, and other legal frameworks emphasize the necessity of transparency and shareholder oversight. This ensures that pay packages are justified by performance and market conditions, preventing excessive remuneration. A vital aspect of executive compensation is motivating executives and protecting shareholder interests. Excessive executive remuneration for business performance may lead to shareholder dissatisfaction and damage the company's brand. Shareholders sometimes attempt to demand executive compensation via mechanisms like "say on pay" votes, augmenting their authority over the salary of top executives. The growing gap between executive remuneration and average employee earnings is an increasing concern in India. Organizations are increasingly mandated to establish regulations that promote equitable compensation and reduce the pay disparity between executives and non-executives.

There are several directions for future research in executive compensation. The significant improvements in data availability, consistency, and quality are expected to result in reduced variation of outcomes in further studies. We anticipate that scholars from other disciplines will capitalize on this opportunity. We assert that a substantial exchange of methodologies between and within disciplines is necessary. Furthermore, we believe the barriers between academic research and practical application should be dismantled broadly, especially in executive remuneration. Furthermore, certain restrictions concerning "Say on Pay" may be interesting. "Say on Pay" refers to the annual, nonbinding opportunity for a company's shareholders to vote on the compensation of the firm's senior executives.

We expect more scrutiny of corporate governance issues, especially concerning the distinction between the roles of the CEO and the Chair of the Board. The intricate issue of severance and "change-in-control" agreements is undergoing further examination by practitioners, and we expect it will garner additional attention from scholars soon. The realm of "risk" is especially appropriate for examination and study by academic experts. We hope that the analysis focused on risk, risk-adjusted remuneration, and related subjects will engage several scholars in the future. We expect executive compensation for executive abilities and education to be paid much more attention. This research analyzes several facets of executive compensation, encompassing pay-for-performance sensitivity, current issues, family firms, managerial power, and the existing state of compensation. Our study aims to aid academics in analyzing these issues while promoting interdisciplinary collaboration among experts in various Human Resources disciplines.

References

Aggarwal, R. K., & Samwick, A. A. (1999). The other side of the trade-off: The impact of risk on executive compensation. *Journal of Political Economy, 107*(1), 65–105. https://doi.org/10.1086/250051

Al Farooque, O., Buachoom, W., & Hoang, N. (2019). Interactive effects of executive compensation, firm performance, and corporate governance: Evidence from an Asian market. *Asia Pacific Journal of Management, 36*(4), 1111–1164. https://doi.org/10.1007/s10490-018-09640-2

Albuquerque, A., Albuquerque, R., Carter, M. E., & Dong, Q. (Flora). (2024). How large is the pay premium from executive incentive compensation? *The Accounting Review, 99*(5), 35–64. https://doi.org/10.2308/TAR-2022-0226

Alolah, O. (2024). CEO compensation complexity: Optimal contracts or agency problems? *Journal of Financial Research, 47*(3), 789–821. https://doi.org/10.1111/jfir.12383

Al-Shaer, H., Albitar, K., & Liu, J. (2023). CEO power and CSR-linked compensation for corporate environmental responsibility: UK evidence. *Review of Quantitative Finance and Accounting, 60*(3), 1025–1063. https://doi.org/10.1007/s11156-022-01118-z

Arya, A., & Mittendorf, B. (2005). Offering stock options to gauge managerial talent. *Journal of Accounting and Economics, 40*(1), 189–210. https://doi.org/10.1016/j.jacceco.2005.02.001

Bebchuk, L. A., & Fried, J. M. (2003). Executive compensation as an agency problem. *Journal of Economic Perspectives, 17*(3), 71–92. https://doi.org/10.1257/089533003769204362

Bebchuk, L. A., & Fried, J. M. (2004). *Pay without performance: The unfulfilled promise of executive compensation.* Harvard University Press.

Bebchuk, L. A., & Fried, J. M. (2010). Paying for long-term performance. *University of Pennsylvania Law Review, 158*(7), 1915–1959.

Bender, R. (2003). How executive directors' remuneration is determined in two FTSE 350 utilities. *Corporate Governance: An International Review, 11*(3), 206–217. https://doi.org/10.1111/1467-8683.00319

Berrone, P., Cruz, C., Gomez-Mejia, L. R., & Larraza-Kintana, M. (2010). Socioemotional wealth and corporate responses to institutional pressures: Do family-controlled firms pollute less? *Administrative Science Quarterly, 55*(1), 82–113. https://doi.org/10.2189/asqu.2010.55.1.82

Bertrand, M., & Mullainathan, S. (2000). *Do CEOs set their own pay? The ones without principals do* (Working Paper 7604). National Bureau of Economic Research. https://doi.org/10.3386/w7604

Bertrand, M., & Schoar, A. (2006). The role of family in family firms. *Journal of Economic Perspectives, 20*(2), 73–96. https://doi.org/10.1257/jep.20.2.73

Bhatia, M., Gulati, R., & Kumar, S. (2024). Are executive pay and firm performance related? Evidence-based on meta-regression analysis. *Asia Pacific Journal of Management*. https://doi.org/10.1007/s10490-024-09979-9

Bhuyan, R., Butchey, D., Haar, J., & Talukdar, B. (2020). CEO compensation and firm performance in the insurance industry. *Managerial Finance, 48*(7), 1086–1115. https://doi.org/10.1108/MF-04-2019-0154

Boschen, J. F., & Smith, K. J. (1995). You can pay me now and you can pay me later: The dynamic response of executive compensation to firm performance. *The Journal of Business, 68*(4), 577–608.

Bouteska, A., & Mefteh-Wali, S. (2021). The determinants of CEO compensation: New insights from United States. *Journal of Applied Accounting Research, 22*(4), 663–686. https://doi.org/10.1108/JAAR-08-2020-0176

Bryson, A., Forth, J., & Zhou, M. (2014). Same or different? The CEO labour market in China's public listed companies. *The Economic Journal, 124*(574), F90–F108. https://doi.org/10.1111/ecoj.12107

Burkart, M., Panunzi, F., & Shleifer, A. (2003). Family firms. *The Journal of Finance, 58*(5), 2167–2201. https://doi.org/10.1111/1540-6261.00601

Cabezon, F. (2024). Executive compensation: The trend toward one-size-fits-all. *Journal of Accounting and Economics, 101708*. https://doi.org/10.1016/j.jacceco.2024.101708

Cai, X., Pan, H., Gao, C., Wang, C., & Lu, L. (2021). Top executive tournament incentives and corporate innovation output. *Accounting & Finance, 61*(5), 5893–5924. https://doi.org/10.1111/acfi.12850

Cennamo, C., Berrone, P., Cruz, C., & Gomez-Mejia, L. R. (2012). Socioemotional wealth and proactive stakeholder engagement: Why family-controlled firms care more about their stakeholders. *Entrepreneurship Theory and Practice, 36*(6), 1153–1173. https://doi.org/10.1111/j.1540-6520.2012.00543.x

Chaidir, M., Santoso, S., & Qalbia, F. (2024). Exploring pay equity: The role of public and employee stakeholders in shaping CEO-worker pay ratios. *International Journal of Management, Accounting & Finance (KBIJMAF), 1*(1), Article 1. https://doi.org/10.70142/kbijmaf.v1i1.216

Chen, G., Chittoor, R., & Vissa, B. (2021). Does nepotism run in the family? CEO pay and pay-performance sensitivity in Indian family firms. *Strategic Management Journal, 42*(7), 1326–1343. https://doi.org/10.1002/smj.3263

Cheng, M., Lin, B., & Wei, M. (2015). Executive compensation in family firms: The effect of multiple family members. *Journal of Corporate Finance, 32*, 238–257. https://doi.org/10.1016/j.jcorpfin.2014.10.014

Conyon, M. J., & He, L. (2016). Executive compensation and corporate fraud in China. *Journal of Business Ethics, 134*(4), 669–691. https://doi.org/10.1007/s10551-014-2390-6

Core, J. E., Holthausen, R. W., & Larcker, D. F. (1999). Corporate governance, chief executive officer compensation, and firm performance. *Journal of Financial Economics, 51*(3), 371–406. https://doi.org/10.1016/S0304-405X(98)00058-0

Core, J. E., & Larcker, D. F. (2002). Performance consequences of mandatory increases in executive stock ownership. *Journal of Financial Economics, 64*(3), 317–340. https://doi.org/10.1016/S0304-405X(02)00127-7

Coughlan, A. T., & Schmidt, R. M. (1985). Executive compensation, management turnover, and firm performance: An empirical investigation. *Journal of Accounting and Economics, 7*(1), 43–66. https://doi.org/10.1016/0165-4101(85)90027-8

Darrough, M. N., & Melumad, N. D. (1995). Divisional versus company-wide focus: The trade-off between allocation of managerial attention and screening of talent. *Journal of Accounting Research, 33*, 65–94. https://doi.org/10.2307/2491375

De Massis, A., Eddleston, K. A., & Rovelli, P. (2021). Entrepreneurial by design: How organizational design affects family and non-family firms' opportunity exploitation. *Journal of Management Studies, 58*(1), 27–62. https://doi.org/10.1111/joms.12568

Dharwadkar, R., George, G., & Brandes, P. (2000). Privatization in emerging economies: An agency theory perspective. *The Academy of Management Review, 25*(3), 650–669. https://doi.org/10.2307/259316

Doucouliagos, C. (Hristos), Graham, M., & Haman, J. (2012). Dynamics and convergence in chief executive officer pay. *SSRN Electronic Journal*. https://doi.org/10.2139/ssrn.2033535

Edmans, A., Gabaix, X., & Jenter, D. (2017). Executive compensation: A survey of theory and evidence✶. In B. E. Hermalin & M. S. Weisbach (Eds.), *The handbook of the economics of corporate governance* (Vol. 1, pp. 383–539). https://doi.org/10.1016/bs.hecg.2017.11.010

Edmans, A., Gosling, T., & Jenter, D. (2023). CEO compensation: Evidence from the field. *Journal of Financial Economics, 150*(3), Article 103718. https://doi.org/10.1016/j.jfineco.2023.103718

Eriksson, T. (1999). Executive compensation and tournament theory: Empirical tests on Danish data. *Journal of Labor Economics, 17*(2), 262–280. https://doi.org/10.1086/209920

Finkelstein, S., & Boyd, B. K. (1998). How much does the CEO matter? The role of managerial discretion in the setting of CEO compensation. *The Academy of Management Journal, 41*(2), 179–199. https://doi.org/10.2307/257101

Frye, M. B. (2004). Equity-based compensation for employees: Firm performance and determinants. *Journal of Financial Research, 27*(1), 31–54. https://doi.org/10.1111/j.1475-6803.2004.00076.x

Gayle, G.-L., Li, C., & Miller, R. A. (2022). Was Sarbanes-Oxley costly? Evidence from optimal contracting on CEO compensation. *Journal of Accounting Research, 60*(4), 1189–1234. https://doi.org/10.1111/1475-679X.12436

Ghosh, A. (2006). Determination of executive compensation in an emerging economy. Evidence from India. *Emerging Markets Finance and Trade, 42*(3), 66–90. https://doi.org/10.2753/REE1540-496X420304

Gomez-Mejia, L. R., Larraza-Kintana, M., & Makri, M. (2003). The determinants of executive compensation in family-controlled public corporations. *The Academy of Management Journal, 46*(2), 226–237. https://doi.org/10.2307/30040616

Gomez-Mejia, L. R., Tosi, H., & Hinkin, T. (1987). Managerial control, performance, and executive compensation. *The Academy of Management Journal, 30*(1), 51–70. https://doi.org/10.2307/255895

Göx, R. F., & Hemmer, T. (2020). On the relation between managerial power and CEO pay. *Journal of Accounting and Economics, 69*(2), Article 101300. https://doi.org/10.1016/j.jacceco.2020.101300

Grossman, S. J., & Hart, O. D. (1983). Implicit contracts under asymmetric information. *The Quarterly Journal of Economics, 98*, 123–156. https://doi.org/10.2307/1885377

Hambrick, D. C., & Finkelstein, S. (1995). The effects of ownership structure on conditions at the top: The case of CEO pay raises. *Strategic Management Journal, 16*(3), 175–193. https://doi.org/10.1002/smj.4250160304

Holmström, B. (1979). Moral hazard and observability. *The Bell Journal of Economics, 10*(1), 74–91. https://doi.org/10.2307/3003320

Islam, E., Rahman, L., Sen, R., & Zein, J. (2022). Eyes on the prize: Do industry tournament incentives shape the structure of executive compensation? *Journal of Financial and Quantitative Analysis, 57*(5), 1929–1959. https://doi.org/10.1017/S0022109021000570

Jaiswal, S. K., & Bhattacharyya, A. K. (2016). The role of operating efficiency and asset productivity in relative performance evaluation and CEO compensation in Indian firms. *Decision, 43*(3), 201–221. https://doi.org/10.1007/s40622-016-0128-2

Jensen, M. C., & Meckling, W. H. (1976). Theory of the firm: Managerial behavior, agency costs and ownership structure. *Journal of Financial Economics, 3*(4), 305–360. https://doi.org/10.1016/0304-405X(76)90026-X

Jensen, M. C., & Murphy, K. J. (1990). *CEO incentives—IT'S not how much you pay, but how.*

Khanna, T., & Palepu, K. (2000). Is group affiliation profitable in emerging markets? An analysis of diversified Indian business groups. *The Journal of Finance, 55*(2), 867–891. https://doi.org/10.1111/0022-1082.00229

Kohli, M., & Gill, S. (2020). Impact of family involvement on strategy and CEO compensation. *Journal of Family Business Management, 10*(3), 189–212. https://doi.org/10.1108/JFBM-09-2019-0060

Lambert, R. A., Larcker, D. F., & Weigelt, K. (1993). The structure of organizational incentives. *Administrative Science Quarterly, 38*(3), 438–461. https://doi.org/10.2307/2393375

Lazear, E. P., & Rosen, S. (1981). Rank-order tournaments as optimum labor contracts. *Journal of Political Economy, 89*(5), 841–864. https://doi.org/10.1086/261010

Lee, K. W., Lev, B., & Yeo, G. H. H. (2008). Executive pay dispersion, corporate governance, and firm performance. *Review of Quantitative Finance and Accounting, 30*(3), 315–338. https://doi.org/10.1007/s11156-007-0053-8

Main, B. G. M., Bruce, A., & Buck, T. (1996). Total board remuneration and company performance. *The Economic Journal, 106*(439), 1627–1644. https://doi.org/10.2307/2235204

Masulis, R. W., & Zhang, S. (2013). *Compensation gaps among top executives: Evidence of tournament incentives or productivity differentials?* (SSRN Scholarly Paper 2275455). https://doi.org/10.2139/ssrn.2275455

Michiels, A., Voordeckers, W., Lybaert, N., & Steijvers, T. (2013). CEO compensation in private family firms: Pay-for-performance and the moderating role of ownership and management. *Family Business Review, 26*(2), 140–160. https://doi.org/10.1177/0894486512454731

Murphy, K. J. (1985). Corporate performance and managerial remuneration: An empirical analysis. *Journal of Accounting and Economics, 7*(1), 11–42. https://doi.org/10.1016/0165-4101(85)90026-6

Murphy, K. J. (1999). Executive compensation. In *Handbook of labor economics* (Vol. 3, Chapter 38, pp. 2485–2563). Elsevier. https://doi.org/10.1016/S1573-4463(99)30024-9

Murphy, K. J., & Sandino, T. (2020). Compensation consultants and the level, composition, and complexity of CEO pay. *The Accounting Review, 95*(1), 311–341. https://doi.org/10.2308/accr-52439

Nawaz, T. (2022). How much does the board composition matter? The impact of board gender diversity on CEO compensation. *Sustainability, 14*(18), Article 18. https://doi.org/10.3390/su141811719

Parthasarathy, A., Menon, K., & Bhattacherjee, D. (2006). Executive compensation, firm performance and governance: An empirical analysis. *Economic and Political Weekly, 41*(39), 4139–4147.

Pollock, S., Switzer, L. N., & Wang, J. (2023). The dynamics of CEO equity vs. Inside debt and firm performance. *Research in International Business and Finance, 64*, 101891. https://doi.org/10.1016/j.ribaf.2023.101891

Raithatha, M., & Komera, S. (2016). Executive compensation and firm performance: Evidence from Indian firms. *IIMB Management Review, 28*(3), 160–169. https://doi.org/10.1016/j.iimb.2016.07.002

Ramaswamy, K., Veliyath, R., & Gomes, L. (2000). A study of the determinants of CEO compensation in India. *MIR: Management International Review, 40*(2), 167–191.

Rehman, A. U., Ali, T., Hussain, S., & Waheed, A. (2021). Executive remuneration, corporate governance and corporate performance: Evidence from China. *Economic Research-Ekonomska Istraživanja, 34*(1), 3092–3118. https://doi.org/10.1080/1331677X.2020.1867214

Rosen, S. (1986a). The theory of equalizing differences. In *Handbook of labor economics* (Vol. 1, Chapter 12, pp. 641–692). Elsevier. https://doi.org/10.1016/S1573-4463(86)01015-5

Rosen, S. (1986b). The theory of equalizing differences. In *Handbook of labor economics* (Vol. 1, Chapter 12, pp. 641–692). Elsevier. https://doi.org/10.1016/S1573-4463(86)01015-5

Shleifer, A., & Vishny, R. W. (1997). A survey of corporate governance. *The Journal of Finance, 52*(2), 737–783. https://doi.org/10.1111/j.1540-6261.1997.tb04820.x

Sridhar, I. (2021). Study of executive compensation of public and private sector enterprises in India—Impact on corporate governance. *International Journal of Public Sector Performance Management, 7*(2), 236–249. https://doi.org/10.1504/IJPSPM.2021.114042

Swami, P. (2005). India's highest paid executives'. *Business India* (p. 62).

Talavera, O., Yin, S., & Zhang, M. (2021). Tournament incentives, age diversity and firm performance. *Journal of Empirical Finance, 61*, 139–162. https://doi.org/10.1016/j.jempfin.2021.01.003

Tosi, H. L., Werner, S., Katz, J. P., & Gomez-Mejia, L. R. (2000). How much does performance matter? A meta-analysis of CEO pay studies. *Journal of Management, 26*(2), 301–339. https://doi.org/10.1177/014920630002600207

van Essen, M., Otten, J., & Carberry, E. J. (2015). Assessing managerial power theory: A meta-analytic approach to understanding the determinants of CEO compensation. *Journal of Management, 41*(1), 164–202. https://doi.org/10.1177/0149206311429378

Wang, Q., & Qiu, M. (2023). Strength in numbers: Minority shareholders' participation and executives' pay-performance sensitivity. *Pacific-Basin Finance Journal, 79*, Article 102015. https://doi.org/10.1016/j.pacfin.2023.102015

Young, M. N., Peng, M. W., Ahlstrom, D., Bruton, G. D., & Jiang, Y. (2008). Corporate governance in emerging economies: A review of the principal-principal perspective. *Journal of Management Studies, 45*(1), 196–220. https://doi.org/10.1111/j.1467-6486.2007.00752.x

Chapter 19
Data as a Corporate Asset: Role of Data Governance Framework in Corporate Governance of MNCs with Reference to Digital Personal Data Protection Act 2023

Manvee and Md Safiullah

Abstract In the twenty-first century, data have emerged as a critical asset for businesses, especially MNCs, as it gives them an edge in the market and provides them with a competitive advantage, innovation, and decision-making. This paper analyzes and explores the role of data as an intangible asset for corporations. Data have the transformative power to revolutionize the business models of MNCs and their operations. This paper also discusses the challenges with the MNCs that they face in data mining and refining, like system interoperability and data fluency; it proposes a solution for all these challenges, like clarifying the accountability of the data and embedding the data fluency across enterprises. This paper also examines the integration of data governance within corporate governance, highlighting the benefits that it poses to MNCs. With the implementation of the Digital Personal Data Protection Act 2023 (DPDPA) in India now, the MNCs must adopt their corporate governance frameworks, ensuring compliance with the DPDPA and fostering a privacy-centric culture within the corporation. This chapter also addressed the need for compliance measures to be taken by corporations after the DPDPA gets implemented what could be the potential challenges for them and how the board can prepare itself for all these challenges have also been suggested. The chapter concludes by stressing the need for continuous monitoring and taking proactive measures to build a robust data protection and privacy framework within the organization, ensuring that MNCs contribute to the long-term success of the digital economy.

Keywords Multinational corporations · Business · Data protection · Data privacy · DPDPA

Manvee (✉) · Md. Safiullah
Chanakya National Law University, Patna, India
e-mail: manvee.nlu.patna@gmail.com

Md. Safiullah
e-mail: safiullah@cnlu.ac.in

V. Kandpal et al. (eds.), *Developments in Corporate Governance*,
Approaches to Global Sustainability, Markets, and Governance,
https://doi.org/10.1007/978-981-96-6366-8_19

19.1 Introduction

In the present-day landscapes of the corporate world, data are an essential ingredient. These days, most of the corporate giants rely on the use of data science and data analytics to operate their business models; in fact, few businesses like *Uber* and *Amazon* are totally data driven. MNCs have made the use of data as an essential element of their businesses. Data (Safiullah & Parveen, 2022) are not just a strategic asset for the business. However, it has become an operational necessity for these MNCs, which helps them to drive innovation, enhance smart decision-making, and give them a competitive advantage (Dexter, 2024). The usage of data-driven decision-making helps these MNCs enhance their business performance and achieve the long-term objectives and goals of the business. This indicates that business needs to accumulate (Reynolds, 2024) a vast amount of data for processing. As a result, there is a need for a robust data governance framework for these MNCs. So, the role of the data governance framework becomes critical here. Recently, India got its full-fledged Data Protection law, i.e., *The Digital Personal Data Protection Act 2023* ("DPDPA"). This increases the compliance for all those MNCs who are willing to operate in India and are currently operating in India. As per the DPDPA, these MNCs need to align their privacy compliance and data governance framework.

The DPDPA is not just a regulatory requirement for corporations, but it is a strategic blueprint that safeguards data integrity and fosters trust in the digital economy. The DPDPA underlines the importance of having a strong and robust data governance (Special, 2023) framework as well as ethical data management for the MNCs and businesses operating in India. DPDPA also ensures that these MNCs should have the highest standards of data protection, privacy, and security within their organizations. The corporate governance practices of the MNCs are linked with their data governance practices. The more effective data governance a corporation has, the more it will ensure that the data are accessible, secure, and accurate, thereby enabling the business to leverage data insights for strategic planning and informed decision-making. The DPDPA makes it mandatory for MNCs to have a comprehensive data governance structure that includes strict consent mechanisms, privacy-by-design principles, and frequent audits. These mandatory requirements of DPDPA compel these MNCs to adopt advanced technologies and practices that safeguard personal data throughout the lifecycle of the data.

As per the DPDPA, data governance is about building (Kuzio et al., 2022) a robust infrastructure, and it extends beyond compliance. This robust infrastructure is needed to stand strong against upcoming cyber threats and adapt to the regulatory landscapes. For MNCs, this means that they need to invest in cutting-edge encryption measures with continuous monitoring systems and robust cybersecurity measures. DPDPA emphasizes transparency, aligning user control with the global best practices, thereby fostering a culture of accountability and trust within the organizations. DPDPA places additional responsibility on MNCs as significant data fiduciaries, especially those that handle large volumes of personal data (Alamy, 2023). These entities are responsible for appointing dedicated data protection officers within the corporation, conducting data protection impact assessments, and implementing advanced security measures for their corporations. The Data Protection Officer shall be responsible for overlooking compliance-related issues. This proactive approach to protecting privacy will position these MNCs as global leaders in data stewardship, enhancing their reputations in the market and providing them with a competitive edge. There are a couple of benefits for these MNCs if they have a strong data governance framework. By adhering to DPDPA strict standards, MNCs can ensure enhanced operational efficiencies and data accuracy and drive innovation. It will also enhance their risk management and help in identifying and mitigating potential data breaches in real time.

This paper aims to analyze Data as a Corporate asset for MNCs, how it is an important asset for a multinational corporation to drive their business globally, how can they mitigate the challenges associated with data governance, how these data will be a key driving factor for the businesses, how MNCs can incorporate data governance within their corporate governance framework and what could be potential challenges associated with the corporate governance in terms of data protection for the MNCs. We will also analyze how the DPDPA will impact the corporate governance practices of all the MNCs operating in India and what could be the way forward for them to overcome the challenges with regard to the DPDPA.

19.2 Literature Review

19.2.1 Data as a Corporate Asset

The perception of data as a corporate asset has evolved rapidly, driven by the immense value it generates for businesses. *Hanhan and Nuwar (2024)* highlight that data are a vital corporate asset, yet many businesses underutilize its potential, impacting innovation and decision-making. Effective data strategies, including governance and integration, are crucial for optimizing AI and machine learning performance. Companies leveraging high-quality data can enhance efficiency and customer experience. *Guy Hanson (2020)* highlights the commercial benefits of effective data management, where high-quality data enhance sales forecasting, customer conversion, and overall productivity. *Firas Sleiman (2022)* further emphasizes that data, akin to oil, is the primary growth driver for today's most dynamic companies. Liberating, understanding, and monetizing data unlock new revenue streams, making it integral to corporate strategy. Case studies by the *British Academy and Royal Society (2021)* further illustrate the governance challenges of harnessing data responsibly across sectors.

19.2.2 Importance of Data Management in MNCs

Professor Dustin Tingley (2023) emphasizes that data science extends beyond tech giants like Facebook and Amazon, and fast-food chains like Chick-fil-A also use data strategies to optimize operations and improve customer experiences. He highlights the importance of collaboration between data scientists and business leaders for the successful execution of data-driven strategies. This relationship enables companies to harness the full potential of data to drive efficiency and revenue. *Molly Reynolds, Joel Ramsey, and Nushrah Amod (2024)* discuss how integrating data governance into corporate governance helps manage data securely and ethically, aligning it with business strategies. Data governance supports innovation, privacy, and cybersecurity by offering a proactive approach to managing risks and ensuring compliance while driving growth through effective data management. *Nisha* Talagala (2022) likens data to oil, highlighting the need for refining it through a solid "Data Practice." She outlines four principles—data provenance, privacy, protection, and preparation—that organizations must follow to derive valuable insights from data and avoid data-related risks.

19.2.3 DPDP Act and Data Governance Issues in MNCs

The Digital Personal Data Protection (DPDP) Act, 2023 has introduced a robust regulatory framework for data governance, posing significant implications for multinational corporations (MNCs) operating in India. As *Deloitte (2024)* notes, the Act emphasizes the protection of personal data, imposing compliance obligations on enterprises to uphold transparency, trust, and consumer rights. MNCs face challenges in adapting their data practices, particularly in relation to third-party data sharing, employee data processing, and safeguarding intellectual property. *Rashmi Pradeep and Arun Prabhu (2024)* further highlight that MNCs must navigate the limitations of consent-based processing and carefully manage the legitimate uses of employee data. This evolving landscape demands that boards of MNCs implement a proactive, top-down governance approach, ensuring adherence to the law while fostering a culture of data privacy compliance.

19.3 Data as a Corporate Asset

"*Data is the new Oil*" (Talagala, 2022). This phrase itself tells the importance of Data in the twenty-first century. In today's fast-paced world, for rapidly evolving businesses data have emerged as the most valuable asset for many organizations. Most Multinational Companies ("MNCs") are (Hanson, 2021) particular about leveraging data for their competitive advantage, driving innovation, and enhancing strategic decision-making. The significance of the data extends beyond its raw form (Smaje, 2020), as it represents an intangible asset with the power to transform business models, enhance operational efficiency, and create new revenue streams. As MNCs recognize the potential of the Data, they are investing (Smaje, 2020) heavily in data-related technologies and strategies. They aim to unlock the secrets and disruptive potential embedded within their vast data reserves.

19.3.1 Data an Intangible Asset

With time, the means and usage of assets (Sleiman, 2022) also saw a change. Initially, corporate assets were predominantly tangible and included physical goods, real estate, machinery, etc. However, with digitalization, intangible assets like intellectual property (Tang, 2020), brand equity, and particularly data, saw a surge in its importance. Data are termed as an intangible asset because it is digital, i.e., not physical, and it cannot be touched or measured in traditional forms. Data include (Corrado et al., 2022) customer information, transaction records, market research, and personal information like contact, email, etc., all of which can be stored, analyzed, and processed to drive the business value. This intangible nature of the data allows data to be infinitely replicable and usable across various applications without degradation.

19.3.2 Importance of Data as an Asset for MNCs

The strategic importance of the data cannot be overstated for the MNCs. The simple reason is that MNCs operate in diverse markets and with varying consumer preferences, regulatory environments, competitive landscapes, and different governance policies. MNCs primarily leverage the data to gain insights into the local markets, optimize the supply chain, and tailor products and services to meet regional demands. For example, *Google* (Atomicwork, 2023) utilizes data as one of its critical assets to power its business. It uses data analytics techniques on its advertising platforms. With billions of daily searches, emails, map queries, and other interactions during browsing, Google collects ample amounts of data about its users on a daily basis. These data also include user preferences, behaviors, and demographics, which Google meticulously analyzes to understand individual and collective trends. These data are primarily used for target advertising. Google provides (Online, 2022) ads by comprehending what users are searching for and are interested in; in this manner, google serves highly relevant ads to specific audiences, enhancing the effectiveness of its ads campaign (Safiullah et al., 2016). This precise targeting serves a dual (Online, 2022) purpose; it not only benefits the advertisers by improving their return on investment but also generates substantial revenue for Google, making advertising its primary income source. Google uses data (Online, 2022) analytics for its Google

Map feature, which uses data from users' mobile devices to predict traffic conditions and suggest an optimal route for its users. *Amazon* (Mpeshev, 2024) leverages data as one of its essential assets because it deploys advanced data analytics techniques to bolster its business operations and maximize profits. Amazon, through its complex algorithms and machine learning techniques, analyzes vast amounts of customer data, customer browsing patterns, purchase history, and customer feedback (Mpeshev, 2024). This analysis of data allows Amazon to create a highly personalized shopping experience, allowing Amazon to offer tailored product recommendations, thereby significantly increasing conversion rates and customer loyalty. This data-driven approach helps Amazon to optimize its inventory management and supply chain operations. Amazon forecasts the demands accurately through historical sales data and external factors, ensuring that popular items are adequately stored, reducing the excess inventory and thus improving order fulfillment efficiency. This results in enhanced customer satisfaction and faster delivery. For marketing and advertising campaigns, Amazon's data analytics enables (Reynolds, 2024) it to deploy highly targeted campaigns reaching out to the right audience with the relevant message for them. This increases the effectiveness of its marketing efforts and enhances the overall customer experience (Reynolds, 2024); continuous customer feedback ensures that Amazon remains at the forefront of innovation, adapting to evolving customer needs and maintaining its industry leadership.

The trends of data analytics (Online, 2022) saw a surge due to 4G and 5G technology, as this enabled the proliferation of connected devices and the rise of extensive sensory networks. This connectivity leads to the generation of unprecedented data volumes, providing rich visualization and thereby facilitating the rapid evolution of data usage and analysis methods. As data storage costs drop and data transfer and processing become more affordable, the opportunities for MNCs to derive actionable insights from data are expanding at a rapid pace.

19.3.3 Data as a Key Driver of New Business Models

The transformative power of data lies (Online, 2022) in its ability to drive new business models and foster innovation. MNCs view data as a core component of their value proposition. Day by Day, MNCs view data as a core component of their value proposition (Online, 2022). For Instance, *Uber* (Pahuja & Cio, 2022) relies heavily on the use of the data to drive its business model. Uber deploys data analytics to enhance its operations and optimize the rider and driver's experience. Uber collects extensive data on rides (Pahuja & Cio, 2022), driver performance, conduct, and traffic patterns; it then uses predictive analytics to match the riders with nearby drivers, reducing the wait time and improving the service quality. These data-driven (Pahuja & Cio, 2022) insights help Uber implement dynamic pricing via surge pricing and adjust the fares on a real-time basis. This is based on the algorithm of demand and supply to maximize the revenue and ensure the availability of the rides during peak times. Uber also utilizes data to reduce idle time (Pahuja & Cio, 2022) and optimize the driver's location. Uber continuously analyses city-specific data to adapt to the strategies for local transportation patterns ensuring that optimal performance is in diverse metropolitan environments. Uber uses Data Visualization (Khune, 2024) techniques to transform complex data into actionable insights, thereby helping both the internal and external teams, as well as the public, to understand the operational efficiency and trends of the market. Whereas the forecasting technique allows Uber to predict future demand and direct the drivers proactively further optimizing the platform.

Hence, these data enable the creation of entirely new revenue sources (Anderson, 2023). Many MNCs have started commercializing or sharing data to generate income. The Data monetization market is expected to grow at a rapid pace, bolstering the economic potential of data as a commodity (Anderson, 2023). MNCs that capitalize on this trend by developing robust data strategies are well-positioned to thrive in the era of the digital economy.

19.3.4 Challenges with Data Mining and Data Refining

Many organizations (British Academy & the Royal Society, 2016) struggle to mine and refine data effectively despite its clear advantages. Several challenges hinder successful data utilization (KPMG, 2024), such as *(i) data fluency* (KPMG, 2024), *(ii) complex and siloed system architectures, (iii) access controls and policies* (KPMG, 2024), *and (iv) interoperability issues* (KPMG, 2024) across the MNCs. These obstacles prevent MNCs from fully harnessing data's potential.

A pervasive challenge is the lack of data literacy among employees. Many workers, including decision-makers, do not fully understand how to interpret data, identify trends, or make data-driven decisions. This gap hinders the organization's ability to use data effectively. Data often reside in disparate systems across various departments, leading to silos that impede data sharing and integration. Different business units using different software systems or databases that are not interoperable often cause these silos.

Organizations must protect sensitive data from unauthorized access and breaches. Implementing robust data security measures, such as encryption and access controls, is necessary but complicates data accessibility and sharing within the organization. It should ensure seamless data flow across different business units and systems that require standardized data formats and protocols. Whereas achieving interoperability often involves overhauling legacy systems and investing in new technology, which can be costly and disruptive.

To address these challenges, leading multinational corporations are reimagining their relationship with data and transforming their IT functions to impact (KPMG, 2024) business outcomes materially. Over the next few years, successful corporations are expected to adopt four key data principles: *(i) clarifying data accountabilities across the enterprise, (ii) embedding data fluency as a strategic imperative, (iii) moving data curation into the business as a core competency, and (iv) reimagining a frictionless data supply chain.*

19.3.4.1 Potential Solutions by the Implementation of Key Data Principles

Clarify Data Accountabilities Across the Enterprise

To mitigate the challenge of *Lack of Clear Accountability and Governance* (KPMG, 2024), corporations should clearly define data ownership and accountability, and organizations can mitigate the risk of data mismanagement. Roles such as Chief Data Officer (CDO), Chief Information Officer (CIO), and Chief Technology Officer (CTO) should be clearly delineated with specific responsibilities and control areas. Corporations should implement a robust data governance framework that outlines data policies, standards, and procedures. This framework should ensure that data-related responsibilities are clearly communicated and enforced across the enterprise. Moreover, corporations shall establish mechanisms to resolve conflicts that arise from overlapping data responsibilities. This may include setting up cross-functional data governance committees that can mediate and make decisions on data-related issues.

If the corporations implement these solutions, then it will lead to enhancement in data quality and trust, leading to more trustworthy and high-quality data; it will also lead to improved decision-making with clear roles and responsibilities, making data-driven decision-making more efficient and reliable; it will also lead to a reduction in data silos that will facilitate better data integration and accessibility.

Embed Data Fluency Across the Enterprise as a Strategic Imperative

To mitigate the challenge of *Low Data Literacy and Fluency* (KPMG, 2024), corporations should develop comprehensive training programs to improve data literacy across the organization for their employees. This includes workshops, online courses, and certifications focused on data analytics, data governance, and data-driven decision-making. Corporations should also encourage the rise of citizen developers who can leverage low-code platforms and tools to create data-driven applications. This will empower non-IT staff to engage with data more effectively. Corporations should foster a culture that values data-driven insights by integrating data fluency into the organization's core values and operational practices.

If corporations implement these solutions, they will have widespread data utilization, leading to more informed decisions. This will also promote innovation and efficiency among corporation employees and play a key role in aligning with the strategic data goals.

Move Data Curation into the Business as a Core Competency

To mitigate the challenge of (KPMG, 2024) *Centralized and Technical Data Management*, corporations should shift data curation responsibilities closer to business units (OECD, 2023). This involves training business staff to manage and curate data relevant to their functions. Create internal data marketplaces that will allow business units to access, curate, and share data easily. These marketplaces should include tools for data discovery, quality assessment, and integration. Corporations should also implement guided self-service models that help users navigate data resources and utilize data effectively without needing extensive technical knowledge.

If corporations implement these solutions, they will have faster and more relevant data access according to the specific needs of the business; by leveraging curated data, business units can enhance their responsiveness. Lastly, it will also lead to greater data utilization across the organization.

Reimagine a Frictionless Data Supply Chain

To mitigate the challenge of (OECD, 2023) *Inefficient Data Movement and Integration,* corporations should start adopting data virtualization techniques and microservices architecture to enable seamless data integration and movement across different systems. This approach reduces dependencies on physical data storage and allows for more flexible data management. Implement Data-as-a-Service (DaaS) models to provide on-demand data access across the enterprise. This will involve creating APIs and services that make data easily accessible and usable by different applications and business units. Corporations should also use AI and machine learning to automate data management tasks, such as data quality monitoring, data cleansing, and metadata management.

If corporations implement these solutions, it will increase the accessibility of data for those who need it and when they need it. It will also help in the scalability of the data and provide the flexibility required to handle large, bulky volumes of data and diverse sources of data. The use of AI-powered data management will improve the data quality and reduce the manual effort required to maintain that data.

19.4 Data Governance vis-à-vis Corporate Governance in MNCs

Data governance is a subset of corporate governance that (Bamberger & Mulligan, 2013) ensures the proper management of data assets. It includes policies, procedures, and controls for collecting, storing, processing, and destroying data within an organization. As a strategic approach, data governance aims (Bamberger & Mulligan, 2013) to maximize the value of data while minimizing associated risks, thereby supporting corporate governance's broader objectives. Corporate governance provides a predominant framework for decision-making, accountability, and advancing the best interests of (Bamberger & Mulligan, 2013) the company. Thus, integrating data governance within corporate governance ensures that data-related decisions align with the organization's overall strategy and processes. This alignment helps maintain regulatory compliance, protect data value, and prevent financial loss and reputational harm.

19.4.1 Why Should MNCs Integrate Data Governance[1] in Their Corporate Governance Model?

19.4.1.1 Enhanced Transparency and Trust

An effective data governance (Tokhadze, 2023) framework within an organization will enhance transparency within the corporation. This will lead to increased trust among investors, customers, and regulators. Thus, transparent data handling within the corporation aligns with corporate governance principles, fostering stakeholder confidence.

19.4.1.2 Improved Compliance

An effective data governance structure will ensure compliance with the updated data protection regulations, helping the (Tokhadze, 2023) MNCs avoid legal and financial ramifications like penalties and suits. A well-defined data governance framework will enable the organization to adhere to data protection regulations by rapidly adapting to new regulations.

19.4.1.3 Effective Risk Management

If an MNC has an effective data governance structure, it will play a crucial role in identifying and mitigating the risks (Tokhadze, 2023) related to data usage, storage, and processing. It will help the corporation minimize potential data breaches and security incidents that are critical for maintaining business continuity and protecting the corporation's reputation.

[1] Tokhadze, A. (2023). The interdisciplinary analysis of institutional role of data protection officer in the system of corporate governance. *Journal of Personal Data Protection Law, 2023*(1), 106–118.

19.4.1.4 Better Decision-Making

It is a well-known fact that high-quality data are essential for accurate reporting and analysis. Thus, a well-framed data governance policy (Tokhadze, 2023) within the corporation will ensure consistency, accuracy, and accountability in the data used across the corporation, facilitating better decision-making.

19.4.1.5 Operational Efficiency and Cost Savings

A robust data governance practice (Tokhadze, 2023) within the corporation will eliminate redundancy and inconsistencies in the data, leading to operational efficiencies and cost savings. This efficiency will support the organization's overall market integrity and economic performance.

19.5 Impact of Digital Personal Data Protection Act 2023 on Corporate Governance

With the enactment of the Digital Personal Data Protection Act 2023 ("DPDPA") now, corporates will have a strict requirement to handle the personal data of customers (Deloitte & Confederation of Indian Industry, 2024), employees, vendors, etc., and this will significantly impact corporate governance. DPDPA will necessitate a standard shift in how the board of the corporation will govern the data privacy inside the organization as well as during its operations, how it is going to manage the risks associated, and how they are going to ensure compliance with the DPDPA. In this part, we will be discussing how DPDPA will (Deloitte & Confederation of Indian Industry, 2024) have a multifaceted impact on the corporate governance of the MNCs, and we shall be highlighting the authoritative role of the board in how it can drive compliance for DPDPA, thereby building a privacy-centric culture and how it will mitigate the associated privacy risks with it.

19.5.1 Familiarity and Self-Awareness with the Law

Corporate boards will now have a primary responsibility (Deloitte & Confederation of Indian Industry, 2024) to stay updated with the regulatory obligations under the DPDPA. They have to develop a thorough understanding of the law and what could be the potential implications for their business. It will also include the fact that corporations have to dive deeper into the specific requirements and obligations posed by the act to the enterprise/corporation with respect to data privacy. Now, the board has to stay abreast (Ciso, 2024) of the evolving industry standards and regulatory changes and accept the emerging trends to adopt for their corporate governance frameworks. The board has to continuously update its knowledge base with regard to the privacy risks associated with the industry. This can be done by attending workshops and training sessions, taking part in Industry consultations and forums, and engaging on a regular basis with the privacy experts. In this way, the board will be in a position to anticipate the potential privacy risks and challenges associated with the business after the DPDPA implementation.

19.5.2 Transformation in Governance

One of the significant impacts that the MNCs (Ciso, 2024) would face after the implementation of DPDPA is the drastic cultural shift in the corporation. To manage this cultural shift, the board must lead the transformation by inculcating data protection and privacy into the organization's corporate culture. This can be done by fostering a privacy-first mindset across the organization at all levels. Regular training sessions on managing privacy in deals and transactions, as well as inside the organizations, should be organized by the board. The board should also make employees aware of the privacy policy and procedures of the organization for all of its employees at all levels. In order to ensure the effectiveness of these initiatives, the board should regularly test the employees' awareness level via periodic tests, quizzes, and scenario-based training programs. In this way, the board can identify the knowledge gap and compliance gaps; this will enable the board to enable targeted interventions to address the deficiencies. This will ensure that employees and (Deloitte & Confederation of Indian Industry, 2024) key managerial personnel are highly aware of compliances under DPDPA, which is essential to mitigate data breach risks.

19.5.3 Transformation of the Corporation

The DPDPA makes it necessary for the corporation (Deloitte & Confederation of Indian Industry, 2024) to have a comprehensive transformation to align with the privacy principles within the corporation. For this, the boards should ensure that privacy concerns are addressed at every stage of data processing. This will involve operationalizing the process and measures to protect data privacy, using advanced technology and automation tools to ensure compliance. By doing this, boards can ensure that data privacy is the core value guiding decision-making and business operations within the corporation. The board should encourage its enterprises to generate value from the data, which can be done by identifying creative and innovative ways to utilize the data collected and its processes. The enterprise should comply with the privacy principles during the collection and processing of the data. If these enterprise corporations balance the need for data-driven insights with robust privacy measures, then they can unlock new growth opportunities without compromising on data protection; this will ensure that the power of data for growth and innovation is contributing to the long-term success of the corporation and its enterprises.

19.5.4 Risk Management

After the DPDPA gets implemented (Deloitte & Confederation of Indian Industry, 2024), the board will have the responsibility to integrate data protection and privacy into the risk management framework of the corporation. It is essential to mitigate the associated risk and prevent the corporation from getting hefty fines from the Data Protection Board of India due to data breaches or non-compliance. The board must implement robust controls like conducting regular risk assessments, preparing comprehensive privacy risk mitigation strategies, developing data privacy risk mapping for the organization, and continuously monitoring the associated risks. This will increase the resilience of the enterprise in managing data privacy risks. One of the most crucial aspects of risk management under the DPDPA is the oversight of the third party involved in data processing. Here, the boards need to ensure that procurement functions give due weight to the privacy obligations when they onboard any supplier. This can be ensured by conducting thorough due diligence, assessing the privacy maturity of the supplier, and maintaining transparency with the customers with respect to the data privacy practices of these suppliers. In this way, the board will be able to ensure that consumers' personal data is safe and that consumer trust is upheld.

19.5.5 Proactive Compliance Measures

In order to ensure compliance with (Deloitte & Confederation of Indian Industry, 2024) the DPDPA, the board must appoint a *Data Protection Officer* ("DPO") and guide the DPO and the stakeholders across the corporation to adopt proactive compliance measures. This includes several major steps such as *(a) obtaining informed consent, (b) maintaining records of the processing activities, (c) securing data transfers, (d) ensuring transparency with the data, and (e) privacy by design.* The board shall ensure that the corporation has obtained informed consent from the data principles, and they haven't retained any personal data which is no longer required as per the lawful basis. This will include proper transparent communication with the data principles regarding the purpose and the duration of the data processing. The DPO should ensure that a comprehensive record of all the data processing activities going on within the corporation is documented and updated properly. This will ensure visibility over all types of personal data collected, stored, disposed and processed across the lifecycle of the data. The DPO will be responsible for the implementation of the security protocols for the transfer of data outside the corporation. He will ensure that robust encryption techniques and secure communication protocols are used within the corporation. This will ensure that the transfer of the data complies with the DPDPA. The board should also ensure that data transparency is maintained within the corporation; if the corporation is asking for any personal data, then it should only do so when it is necessary for the intended purposes. This will ensure the transparency and trust factor within the corporation. The board should adhere to the data minimization principles so that data principles have clear explanations about what kind of data is being collected and that only those data are being collected that are absolutely necessary for the activity and not any other irrelevant data. The board can also adopt a privacy-by-design approach whereby they can deploy robust security control at all points of the collection of the data, usage, transfer, and storage. In this approach, the organization will ensure that since the inception of its products or services or business process, the proactive data protection approach is being catered to all the data principles or the consumers.

19.5.6 Creating a Data Privacy Framework

When developing a data privacy framework (Deloitte & Confederation of Indian Industry, 2024) within the corporation and then flagging areas requiring immediate attention, the board must establish an escalation matrix. This matrix will provide a framework for the process of escalating privacy-related issues to appropriate stakeholders within the corporation, thereby ensuring a timely resolution. The board should also ensure that the matrix is well-defined, well-communicated, and followed across the corporation. Another important aspect of creating a data privacy framework would be the formation of a risk committee on the board. The duty of this committee will be to track real-time compliance metrics. This will also focus on the misses to identify the gaps in the framework. This committee will regularly meet with the board to review the existing initiative and discuss the potential privacy risk, thereby developing further action plans to mitigate the identified gaps. If the board keeps closely monitoring these data privacy risks, then the board can ensure continuous improvement in privacy practices.

19.5.7 Privacy Compliance Responsibility for the Board

Once the DPDPA comes into force, the board member's (Deloitte & Confederation of Indian Industry, 2024) role commences with identifying who is responsible for ensuring privacy compliance within the organization. This will include the assignment of a clear job profile and responsibilities to the key personnel. This will ensure that there is accountability for the privacy compliance activities. Ideally, it is suggested for all corporations to either appoint a Data Protection Officer or a Privacy Officer who will be responsible for all the compliances related to consent management, data protection, privacy, processing, and transferring of the data within and outside the organization. The Board should also have time to time assess the corporation's compliance posture, identify the areas of non-compliance, and develop a strong action map to address these privacy gaps. This will involve conducting regular internal audits by an internal auditor, reviewing the privacy compliance framework from time to time, and also engaging external auditors to provide an independent assessment of the privacy compliance framework of the corporation.

19.5.8 Future of Privacy for the Board

As the enforcement of DPDPA comes (Deloitte & Confederation of Indian Industry, 2024) near, it raises dozens of compliance-related concerns for the corporation's continuity, network, and infrastructural security. Thus, building a robust privacy framework will require ongoing vigilance, thereby enabling a strong privacy framework within the organization. Boards have to realize now that data privacy is an essential strategic requirement to drive consumer trust within the organization and it is not merely a regulatory requirement anymore. This will drive the success of the business. Some of the key focus areas for the board should be (a) integrating risk management within the corporation, (b) having a strong oversight mechanism, (c) inculcating the privacy culture within the corporation, and (d) continuous monitoring and reporting.

Corporations should adopt risk assessment and mitigation as enterprise-wide activities. Integration of risk management will include hiring privacy professionals who are fluent in law technology and operations of the corporation. Privacy risks should be delineated across corporations with well-defined roles and responsibilities to manage DPDPA compliance. If the corporation hires external professionals to deal with matters like impact assessments, data discovery, privacy incident remediation, etc., then it can have a powerful oversight mechanism within the organization. The board should start building a privacy culture inside the corporation by regularly engaging employees and key managerial personnel in training and workshops led by the senior management and the board members of the corporation.

19.5.9 Practical Challenges that MNCs Will Face in the Implementation of DPDPA Compliances and Mitigation Recommendations

19.5.9.1 Overlap of Global Privacy Laws (CCPA, GDPR, etc.)

Since MNCs operate globally, there is a risk of overlapping laws. MNCs will face operational difficulties in data processing activities with the DPDPA while ensuring compliance with other regulations like GDPR (EU) and CCPA (California). Under the

DPDPA, data localization and cross-border data transfer requirements will certainly hamper the operations of these corporates.

MNCs can implement a unified global privacy framework where region-specific compliance checklists and modules can be prepared. They can also conduct Data Transfer Impact Assessments (DTIAs) when data are transferred from one country to another. Contractual safeguards such as Standard Contractual Clauses may be used for lawful data transfers.

19.5.9.2 Cross-Border Data Transfers and Data Localization Requirements

DPDPA mandates obtaining explicit consent from users and provides for government notification for the transfer of data across borders. However, this would disrupt the operations of MNCs, increasing the compliance cost and disrupting global data flows because these MNCs rely on centralized cloud storage or shared service centers outside India.

MNCs can adopt a hybrid cloud strategy. They can store sensitive personal data locally and use global cloud services to process non-personally sensitive data. MNCs would now have to invest in establishing localized data centers and engaging with the Data Protection Board of India to clarify the transfer mechanisms of data cross-border.

19.5.9.3 Notice and Consent Management Requirements

DPDPA mandates that privacy notices be served to obtain consent, which should be clear, specific, and granular; however, this will create obstacles in handling large-scale consumer data. MNCs operating in multiple languages and sectors may struggle with obtaining valid consent across different platforms. The DPDPA mandates that notice be served in either English or in any of the 22 languages mentioned in the VIIIth schedule of the Indian constitution.

MNCs may invest in technologies that can deploy automated consent management systems with real-time preference tracking. They can also use PETs (privacy-enhancing technologies) like consent dashboards and AI-driven adaptive notices for transparency in the context.

19.5.9.4 Allocation of Resources and Compliance Expenditure

Small and medium-sized subsidiaries of MNCs will face challenges with the resources to comply with the strict obligation of DPDPA, which includes appointing a data protection officer (DPO), conducting data protection impact assessments (DPIA), and implementing security measures and safeguards.

MNCs can implement a tiered compliance model in which all the high-risk data processing activities receive a priority compliance investment, and the low-risk functions leverage the existing frameworks. MNCs may also try to establish a regional privacy hub that will provide centralized support to all the business units in that region.

19.6 Implications and Scope of Future Work

19.6.1 Improved Corporate Governance in MNCs and Compliance

Under the DPDPA integrating data, governance will reshape the corporate strategies of the MNCs. Now, the board will need to enhance its governance structure by embedding data protection principles into its core operational and decision-making process. This will ensure better compliance, risk management, and transparency, fostering trust among the stakeholders (Deloitte & Confederation of Indian Industry, 2024), which includes customers, investors, and regulators. Now, the board has to promote a data-driven culture within the organization, which will eventually lead to improved decision-making by the directors and promoters of the MNCs. By clarifying the data responsibilities, embedding data curation as a core competency, and promoting data fluency, MNCs will enable data-driven insights to take up all levels of decision-making. This transformation will lead to a data-driven economy.

19.6.2 Operational Efficiency and Enhanced Data Security

The operational efficiency of the MNCs can now be increased (Deloitte & Confederation of Indian Industry, 2024) by the implementation of Data-as-a-Service (DaaS) models, AI-powered data management, and data virtualization techniques, which will enhance the operations of the corporation. By incorporating these practices, the firms will reduce the data redundancy, minimize the integration efforts, and thereby streamline the access to the data across all their business units. This not only ensures an agile data supply but also helps in saving the costs of the company chains within the MNCs. DPDPA vouches for a Privacy by Design Approach, strict data protection measures, continuous risk assessments, and informed consent mechanisms. Now, the boards have to be proactive in overseeing data privacy compliance, including the mitigation of associated risks with data breaches and regulatory penalties, thereby protecting the MNC's reputation and financial stability.

19.6.3 Cultural Shifts and Strategic Alignment with Business Goals

As the DPDP Act has introduced a privacy-centric governance, it will now necessitate a cultural shift within the organization. The board should now start leading by setting an example of a privacy-first mindset and embedding privacy values into corporate ethics. The board now has to conduct regular training, awareness programs, and scenario-based learning to bridge the knowledge gaps and ensure that all employees are aligned with the data protection goals within the organization. Moreover, the integration of data governance with corporate governance will ensure that data strategies are in alignment with the business objectives. This alignment will drive new business opportunities, support sustainable growth, and facilitate innovation by leveraging data as a strategic asset, thereby maintaining robust compliance standards. The evolving nature of data governance will now create an obligation upon the board to create a specialized data governance committee, appoint Data Protection officers and privacy officers, and create specialized Data Privacy and management roles within the organization.

19.6.4 DPDP Act Impact Assessment on Corporate Strategies

One of the critical areas of focus will be the future (Ciso, 2024) involving the assessment of the long-term impact of the DPDP Act on the corporate governance practices in MNCs, which will include an evaluation of how regulatory compliance shaped the decision-making of the board of directors, how effective the operational strategies were after the application of privacy principles in the organization, how the overall landscape of the business benefitted or not from the implementation of the principles of DPDPA. How corporations can balance compliance with innovation, especially in data-intensive industries can also be explored and be a part of areas of focus.

19.6.5 Exploring Machine Learning and AI in Data Governance

Another critical area of concern is the training of employees across the organization for implementing technologies that automate data discovery, classification, and quality control, ensuring accurate and consistent data handling across complex environments. Machine Learning models detect errors, while AI-driven tools monitor the access controls and detect security threats. AI also plays a crucial role in regulatory compliance by mapping data processes against legal requirements, adapting to regulatory changes, and automating policy enforcement. Predictive analytics powered by AI can forecast governance challenges and identify compliance violations, enabling proactive adjustments. However, challenges include high implementation costs and the need for transparent and explainable AI decisions.

19.7 Conclusion

In the current landscape of corporates and businesses, data have emerged as an essential component similar to oil. That is why it is often said, "Data is the new oil." The reason is that it is essential today to provide a competitive advantage to the business and help the board in strategic decision-making for MNCs. It is an intangible asset because it holds the ability to transform business models, enhance operational efficiencies, and generate new revenue streams. In fact, leading MNCs like Google, Uber, and Amazon are using data analytics to personalize the user experience of their customers and enhance their operations. However, this effective utilization of data as an asset is caught up with some challenges, too, such as siloed systems, strict access controls for the data, and, most importantly, a lack of data literacy among the people. To overcome these challenges, MNCs need to adopt key data principles like accountability, transparency, fluency in data, etc., which is also one of the essential components of corporate governance principles. Integrating the data governance inside the MNC corporate governance framework will help these MNCs maximize their data value while minimizing their risks. Effective data governance will enhance user trust and compliance framework and will help manage the risks for the organization. The enactment of the Digital Personal Data Protection Act 2023 ("DPDPA") further mandates the need for a robust data governance framework for the MNCs. It compels the MNCs to have a privacy-centric approach to their corporate governance practice. DPDPA also makes it mandatory for MNCs to have a cultural and operational shift within the organization, thereby emphasizing the role of the board in creating a privacy-first mindset to uphold consumer trust and compliance with the law. Thus, as data continues to grow in the digital economy, MNCs must keep adapting and integrating robust data governance practices and privacy frameworks into their corporate strategies.

References

Alamy. (2023). The financial consequences of inadequate data governance. (2024). *ITPro Today*. https://www.itprotoday.com/regulatory-compliance/the-financial-consequences-of-inadequate-data-governance

Anderson, M. (2023). How do different industries use data science? *SACS Meharry*. https://sacsmeharry.org/blog/how-do-industries-use-data-science/

Atomicwork. (2023). How Google uses data and people analytics to build a great workplace. *LinkedIn*. https://www.linkedin.com/pulse/how-google-uses-data-people-analytics-build-great-workplace/

Bamberger, K. A., & Mulligan, D. K. (2013). Privacy in Europe: Initial data on governance choices and corporate practices. *George Washington Law Review, 81*(5), 1529–1664. https://www.gwlr.org/wp-content/uploads/2013/09/Bamberger.pdf

British Academy and the Royal Society. (2016). Data management and use: case studies of technologies and governance. *Royal Society*. https://royalsociety.org/-/media/policy/projects/data-governance/data-governance-case-studies.pdf

Ciso, E. (2024). Company boards must navigate complexities of data privacy in DPDP era: Report. *ETCISO.in*. https://ciso.economictimes.indiatimes.com/news/grc/company-boards-must-navigate-complexities-of-data-privacy-in-dpdp-era-report/108993591

Corporate Data Management Challenges. (2024). *Casepoint*. https://www.casepoint.com/blog/corporate-data-management-challenges/

Corrado, C., Haskel, J., Iommi, M., & Jona-Lasinio, C. (2022). Measuring data as an asset. In *OECD economics department working papers*. https://doi.org/10.1787/b840fb01-en

CX Network. (2024, July 25). *Why data is the most underappreciated asset in your company*. https://www.cxnetwork.com/customer-insights/articles/why-data-is-the-most-underappreciated-asset-in-your-company

Deloitte & Confederation of Indian Industry. (2024). The DPDP act and enterprises in India: Privacy for the board. *Deloitte*. https://www2.deloitte.com/content/dam/Deloitte/in/Documents/risk/in-ra-the-dpdp-act-and-enterprises-in-india-final-19.04-noexp.pdf

Dexter, C. (2024). How does data governance fit within the framework of corporate governance? *Secoda*. https://www.secoda.co/blog/data-governance-in-corporate-governance

Hanson, G. (2021). Why data should be considered a corporate asset. *Dynamic Business*. https://dynamicbusiness.com/featured/why-data-should-be-considered-a-corporate-asset.html

Khune, A. (2024). Modernizing Uber's batch data infrastructure with Google cloud platform. *Uber Blog*. https://www.uber.com/en-IN/blog/modernizing-ubers-data-infrastructure-with-gcp/

KPMG. (n.d.). (2024). Data as an asset. *KPMG*. https://assets.kpmg.com/content/dam/kpmg/xx/pdf/2019/10/data-as-an-asset.pdf

Kuzio, J., Ahmadi, M., Kim, K. C., Migaud, M. R., Wang, Y. F., & Bullock, J. (2022). Building better global data governance. *Data & Policy, 4*.https://doi.org/10.1017/dap.2022.17

Mpeshev. (2024). The power of data: How Amazon utilizes big data to drive sales. *CEO Hangout*. https://ceohangout.com/the-power-of-data-how-amazon-utilizes-big-data-to-drive-sales/

OECD. (2023). OECD Legal Instruments. (n.d.). *Organization for economic co-operation and development*. https://legalinstruments.oecd.org/en/instruments/OECD-LEGAL-0413

Online, H. (2022). How data science can benefit your business decisions. *Harvard Online*. https://www.harvardonline.harvard.edu/blog/how-data-science-can-benefit-your-business-decisions

Pahuja, R., & Cio, E. (2022). Uber's approach towards data governance. *ETCIO.com*. https://cio.economictimes.indiatimes.com/news/big-data/ubers-approach-towards-data-governance/90632469

Reynolds, M. (2024). Why good data governance makes for good corporate governance. *Mondaq*. https://www.mondaq.com/canada/corporate-governance/1474748/why-good-data-governance-makes-for-good-corporate-governance

Safiullah, M., & Parveen, N. (2022). Big data, artificial intelligence and machine learning: A paradigm shift in election campaigns. In *The new advanced society: Artificial intelligence and industrial internet of things paradigm* (pp. 247–261). https://doi.org/10.1002/9781119884392.ch11

Safiullah, M., Pathak, P., & Singh, S. (2016). View of emergence of social media and its implications for public policy: A study of Delhi assembly election 2013. (n.d.). *XII*(1), *Management Insight Journal, School of Management Sciences,* Varanasi. https://journals.smsvaranasi.com/index.php/managementinsight/article/view/394/373

Sleiman, F. (2022). Data—The game-changing asset. (n.d.). *PwC*. https://www.pwc.com/m1/en/blog/data-the-game-changing-asset.html

Smaje, K. (2020). *How six companies are using technology and data to transform themselves*. McKinsey & Company. https://www.mckinsey.com/capabilities/mckinsey-digital/our-insights/how-six-companies-are-using-technology-and-data-to-transform-themselves

Special, E. S. (2023). Data-driven decision making: Navigating the analytics landscape product decision making. *The Economic Times*. https://economictimes.indiatimes.com/jobs/c-suite/data-driven-decision-making-navigating-the-analytics-landscape-product-decision-making/articleshow/105877287.cms?from=mdr

Talagala, N. (2022). Data as the new oil is not enough: Four principles for avoiding data fires. *Forbes*. https://www.forbes.com/sites/nishatalagala/2022/03/02/data-as-the-new-oil-is-not-enough-four-principles-for-avoiding-data-fires/

Tang, D. (2020). New WIPO data show intangible asset investment growing fast, spreading globally. *Office-russia*. https://www.wipo.int/web/office-russia/w/news/2024/new-wipo-data-show-intangible-asset-investment-growing-fast-spreading-globally

Tokhadze, A. (2023). The interdisciplinary analysis of the institutional role of data protection officer in the system of corporate governance. *Journal of Personal Data Protection Law, 2023*(1), 106–118.

Chapter 20
Corporate Custodianship and Data Sovereignty: An Indo-Indonesian Perspective

Siddharth Kanojia and Inas Afifah Zahra

Abstract In the present-day landscape, the implementation of corporate governance and data privacy is deeply interconnected. These two facets of a business are critical for regulating compliance, building the trust of stakeholders, and making sure that businesses are operated responsibly and efficiently. Accordingly, this chapter explores the delicate balance that these two of Asia's most dynamic economies, i.e., India and Indonesia, manage to maintain between the requirements for data privacy and accountability as well as transparency. The legislative frameworks and enforcement tactics that have influenced corporate governance and data privacy practices in these nations are significant factors to be taken into consideration for focus areas. The chapter will provide insight into how organizations employ these regulations, prevalent problems that emerge with them, and innovative solutions that are implemented to uphold compliance and safeguard individuals' privacy. The chapter also provides comparative viewpoints that can be used to identify common issues and viable solutions. It also includes policy recommendations that can be implemented to improve corporate custodianship towards privacy of data.

Keywords Corporate Governance · Data Governance · Privacy · Regulatory Policies · Data Regulation · Legal Landscape · Corporate Responsibility · Data Sovereignty

S. Kanojia (✉)
O. P Jindal Global University, Sonipat, India
e-mail: Siddharthkanojia.fol@gmail.com

I. A. Zahra
Politeknik Negeri Jakarta, Depok, Indonesia
e-mail: Inas.afifahzahra@akuntansi.pnj.ac.id

V. Kandpal et al. (eds.), *Developments in Corporate Governance*,
Approaches to Global Sustainability, Markets, and Governance,
https://doi.org/10.1007/978-981-96-6366-8_20

20.1 Introduction

The interplay of corporate governance with data privacy is, therefore, a strict necessity for every business in the face of unrestricted international flows of intelligence and the fast-paced development of technologies today. Proper corporate governance is essential in the sense that it forms the very basis of how a business runs and principally dictates how businesses choose to handle sensitive data. Integrity, honesty, accountability, and responsibility assume a central role in corporate governance. It promotes the development of relations with credible stakeholders. Corporate governance can no longer afford to remain divorced from data protection processes. The information can offer ways in which companies can be more innovative, effective, and faster. The need is felt for integrating corporate governance with stringent data privacy. Concerning the regulatory frameworks of Corporate Governance, India is particularly concerned about corporate governance and is adept at managing investor confidence and promoting sustainable economic growth (Chakrabarti & Megginson, 2009). India has a comprehensive legal system alongside the rapidly expanding economy. The Companies Act 2017 is a new Act that makes a more improved concept that can be employed in improving investor protection, accountability, and transparency. Some of the main requirements include compulsory board structures, audit committee requirements, and liable counselors. At this junction, it is worth noting that India's corporate governance goes far beyond the legal framework of its two regulatory bodies (Gatti et al., 2019). These two bodies are the Reserve Bank of India (RBI) together with the Securities and Exchange Board of India (SEBI). The agglomeration of RBI is the non-banking sector while that of SEBI is listed companies. Some of the specific objectives of these regulators include ensuring that there is transparent disclosure, public board duty, and a compensation package that also focuses on risk- and quality-based incentives for management. This might permit the business to run under morally correct business principles and thus help overcome flaws such as fraud and bribery involved in any corporate schemes. The fact that the Ministry of Corporate Affairs (MCA) ultimately accepted an initiative of self-regulation also makes it rather clear that India is willing to do more on governance standards than what the law requires. Thus, by developing an enabling environment within which businesses can operate fairly, with optimum risk management strategies, and a well-diversified and representative board membership, these practices show that the principles are doing something right in putting sustainable business principles into practice. On the contrary, Indonesia has one of the most robust corporate governance frameworks, restyling its economy into a stable haven for investors. Corporate governance standards actually get their roots from Indonesian Companies Law itself, especially Law No. 40/2007 on Limited Liability Companies, which lays down important provisions related to shareholder rights, board duties, and transparency obligations. Like India, Indonesia is also subject to supervision during corporate governance by any agency like the Financial Services Authority (Otoritas Jasa Keuangan, OJK), popularly called OJK, and the Indonesia Stock Exchange, or IDX. These agencies are responsible for implementing governance laws on the part of companies listed on their main stock exchange, making sure they adhere to rules related to fair corporate behavior, board independence, and periodic reporting. Indonesia has currently been working on several initiatives to improve corporate governance behavior, amongst others, the establishment of the Good Corporate Governance (GCG) mechanism (Basri et al., 2017). It highlights postulations like the transparency of information, accountability of management, and equitability of the decision-making process solely to establish it as the most dependable source of irritants and, in due course of time, one of the tools fostering economic growth. It is also evident that the interlinkages between corporate governance in both countries are further nailed by the increasing interdependence of markets across the globe and the emergence of Multinational Companies (MNCs) that often straddle both these countries.

As observed by Solomon, 2020, corporate governance should now be holistic rather than just a purely financial affair, and confidentiality protection should be embraced with other objectives relevant to an organization. The implication of governance-privacy interplay becomes crucial in light of continuous scrutiny from the agencies of regulatory bodies on the rise. For example, the General Data Protection Regulation (GDPR) in Europe has become the benchmark of stricter data privacy rules globally, subsequently informing corporate governance practices across various sectors (Voigt & von dem Bussche, 2017). Existing literature suggests that those companies that incorporate privacy measures into their governance structures are better placed to deal with compliance risks, data breaches, or reputational damage, therefore entrenching greater investor confidence (Brown & Marsden, 2023). Moreover, the world is rapidly adopting sustainability and socially responsible business operations, which call for a holistic approach to corporate governance, wherein data privacy needs to form the centerpiece of it. In this respect, The Organization for Economic Cooperation and Development (OECD) claimed that in order to establish and maintain stakeholder trust, many elements, including aspects such as transparency, fairness, and ethical considerations, have now become essential to corporate actions, especially in light of increasing stakes about corporate misuse of personal data. The chapter, therefore, identifies the exact regulatory developments within these two of the most voluminous Asian economies, Indonesia and India, to relate changes in the laws on data privacy with the paradigm of corporate governance. The chapter will be a useful resource for the global audience in the sense that it will inform about and update trends in regulatory developments in two of the leading emerging economies in Asia with important corporate governance reforms. Beyond highlighting the best practices about the governance and privacy framework for both countries, the chapters contain lessons applicable to other jurisdictions. They could help companies negotiate such complex environments while optimizing their operational effectiveness. These findings from the chapter will be of special relevance to MNCs cutting across borders as guidelines for how best to design governance structures that meet international standards and a framework for compliance with local privacy rules. In integrating data privacy into corporate governance, the chapter gives companies a roadmap for the innovation and sustainability of their competitive advantage in the global marketplace under the emergent laws of data protection.

20.2 Data Privacy and Corporate Governance Framework

Data privacy concerns the rights and responsibilities individuals have with regard to the gathering, utilizing, and disclosing of personal data. The need to safeguard personal data from potential breaches, unauthorized access, and misuse is growing in a world where businesses are continuously using data-driven strategies to boost profitability, optimize performance, and improve customer experience (Kanojia, 2024). Organizations are legally obliged to protect individuals' privacy rights under data protection laws, such as the General Data Protection Regulation (GDPR) in Europe, the California Consumer Privacy Act (CCPA) in California, the Personal Data Protection Act (PDPA) in Singapore and India's Digital Personal Data Protection Act (DPDPA) 2023. These regulations impose strict requirements on organizations for fetching express consent before collecting data, putting security measures in place, promptly reporting data breaches, and enabling data subjects to access, edit, or remove their personal information. Further, there is a need to prioritize really strong data governance and compliance initiatives because non-compliance with data privacy laws not only carries significant financial penalties but also demonstrates evidence of legal liabilities and reputational damages (Lancieri, 2022).

Recognizing that privacy protection involves ethical imperatives and legal commitments, dealing effectively with corporate privacy is by no means a simple matter. In this framework, corporate privacy protection is complicated by the variety and complexity of the global legal environment, leading to different levels of regulatory requirements and applying different scopes, requirements, and enforcement mechanisms (Bennett & Raab, 2017). It is also observed that multinational corporations need to navigate border concepts by trying to reconcile different regulatory requirements and still keep up with global data protection standards (Hoofnagle et al., 2019). Above all, it is the rapidly changing technological environment, especially with the latest areas of innovation, Artificial Intelligence (AI), big data analytics, and the Internet of Things (IoT), that new issues and exposures in the governance of data privacy are brought out (Sollins, 2019). A further drive to innovation requires the adoption by corporations of privacy-by-design, carrying out a privacy impact assessment, and undertaking robust technical safeguards against the risks of breach and unauthorized access. Furthermore, as cyber threats become more sophisticated, organizational inertia, cultural attitudes towards privacy, resource limitations, and resource constraints themselves become more significant obstacles to providing comprehensive protection of data privacy (Mikalef et al., 2021). To help overcome these obstacles, organizations should make investments in cybersecurity measures, teach staff members best practices for protecting information resources, and cultivate a culture of compliance and privacy awareness within their walls (McIlwraith, 2021).

In this regard, corporate governance may offer the structure that enables an organization to maintain the required ethical perspective while addressing the interests of all stakeholders, be transparent in its operations, and hold accounting (Solomon, 2020). All the while making sure that the processing of the personal data on the data subjects is being done. A culture of compliance with data privacy laws and regulations is stimulated by well-established corporate governance mechanisms, which lowers all potential risks of a data breach and increases organizational resilience against changing cyber threats (Schinagl & Shahim, 2020). Nonetheless, there are few obstacles to creating corporate governance structures that guarantee strong data privacy protection. One of the most significant challenges will be reconciling data privacy regulations with business goals, more so in highly data-driven or data-centric industries where operations and continuous innovation rely on constant feeds of data to feed the pipelines of operations and innovation alike. Appropriate corporate governance should support privacy and balance value in data through thoughtfully designed policies, moral guides, and oversight mechanisms that reassure stakeholders (Scheibner et al., 2020). Meanwhile, the global transfers of data also create challenges in adhering to the various statutes related to data privacy that have been enacted by different jurisdictions. Among other factors, MNCs that operate in different countries move within complex regulatory frameworks and cultural differences but operate at a consistent level of data privacy (Ebert, 2021). This can be attained through harmonization of governance principles, collaborative functions, and technology use in promoting accountability and compliance. In this regard, the succeeding part of the chapter will discuss the efficiency of the regulatory framework of Corporate Governance in India and Indonesia to address the mandate of Corporations to maintain the secrecy of data.

20.2.1 India

The convergence of data privacy and corporate governance has lately attracted much attention in the Indian subcontinent. Here, businesses depend heavily on data-driven strategies to drive growth and efficiency. The regulatory structure that oversees corporate governance in India is essential to guaranteeing that businesses respect moral principles and abide by strict data privacy laws. The Information Technology (IT) Act, 2000, which was amended in 2008 to include new provisions pertaining to the protection of personal data and information, serves as the primary legislative foundation for data privacy laws in India (Burman, 2022). Nevertheless, data privacy issues were not fully addressed by the IT Act. Thus, the Digital Personal Data Protection Act (DPDPA), which was implemented in 2023, marked a significant change in the landscape. In line with global norms like the General Data Protection Regulation (GDPR) of the European Union, the DPDP seeks to create a strong framework for the protection of personal data (Sundara & Narendran, 2023).

As a result of the Supreme Court's historic rulings establishing the right to privacy as a fundamental right guaranteed by the Indian Constitution, this right should currently be protected by legislative frameworks that both ensure compliance with international norms and take into account the particular opportunities and challenges posed by India's diverse population and quickly evolving technological landscape (Khosla, 2017). The first and perhaps one of the most convincing arguments why India needs to set high standards of data protection is to protect the rights and freedoms of the people. Based on the availability and increased use of digital platforms and services, large amounts of personal data are produced, transmitted, and archived on a daily basis. When there is no proper protection in place, such data may be abused or violated in some way, thus violating the privacy of individuals. This demand would mean that there would be a clear understanding of how personal data should be collected, processed, and used, thus enabling individuals to have more control over their data. In addition, the protection of data privacy is important in building confidence in the digital economy hence the need for strong legal protection mechanisms. To a certain extent, data are a global commodity in a world where information crosses borders; consumers and corporations require certainty of the appropriate handling of their data (Tene & Polonetsky, 2012). Thus, the adoption of laws that would comply with international standards would positively affect the Indian market's image as a reliable place for investments in digital technologies, as well as developing domestic and international businesses to boost the country's economy.

Over the last few years, many Indian corporations have gone out of their way to increase the privacy and security of data, which shows increased consciousness and responsibility towards the customers' information. A good example is banking services, where institutions such as the HDFC (Housing Development Finance Corporation) Bank and the ICICI (Industrial Credit and Investment Corporation of India) Bank have set up stringent measures regarding the privacy of customer information. These banks have put into practice measures such as encryption and the use of passwords, tokens, etc., to ensure the safety of Internet transactions and customer information. Further, they have implemented state-of-the-art cybersecurity tools, and they perform vulnerability tests to determine the insecure areas in the company's software systems (Yelikar, 2019). Another industry that can be noted as demonstrating high levels of activity regarding data privacy is telecommunications. Currently, Indian companies like Reliance Jio and Bharti Airtel have developed privacy policies and measures that would ensure that subscriber's information is safe (Dhar, 2017). These are to ensure that the individual has consented to the collection and processing of their information and to ensure that the information is protected from unauthorized users. Moreover, corporations in India are gradually implementing international standards in the protection of data. The majority of IT companies, such as Infosys and TCS (Tata Consultancy Services), have also set up their independent privacy offices and recruited Chief Privacy Officers to ensure that they comply with national and international laws on the privacy of data (Attili et al., 2018). To protect privacy, these firms conduct periodic privacy impact assessments and also offer training to the employees regarding the firm's standard practices concerning data handling. Besides top management and organization programs, industry associations like NASSCOM (National Association of Software and Service Companies) have also significantly contributed to raising awareness of data protection in Indian organizations (Godse, 2010). They have created policies and procedures that companies can use to follow the standard measures of data protection and cyber security.

On the other hand, there are some examples of failure on the part of Indian corporations to protect personal data, which have raised serious questions on the issue of data protection and the necessity for organizations to begin implementing better security measures. For instance, in 2020, India's biggest online grocery delivery service, BigBasket, was hacked, which put more than 20 million of its users' data at risk. These include the names, email addresses, and phone numbers, as well as hashed passwords to some of the users' accounts (Mukhopadhyay, 2022). This event also exposed the lack of security in BigBasket's technological framework in addition to raising the threat implications for consumers when their information is not protected. Likewise, in the same year, Domino's India faced a data breach through which the personal details of more than 180 million orders were leaked (Dewan, 2022). Among those were information that should never have been exposed to the public, such as customers' names, phone numbers, and addresses, which were found on the dark web for sale. These breaches prove a need to adopt privacy as a top priority in corporate processes, at least for those that are illustrated here. In particular, data protection is critical not only in legal terms but also in terms of customer trust. Therefore, if personal data are not protected sufficiently enough, it can lead to damage, fines, and other losses, including the potential for legal action to be taken against the company (Voss & Bouthinon-Dumas, 2021). Furthermore, the ability to guarantee excellent data protection measures is gradually becoming one of the sources of competitive advantage in the modern world where communication and data sharing are tight, and the customer is becoming more sensitive to how their information is treated.

20.2.2 Indonesia

The Indonesian case of data privacy and corporate governance issues are well-coordinated because this is where the legal, ethical, and business aspects meet. Since the digital economy is growing rapidly across the largest economy in Southeast Asia, the need for bona fide personal data protection has become more evident. At the same time, Indonesian corporations are operating in an environment where commitment to strict standards of governance improves business processes' disclosure and cultivates trust among investors (Rustam & Narsa, 2021). This part of the chapter discusses the extent to which Indonesian business organization's data privacy and corporate governance issues have become intertwined to influence business operations, legal requirements, and social norms.

The structure of corporate governance is primarily guided by the Indonesia Corporate Governance Manual (ICGM) and the Financial Services Authority (OJK) regulations (Helmi & Mulyany, 2020). As the ICGM outlines, transparency, accountability, responsibility, independence, and fairness are the cornerstones of corporate operations (OECD, 2015). The OJK, established by Law No. 21/2011, is responsible for supervising and regulating financial services, ensuring that companies adhere to the principles of good governance (OJK, 2019). Moreover, the Indonesian data privacy framework is undergoing a significant evolution, with the primary legislation being the Electronic Information and Transactions Law (EIT Law) and the Government

Regulation No. 71/2019 on the Implementation of Electronic Systems and Transactions (GR 71/2019). These legislative instruments mandate the protection of personal data, the implementation of security measures, and the rights of data subjects (Lestari, 2020).

Indonesia has recently become concerned with the question of data privacy because of the increasing trend of Internet and mobility. Due to the active use of digital services, e-commerce platforms, and social networks, personal data are accumulated, processed, and shared massively (Prastyanti & Sharma, 2024). However, threats such as data spillage, identity theft, and unauthorized use of the information have created awareness of the need to protect data privacy. In this regard, Indonesia passed the Personal Data Protection Bill (RUU PDP) in 2020 to govern the data processing activities and protect the personal data of individuals as per global standards (Riyadi, 2021). This legislative attempt signals Indonesia's willingness to conform to international data protection standards and to adapt to the complexity of its population and digital economy. Whereas Indonesian corporate governance structures are still in the process of being developed, integrating the principles of accountability, transparency, and ethical behavior in business is an emerging concept associated with the concept of sustainable and responsible business (Tjahjadi et al., 2021). Good governance also reduces risks and maintains legal compliance but also strengthens the organizations' ability to adapt to future conditions and build sustainable value. There is a growing trend of Indonesian corporations to internalize governance practices that enhance the integrity, fairness, and accountability of decisions made in the corporation, hence boosting investor confidence and stakeholder confidence.

Indonesia had initiated broad data privacy legislation as far back as 2008 when the basic law titled the Electronic Information and Transactions Law came into being (Walters et al., 2019). This was a pathbreaker to set norms for e-commerce or electronic transactions and other related issues, including personal data protection and cybersecurity/privacy in cyberspace. Through all these years, UU ITE has gone through a number of revisions to cope with the fast-changing digital world and new challenges to data protection (Setyardi, 2023). The need for a specific and comprehensive law on data protection became more patent with the growth in Indonesia's digital economy, against the foundational role played by UU ITE. Influenced by global trends and best practices in data privacy, Indonesia then began drafting the Personal Data Protection Bill. The RUU PDP's objective is to provide clear guidelines and principles on the collection, processing, and storage of personal data; these standards must be in line with international law, specifically the GDPR of the European Union (Simbolon & Juwono, 2022). At the core of the RUU PDP principles lies respect for the rights of natural persons over their personal data: the free and prior informed consent principle obtained before collection, the principle of processing legality and fairness, and all the more, putting rigorous obligations on controllers and processors to take every measure possible to guarantee the security and confidentiality of personal data.

Notwithstanding these improvements, Indonesia has a number of challenges in the implementation of its data privacy laws. One critical challenge is building the level of awareness and capacity of businesses, government agencies, and the general public to perform their duties and exercise their rights under the RUU PDP (Prawiraharjo et al., 2023). It would involve education for the different stakeholders on data protection, good practices for handling data, and robust mechanisms for reporting data breaches and acting upon them. Furthermore, compliance enforcement and mitigation of new threats will call for adequate resourcing and capabilities within the regulatory bodies. Another feature of Indonesia's process of strengthening its legislative and regulatory framework on data protection is how it aligns its law with international requirements. This will involve supporting cooperation with international counterparts by engaging in forums on data protection and regulation in a manner that allows cross-border data flows with appropriate safeguards for personal data. Indonesia can thus continue to strengthen consumer trust in Indonesia, encourage the growth of its digital economy, and establish itself as a responsible actor in the global landscape of data governance and privacy protection.

In recent years, Indonesian corporations have been more conscious of the need to protect personal data and have been proactive in efforts towards better practices in data privacy. As an illustration, PT Telekomunikasi Indonesia Tbk, better known as Telkom, is the largest telecommunications and network service provider in Indonesia. It has instituted rigid measures for data protection to ensure information security and customer information confidentiality (Rudito, 2010). This is achieved by robust encryption protocols, secure data storage systems, and frequent audits to enhance their cybersecurity. Enhanced data privacy at Telkom ranks this firm among the few hundred companies eyeing compliance with regulatory provisions in a bid to find its way to millions of customers' hearts who use its services daily. Another example might include Gojek, a leading technology company providing ride-hailing services, food delivery, and digital payment across Southeast Asia. Gojek has institution-wide policies and practices towards data protection for its users' personal information. Incorporated measures include advanced cryptographic methods, periodic security testing processes, and training programs on the best practice of data privacy for employees. Gojek regards the requirements as the minimum threshold for how sure and transparent their dealings are in treading the fast digital transformation pace in an era dominated by burgeoning cyber threats (Dan & Dewi, 2023). Besides that, Indonesian financial institutions have progressed in terms of data privacy. Bank Mandiri, one of Indonesia's biggest banks, has been trying to provide tight security in storing customers' financial data. Here, multi-factor authentication in online transactions and monitoring against suspicious activity are assured with regulations about personal data processing and storage protection (Jameaba, 2024). Bank Mandiri's integration of robust cybersecurity practices assures customers of the commitment to the confidentiality of private information while maintaining the integrity of financial transactions in this very vibrant Indonesian banking sector. These are examples of how Indonesian corporations are more proactive in entrenching data privacy as an aspect of their corporate governance. It would then be possible to enhance binding trust and credibility between the companies and their customers by improving their cybersecurity posture, investing in cutting-edge technologies, implementing comprehensive policies, and establishing a culture of privacy awareness.

However, corporations in Indonesia have been few and far between in breaching personal data, underlining the critical need to maintain data privacy and put secure, robust security measures in place. Among larger cases was one in 2021 when Tokopedia, one of Indonesia's largest e-commerce platforms, was hit by what turned out to be a colossal data breach affecting millions of its users. The leak showed personal details, including name, email addresses, phone numbers, and hashed passwords, which indicated gaps in Tokopedia's cyber defenses (Almeyda et al., 2024). The breach not only violated the privacy of millions of customers but also threw up questions regarding the preparedness of Indonesian digital platforms to protect the data of their users (Susanto et al., 2021). Likewise, in 2020, Bhinneka, one of the major online electronic sellers in Indonesia, faced a data breach through which confidential customer data, like its name, full address, and payment particulars, got compromised (Prastyanti & Sharma, 2024). It demonstrated potential threats arising from weak cyber hygiene, such as the consequences that could be faced by both individuals and corporations in the event of a data breach. This has further triggered calls for more regulatory control and sensitization among corporations to invest in robust measures of safeguarding data. These breaches underline the immediate call for Indonesian corporations to take privacy seriously and have comprehensive means to protect personal data. These technologies, coupled with strict policies about data protection and regular security audits, can go a long way towards making Indonesian companies more resilient against cyber threats and showcasing commitment to ethical habits of data handling. These abovementioned instances point to the role played by regulatory frameworks in setting clear guidelines on data privacy and making corporations liable in the case of breaches, such as Indonesia's developing Personal Data Protection Bill, RUU PDP. It foresees the establishment of principles of lawfulness regarding processing, enhancing the position of data subjects vis-à-vis control over their personal data, and imposes sanctions for non-compliance (De Hert, 2021).

Hence, it can be further argued that data privacy and corporate governance in Indonesia are on a convergence course that will bring about transformative change in all sectors. Organizations are steadily realizing that protecting personal data is not merely about adhering to legal requirements but is more of a strategic interest in maintaining competitive advantage and gaining consumer confidence. Developing robust measures to protect the concerned data through encryption, access controls, and data minimization lowers the threats associated with data breaches. It gives companies the perception that they respect the privacy rights of individuals. Moreover, building privacy considerations into these corporate governance frameworks will ensure that their handling practices are in compliance with ethical standards and regulatory requirements in order to improve the resilience and sustainability of organizations. In that respect, data privacy and corporate governance are coming together in Indonesia as an important trajectory of regulatory and industry standard-setting (Denis, 2021). Government agencies, industry associations, and civil society organizations now collaborate on the development of guidelines, codes of conduct, and certification schemes in ways that diffuse good practice with respect to data protection and governance. These can really help businesses operate more clearly and give clear guidance in establishing a responsible culture as regards data stewardship and ethical corporate conduct.

20.3 Challenges and Redressal in Balancing Governance and Privacy

Organizations in Indonesia must address several challenges in balancing corporate governance and data privacy. The first challenge that exists is regulatory complexity. Navigating through the web of complex data privacy regulations and corporate governance, of course, is not very easy. There is no doubt that the existence of local and international laws adds to this complexity since they lack harmonization with each other (Taylor, 2019). The second challenge is a lack of resources. Implementing complete data privacy measures requires financial and human resources, which are insufficient measures in many organizations. These demands cannot be met by Small and Medium-Sized Enterprises (SMEs) (Peters, 2021). Apart from that, the fast pace of technological change silently surpasses the development of regulatory frameworks, in which it is difficult for an organization to keep up with compliance (Kshetri, 2020). Organizations need to continue updating their governance and privacy-related practices to remain compliant with the continuous emergence of new technologies. It is pretty resource-intensive and needs sharp vigilance.

Despite these challenges, some organizations in Indonesia have been able to balance between corporate governance and data privacy. For example, one of the e-commerce companies had a compliance program that was integrated to address issues of streamlining efforts and reducing redundancies through comprehensive data protection policies. The company also provided training on data privacy to its staff frequently, which has kept the company compliant and trusted by customers (Smith & Dinev, 2018). Innovative approaches, such as adhering to privacy by design and harnessing newer technologies like blockchain for secure data storage, have been equally efficient. Some of the tools that go into making a vehement data security system are encryption, access controls, and monitoring systems (Bélanger & Crossler, 2019). These would not only help in matters of compliance issues but also add to the overall governance framework of organizations. Harmonization or convergence of local and international laws on data privacy can simplify compliance for multinationals. It is in such cooperation between the regulators that cohesive regimes can be built and result in more practical and effective frameworks.

An overview of India's approach may be an interesting learning curve. Broadly, the two nations stand comparable in the prevailing challenge of reconciliation between corporate governance and data privacy because both are on a rapid growth path of their digital landscapes that warrants a robust regulatory framework for safeguarding personal data by bringing accountability at the corporate level (Basu, 2020). Regulatory frameworks and strategies differ in both. Underpinning the landscape of data privacy in India are the Information Technology Act of 2000 and the Personal Data

Protection Bill of 2019. The second is much more far-reaching in its reach and redolent to the General Data Protection Regulation compared to Indonesia's Electronic Information and Transaction Law (Srinivasan, 2020). India represents a tight regulatory environment, whereas Indonesia's regulations have been at a developing stage (Pathak, 2020).

20.4 Policy Recommendation

Some policy recommendations, following this study's basic structure, could be made to improve corporate governance and data privacy in India and Indonesia. First, strengthen data privacy regulations. Indonesia should continue to develop its data privacy laws to align with global standards like the GDPR. It will enhance data protection and build international trust (Chander, 2020). Second, enhance corporate governance frameworks. Incorporating data privacy into corporate governance codes and guidelines will ensure a holistic approach to organizational management (Solove & Schwartz, 2020). Third, promote regulatory collaboration. Encouraging collaboration between regulatory bodies within Indonesia and international counterparts can lead to more consistent and effective data privacy enforcement and corporate governance standards (Hoffman, 2021). Fourth, providing resources and support to SMEs will help them implement necessary data privacy measures without undue financial burden. Government grants and training programs can be beneficial (Chander, 2020).

Furthermore, the policymakers must align national frameworks with international standards. This would ensure that India and Indonesia integrate their data governance policies with global best practices, such as the General Data Protection Regulation (GDPR), the OECD guidelines, and the Asia–Pacific Economic Cooperation (APEC) Cross-Border Privacy Rules. Such integration would enhance data security and regulatory compatibility (Hoofnagle et al., 2019). Furthermore, the implementation of more stringent policies mandating the establishment of Chief Data Protection Officers (CDPOs) and data security committees within businesses will facilitate enhanced oversight of compliance and risk management (Bélanger & Crossler, 2019). Secondly, corporations should adopt AI-driven compliance tools to implement artificial intelligence (AI) and machine learning solutions for automated risk assessment, anomaly detection, and compliance monitoring, thus improving data protection mechanisms (McIlwraith, 2021). Organizations must also conduct periodic Privacy Impact Assessments (PIAs) to identify and mitigate risks associated with data processing activities (Mikalef et al., 2021). Furthermore, companies should develop cross-border data governance strategies to ensure compliance with extraterritorial regulations such as the EU GDPR and the California Consumer Privacy Act (CCPA) (Ebert, 2021). Lastly, regulators must implement standardized reporting requirements. Such bodies as SEBI, RBI, OJK, and IDX must introduce corporate data governance reporting obligations to improve oversight and accountability (OECD, 2015). Additionally, stricter accountability measures must be enforced. These measures include sector-specific penalties for noncompliance, financial sanctions, reputational penalties, and enhanced monitoring of corporate data custodianship (Solomon, 2020).

20.5 Conclusion

The interplay between corporate governance and data privacy in the digital age poses both potential risks and opportunities for any organization operating in any country in Asia. This chapter has attempted to underscore the critical need for balance between transparency, accountability, and protection of personal data. It is evident, by taking the case of the legislative framework and enforcement strategies in two dynamic Asian economies, how an organization handles these complexities to balance the maintenance of regulatory compliance and stakeholder trust with the assurance of safeguards against the diminution of privacy rights. Such common challenges have been underlined as regulatory ambiguity, that the pace of technological development has outpaced that of the related legal framework, and the global nature of data flows. At the same time, innovative approaches have been taken by businesses to solve these issues from a comparative perspective. The chapter thus can be instrumental in teasing out effective practices and associated policy recommendations for strengthening data privacy protections and corporate governance standards. Accordingly, such steadfast dialogue among policymakers, the business community, and civil society would be necessary for adapting regulatory frameworks and enforcement mechanisms to address the fast-changing landscape in areas as divergent as corporate governance and data privacy. By drawing on each other's experiences and thereby being enabled to adopt forward-looking policy guidelines, the Asian economies will have created an enabling environment in which innovation will thrive alongside stringent safeguards for personal data protection and responsible corporate practice.

References

Almeyda, I. T., & Prasetyawati, E. (2024). Consumer Protection for The Hacking of Personal Data of Tokopedia Marketplace Users. *Journal Evidence of Law, 3*(2), 93–106.

Attili, V. S., Mathew, S. K., & Sugumaran, V. (2018). Understanding information privacy assimilation in IT organizations using multi-site case studies. *Communications of the Association for Information Systems, 42*(1), 4.

Basri, H., Ulfah, A. K., & Majid, M. (2017). The implementation of good corporate governance (GCG) to improve service quality: the case of state-owned electricity company in Indonesia.

Basu, S. (2020). Data Protection in India: The New Personal Data Protection Bill. *Journal of Data Privacy and Security, 12*(3), 45 59.

Bélanger, F., & Crossler, R. E. (2019). Privacy in the Digital Age: A Review of Information Privacy Research in Information Systems. *MIS Quarterly, 43*(1), 123–168.

Chambers, C. (2020). Supporting SMEs in Data Privacy Compliance. *International Journal of Small Business Management, 28*(4), 297–313.

Bennett, C. J., & Raab, C. D. (2017). *The governance of privacy: Policy instruments in global perspective*. Routledge.
Brown, I., & Marsden, C. T. (2023). *Regulating code: Good governance and better regulation in the information age*. MIT Press.
Burman, A. (2022). *Will India's Proposed Data Protection Law Protect Privacy and Promote Growth?* Carnegie Endowment for International Peace.
Chakrabarti, R., & Megginson, W. I. (2009). Corporate governance in India. In *Global Corporate Governance* (pp. 151–176). Columbia University Press.
Chander, A. (2020). How Law Made Silicon Valley. *Harvard International Law Journal, 61*(2), 273–327.
DAN, S. S., & Dewi, E. F. (2023). *The Future of Digital Finance in Southeast Asia: A Legal Perspective*. Prenada Media.
De Hert, P. (2021). EU sanctioning powers and data protection: New tools for ensuring the effectiveness of the GDPR in the spirit of cooperative federalism. In *EU Law Enforcement* (pp. 291–324). Routledge.
Denis, E. (2021). The promises and pitfalls of SupTech for corporate governance-related enforcement.
Dewan, Y. (2022). An Analysis of Digital Privacy Laws. *Issue 6 Indian JL & Legal Rsch., 4*, 1.
Dhar, S. (2017). Digitalisation & Data Protection-Changing Landscape of Indian Telecom Sector. *Int'l. in-House Counsel J., 11*, 1.
Ebert, I. L. (2021). The tech company dilemma: navigating government requests for user data.
Gatti, L., Vishwanath, B., Seele, P., & Cottier, B. (2019). Are we moving beyond voluntary CSR? Exploring theoretical and managerial implications of mandatory CSR resulting from the new Indian Companies Act. *Journal of Business Ethics, 160*, 961–972.
Godse, V. (2010, January). Building an ecosystem for cyber security and data protection in India. In *International Conference on Ethics and Policy of Biometrics* (pp. 138–145). Berlin, Heidelberg: Springer Berlin Heidelberg.
Greenleaf, G. (2018). Global Data Privacy Laws 2018: 132 National Laws and Many Bills. *Privacy Laws & Business International Report, 157*, 14–18.
Helmi, I. D., & Mulyany, R. (2020). Does Islamic Corporate Governance Contribute to the Performance of Islamic Banks? Evidence from Indonesia and Malaysia. In *Proceedings Aceh Global Conference-Business, Economics, and Sustainable Development Trends* (Vol. 2, pp. 129–143).
Hoffman, D. A. (2021). The Promises and Perils of Regulatory Collaboration. *Regulation & Governance, 15*(1), 103–123.
Hoofnagle, C. J., Van Der Sloot, B., & Borgesius, F. Z. (2019). The European Union general data protection regulation: What it is and what it means. *Information & Communications Technology Law, 28*(1), 65–98.
Jameaba, M. S. (2024). Digitalization, emerging technologies, and financial stability: challenges and opportunities for the Indonesian banking sector and beyond. *Emerging Technologies, and Financial Stability: Challenges and Opportunities for the Indonesian Banking Sector and Beyond (April 26, 2024)*.
Kanojia, S. (2024). Digitalization in Corporations: Integrating Utility of Digital Technology With Accessibility and Privacy of Data. In *Balancing Human Rights, Social Responsibility, and Digital Ethics* (pp. 227–245). IGI Global.
Khosla, S. (2017). The Right to Privacy: Constitutional Perspective in India. *Innovative Research Thoughts, 3*(2), 121–127.
Kshetri, N. (2020). The Economics of Internet of Things (IoT) Security. *Journal of Cybersecurity, 6*(1), 1–12.
Lancieri, F. (2022). Narrowing data protection's enforcement gap. *Me. l. Rev., 74*, 15.
Lestari, Y. (2020). The Evolution of Data Privacy Laws in Indonesia. *Asian Journal of Law and Society, 7*(2), 245–267.
McIlwraith, A. (2021). *Information security and employee behaviour: How to reduce risk through employee education, training and awareness*. Routledge.

Mikalef, P., van de Wetering, R., & Krogstie, J. (2021). Building dynamic capabilities by leveraging big data analytics: The role of organizational inertia. *Information & Management, 58*(6), Article 103412.

Mukhopadhyay, I. (2022). Cyber threats landscape overview under the new normal. In *ICT Analysis and Applications* (pp. 729–736). Springer Singapore.

Organization for Economic Co-operation and Development (OECD). (2015). *Indonesia Corporate Governance Manual*. OECD Publishing.

Pathak, P. (2020). Data Privacy Laws in India: Challenges and Prospects. *Indian Journal of Law and Technology, 16*(2), 98–120.

Peters, R. (2021). Data Privacy Challenges for SMEs in the Digital Economy. *Journal of Small Business Strategy, 31*(1), 83–101.

Prastyanti, R. A., & Sharma, R. (2024). Establishing Consumer Trust Through Data Protection Law as a Competitive Advantage in Indonesia and India. *Journal of Human Rights, Culture and Legal System, 4*(2), 354–390.

Prawiraharjo, B. S., Priyono, F. X., & Trihastuti, N. (2023, January). The Jurisprudence Regarding the Protection of Personal Data for the Communities and Business Actors in Indonesia. In *Proceedings of the 1st International Workshop on Law, Economics and Governance, IWLEG 2022, 27 July 2022, Semarang, Indonesia.*

Riyadi, G. (2021). *Data Privacy in the Indonesian Personal Data Protection Legislation* (No. 7). Policy Brief.

Rudito, P. (2010). An investigation into the role of moderating variables on mobile broadband technology acceptance in Indonesia. *RMIT University.*

Rustam, A. R., & Narsa, I. M. (2021). Good corporate governance: A case study of family business in Indonesia. *The Journal of Asian Finance, Economics and Business, 8*(5), 69–79.

Sarma, G. D., Choudhury, S., Bharadwaj, P., & Sarma, M. (2024). Navigating Corporate Governance and Ethics: The Cornerstones Of Sustainable Business Practices. *Educational Administration: Theory and Practice, 30*(5), 5442–5454.

Schinagl, S., & Shahim, A. (2020). What do we know about information security governance? "From the basement to the boardroom": Towards digital security governance. *Information & Computer Security, 28*(2), 261–292.

Scheibner, J., Ienca, M., Kechagia, S., Troncoso-Pastoriza, J. R., Raisaro, J. L., Hubaux, J. P., … & Vayena, E. (2020). Data protection and ethics requirements for multisite research with health data: a comparative examination of legislative governance frameworks and the role of data protection technologies. *Journal of Law and the Biosciences, 7*(1), lsaa010.

Setyardi, H. U. (2023). Indonesian Legal Readiness: Regulation of Right to be Forgotten in Relation to Big Data. *Jurnal Kewarganegaraan, 7*(1), 1068–1079.

Simbolon, V. A., & Juwono, V. (2022). Comparative Review of Personal Data Protection Policy in Indonesia and The European Union General Data Protection Regulation. *Publik (Jurnal Ilmu Administrasi), 11*(2), 178–190.

Smith, H. J., & Dinev, T. (2018). Information Privacy Research: An Interdisciplinary Review. *MIS Quarterly, 42*(4), 889–910.

Sollins, K. R. (2019). IoT big data security and privacy versus innovation. *IEEE Internet of Things Journal, 6*(2), 1628–1635.

Solomon, J. (2020). *Corporate governance and accountability*. John Wiley & Sons.

Solove, D. J., & Schwartz, P. M. (2020). *Information Privacy Law*. Aspen Publishers.

Srinivasan, S. (2020). Comparative Analysis of Data Privacy Laws: India and GDPR. *International Data Privacy Law, 10*(3), 245–257.

Sundara, K., & Narendran, N. (2023). Protecting Digital Personal Data in India in 2023: Is the lite approach, the right approach? *Computer Law Review International, 24*(1), 9–16.

Susanto, H., Yic, L. F., Setiana, D., Asih, Y., Yoganingrum, A., Riyanto, S., & Saputra, F. A. (2021). Digital ecosystem security issues for organizations and governments: Digital ethics and privacy. In *Web 2.0 and cloud technologies for implementing connected government* (pp. 204–228). IGI Global.

Taylor, C. R. (2019). Global Data Protection Regulations and Corporate Compliance. *Journal of International Business Studies, 50*(5), 867–889.
Tene, O., & Polonetsky, J. (2012). Big data for all: Privacy and user control in the age of analytics. *Nw. J. Tech. & Intell. Prop., 11*, 239.
Tjahjadi, B., Soewarno, N., & Mustikaningtiyas, F. (2021). Good corporate governance and corporate sustainability performance in Indonesia: A triple bottom line approach. *Heliyon, 7*(3).
Utama, C. A., Utama, S., & Amarullah, F. (2017). Corporate governance and ownership structure: Indonesia evidence. *Corporate Governance: The International Journal of Business in Society, 17*(2), 165–191.
Voigt, P., & Von dem Bussche, A. (2017). The EU General Data Protection Regulation (GDPR). *A Practical Guide, 1st Ed., Cham: Springer International Publishing, 10*(3152676), 10–5555.
Voss, W. G., & Bouthinon-Dumas, H. (2021). EU general data protection regulation sanctions in theory and in practice. *Santa Clara High Tech. LJ, 37*, 1.
Walters, R., Trakman, L., & Zeller, B. (2019). Data protection law. *Springer Nature.*
Yelikar, B. (2019). Internet Banking Trends in India: Evolution, Challenges and Opportunities. *International Journal of Management, Technology and Engineering, IX (V)*, 2452–2462.

Chapter 21
Redefining Corporate Governance in Digital Era: Strategies for Mitigating Organizational Risk

Siddharth Dhongde, Sunil Kumar, Sujit Kumar Patra, and Subhendu Kumar Mishra

Abstract Corporate governance refers to a set of rules, norms, and guidelines to determine how corporations are controlled. It also encompasses the relationship between the company's shareholders, other stakeholders, and the management board. This chapter considers the fast-evolving context of corporate governance today with a specific focus on the rigor of digital penetration. It also highlights the need for anti-organizational risk frameworks and strategies. We analyze how digital transformation alters the governance designs, increases the boundaries of supervision, and increases suspicion. The advancement of technology has not only changed business processes but has also advanced new problems, many of which can be solved. So, we discuss the pros and cons of these technologies to see how they can be used to improve the processes of governance. The area of focus covers data governance, cyber security, and ethical aspects. We then set forth some measures that deal with the risks emerging from digital transformation, such as improving digital know-how, data security, and the establishment of agile governance frameworks. Moreover, we point out the importance of collaboration between different participants in the process, such as IT units, risk managers, and even regulators. Lastly, this chapter suggests a more sophisticated view of new and future potential developments in corporate

S. Dhongde (✉)
Department of Finance, GITAM School of Business, GTAM (Deemed to Be University), Visakhapatnam, India
e-mail: sdhongde@gitam.edu

S. Kumar
Department of Entrepreneurship, GITAM School of Business, GTAM (Deemed to Be University), Visakhapatnam, India
e-mail: skumar6@gitam.edu

S. K. Patra
Department of Marketing, Jaipuria Institute of Management, Lucknow, India

S. K. Mishra
Department of Entrepreneurship, GITAM School of Business, GITAM (Deemed to Be University), Visakhapatnam, India
e-mail: smishra2@gitam.edu

V. Kandpal et al. (eds.), *Developments in Corporate Governance*,
Approaches to Global Sustainability, Markets, and Governance,
https://doi.org/10.1007/978-981-96-6366-8_21

governance. By reconstructing the digital age governance frameworks, organizations can have more protection against the erosion of assets and reputation, as well as increase the chances of long-term sustenance.

Keywords Digital transformation · Corporate governance · Data protection · Data governance · Risk management

21.1 Introduction

In this day and age when technology is evolving at a rapid pace, a company needs to adopt digital transformation in order to grow and succeed in the economy and digitized world (Li et al., 2022; Pandita & Vapiwala, 2023). Recent corporate restructurings have underscored the critical importance of digital transformation for businesses. Increasingly, organizations are expanding "digital technologies" to enhance efficiency, profitability, and competitiveness (Zeng et al., 2023). The concept of "digital transformation" encompasses a variety of activities, including using new technologies, reengineering corporate processes, and innovating new business practices and models (Machado et al., 2022). In today's twenty-first-century business environment, it is imperative for every company to function as a digitally empowered entity (Arbabi, 2022). Harnessing digital capabilities to enhance products and services has become essential for increasing effectiveness. Organizations embracing this approach will experience a gradual erosion of their competitive standing. The convergence of digital transformation and technological advancements not only yields unforeseen outcomes but also serves as pivotal catalysts for market trends. The digital revolution has sparked an entirely new age of corporate streamlining. Adapting to the evolving business landscape is imperative for businesses striving to remain competitive. In this context, effective corporate governance plays a critical role by providing directives on how businesses should handle their operations and answer to external environmental changes.

Digital revolution states the combination of digital technologies across various phases of an organization, fundamentally reshaping its operations and enhancing value delivery to clients (Kumar & Gupta, 2021; Kotni et al., 2024; Riaz et al., 2022). This evolution brings numerous benefits such as increased productivity, reduced costs, and enhanced customer satisfaction. However, digital transformation also presents challenges, particularly in the context of restructuring businesses in developing countries (Seete, 2022). The growing dependence of individuals, businesses, and society on digital technologies is increasingly evident. Activities like online information retrieval, communication with loved ones, and urban navigation exemplify this reliance. Through task automation, digital transformations reshape work processes and offer organizations new opportunities to review and streamline their operations—a concept termed "informatization" (Zuboff, 1988). Digital technologies introduce enhanced visibility, expanded editing capabilities, information durability, and broader connectivity in both everyday life and organizational settings (Flyverbom et al., 2019; Treem & Leonardi, 2012). The foundational framework guiding how businesses operate has always been dynamic, shaped by various internal and external factors.

The advancement of technology has profoundly reshaped the framework of corporate governance, necessitating a comprehensive evaluation. The corporate landscape is undergoing disruption due to the rapid and exponential progress in technological advancements over recent decades, including "big data, blockchain, artificial intelligence, and the Internet of Things." Therefore, values and structures of corporate governance have undergone significant change. Here is an important change in business operations and strategies through technology implementation. As an outcome, corporate governance has moved far from its historical emphasis on management to a more complex field intensely related to technology. Most of the important companies are growing and prospering through using digital technology and making their operations inclusive. These technological platforms and their network technology are inter-related. These platforms use network technologies to promote relationships and make information sharing smooth and easy. It also helps with easier economic transactions. Companies are using a more expansive and inclusive organizational framework to promote better cooperation among all stakeholders. It was possible to improve various features and services in digital practices (Fenwick et al., 2019).

The objective of this study is to explore the effect of technology in improving "corporate governance" practices. The study emphasizes the important role of digital technology, which is playing a critical role in the practices of corporate governance. Disruptive technology, like AI, blockchain, and big data, is used in business governance practices. It highlights the integration of digital technology and today's organizations to understand various dynamics of corporate governance, which are being impacted by digital transformation. It also explains how good governance practices can reduce organizational risks in the context of digital revolution.

21.2 Landscape of Corporate Governance in the Digital Era

Corporate governance is not a new coinage; it has existed for a long time, but over a period of time, it has evolved as per the needs of the corporations and the governing bodies, otherwise called the government watch dogs. It has received strategic importance after the Asian financial crisis and many corporate scams during the period of the new millennium. Corporate governance in countries like India and China has grown with the world trend. Since both emerging nations have seen tremendous economic growth and gave birth to several multinational and global companies, it has attracted the attention of several social scientists who are carrying out research in this regard. Corporate Governance (CG from now onwards) can be defined as "the framework that determines how the company's goals are established, how to achieve them, and how to evaluate its performance." Further, it defines a "company's management, board, shareholders, and other stakeholders who are an integral part of it (OECD, 2015). CG, which is defined as compensating for the sometimes-divergent interests of stakeholders, is a system of regulations and organizational structures that form the foundation for proper corporate functioning (Du Plessis et al., 2018). Traditionally, CG was an instrument that safeguarded investors' interests from the greed of managers or so-called custodians (Roberts & Van den Steen, 2000). With the growth of technology and digitization, CG has seen its evolution, business models based on digital practices are changing the dynamics of rivalry and relations between companies and their stakeholders. Due to various digital platforms and artificial intelligence (AI), there are now much fewer restrictions on scale, scope of operations, and learning (Iansiti & Lakhani, 2020). The roles and responsibilities which stakeholders can play and the intensity with which they can integrate, add to value proposition, utilize optimum capabilities, and infuse trust are now looked through a new lens in digital business models (Lund & Nielsen, 2018).

21.2.1 Frameworks

Two important theories through which we can see the development in CG are broadly classified as Agency theory and stewardship theory, which have evolved in contemporary organizational environments in terms of governance and principal-agent relationships (Puyvelde et al., 2012). Agency theory postulates that agents are inspired by their personal goals and will not leave a single opportunity to cash in on their individualism, sacrificing the interests of principals (Jensen & Meckling, 1976). Stewardship theory basically resides in the fact that the literal meaning of steward is to take care and nurture, and tries to solve the issues or limitations of the principal–agent relationship. "Stewardship theory" comes with an assumption of being responsible for a greater cause (Donaldson & Davis, 1991; Fox & Hamilton, 1994). To proactively protect organizational perpetuity, stewardship theory advocates the responsibilities and roles of top-notch office bearers as stewards, who are committed to collaborating with the individual objectives of leaders and managers with those of their organization and society. The importance of good stewardship and its best practices can be diffused across the company so that organizational members at all levels can infuse a stewardship ethic and promote it within their departments, senior officers of the organization are the ones who exemplify stewardship in their respective organizations. As a result, there is a cumulative effect across the entire organization, which can effectively guarantee that good stewardship is promoted even if founding leaders go (Sama et al., 2022).

The Balanced Digital Governance Model is designed to synthesize two foundational theoretical perspectives—Agency Theory and Stewardship Theory—with modern digital technologies to enhance corporate governance. Agency Theory, which emphasizes the alignment of managerial actions with shareholder interests through control mechanisms (Jensen & Meckling, 1976), is paired with Stewardship Theory, which stresses intrinsic motivation and collective responsibility for long-term organizational success (Donaldson & Davis, 1991). This hybrid approach leverages digital tools to both monitor compliance and empower stakeholders, thereby ensuring a robust and ethically sound governance framework. The foundational principle provides the necessary (Fig. 21.1).

Governance Foundations

At the core of the model are the governance foundations that reflect the dual perspectives:

Agency Perspective:

This perspective emphasizes monitoring through digital means such as Artificial Intelligence-driven analytics and automated controls. Further, the deployment of blockchain for audit demonstrates how immutable digital records can reduce information irregularity and mitigate agency costs (Jensen, October 1976).

Stewardship Perspective:

In the view of stewardship, it fosters a culture of shared responsibility and decentralized decision-making. In addition to those collaborative platforms, ethical guidelines, and training initiatives, they serve to enhance fundamental motivation to ensure that managers act as stewards of the organization (James H. Davis, 1997). These core principles strike the right balance between strict oversight and employee freedom, creating a workplace where control and autonomy complement each other.

Technology Enabler

This model identifies that the technology functions play a critical enabler in modern corporate governance practices

Agency Enablers:

This model uses automated controls, such as access restrictions and role-based access controls (RBAC), which facilitate real-time monitoring and compliance. On the other hand, tools that provide digital audit trails, like blockchain, ensure that managerial actions are transparent and supportable (Alnoor Bhimani, 2014).

Stewardship Enablers:

In this, the enablers cover digital collaborative tools, including open innovation platforms, to foster a participatory environment. Further, Data democratization initiatives, such as self-service analytics, empower employees to make informed decisions and accept accountability for digital assets. It covers the dual approach to ensure that technology is leveraged not only to enforce compliance but also to promote a culture of continuous improvement and ethical innovation in the organization's corporate governance.

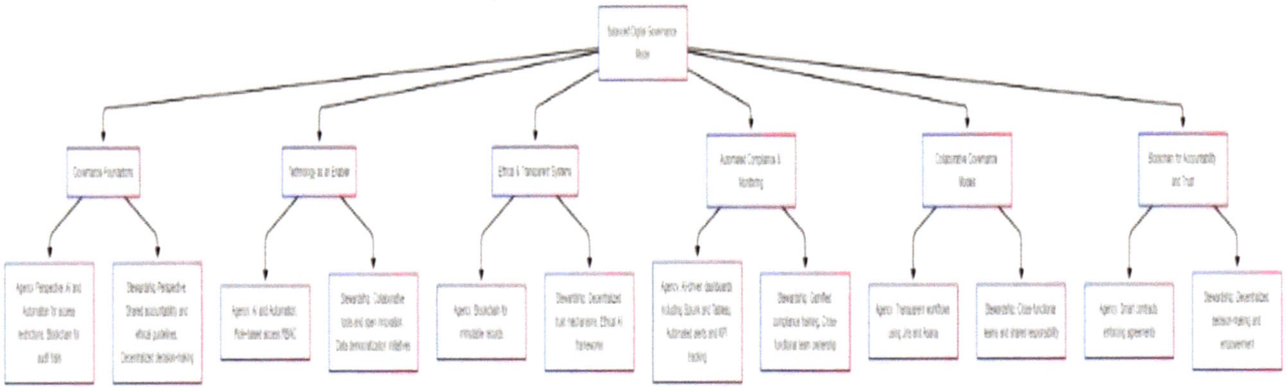

Fig. 21.1 Balanced digital governance model. (*Source* Authors)

Ethical and Transparent Systems

An ethical code of conduct and transparency play pivotal roles in corporate governance. The corporate governance rules are important because they outline the organization's ethical beliefs and roadmap. They cover both approaches, like agency and stewardship mechanisms.

Agency Mechanisms:

It is the integration of blockchain technology, immutable record-keeping, ensuring that all transactions and decisions are recorded transparently. This aligns with the agency's need for control and verification of managerial actions.

Stewardship Mechanisms:

Decentralized trust mechanisms and ethical AI frameworks support accountable innovation while allowing for a more flexible decision-making process. These systems foster trust by engaging all stakeholders in ethical governance practices. Lastly, the combination of both approaches leads to a governance structure that is transparent and ethically robust, promoting stakeholder trust and long-term organizational sustainability.

Automated Compliance and Monitoring

In corporate governance, a management system plays a key role in organizations' following the rules and regulations. So automated compliance management reduces the business risks. In addition to that, automating the compliance management systems soothes the flow of work and monitoring as well. So, this model incorporates automated tools to maintain compliance and monitor organizational performance:

Agency Tools:

In current practices, the real-time dashboards powered by AI (e.g., Splunk, Tableau) track key performance indicators (KPIs), IT spending, and data usage. As far as Splunk is a powerful tool for searching, monitoring, and analyzing machine-generated data to understand and take operational decisions effectively. Further, automated alerts and audit trails are critical for ensuring that deviations from set standards are promptly identified and addressed.

Stewardship Initiatives:

Making compliance training more engaging through gamification techniques and building cross-functional teams fosters a shared sense of responsibility for cybersecurity and data ethics. Every organization stays actively involved in maintaining strong governance standards. Lastly, by integrating these mechanisms, organizations can simultaneously reduce agency costs and build an environment that encourages shared accountability.

Collaborative Governance Models

In this framework, it emphasizes the importance of collaboration in digital governance for improved governance:

Agency Approach:

In case of use of transparent workflow methods through project management tools like e.g., Jira, Asana, which provides clear oversight and documentation of digital initiatives, ensuring that all actions are aligned with organizational policies and goals.

Stewardship Approach:

Cross-functional teams promote collaborative decision-making and empower employees to take an active role in shaping organizations' governance practices. This will foster innovation and align individual efforts with long-term organizational goals. Lastly, such models will ensure that the governance process is both comprehensive and inclusive, balancing control with collaborative engagement.

Blockchain for Accountability and Trust

Blockchain technology enhances transparency and trust in processes and styles, enabling well-informed decision-making. It also enables secure and efficient operations with improved traceability, facilitating better governance in organizations.

Agency Implementation:

In blockchain smart contracts platforms, the document enforcement of agreements is automated, which ensures the contractual obligations are fulfilled with oversight and other challenges. So, this will reinforce the agency's focus on control and compliance.

Stewardship Implementation:

Blockchain technology, which uses decentralized decision-making mechanisms via distributed ledgers, will empower stakeholders by democratizing governance processes. This enables a stakeholder's participatory approach that supports the organization's ethical and long-term strategic objectives to achieve corporate governance practice. Lastly, the integration of blockchain has been showcased as a robust foundation for both ensuring compliance and fostering trust within the organization.

Lastly, in the above Balanced Digital Governance Model it effectively integrates the perspectives of Agency theory and Stewardship theories. In the model, by leveraging advanced digital technologies such as AI, blockchain, and collaborative platforms, the model emphasizes a comprehensive framework that enhances oversight, accountability, and ethical decision-making. This dual approach not only mitigates risks associated with information irregularity and managerial self-interest but also promotes a culture of shared responsibility and continuous improvement. Therefore, organizations that adopt this model are better equipped to address the complexities of modern digital governance while achieving sustainable long-term goals and success.

The integration of each component, as depicted in the flowchart, is explained here, showing how the components function in combination to create an enhanced governance framework that makes use of digital instruments that enable control and empowerment in a contemporary business setup (Malik, 2021). As a rule, corporate governance literature views audit as a tool of control or governance that aims to disclose credible accounting information and prevent possible self-interested actions of stewards and investors (Carcello et al., 2011).

However, the audit procedure needs to be changed primarily for three reasons. Firstly, the audit is primarily designed in the interest of shareholders, who consider it a shield that offers them a decent level of security and assurance. The past performance of the company should most appropriately represent the company and stick to legal regulations and framework. Secondly, because the audit report is based on past data and was generated many months after the fiscal year ended. It does not include any elements that guide ahead. Thirdly, since most reports do not include recommendations, managers view audits as unnecessary extra costs rather than assets (Manita et al., 2022).

21.2.2 Digital Tools and Platforms

In the world of technological dominance and the digital business world, most companies operate through open platforms, which are structured and secure, and they report to the government smoothly (Parker & Alstyne, 2018). Most platforms of the business world use networked technologies to add people, information, and currency exchange (Parker et al., 2016; Rochet & Tirole, 2003). Technology giant companies like the Big Four see it as a business that provides platforms to create relationships between value producers and extractors in order to produce income for the platform's owner (Evans, 2006). Platform companies are not just about using new technology to let interested parties engage socially or economically. Additionally, it facilitates cooperation across all parties, such as stakeholders and their business structure, along with their internal operations, in a heavier and more inclusive manner (Kiron & Unruh, 2018). Policymakers and the government together should see how it can integrate companies and various CG platforms effectively for better coordination. Currently there is indication of mismatch between the demands of firms trying to become successful platforms and the regulatory pressures brought about by corporate governance (Fenwick et al., 2019).

21.3 Digital Transformation in Corporate Governance

Corporate governance digitalization is the adoption of sophisticated digital technologies into age-old governance regimes to promote openness, accountability, and decision-making. Through tools like AI-enabled analytics, blockchain for tamper-proof record maintenance, and interactive digital platforms, organizations can fill the gap between agency and stewardship views to a large extent. Not only do these make compliance procedures automatic and risk assessments seamless, but they also drive a culture of collective responsibility and moral conduct. These technological innovations minimize information asymmetry and enhance stakeholder participation, allowing organizations to manage risks more effectively while facilitating sustainable growth. As digital technologies advance, they transform corporate governance structures to make them more responsive to the changing business environment. Adrian Cadbury, while chairing a committee in the 1990s to develop corporate governance standards following UK scandals, did not have access to contemporary technical tools.

Similarly, post-boards faced heightened corporate governance responsibilities without the technological tools available today. Recognizing the profound impact of technology on the evolution of corporate governance processes and the roles of traditional stakeholders is crucial. As noted by Cross (2004), information technology (IT) has accelerated the development or transformation of new cultural influences, methodologies, business practices, and its models across various domains, including corporate governance (Kumar & Gupta, 2019; Mishra & Kumar, 2023). The next wave of challenges has pushed outdated practices into the hands of corporate administrations and regulatory bodies. However, advancements in data transfer speed and processes, supported by a growing ecosystem, have effectively eliminated these inefficiencies. This transformation underscores the importance for businesses to continuously monitor and disclose non-financial metrics, and to standardize operations across all facets, including supply chains, distribution channels, and manufacturing processes, to meet public expectations (Eng & Mak, 2003; Gupta et al., 2021; Perrini, 2006; Sun et al., 2010). In addition to developments in information technology, technologies such as blockchain and "artificial intelligence" (AI) have created opportunities for startups whose business models are shaped by these innovations. The impact of technology-driven businesses has converted the expectations and demands of corporate governance, ushering in an era focused on technology-driven governance to cultivate stronger relationships with stakeholders (Fenwick et al., 2018).

According to Marcoux (2003), digitalization challenges claim that it is impossible to serve many participant groups simultaneously while providing real answers to remove irregular information. Abraham (2012) has explained that it is sensible to evaluate new technologies using this framework. To understand its function as a facilitator to stakeholder relations and practices of corporate governance. According to Fenwick et al. (2018), two-way communication with transparency is possible through technology. Many firms are using these changes to their benefit and embracing proactive steps to reduce risks. It jeopardizes its sustainability, such as technological complications, to ensure that management and shareholders' objectives are in one direction (Flyverbom et al., 2019; Lin et al., 2020a, 2020b). According to Fenwick et al. (2019), to strategically use emerging technology, companies can now create a direct governance structure that promotes digitalization.

In order to control this temporary period and respond to digital changes, good corporate governance is becoming increasingly important. Here, technology infuses every aspect of business operations in the Indian corporate sector, from decision-making to stakeholder interaction. Firms need to use digital channels to get feedback, solve issues, and build networks with stakeholders. It also improves information accessibility to stakeholders. Such interactive communication fosters transparency and accountability in company practices, which can validate industry standards and stakeholder expectations. It ensures compliance with regulatory obligations about data collection and storage. It also prevents mistakes and misuse during the implementation processes and develops strong corporate governance measures.

For digital economy data centers and strong computer infrastructure is needed and it is called as "oil of the future." The surge of digital technology has majorly changed the practices within companies. It easily engages with the company's stakeholders and helps them communicate. It is also helpful in corporate governance practices and meets standards. In order to increase operations in various functions of the company and improve efficiency to expedite decision-making processes, digital technology is important. Companies should take advantage of different digital technologies such as "cloud computing, data analytics, artificial intelligence, blockchain," and other technologies as well, used in current business practices. The way companies are practicing the use of this technology in their boardroom operations will help in strengthening corporate governance practices. People working in companies have to learn new practices regarding new technology and digital practices. Digital technologies are revolutionizing company practices and are cost-saving in all types of business organizations.

Corporate governance will benefit from such technological development. It will improve transparency, openness, efficiency, and their responsibility towards stakeholders. Changing business practices and environment in such a volatile situation, it is important to adopt digital practices in companies. Corporate governance practices must include digital technology. Business houses to achieve their goals in the long run, business organizations need to practice digital technology integration and foster innovation to provide a flexible work environment. It also helps in practicing mature corporate governance at a fast pace in the current business environment. Implementation of new technology is the judicious manner to be responsible and accountable for corporate governance practices.

21.3.1 Big Data

Big Data offers a fresh viewpoint by providing insights into potential futures and facilitating the use of cutting-edge business analytics techniques to forecast upcoming trends (Fotaki et al., 2021). This makes it possible to develop analytical models that mimic organizational dynamics, highlighting the demand for cutting-edge methods for the processing, storing, and improvement of big data. Businesses embracing this mind-set, known as "Data-Driven Businesses," prioritize strategic management and decision-making based on comprehensive information (Sadasivam et al., 2016). In recent years, the financial sector has experienced significant growth in the adoption and utilization of Big Data (Ye & Li, 2017). Despite advancements, there remains limited understanding of how Big Data influences strategic accountability, its governance at the board level, and its impact on financial decision-making (Nutt & Wilson, 2010). Nonetheless, big data has the potential to provide senior leadership teams with a more holistic view, potentially reducing risks and enhancing strategic decision-making capabilities (Lin et al., 2020a, 2020b).

21.3.2 Artificial Intelligence

Artificial intelligence (AI) represents a significant opportunity for companies across various scales, promising enhanced profitability, decision-making process, and operational efficiency (Hilb, 2020). However, integrating AI poses challenges such as the need for investments in specialized skills and ensuring ethical and transparent deployment. In order to ensure the appropriate and efficient use of AI, this section examines basic ways for integrating it into corporate plans (Tokmakov, 2021). Companies providing financial services and using AI to automate the process and compliance will be able to do fraud detection. Business organizations should be able to determine the uses of AI and have a sound strategic plan for cost-effective implementation. In companies within the framework of corporate governance, artificial intelligence includes collection, Creation, use, and control, planning, dissemination, and removal of information. At the expense of companies, business processes use such practices to ensure the important information to support business operations (Hickman & Petrin, 2021).

It can be better understood as "information about information," it entails a closer examination of "data in context" and "information in context." Using AI in the company's business practices helps in data mining, loading, transformation, repository management, and modeling, which are important tasks related to this. Information security through encryption, and upholding qualities, and handling metadata-related tasks such as "architecture" are all included in information management practices in a business organization (Albalawee & Fahoum, 2024). To attain desired success in business, a number of important mechanisms are essential. These include in companies optimizing the value taken from the data and ensuring AI based access through different channels. Furthermore, consideration is given to set measurable high standards and enabling data sharing with external bodies and within the organization to strengthen the corporate governance practices.

21.3.3 Blockchain

Blockchain technology is crucial in managing various aspects of business organizations. It plays an important role in improving transparency. In corporate securities and stocks, blockchain technology has the ability to streamline various practices (Akhtar, 2024). It also reduces intricacies in business organizations. But changing from a traditional system to blockchain-based solutions is difficult for an organization. Careful implementation should be done to achieve the desired outcome in corporate governance.

Blockchain can improve shareholder and stakeholder rights, but its uses and restrictions need to be carefully considered. The advent of blockchain technology offers a substitute for current corporate governance frameworks, utilizing peer-to-peer networking, crowdsourcing, coding, and cooperation to possibly minimize middlemen. By incorporating blockchain-based guarantees into the code, any party engaged in business transactions or agency interactions cannot avoid governance principles and can assist in ensuring adherence to them (Ronaghi, 2022). Transactions in blockchain-based contract execution between a principal and an agent only happen when all contract requirements are met, as confirmed by a consensus procedure. Because there is no longer as much need for close supervision and monitoring of agents under blockchain architecture, this modifies the principal–agent relationship's cost structure. Blockchain legal frameworks have made it possible for decentralized autonomous organizations to develop, which calls into question the conventional wisdom on the need for agency in governance. Conventional business structures still incorporate some elements of blockchain-driven business governance. Additional regulatory actions may be required due to worries about blockchain's connection to speculative activity and volatile cryptocurrencies. Still, there is a good chance that distributed ledger technologies—like blockchain—will have a big influence on corporate governance (Ezzi et al., 2023).

In conclusion, blockchain technology keeps a duplicate ledger on each user's computer, allowing users to instantaneously view transaction data. Blocks of transactions are added to the ledger so as to maintain the chronological sequence without erasing earlier entries. This guarantees the data's openness, verifiability, and immutability—qualities that are anticipated to improve network transparency and trust. In today's digital era, corporate governance faces a number of dangers that can have serious consequences for an organization's integrity, reputation, and efficiency. As businesses rely more on digital technological penetrations, they are more open to cybersecurity attacks, data breaches, and complicated compliance challenges (Mutamimah et al., 2023). Cybersecurity dangers such as hacking, phishing, and malware assaults are continually threatening critical corporate and customer data. Data breaches may cause substantial financial losses, legal issues, and major reputational harm.

21.4 Cybersecurity Threats

Cybersecurity is the umbrella term for the collection of preventive measures that can be used to safeguard user assets and cyberspace against intrusions and illegal access. It has always been a difficulty that has impacted the corporate governance and compliance procedures now in place at the organization. Conversely, one of the main challenges during the COVID-19 epidemic was corporate governance. Cyber-attacks and associated risks surged dramatically at that time, presenting a new set of difficulties for businesses and ultimately hurting the economy. (Aguilar, 2015) It is helpful to keep in mind the board's general responsibilities to the company as well as its roles in corporate governance and risk management when thinking about the board's involvement in cybersecurity matters (Christopher Nolan, 2019). Because cybersecurity presents serious financial threats to organizational value, it is appropriate to classify it as a management quality indicator and a governance concern. Prioritizing risk management, investor agendas, and senior management are all impacted by cybersecurity in the environment, social context, and governance framework. Companies are being forced by important regulatory oversight agencies to handle cyber risk in a way that is comparable to other macro-level business hazards (Gorejena Nyasha, 2024).

21.4.1 Data Breaches

A data breach is defined as "a security compromise that results in unintentional or deliberate loss, alteration, disclosure, or access to protected data that is transmitted, stored, or processed in any other way." Data breaches include a wide range of activities, such as someone breaking into a company website and publishing pictures of celebrities, taking credit and debit card information, or carelessly discarding confidential documents that an outsider discovers in a public trash can. A management team with a long-term vision and a willingness to safeguard stakeholders is necessary for organizations investing in information security to prevent data breaches, since many companies have experienced them (Lending et al., 2018). Compliance issues in the corporate governance compliance issue cover the legal rules and cover the company or the organization. So the regulatory part and the standard are applied to the company. As far as the issues are concerned, consider following the regulation so that there will not be any action against the organization (Markets Compliance, 2024). Thus, the primary function of regulatory compliance, which is a component of corporate governance, is to focus efforts on preventing employees and stakeholders from breaking laws and regulations. Compliance shields the business against meddling from the government.

21.5 Mitigation Strategies of Potential Risk

An essential component of any organization's cybersecurity strategy is risk minimization. This chapter explores the foundations of comprehending and creating successful risk mitigation plans, from selecting the best controls to putting them into practice. It looks at how to apply user security rules and reduce the dangers associated with interconnectivity. Policies, standards, and procedures articulate an organization's expectations concerning its security posture. Asset management processes are crucial in maintaining the asset inventory and classification. Asset risk assessment is a vital part of the asset management process. Interconnectivity risks extend to using Application Programming Interfaces, as they form bridges connecting disparate systems and services. User controls are vital to an organization's overall security framework as they manage the potential risks arising from users of its systems and data. A robust change management process is a critical risk mitigation strategy that can have profound implications for the organization's resilience and long-term success (Jason Edwards, 2024).

21.5.1 Mitigation Strategies

Put strong cybersecurity measures in place, including firewalls, multi-factor authentication, encryption, intrusion detection systems, and frequent security assessments. Regarding the cyber. A comprehensive cybersecurity plan called cybersecurity governance works with the organizational operations to avoid activities from being disrupted by cyberthreats or assaults (Bauch, 2024).

In the current digital age, data privacy has become a top priority due to the major ethical, legal, and societal problems that have been generated by the widespread use of data gathering and utilization technology. The organization implements strict procedures for protecting data, uses data encryption, regularly evaluates its susceptibility, and makes sure that data protection laws like the California Consumer Privacy Act (CCPA) and the General Data Protection Regulation (GDPR) are followed (Bauch, 2024). The General Data Protection Regulation (GDPR), which is enforced by the European Union (EU), seeks to improve individual control over their personal data and harmonize data protection rules among EU member states.

Organizations managing personal data must adhere to stringent regulations, which include clauses pertaining to permission, data subject rights, reporting of data breaches, and sanctions for noncompliance. Additional California Consumer Privacy Act Information. The California Consumer Privacy Act (CCPA), which was passed into law in the US, gives Californians certain rights about their personal information. These rights include the ability to access and remove their data that is kept by businesses, the right to know what data is being collected, and the right to refuse to have their data sold. Other states contemplating enacting laws akin to the CCPA have looked to it for inspiration.

The compliance risk program is a process of identifying the risk and mitigating the financial, legal, and operational impact on the noncompliance of the regulatory and misleading the information of the organization, which has led to an increase in the level of risk. So, to effective management of risk, it requires key steps. First is to identify the gaps of applicable regulation and assess the risk according to the objective of the organization. Second is assigned the role to manage and control the risk. So, for the specific task, the organization appoints the compliance office. His role is to check and confirm whether a company follows important regulations, laws, and government guidelines. An organization needs to establish a comprehensive compliance program, stay updated with regulatory changes, conduct regular compliance audits, and provide training for employees on regulatory requirements. Lastly, conduct thorough planning and testing before integration, employ experienced IT professionals, and ensure robust backup and recovery plans are in place. In addition to that, it has to perform due diligence when selecting vendors, include security requirements in contracts, regularly assess third-party security practices, and establish clear communication channels for incident response.

Another risk mitigation in the internal level has to be implemented, which covers Insider threats, employees or contractors who misuse their access to corporate data for malicious purposes. It can cause significant financial and reputational damage, as well as legal complications. To mitigate the mitigation Strategies, Organizations Implement strict access controls, monitor employee activities, conduct background checks, and foster a culture of security awareness within the organization. Below is a diagram illustrating the framework for mitigating risks associated with digital corporate governance (Fig. 21.2):

Frameworks and Models for Managing and Mitigating Risks in Digital Corporate Governance.

To effectively manage and mitigate risks associated with digital corporate governance, organizations can adopt various frameworks and models that provide structured approaches and best practices. Here are some widely recognized frameworks and models:

1. ***NIST Cybersecurity Framework (CSF)***

The National Institute of Standards and Technology (NIST) Cybersecurity Framework provides a comprehensive guide for organizations to manage and reduce cybersecurity risk. It consists of five core functions:

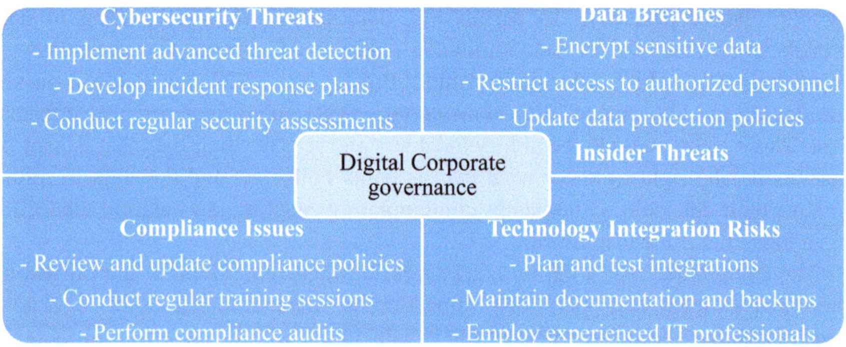

Fig. 21.2 Digital corporate governance risk mitigation framework. (*Source* Authors)

- To Develop an understanding of the organization's cybersecurity risks, including assets, systems, data, and capabilities.
- To Implement appropriate safeguards to ensure the delivery of critical services.
- To Develop and implement the ability to identify cybersecurity events.
- To Implement plans to take action regarding a detected cybersecurity event.

2. ***Iso/iec 27,001***

ISO/IEC 27,001, this is an international standard for managing information security. It provides a systematic approach to managing sensitive company information, ensuring it remains secure. Key components include:

- Risk Assessment: Identify and assess information security risks.
- Security Controls: Implement a set of controls to mitigate identified risks.
- Management Commitment: Ensure top management's commitment to information security policies and practices.
- Continuous Improvement: Regularly review and improve the information security management system (ISMS).

3. ***COSO Enterprise Risk Management (ERM) Framework***

This is a Committee of Sponsoring Organizations of the Treadway Commission (COSO) that provides the Enterprise Risk Management (ERM) Framework. This helps organizations identify, assess, and manage risks to achieve not only their objectives but Key elements which include:

- Governance and Culture: Establish risk governance structures and promote a risk-aware culture.
- Strategy and Objective-Setting: Align risk management with organizational strategy and objectives.
- Performance: Identify and assess risks that could impact performance.
- Review and Revision: Monitor and revise risk management processes to adapt to changing conditions.
- Information, Communication, and Reporting: Ensure effective communication and reporting of risk information.

4. ***FAIR Model (Factor Analysis of Information Risk)***

The Factor Analysis of Information Risk (FAIR) paradigm assists companies in understanding, quantifying, and analyzing cybersecurity risk. The FAIR risk assessment technique assists businesses in making timely and educated choices about how to avoid and remediate various types of cyberattacks on vital data and systems. The FAIR methodology's most significant feature is its ability to quantify multiple types of risk using monetary or dollar values. This enables organizations to transform cyber risk into practical, financially sensible decisions. It is a Model is a quantitative risk management framework that provides a structured approach for understanding, analyzing, and measuring information risk. It includes:

- Risk Assessment: Identify and quantify potential risks in terms of frequency and impact.
- Risk Analysis: Analyze the potential loss magnitude and likelihood of risk scenarios.
- Risk Management: Implement strategies to mitigate identified risks based on quantitative analysis.

5. ***COBIT (Control Objectives for Information and Related Technologies)***

COBIT is a framework to help the organization to develop, implement, monitor, and improve IT governance and information management. In addition, it is also for managing and governing enterprise IT. It provides guidelines for aligning IT with business goals and managing IT-related risks. Key components include:

- Governance Framework: Establish governance structures and principles.
- Process Model: Define and manage IT processes to ensure effective governance and management.
- Performance Measurement: Use metrics and KPIs to measure and monitor IT performance.
- Maturity Models: Assess the maturity of IT processes and practices.

6. ***CIS Controls (Center for Internet Security Controls)***

Implementing prescriptive, streamlined, and prioritized CIS Critical Security Controls (CIS Controls) may improve your cybersecurity posture. Thousands of cybersecurity professionals worldwide utilize CIS Controls now and/or participate in their development through a community consensus process.

Further CIS Controls are a set of best practices for securing IT systems and data. They provide a prioritized approach to implementing cybersecurity measures. Key controls include:

- Inventory and Control of Hardware Assets: Maintain an accurate inventory of all hardware assets.
- Inventory and Control of Software Assets: Maintain an inventory of all software assets.
- Continuous Vulnerability Management: Continuously acquire, assess, and take action on information regarding new vulnerabilities.
- Controlled Use of Administrative Privileges: Manage and control the use of administrative privileges.

21.6 Framework for Risk Management into Practice

Make Thorough Risk Assessments: Determine and assess risks regularly to learn about their likelihood and possible effects. For a comprehensive evaluation, combine qualitative and quantitative methodologies.

Create and Put into Practice Security Policies: Make comprehensive security rules and procedures based on the selected model or framework. Make sure that these rules address every facet of risk management, including compliance, incident response, access control, and data security.

Involve the Parties: Involve all relevant parties in the risk management process, such as employees, IT staff, and top-level management. Ensure that each of them knows about their responsibilities and liabilities.

Provide Training and Awareness Programs: Conduct regular training sessions to educate staff members on cybersecurity risks, risk management processes, and legal requirements. Foster a security-conscious culture.

Review and Monitor: Review and Monitor the effectiveness of the controls and risk management processes.

Leverage Technology to enhance risk identification, analysis, and response capabilities, utilize advanced technology such as artificial intelligence (AI), machine learning, and automated monitoring systems.

Check Compliance: Stay current with all applicable laws, regulations, and industry standards.

To ensure compliance with all applicable regulations, perform audits on a regular basis.

References

Akhtar, T. (2024). Blockchain technology: The beginning of a new era in reforming corporate governance mechanisms. *Journal of the Knowledge Economy, 15*(1), 3059–3084.

Albalawee, N., & Fahoum, A. A. (2024). A novel legal analysis of Jordanian corporate governance legislation in the age of artificial intelligence. *Cogent Business & Management, 11*(1), 2297465.

Bel, I., & Grau, A. (2020). Big data in corporate governance decision. In *Congress UPV* (p. 103).

Carcello, J. V., Hermanson, D. R., & Zhongxia, Y. (2011). Corporate governance research in accounting and auditing: Insights, practice implications, and future research directions. *Auditing, 30*(3), 1–31.

Christopher Nolan, G. L. (2019). Cybersecurity: today's most pressing governance issue. *Journal of Cyber Policy*. Corporate Governance in India: The Changing Paradigm. NSE Centre for Excellence in Corporate Governance.

Cross, S. R. (2004). Corporate governance, information technology and the electronic company in the United Kingdom. *Information and Communications Technology Law, 13*(2), 117–128.

Deloitte India. (2021). Digital disruption and corporate governance. *Digital India*. Ministry of Electronics & Information Technology, Government of India.

Donaldson, L., & Davis, J. H. (1991). Stewardship theory or agency theory: CEO governance and shareholder returns. *Australian Journal of Management, 16*(1), 49–64.

Du Plessis, J. J., Hargovan, A., & Harris, J. (2018). *Principles of contemporary corporate governance*. Cambridge University Press.

Elizabeth Abraham, S. (2012). Information technology, an enabler in corporate governance. *Corporate Governance: The International Journal of Business in Society, 12*(3), 281 291.

Eng, L. L., & Mak, Y. T. (2003). Corporate governance and voluntary disclosure. *Journal of Accounting and Public Policy, 22*(4), 325–345.

Evans, D. S., Hagiu, A., & Schmalensee, R. (2006). *Invisible engines: How software platforms drive innovation and transform industries*. The MIT Press.

Ezzi, F., Abida, M., & Jarboui, A. (2023). The mediating effect of corporate governance on the relationship between blockchain technology and investment efficiency. *Journal of the Knowledge Economy, 14*(2), 718–734.
Fenwick, M., Kaal, W. A., & Vermeulen, E. P. M. (2019a). The "unmediated" and "tech-driven" corporate governance of today's winning companies. *New York University Journal of Law & Business, 16*(1), 75–120.
Fenwick, M., McCahery, J. A., & Vermeulen, E. P. (2019b). The end of 'corporate ' governance: Hello 'platform ' governance. *European Business Organization Law Review, 20*, 171–199.
Fenwick, M., Mccahery, J. A., Vermeulen, E. P. M., Fenwick, M., & Mccahery, J. A. (2018, December). *The end of "Corporate" governance (Hello "Platform" Governance).*
Flyverbom, M., Deibert, R., & Matten, D. (2019). The governance of digital technology, big data, and the internet: New roles and responsibilities for business. *Business and Society, 58*(1), 3–19.
Fotaki, M., Voudouris, I., Lioukas, S., & Zyglidopoulos, S. (2021). More accountable, more ethical, yet less trusted: Misplaced corporate governance in the era of big data. *British Journal of Management, 32*(4), 947–968.
Fox, M. A., & Hamilton, R. T. (1994). Ownership and diversification: Agency theory or stewardship theory. *Journal of Management Studies (Wiley-Blackwell), 31*(1).
Gorejena Nyasha, L. I. (2024). A systematic literature review on the impact of cybersecurity threats on corporate governance during the covid-19 era. *ICABR Conference* (pp. 157–174). Springer Proceedings in Business and Economics.
Gupta, S., Mishra, O. N., & Kumar, S. (2021). Citizen empowerment and adoption of E-governance services: The role of online citizen skills, awareness, and engagement. *International Journal of Electronic Governance, 13*(4), 386–407.
Hickman, E., & Petrin, M. (2021). Trustworthy AI and corporate governance: The EU's ethics guidelines for trustworthy artificial intelligence from a company law perspective. *European Business Organization Law Review, 22*, 593–625.
Hilb, M. (2020). Toward artificial governance? The role of artificial intelligence in shaping the future of corporate governance. *Journal of Management and Governance, 24*(4), 851–870.
Iansiti, M., & Lakhani, K. R. (2020). *Competing in the age of AI: Strategy and leadership when algorithms and networks run the world*. Harvard Business Press.
Jason Edwards, G. W. (2024). Risk mitigation. In G. W. Jason Edwards (Ed.), *The cybersecurity guide to governance, risk, and compliance* (p. 26). Wiley Online Library.
Jensen, M. C., & Meckling, W. H. (2019). Theory of the firm: Managerial behavior, agency costs and ownership structure. In *Corporate governance* (pp. 77–132).
Kiron, D., & Unruh, G. (2018, April). Corporate 'purpose' is no substitute for good governance. *MIT Sloan Management Review*. https://sloanreview.mit.edu/article/corporate-purpose-is-no-substitute-forgood-governance/
Kotni, V. D. P., Patra, S. K., & Kumar, S. (2024). Bibliometric analysis on digital wallets: Current research status and future research directions. *Digital currency assets and challenges to financial system stability*, 1–21.
Kumar, S., & Gupta, A. (2019). Impact of IT support system and organizational culture on innovation and job performance: Mediating role of KM attitude. *Ramanujan International Journal of Business and Research, 4*, 137–160.
Kumar, S., & Gupta, A. (2021). Factors affecting adoption of M-Wallets: Moderating role of financial incentives. *Ramanujan International Journal of Business and Research, 6*, 132–143.
Lin, R., Xie, Z., Hao, Y., & Wang, J. (2020). Improving high-tech enterprise innovation in big data environment: A combinative view of internal and external governance. *International Journal of Information Management, 50*(November 2018), 575–585.
Lin, R., Xie, Z., Hao, Y., & Wang, J. (2020b). Improving high-tech enterprise innovation in big data environment: A combinative view of internal and external governance. *International Journal of Information Management, 50*, 575–585.

Manita, R., Elommal, N., Baudier, P., & Hikkerova, L. (2020). The digital transformation of external audit and its impact on corporate governance. *Technological Forecasting and Social Change, 150*, Article 119751.
Marcoux, A. M. (2003). A fiduciary argument against stakeholder theory. *Business Ethics Quarterly, 13*(1), 1–24.
Markets Compliance, S. C. (2024, May 30). *financialcrimeacademy*. Retrieved from financialcrimeacademy: https://financialcrimeacademy.org/compliance-and-corporate-governance/#:~:text=Corporate%20governance%20refers%20to%20the,participate%20in%20the%20business%20environment
Mishra, M. K., & Kumar, S. (2023). Impact of digitalisation on youth consumer dynamics: A review of literature using R-software. *International Journal of Public Sector Performance Management, 11*(2), 177–190.
Mutamimah, M., Alifah, S., & Adnjani, M. D. (2023). Corporate governance innovation framework to reduce credit risk in MSMEs using blockchain technology. *Cogent Business & Management, 10*(3), 2250504.
Tokmakov, M. A. (2021). Artificial intelligence in corporate governance. In *Digital economy and the new labor market: Jobs, competences and innovative HR technologies* (pp. 667–674). Springer International Publishing.
Sadasivam, G. S., Subrahmanyam, M., Himachalam, D., Pinnamaneni, B. P., & Lakshme, S. M. (2016). Corporate governance fraud detection from annual reports using big data analytics. *International Journal of Big Data Intelligence, 3*(1), 51–60.
Nielsen, C., Lund, M., Montemari, M., Paolone, F., Massaro, M., & Dumay, J. (2018). *Business models: A research overview*. Routledge.
Nutt, P., & Wilson, D. (2010). *Handbook of decision making*. Wiley.
OECD. (2015). *G20/OECD Principles of corporate governance*. OECD Publishing.
Parker, G., & Van Alstyne, M. (2018). Innovation, openness and platform control. *Management Science, 64*(7), 3015–3032.
Parker, G. G., Van Alstyne, M. W., & Choudary, S. P. (2016). *Platform revolution: How networked markets are transforming the economy and how to make them work for you*. Norton.
Perrini, F. (2006). The practitioner's perspective on non-financial reporting. *California Management Review, 48*(2), 73–103.
Roberts, D. J., & Van den Steen, E. (2000). Shareholder interests, human capital investment and corporate governance. *Stanford GSB Working*. https://doi.org/10.2139/ssrn.230019
Rochet, J., & Tirole, J. (2003). Platform competition in two-sided markets. *Journal of the European Economic Association, 1*(4), 990–1029.
Ronaghi, M. H. (2022). Contextualizing the impact of blockchain technology on the performance of new firms: The role of corporate governance as an intermediate outcome. *The Journal of High Technology Management Research, 33*(2), Article 100438.
Sama, L. M., Stefanidis, A., & Casselman, R. M. (2022). Rethinking corporate governance in the digital economy: The role of stewardship. *Business Horizons, 65*(5), 535–546.
Sun, N., Salama, A., Hussainey, K., & Habbash, M. (2010). Corporate environmental disclosure, corporate governance and earnings management. *Managerial Auditing Journal, 25*(7), 679–700.

Chapter 22
Restrictive Approach to Say on Pay Vote in India

Yash Pandey and Hartej Singh Kochher

Abstract 'Say on Pay' is a mechanism that allows shareholders to vote on executive compensation. Unlike other countries, India lacks a mandatory or advisory 'Say on Pay' vote. Instead, Section 197 of the Companies Act 2013 imposes a trigger limit, requiring shareholder approval for executive compensation exceeding specified thresholds: shareholders' approval is required where executive remuneration exceeds 5% of net profits for a single manager, or whole-time director or managing director or 10% for such multiple directors. This restrictive approach limits shareholder power to express opinions on compensation schemes below these thresholds, and it creates an artificial benchmark, making higher compensation appear excessive. Consequently, shareholders are hindered from challenging overgenerous packages or weak performance-linked pay. This paper analyses global 'Say on Pay' structures and proposes an optimal model for India. The authors have also emphasized on case studies to explain the practical implications of the structures. By examining case studies from different jurisdictions, this paper illustrates the effectiveness of various 'Say on Pay' models and their impact on corporate governance. The analysis includes the UK's binding vote model, the advisory approach in the US, and Australia's two-strike rule, providing a comparative perspective on how these mechanisms enhance shareholder engagement and accountability. The paper recommends changes to India's legal framework and corporate governance regime on executive compensation. The proposed changes aim to empower shareholders, foster transparency, and align executive remuneration with long-term company performance and shareholder value.

Keywords Executive remuneration · 'Say on Pay' vote · 'Say on Pay' vote in India · Corporate governance and executive compensation

Y. Pandey (✉)
Galgotias University, Greater Noida, India
e-mail: yashpandeygnlu@gmail.com

H. S. Kochher
Vinayaka Missions Law School, Vinayaka Mission's Research Foundation (DU), Chennai, India
e-mail: hartejsinghkochher@gmail.com

V. Kandpal et al. (eds.), *Developments in Corporate Governance*,
Approaches to Global Sustainability, Markets, and Governance,
https://doi.org/10.1007/978-981-96-6366-8_22

22.1 Introduction

Shareholder activism has become a central theme in corporate governance. Shareholders vote on directors' appointments, amendments to the Memorandum or Articles of Association, auditor appointments, etc., and influence the company's decisions. Additional tools like meeting requisition powers and class action suits further empower shareholders in the governance of companies in India. Shareholder democracy is embedded in the corporate governance framework globally, and it is facilitated by voting rights and adequate disclosure requirements. These mechanisms enable shareholders to voice their concerns effectively. However, shareholder activism is sometimes criticized, particularly by advocates of the stakeholder theory of corporate governance (Stout, 2012).

The right of the shareholders to participate in corporate governance is globally recognized. The extent of this influence is determined by two theories: shareholder primacy theory and stakeholder theory or stakeholderism. Shareholder primacy advocates maximizing returns for shareholders as the primary goal of corporate governance (Bebchuk & Tallarita, 2020). In contrast, stakeholder theory emphasizes the welfare of all stakeholders affected by the company's actions. Lynn Stout, in *The Shareholder Value Myth* (2012), argues that focus on shareholder returns can be detrimental, and to support this idea, she cites the BP Plc oil spill as a prime example. According to her, BP's focus on maximizing shareholder returns led to safety compromises, resulting in a catastrophic explosion in the Gulf of Mexico, which resulted in environmental damage and reputational harm to BP. Stout contends that corporate focus should shift from value maximization to stakeholder welfare (2012). Proponents of stakeholderism argue that excessive shareholder activism can prioritize returns over broad stakeholder interests (Stout, 2012). However, well-intended shareholder activism can enhance corporate governance (Flammer et al., 2021). Recent research indicates that shareholder activism can lead to increased disclosures, demonstrating its potential benefits for all stakeholders (Flammer et al., 2021).

Despite the noble cause advocated by Lynn Stout (2012) through stakeholderism, this paper presents objections to her views. The authors argue that corporate purposes that are not centered on shareholder primacy are likely to fail. Bebchuk and Tallarita (2020), in *The Illusory Promise of Stakeholder Governance*, highlight two key reasons for this stance. First, prioritizing goals other than shareholder value maximization gives excessive discretion to directors and managers, potentially leading to superficial efforts that may harm stakeholders themselves. A purpose, even related to ESG if aligned with shareholders' value maximization, can be better supervised by the shareholders in comparison to a purpose that is kept out and beyond shareholders' scrutiny. Although issues like corporate social responsibility (CSR) and environmental, social, and governance (ESG) criteria may be pursued through legislative mandates, stakeholderism alone is insufficient to ensure that these goals are met. Second, wide discretion given to corporate directors can undermine shareholder power to oversee and direct companies, potentially eroding investor confidence, which can lead to investment withdrawal from the corporation.

Executives often receive substantial compensation. This compensation is determined by the remuneration and nomination committee of the board. The committee is part of the board, and there is potential that it can be manipulated to provide excessive remuneration to executives. Therefore, shareholders use activism to monitor executive remuneration, employing mechanisms like the 'say on pay' vote. This vote allows shareholders to express their opinions on executive pay packages and exists in two forms: advisory and binding. An advisory vote is non-binding, requiring companies to disclose how shareholder feedback influences future remuneration schemes. A binding vote mandates compliance with shareholder decisions on executive compensation.

In India, there is no mandatory or advisory 'say on pay' vote, and therefore, the position in India is significantly different from that of its foreign counterparts. Instead, Section 197 of the Companies Act 2013 imposes a trigger point requiring shareholder approval for remuneration exceeding a certain threshold: executive remuneration in excess of 11% of net profits for managerial remuneration requires shareholders' approval; similarly, if executive remuneration exceeds 5% for a single managing director/whole-time director/ manager and 10% for multiple directors, then it requires shareholders' approval. This structure is in contrast with global practices and has notable disadvantages, as illustrated in Table 22.1.

India's current 'Say on Pay' vote structure prevents shareholders from determining whether a remuneration package is aligned with the organization's needs or not, as shareholders can only vote when trigger thresholds are breached by a remuneration scheme. This decision should rest with the shareholders. Additionally, high remuneration packages that breach the statutory limit require shareholders' approval, creating a perception of excessiveness. Paying more does not necessarily mean that the pay is excessive. It is only a disproportionate remuneration scheme that should be looked down upon and not a highly competitive remuneration package.

Determining what is appropriate or excessive should be shareholders' prerogative, especially in order to stay competitive globally. Therefore, the authors herein believe that Section 197 of the Companies Act 2013 requires an overhaul to include at least an advisory 'say on pay' vote, allowing shareholders to voice their opinions. Furthermore, it is crucial to inform shareholders about the design of remuneration schemes. Current managerial remuneration rules provide some guidance; however, additional regulations on remuneration consultants are required in India. Companies often rely on these consultants to develop compensation schemes, but without proper regulation, consultants may support higher executive pay in exchange for personal favors. Therefore, comprehensive rules regulating remuneration consultants in India are essential to ensure fair and appropriate executive compensation.

Table 22.1 Comparison table of different say on pay vote models

Say on pay vote model	Autonomy to shareholders to decide remuneration of senior executives	Helps in determining appropriateness of a remuneration scheme	Promotes competitive remuneration scheme
Binding vote	Yes	Yes, shareholders can vote on the remuneration amount and design of the remuneration scheme	Yes
Advisory vote	Yes, though these votes are not binding in nature, companies are required to justify how the opinion of shareholders was considered while formulating the remuneration contract	Yes, shareholders can vote on the remuneration amount and design of the remuneration scheme	Yes
Restrictive say on pay vote	No. Autonomy is absent under the Indian say on pay vote structure. The shareholders can only express their opinion when the trigger limit is reached	Shareholders can vote on the remuneration amount only, and that is also true in cases where the trigger limit is reached	No, it does not promote a competitive remuneration scheme. First of all, it allows companies to decide upon remuneration schemes up to the ceiling limits as provided under the law. Second, it creates an illusion that remuneration above 11% is excessive

22.2 Literature Review

22.2.1 Prior Research on 'Say on Pay'

A considerable amount of research on the 'Say on Pay' vote has already been conducted, offering significant insights into various aspects of executive compensation. For instance, scholars such as Konstantinos Stathopoulos and Georgios Voulgaris (2015) have discussed how the 'Say on Pay' vote can be made more effective. Similarly, Mason et al. (2016) have provided a comprehensive perspective on the origin and effectiveness of the 'Say on Pay' vote. Researchers like Shyam Kishore (2021) have studied the Indian 'Say on Pay' vote. However, it should be noted that all previous research on the 'Say on Pay' vote in India has primarily focused on the overall concept of executive compensation rather than specifically exploring the concept of the 'Say on Pay' vote in India. Randall S. Thomas and Christoph Van Der Elst (2015) have conducted comprehensive research on the structures of 'Say on

Pay' around the world in their paper titled '*Say on Pay Around the World.*' However, their seminal work does not discuss the unique 'Say on Pay' vote structure present in India. Additionally, researchers have examined the effect of 'Say on Pay' vote on the quantum of managerial remuneration (Ferri & Gox, 2018) but most of the previous research was not based on case studies.

22.2.2 Gaps in Existing Literature

Previous research has primarily focused on either the various structures of 'Say on Pay' vote or its impact on managerial remuneration. However, none of the existing literature has specifically addressed the unique structure of the Say on Pay' vote in India, nor have they incorporated any case studies into their analysis. This current research is therefore groundbreaking, as it not only explores the restrictive nature of India's Say on Pay' framework, but it also provides practical insights through case studies that illustrate the application of this mechanism in corporate governance.

22.3 Research Methods

This paper adopts a qualitative research approach, primarily based on legal analysis and comparative study. The research draws on both primary and secondary data sources. Primary sources include legal texts such as the Companies Act, 2013 and other relevant statutes covering executive remuneration in India and abroad. Secondary sources include scholarly articles, reports, and case studies from countries like the USA, UK, Australia, and Germany. Comparative legal analysis was employed to examine the varying 'Say on Pay' mechanisms globally and identify how these models influence executive compensation and shareholder engagement. Case studies were selected to illustrate the practical implications of these models and to offer a comparative lens for India's approach.

22.4 Analysis

Shareholders' power in India is restricted by the 'Say on Pay' vote structure, which is regulated by Section 197 of the Companies Act 2013. The 'Say on Pay' vote in India only comes into effect when executive remuneration exceeds predetermined thresholds. This approach stands in stark contrast to international practices, where advisory or binding votes enable shareholders to participate more effectively in corporate governance. Case studies from the US and the UK reveal that shareholder activism, facilitated by 'Say on Pay' votes, can directly impact executive compensation. When compensation does not align with company performance, it can result in an increase in agency costs, which in turn results in a loss to shareholders. Improper remuneration might also lead to brain drain from organizations. India's 'Say on Pay' vote is antithetical to shareholders' activism. It promotes low levels of shareholder involvement and impedes shareholder democracy. The authors believe that having an advisory 'Say on Pay' vote in India could enhance shareholder empowerment. It will also help in aligning executive pay with performance, and it will also increase transparency.

22.5 Executive Remuneration and Corporate Governance

The Cadbury Report defines corporate governance as '*the system by which companies are directed and controlled*' (Cadbury, 1992). Over time, this concept has gained global legislative attention, leading to the development of various practices, rules, and regulations aimed at improving corporate governance. Corporate governance encompasses practices that ensure effective oversight over a corporate organization. A key issue within this framework is executive remuneration. Different jurisdictions have developed unique practices to determine appropriate executive remuneration. In the US and UK, there are advisory/mandatory 'say on pay' votes present in order to allow shareholders to have a say in governance regarding executive remuneration. However, in India, there is a ceiling on executive remuneration under Section 197 of the Companies Act 2013 that limits total managerial pay to 11% of the company's net profits. Shareholders in India only get to voice their opinions regarding executive remuneration when this 11% cap is breached. No similar cap exists in the US or UK.

22.5.1 Defining Executive Remuneration or Executive Compensation

Grossman and Hart define executive compensation as '*a way of designing contracts which will incentivize managers to choose actions that maximize the firm's value and make them exert the required effort to achieve the expected outcome*' (as cited in Sarkar & Sarkar, 2012). Executive compensation typically includes a fixed salary and incentive-based payments such as bonuses or equity. It may also encompass stock options, restricted share plans, pensions, and other benefits like cars or healthcare (Mallin, 2016). Most companies follow a pay-performance relationship for bonuses. For example, Mr. Salil Parekh, CEO and managing director of Infosys Limited, has a contract that includes fixed pay, variable pay, stock compensation, and other benefits (Infosys, 2018). His variable pay is contingent upon the company's achievement of specific milestones as determined by the board or its committee in consultation with him. The contract also includes stringent clawback provisions, allowing for forfeiture, cancellation, recoupment, or clawback of compensation.

22.5.2 Agency Cost and Executive Remuneration

Investors bring capital to companies, and their confidence is crucial for creating an environment conducive to investment, benefiting both companies and investors. However, the structure of a company often separates ownership from management. Jensen and Meckling (1976) explain the concept of agency cost as the cost born by a company by the opportunistic tendency of agents/managers. They explain that agents/managers may fill their own pockets at the expense of the company and shareholders, and the cost borne by the company by such actions of managers/agents is known as agency cost. While agency costs cannot be completely avoided, they can be minimized. The optimal way to align the interests of both shareholders and managers, who are both utility maximizers, is through executive remuneration. On the other hand, if we seek to regulate agency costs through penal provisions, it can lead to lackluster performance or aversion to risk on the part of the managers/agents. A properly regulated executive remuneration scheme would incentivize managers to act in the best interest of the company, and it will also balance utility for both parties.

22.5.3 Say on Pay

The demand for 'Say on Pay' votes regarding executive compensation gained momentum after the 2008 global financial crisis. In 2009, the collapse of major banks in the UK highlighted the issue of executive remuneration, where excessive compensation led to reckless risk-taking. These companies took on more risk than they could manage, resulting in their inability to repay debts. Similarly, in 2008, the US introduced an advisory 'Say on Pay' vote under the Emergency Economic Stabilization Act of 2008 (EESA) in response to public outcry over the financial crisis (Thomas & Elst, 2015). Today, many countries, including Australia, France, Germany, Sweden, and the Netherlands, have implemented 'Say on Pay' votes for executive remuneration (Thomas & Elst, 2015). However, these models and implementation vary globally. Understanding why executive remuneration is a significant corporate governance issue, and its impact on shareholders and the company is crucial before delving further into the 'Say on Pay' vote.

In a corporate structure, shareholders own the company, while professional managers handle its management. This separation can lead to conflicts of interest, where managers may prioritize personal gain over shareholder interest, leading to disproportionate compensation and benefits. Managers might make decisions that benefit them at the expense of the company's long-term prospects or shareholder interests, a phenomenon known as the cost of separation of ownership and management (Sarkar & Sarkar, 2012). To mitigate this agency problem, it is essential to align executive remuneration with shareholder interest. There is a growing demand to link executive pay with company performance. An inadequate remuneration package fails to incentivize senior executives to work harder, while an excessive one can lead to complacency. Therefore, executive remuneration is a critical aspect of corporate governance. A remuneration package should reward executives and reflect the value they create without being excessive (Ferri & Gox, 2018). The 'Say on Pay' vote was developed to address these concerns, allowing shareholders to influence executive compensation decisions.

22.5.4 Say on Pay in Various Countries

The 'Say on Pay' vote allows shareholders to express concerns about executive compensation determined by the board or its committee and these votes can be either advisory or binding. In the US, 'Say on Pay' votes are advisory, meaning they do not compel the company to adjust executive compensation. This raises concerns about their effectiveness (Ferri & Gox, 2018). However, US companies must disclose in their annual proxy statements as to how shareholders feedback from these votes has influenced their compensation policies, which encourages companies to take these votes seriously (SEC, 2011). In the UK, 'Say on Pay' votes are binding, requiring companies to act according to shareholder decisions. This helps to align the company's compensation policies with shareholder interests, and it discourages excessively generous remuneration for the executives.

The US Securities and Exchange Commission (SEC) mandates that most public companies must periodically, at least every three years, provide shareholders with an advisory vote on the compensation of top executives. Companies must disclose, typically in a proxy statement, how their compensation policies and decisions have incorporated the results of the most recent 'Say on Pay' votes. Randall S. Thomas and Christoph Van Der Elst (2015) offer a comprehensive definition in their article '*'Say on Pay' Around the World.*' They define 'Say on Pay' as '(1) a recurring, mandatory, (2) binding or advisory shareholder's vote, (3) provided by law, that (4) directly or indirectly through the approval of the remuneration system, remuneration report or remuneration policy, (5) governs the individual or collective global remuneration package of the executives or managing directors of the corporation' (Thomas & Elst, 2015).

22.5.4.1 Say on Pay in the U.S.

In the US, the 'Say on Pay' vote is advisory. Section 951 of the Dodd–Frank Act of 2010 mandates these advisory votes for public companies providing remuneration to top executives (Thomas & Elst, 2015). Companies with a public float equity of $75 million or more are required to conduct these votes. However, these advisory votes do not allow shareholders to directly vote on specific elements of compensation, such as options, bonuses, stocks, et cetera (Thomas & Elst, 2015).

22.5.4.2 Say on Pay in the UK

Previously, the UK practiced a non-binding advisory vote under the Directors' Remuneration Report Regulation 2002 (Gerner-Beuerle & Kirchmaier, 2018). In 2013, Section 439 of the Companies Act 2006 was amended, adding Section 439A. This section introduced a binding vote on future remuneration policies for directors. Consequently, English law includes both an advisory vote on the annual remuneration report (backward-looking) and a binding vote on future remuneration policies for directors (forward-looking).

22.5.4.3 Say on Pay in Australia

The Australian government has taken a more stringent approach to address excessive executive compensation. Initially, the Corporate Law on Economic Reform Act, 2004 provided for a non-binding advisory vote on executive remuneration (Monem & Ng, 2013). However, in 2011, the Remuneration Amendment Act introduced the 'Two-Strike Rule,' empowering shareholders to tackle excessive pay. This rule stipulates that if 25% or more shareholders vote 'no' on a company's remuneration report at two consecutive annual general meetings, then all directors, except the managing director, must stand for re-elections within 90 days (Monem & Ng, 2013).

22.5.4.4 Say on Pay in Sweden

In Sweden, director remuneration is determined by shareholders at the general meeting, where each director's fee is decided individually. Directors receive a fixed fee, with additional fees for committee work and chairing the board (Thomas & Elst, 2015). Shareholders also vote on the board's proposed remuneration guidelines for the managing director and senior management, which are approved annually and are forward-looking, limited to the period until the next annual meeting (Thomas & Elst, 2015). The Swedish Companies Act stipulates that compensation packages involving shares, convertible, or warrants do not require remuneration guidelines but must be approved by a 90% supermajority of represented votes and shares (Thomas & Elst, 2015). This rule complicates the use of share-based incentives, leading many companies to adopt alternatives, such as Phantom stocks (Thomas & Elst, 2015).

22.5.4.5 Say on Pay in Germany

In Germany, companies have a two-tier board structure consisting of a management board that handles business operations and a supervisory board that oversees the management board. Members of the management board are elected by the supervisory board for a five-year term. Since 1937, the German Stock Corporation Act has required the supervisory board to ensure that management board compensation is reasonable, considering their duties and the company's financial condition (Thomas & Elst, 2015). Historically, shareholders had no voting rights on executive pay, except for share options that could dilute shareholding, which required a three-fourths majority of shareholders representing more than 50% of the company's capital (Thomas & Elst, 2015).

The 1998 Control and Transparency Act, KonTraG, mandated shareholder approval for share buybacks used for management board stock option plans (Thomas & Elst, 2015). In 2009, the German law on the Appropriateness of Directors' Compensation, VorstAG, introduced a non-mandatory advisory vote for shareholders on management board remuneration (Thomas & Elst, 2015). The German Corporate Governance Code allows shareholders to approve the remuneration system, but the supervisory board must approve the total remuneration of individual management board members. Thus, while an advisory vote is encouraged, it is not mandatory under German law.

22.6 Restrictive Approach to Say on Pay Vote in India

The origins of Company Law in India can be traced back to the Indian Companies Act of 1866, which was modeled on Britain's Companies Act of 1862. The 1866 Act did not impose a statutory cap on overall managerial remuneration; instead, it focused primarily on disclosure requirements related to directors' remuneration. As a result, the Act left it to companies and shareholders to determine managerial pay rather than enforcing a legal limit. Similarly, the Companies Act of 1913, under Section 69, stipulated that directors' remuneration would be decided by the company at its general meeting. A significant shift in the approach toward capping managerial remuneration occurred when the C.H. Bhaba Committee Report of 1952 first introduced discussions on the subjcct. (Ministry of Law, 1952).

Although the C.H. Bhaba Committee Report did not propose a statutory cap on directors' remuneration, it recommended capping managing agents' remuneration at a maximum of 12.5% of the company's net profits. In Paragraph 126, the report highlighted several loopholes in the regulation of managerial remuneration, leading to the proposal of an upper limit of 12.5%. Notably, this was the first instance where central government approval was mandated, in addition to a special resolution by the company, for any managerial remuneration exceeding 12.5% of net profits (Ministry of Law, 1952).

It is also worth noting that the early post-independence era was influenced by socialist economic principles, which may have contributed to the introduction of government intervention in managerial remuneration approvals. However, at this stage, we choose not to explore this perspective further. The Bhabha Committee's recommendations led to the enactment of Section 198 of the Companies Act, 1956, which imposed an overall cap of 11% of net profits on managerial remuneration.

Additionally, Section 309 of the 1956 Act provided specific limits on remuneration for Managing and whole-time directors—5% of net profits for an individual director and 10% collectively for multiple managing or whole-time directors. For other directors, Section 309 imposed a 1% ceiling on managerial remuneration, which could increase to 3% if the company had no managing or whole-Time Director. To exceed these limits, a special resolution was required, along with central government approval. The Companies Act, 1956, in this regard, bears a close resemblance to the Companies Act, 2013 concerning managerial remuneration provisions. Notably, the pre-independence era lacked any statutory cap on managerial remuneration, and it was only after the Bhabha Committee's recommendations that such limitations became a fundamental aspect of Indian corporate law jurisprudence.

In the current regime, Section 197 of the Companies Act 2013 provides for a 'Say on Pay' vote by shareholders. According to Section 197, managerial remuneration should not exceed 11% of the net profits of the company for that financial year. If a company wishes to pay remuneration exceeding 11%, it must get approval in a general meeting, subject to provisions under Schedule V of the Companies Act, 2013. The second provision to Section 197(1) states that, except with special resolution approval in a general meeting, remuneration paid to a single managing director or whole-time director should not exceed 5% of net profits, and if there are multiple such directors, then their combined remuneration should not exceed 10% of net profits. Additionally, directors who are neither managing nor whole-time directors should not receive more than 1% of net profits if the company has a managing or whole-time director or manager. If the company lacks such managing and whole-time directors, then the remuneration of other directors should not exceed 3% of net profits. Under Schedule V of the Companies Act 2013, there are restrictions on paying remuneration to Key Managerial Personnel in companies with inadequate or no profit.

Under the current legislative framework for 'Say on Pay' vote in India, shareholders do not have direct authority to determine executive remuneration. Instead, the 'Say on Pay' vote is triggered only when the cap under Section 197 is breached, which significantly limits shareholder activism. Notably, the Standing Committee on Finance (2010), in its Company Law Bill 2009 report (page 205, para 13.18), recommended retaining an overall cap on managerial remuneration while suggesting that the Ministry may develop a rational formula based on corporate profit growth and other relevant factors to determine an outer ceiling. The Committee also proposed that within this ceiling, remuneration decisions should be left to the Board's Remuneration Committee or the shareholders, as outlined in the Bill.

However, no progressive formula was ultimately adopted to adjust the ceiling based on the company's profitability and growth. Instead, managerial remuneration decisions were handed over to the Nomination and Remuneration Committee under the Companies Act 2013. Therefore, we can clearly see that in the pre-independence era, companies had the discretion to determine directors' remuneration through a special resolution, suggesting that shareholder democracy was, in some ways, stronger during that period. To compensate for the lack of say to shareholders in regard to executive remuneration, the legislators in India have sought a disclosure-based regime to ensure transparency in executive remuneration.

The Companies (Appointment and Remuneration of Managerial Personnel) Rules, 2014, under Rule 5, mandate detailed remuneration disclosures for listed companies. These disclosures can prevent arbitrary executive compensation schemes while empowering shareholders with the necessary information to evaluate the adequacy and fairness of pay packages.

Rule 5 of The Companies (Appointment and Remuneration of Managerial Personnel) Rules 2014 provides for elaborate disclosures that are required to be made by every listed company; the following are the disclosures required under rule 5:

1. The ratio of remuneration given to each director compared to the median remuneration of employees.
2. The percentage increase in remuneration of each director, CFO, CEO, company, secretary, or manager.
3. The percentage increase in the median remuneration of employees.
4. An explanation of the relation relationship between the average increase in remuneration and company performance.
5. A Comparison of remuneration to Key Managerial Personnel and company performance.
6. The key parameters for any variable component of remuneration availed by directors.

The authors suggest that the current provisions on managerial remuneration in India should be revised in line with the best international practices to enhance shareholder empowerment. As demonstrated in the Table 22.2, several jurisdictions, including the UK and US, have implemented either a mandatory or an advisory 'Say on Pay' vote to strengthen shareholder oversight. Some jurisdictions, such as Australia, have granted an even higher level of shareholder autonomy in remuneration decisions. Notably, in Australia, if 25% or more shareholders disapprove of managerial remuneration at two consecutive annual general meetings, then the entire board of dircctors is required to stand for re-election.

While recognizing the merits of the Australian model, the authors acknowledge that a sudden shift from India's current restrictive 'Say on Pay' framework to such a stringent model would be too extreme and impractical. Therefore, to strengthen shareholder democracy in India, we propose that an advisory 'Say on Pay' vote should be introduced under Section 197 of the Companies Act, 2013, while retaining the existing mandatory 'Say on Pay' vote for breaches of ceiling limits. Additionally, as recommended by the Standing Committee on Finance, the government should develop a comprehensive formula for determining the outer ceiling limit, incorporating corporate profit growth and overall business development.

Table 22.2 Comparative analysis of say on pay frameworks

Country	Voting mechanism	Shareholder empowerment	Transparency & disclosure	Executive accountability
India	Restricted SoP Vote (Triggered at 11% cap)	Weak—Shareholders vote only if the cap is breached	Strong—Mandatory disclosure under Rule 5 of Managerial Remuneration Rules, 2014	Weak—No action is required if shareholders disapprove
UK	Binding Vote (Future Pay) & Advisory Vote (Past Pay)	Strong—Shareholders have a binding vote on future executive pay policies	Strong—Companies must disclose how feedback shaped pay structures	Strong—Boards must revise pay plans if rejected
US	Advisory SoP Vote	Moderate—Non-binding vote, but companies must justify compensation	Strong—SEC mandates disclosures in proxy statements	Moderate—Shareholders can pressure the board but cannot block pay
Australia	Two-Strike Rule	Very Strong—If 25% of shareholders vote 'No' twice, the Board must resign	Strong—Detailed remuneration reports	Very Strong—The board is removed if shareholder concerns persist
Germany	Advisory SoP Vote	Moderate—Non-binding vote on executive pay	Strong—Disclosure required under German Corporate Governance Code	Moderate—The supervisory board has the final say but considers shareholder input

22.7 Case Studies from Around the World

22.7.1 The Aviva CEO Resignation: A Case Study in Executive Compensation and Shareholder Activism

The resignation of Andrew Moss as CEO of Aviva in May 2012 exemplifies the intricate relationship between executive compensation, shareholder activism, and corporate governance.

Moss's departure followed shareholders' rejection of Aviva's executive pay packages at the company's Annual General Meeting (AGM), underscoring the increasing influence of shareholder input in determining appropriate executive remuneration.

Shareholders at Aviva expressed significant dissatisfaction with Moss's compensation package, raising concerns about excessive pay in light of the company's weak share price performance during his tenure (BBC, 2012). Earlier that year, Moss attempted to address shareholder unrest by declining an annual salary increase

(The Guardian, 2012b). However, this gesture was insufficient to appease investor discontent.

The rejection of Aviva's pay deals at the AGM clearly indicated shareholders' unwillingness to tolerate perceived misalignments between executive compensation and performance. This case highlights the critical role of the 'Say on Pay' Vote, a mechanism that empowers shareholders to express their opinions on executive compensation schemes, thereby promoting greater alignment between executive rewards and company performance.

22.7.2 Barclays and the Shareholder Revolt, a Case Study and Executive Compensation and Corporate Governance

In April 2012, Barclays faced significant shareholder backlash when approximately 27% of investors voted against the bank's remuneration report at its Annual General meeting (Reuters, 2012). This discontent was primarily directed at the £17.7 million pay package awarded to CEO Bob Diamond, which many shareholders deemed excessive given the bank's performance and the broader context of the financial crisis. This dissatisfaction was heightened by Barclays' involvement in the Libor rate-rigging scandal, which had emerged around the same time (Rose & Sesia, 2014).

According to Alan Dignam (2013), the shareholder activism in 2012 in the UK had concentrated on the poor performance of executives in relation to shareholder interests rather than opposing high pay levels, simplified by Barclays, where despite employee bonuses totaling £2.15 billion and shareholder returns of only £730 million, 31.4% of the shareholder did not support the CEO's remuneration package, even made intense efforts to gain shareholder approval.

22.7.3 Vikram Pandit and Citigroup's 'Say on Pay' Vote

Vikram Pandit assumed the role of CEO of Citigroup in December 2007, at a time when the financial sector was in significant turmoil. Under his leadership, Citigroup received multiple government bailouts due to substantial losses during the financial crisis. In a symbolic gesture to address concerns over executive compensation, Pandit accepted a salary of one dollar for 2010, pledging to forgo bonuses until the bank returned to profitability (The Wall Street Journal, 2012).

Despite Pandit's efforts, Citigroup's financial performance remained stagnant, and shareholders grew increasingly concerned about the disconnect between executive compensation and the bank's performance. In 2012, Citigroup proposed a $15 million pay package for Pandit, which drew significant criticism, given that the bank's stock had declined by 44% in the previous year (The Guardian, 2012a).

In the 2012 Citigroup annual meeting, more than 55% of the shareholders voted against the proposed executive compensation plan for Mr. Vikram Pandit (CNBC, 2012). This highlighted significant dissatisfaction with the alignment between executive pay and the bank's performance. A major criticism from shareholders was the absence of performance metrics in the compensation plan. In the wake of the shareholder revolt, Citigroup's Board of Directors acknowledged the vote and committed to engaging with investors to better understand their concerns. The bank also announced plans to explore a more formula-based approach to executive compensation, moving away from the discretionary nature of Pandit's payback (The Guardian, 2012a). The failed 'Say on Pay' vote significantly impacted Vikram Pandit's leadership. Just six months after the shareholder revolt, Pandit unexpectedly resigned as CEO of Citigroup. (Business Standard, 2012).

22.7.4 The Elon Musk-Tesla Compensation Saga

In January 2018, Tesla's board of directors approved a compensation plan for Elon Musk that was unprecedented in scale, potentially valued at up to almost $56 billion. The plan was structured around ambitious performance and milestones, including market, capitalization, revenue, and adjusted earnings, which Musk was required to achieve over a decade to receive stock options. This performance-based structure aimed to align Musk's incentives with the long-term success of Tesla, particularly during a period of financial challenges and skepticism regarding the company's viability (Barron, 2023).

Some shareholders supported Elon Musk's pay package due to concerns he might leave Tesla for another venture if not adequately compensated. Musk himself hinted at leaving unless he gained 25% voting control of Tesla. Tesla chairwoman, Denholm, stressed the importance of retaining Musk's dedication, stating that rewarding him is crucial to motivate his continued contribution to Tesla's success (CBS NEWS, 2018).

Conversely, some shareholders, including the California State Teachers Retirement System, opposed the pay package due to its magnitude and potential dilution of shareholder value. They also criticized the lack of focus on profitability in the compensation plan (CBS NEWS, 2018).

A Tesla shareholder named Richard Tornetta filed the lawsuit in June 2018, alleging that by approving the compensation plan without appropriate negotiation or independent oversight, Musk and other directors had violated their fiduciary duties. The Delaware Court of Chancery rendered a decision in the *Richard J. Tornetta v. Elon Musk* case on January 30, 2024, addressing the contentious compensation plan for Musk. The court determined that Musk was a controlling shareholder even though he only owned 21.9% of Tesla, subjecting the compensation plan to rigorous judicial scrutiny in Delaware corporate law, known as the '*entire fairness*' standard of review. The Delaware law applied because Tesla was incorporated in Delaware.

The 'entire fairness' standard of review is a stringent level of judicial scrutiny in Delaware Corporate law applied primarily in cases involving conflicts of interest, such as transactions with controlling shareholders. It requires the defendant to demonstrate both fair dealing and fair price. Fair dealing pertains to the process by which the transaction was approved, including timing, structure, and negotiations. Fair price relates to the economic and financial considerations of the transaction, ensuring it is substantively fair to the minority shareholders.

There are three major grounds on which the court ruled against this compensation scheme in this case: (i) **Lack of independence:** The court determined that several directors involved in approving the plan lacked independence from Musk, which compromised the fairness of the process. The compensation committee was described as acting more like an advisory body to Musk rather than an independent entity negotiating on behalf of the shareholders. (American Bar Association, 2024) (Dentons, 2024). (ii). **Informed shareholder vote:** the court ruled that the shareholder vote approving the compensation plan was not fully informed. The proxy statement provided to shareholders omitted critical information, including the relationships between Musk and majority of the board members, which raised questions about their independence. Additionally, internal projections indicated that the company was likely to meet the performance milestone for Musk's compensation shortly after the grant. (Harvard Law School Forum on Corporate Governance, 2024). (iii). **Failure to prove fairness:** the defendants failed to demonstrate that the compensation plan was the result of fair dealing or at a fair price. The court criticized the compensation committee for not considering alternative compensation structures or conducting a customer benchmarking analysis. (American Bar Association, 2024) (Dentons, 2024).

As a consequence of this judgment**,** the Board of Tesla made the decision to reapply for shareholder approval of the compensation package in light of the court's decision (Fortune India, 2024). This action was taken in an attempt to allay worries about the preliminary approval process's lack of independence and transparency.

On June 13, 2024, at the company's annual meeting, Tesla shareholders approved Elon Musk's compensation package.

The matter went to court again, and the Delaware Court of Chancery denied Tesla's motion to revise its post-trial ruling, reaffirming its January 2024 decision that rescinded Elon Musk's $55.8 billion compensation package (Tornetta v. Musk, 2024). Despite Tesla's attempt to ratify the package through a shareholder vote in June 2024, the court ruled that a stockholder vote cannot retroactively validate a conflicted-controller transaction, particularly when the original approval process failed to meet the 'entire fairness' standard. Additionally, the court found that the proxy statement for the shareholder vote contained material misstatements, further undermining its legitimacy and precluding its use as a ratification defence (Tornetta v. Musk, 2024).

We should understand from this case study that even though shareholders might have approved a compensation scheme, however, it does not bar scrutiny by the judicial authorities into such compensation scheme. Therefore, a compensation scheme should be appropriate, aligned with interest of the company and it should have clear and transparent Key Performance Indicators. Otherwise, it can be struck down by the shareholders or the judicial authorities.

22.7.5 *Vishal Sikka's Compensation Saga at Infosys*

Vishal Sikka was appointed as the MD & CEO of Infosys in August 2014 with high hopes to take Infosys into its glorious prime (Firstpost, 2017). However in February 2016, disagreements started brewing up between the founders and Vishal Sikka over hike in his compensation (Firstpost, 2017). The Infosys board of directors approved a 55% pay hike for Vishal Sikka. N. R. Narayana Murthy, one of the founders of Infosys, was especially vocal about the excessive compensation being paid to Vishal Sikka. This resulted in an all-out public spat between Vishal Sikka and Infosys. Vishal Sikka finally resigned from Infosys in August 2017, during this chaos Infosys stocks plunged in the stock market (Firstpost, 2017).

This public debacle could have been avoided if there was an advisory 'Say on Pay' vote in India. Shareholders/promoters like N. R. Narayana Murthy could have raised these concerns in the beginning when Vishal Sikka was appointed as the MD & CEO of Infosys or when the remuneration payable toVishal Sikka was increased, an advisory 'Say on Pay' vote could have helped to control the situation. As there was no breach of trigger limits under Section 197 of Companies Act 2013, no 'Say on Pay' vote was cast and hence a full-blown public spat was witnessed by Infosys, which even resulted in legal battles between Vishal Sikka and Infosys. Therefore, this case study helps us to understand the importance of advisory 'Say on Pay' vote.

22.8 Suggestions

1. Introduction of either advisory 'Say on Pay' vote or a mandatory 'Say on Pay' without requirement of touching ceiling limits under Section 197 of Companies Act 2013

Rationale: Implementing mandatory or advisory 'Say on Pay' vote would allow shareholders to express their opinions on executive compensation, annually. If a mandatory 'Say on Pay' vote is not viable then having at least an advisory 'Say on Pay' vote would compel companies to consider shareholder feedback seriously and disclose how these inputs have influenced remuneration policies in the future.

Alternatively, we can adopt the Australian model of two-strike rule that if 25% or more shareholders vote 'no' on a company's remuneration report at two consecutive annual general meetings, then all directors, except the managing director, must stand for re-elections within 90 days (Monem & Ng, 2013). Such an advisory 'Say on Pay' vote will make the exiting directors of a company to take shareholders advisory 'Say on Pay' vote seriously. However, it can lead to short-termism and might affect overall long-term growth of a company. Therefore, if we have to adopt two-strike rule then meticulous research should be conducted beforehand in order to prevent short-termism by any such change.

Impact: This approach can enhance transparency, and it will ensure that the executive remuneration reflects shareholders interest, and it would also promote better alignment of executive remuneration with corporate performance.

2. Strengthening disclosure requirements

Rationale: Enhance the existing disclosure norms to include detailed information about the rationale behind the executive pay packages, the criteria used for performance-based incentives. The disclosure should also include the details of findings of remuneration consultants.

Impact: Greater transparency will empower shareholders with the necessary data to make informed decisions, reducing the knowledge gap, and fostering a culture of accountability.

3. Regulation of remuneration consultants

Rationale: Introduce stringent regulations governing the activities of remuneration consultants to prevent conflicts of interest and ensure that their recommendations are in the best interest of the company and shareholders.

Impact: this will mitigate the risk of consultants, promoting excessive pay packages in exchange for personal favors, ensuring fair and justified executive compensation.

4. Ex-post facto advisory 'Say on Pay' vote

Rationale: We can also have an Ex-post facto (backward-looking) advisory 'Say on Pay' vote like one present in the UK (Gerner-Beuerle & Kirchmaier, 2018). This empowers the shareholders to express their opinion regarding an executive remuneration package after the performance of an executive is assessed in the next annual general meeting for the past year. This Ex-post facto (backward-looking) advisory 'Say on Pay' vote coupled with clawback provisions can prove to be a very effective tool to regulating the working of executives of a company.

Impact: This Ex-post facto (backward-looking) advisory 'Say on Pay' vote coupled with clawback provisions can help a company to recover remuneration paid to an executive for wrongful actions or wrongful conduct, without going to the judicial authorities for enforcement of such an order. This will also help in reducing unnecessary agency cost.

22.9 Conclusion

The restrictive approach to the say on the vote in India as provided by Section 197 of the Companies Act 2013 highlights significant limitations in empowering shareholders to influence executive remuneration. The current framework, which triggers shareholder approval, only when remuneration exceeds certain threshold, fails to address scenarios where shareholders may have legitimate concerns about executive compensation, even if it falls within the prescribed limits. This mechanism not only stifles shareholder activism, but also perpetuates a governance structure that can be seen as opaque and unresponsive to investor sentiments.

In contrast, global practices provide a range of models—from advisory to binding vote—that actively involve shareholders in determining executive pay, fostering a more transparent corporate governance environment. The analysis of various international 'Say on Pay' frameworks underscores the benefits of these approaches in promoting alignment between executive remuneration and company performance, thereby enhancing shareholder trust and corporate accountability. Therefore, Indian 'Say on Pay' Structure should also be reformed now in line with best practices followed around the world and it should at least include an advisory 'Say on Pay' vote.

We suggest that it would be in the interest of shareholders democracy in India that an advisory 'Say on Pay' vote is provided under Section 197 of Companies Act 2013 and the current mandatory 'Say on Pay' vote over ceiling limits should be retained. Further, as suggested by the standing committee, comprehensive formulae should be devised by the Government to determine outer ceiling limit taking into consideration increase in profits and overall growth of a company.

22.10 Scope for Further Research

This paper explores the 'Say on Pay' vote globally, with a particular focus on the Indian Model. The Indian 'Say on Pay' vote is distinctive as it only comes into effect when the managerial remuneration exceeds the trigger limits outlined in Section 197 of the Companies Act 2013. Further research could examine how this statutory ceiling functions as an artificial limit and investigate the frequency with which shareholders in Indian companies have rejected remuneration that surpasses the trigger limit. Such analysis would enhance our understanding of whether shareholders consider compensation exceeding the statutory threshold limit to be excessive or not. Another area worth exploring is whether India should adopt a mandatory 'Say on Pay' vote for executive remuneration that remains within the statutory limits of Section 197 of the Companies Act 2013. The influence of remuneration consultants and proxy advisory firms on shareholders' voting behavior in regard to the 'Say on Pay' vote is another interesting area that can be explored.

22.11 Implications

There are huge implications for the research work. This research will help shareholders and corporations to decide on appropriate remuneration for their executives in India. If India wants to compete in the global corporate world, then the standards of compensation that is paid in India should match the global standards. Therefore, the existing artificial legislative ceiling should be removed. On the other hand, it should be left open to the shareholders to choose how they want to pay the executives of a company on the basis of the existing needs of the company. The peril of excessive compensation can be regulated by an advisory 'Say on Pay' vote, which can help the shareholders and the company to decide how a remuneration policy is to be made. This will also help the company to know how shareholders feel about the compensation offered to an executive. In this manner, executive compensation in India will become competitive, and at the same time, we will be able to prevent instances of excessive executive remuneration.

References

American Bar Association. (2024, March). Delaware Court of Chancery orders rescission of Musk's Tesla compensation plan. *Business Law Today*. https://www.americanbar.org/groups/business_law/resources/business-law-today/2024-march/del-court-chancery-orders-rescission-musks-tesla-compensation-plan/

Barron's. (2023, May 16). Elon Musk's pay package explained. *Barron's*. https://www.barrons.com/livecoverage/tesla-shareholder-vote-results-meeting-today/card/elon-musk-s-pay-package-explained-0w9Z5dIizpn4P1pyo730

BBC News. (2012, May 3). Aviva CEO Andrew Moss resigns after pay revolt. *BBC News*. https://www.bbc.com/news/business-17987271

Bebchuk, L. A., & Tallarita, R. (2020). The illusory promise of stakeholder governance. *Cornell Law Review*, 106. https://www.cornelllawreview.org/wp-content/uploads/2021/02/The-Illusory-Promise-of-Stakeholder-Governance.pdf

Business Standard. (2012, October 16). Citigroup CEO Vikram Pandit resigns. *Business Standard*. https://www.business-standard.com/article/international/citigroup-ceo-vikram-pandit-resigns-112101600235_1.html

Cadbury, A. (1992). *The financial aspects of corporate governance*. https://ecgi.global/sites/default/files//codes/documents/cadbury.pdf

CBS News. (2018, March 21). Elon Musk pay package vote: CBS News explains. *CBS News*. https://www.cbsnews.com/news/elon-musk-pay-package-vote-cbs-news-explains/

CNBC. (2012, April 18). Citigroup's chief rebuffed on pay by shareholders. *CNBC*. https://www.cnbc.com/2012/04/18/citigroups-chief-rebuffed-on-pay-by-shareholders.html

Dignam, A. (2013). The future of shareholder democracy in the shadow of the financial crisis. *Seattle University Law Review, 36*(2), 639–668. https://digitalcommons.law.seattleu.edu/sulr/vol36/iss2/12/

Dentons. (2024, February 21). Delaware Court of Chancery strikes down Elon Musk's Tesla compensation plan. *Dentons*. https://www.dentons.com/en/insights/alerts/2024/february/21/delaware-court-of-chancery-strikes-down-elon-musks

Ferri, F., &Göx, R. F. (2018). Executive compensation, corporate governance, and say on pay. *Foundations and Trends® in Accounting, 12*(1), 1–103. https://doi.org/10.1561/1400000043

Firstpost. (2017, August 18). Infosys CEO Vishal Sikka resigns highlights: Company stocks fall steepest in over four years. *Firstpost.* Infosys CEO Vishal Sikka resigns highlights: Company stocks fall steepest in over four years—Firstpost.

Flammer, C., Toffel, M. W., & Viswanathan, K. (2021). Shareholder activism and firms' voluntary disclosure of climate change risks. *Strategic Management Journal, 42*(10), 1850–1879. https://doi.org/10.1002/smj.3313

Fortune India. (2024, June 13). Tesla shareholders re-approve $56 bn package for Elon Musk. *Fortune India.* https://www.fortuneindia.com/enterprise/tesla-shareholders-re-approve-56-bn-package-for-elon-musk/117173

Gerner-Beuerle, C., & Kirchmaier, T. (2018). *Say on Pay: Do shareholders care?* FMG Discussion Paper DP751, European Corporate Governance Institute (ECGI)—Finance Working Paper No. 579/2018. https://doi.org/10.2139/ssrn.2720481

Harvard Law School Forum on Corporate Governance. (2024, February 27). Tornetta v. Musk: Is the rule of law at work? *Harvard Law School Forum on Corporate Governance.* https://corpgov.law.harvard.edu/2024/02/27/tornetta-v-musk-is-the-rule-of-law-at-work/

Indian Express. (2024, June 13). Elon Musk wins shareholders' approval for Tesla pay package. *Indian Express.* https://www.indianexpress.com/article/world/elon-musk-shareholders-approval-tesla-pay-package-9393178/

Infosys. (2018). *CEO executive employment agreement.* https://www.infosys.com/investors/reports-filings/Documents/CEO-executive-employment-agreement2018.pdf

Jensen, M. C., & Meckling, W. H. (1976). Theory of the firm: Managerial behavior, agency costs, and ownership structure. *Journal of Financial Economics, 3*(4), 305–360.https://doi.org/10.1007/978-94-009-9257-3_8

Kishore, V. Shyam (2021). Say what on pay?—A comparative evaluation of the impacts of the regulatory reforms and COVID-19 on executive compensation in the UK, US and India. *NMIMS Law Review*1, 12–39, Available at SSRN: https://ssrn.com/abstract=3846808

Mallin, C. A. (2016). *Corporate governance* (3rd ed.). Oxford University Press.

Mason, S. A., Medinets, A. F., & Palmon, D. (2016). Say-on-Pay: Is anybody listening? *Multinational Finance Journal, 20*(4), 273–322. https://doi.org/10.17578/20-4-1

Ministry of Law. (1952). Report of the Company Law Committee. Government of India. http://oldopac.nls.ac.in:8081/xmlui/handle/123456789/559

Monem, R. M., & Ng, C. (2013). Australia's 'two-strikes' rule and the pay-performance link: Are shareholders judicious? *Journal of Contemporary Accounting & Economics.* https://ssrn.com/abstract=2334285

Reuters. (2012, April 27). Barclays pay opposed by 27 pct of votes cast. *Reuters.* https://www.reuters.com/article/markets/barclays-pay-opposed-by-27-pct-of-votes-cast-idUSWLA7153

Rose, C. S., & Sesia, A. (2013). Barclays and the LIBOR scandal (Harvard Business School Case 313–075). *Harvard Business School.* https://www.hbs.edu/faculty/Pages/item.aspx?num=43888

Standing Committee on Finance. (2010, August). Finance Committee Report (Report No. 21). Lok Sabha Secretariat, Parliament of India. https://eparlib.nic.in/bitstream/123456789/64114/1/15_Finance_21.pdf

Sarkar, J., & Sarkar, S. (2012). *Corporate governance in India.* SAGE Publications Pvt.

SEC. (2011). SEC adopts rules for say-on-pay and golden parachute compensation as required under Dodd-Frank Act. *U.S. Securities and Exchange Commission.* http://www.sec.gov/news/press/2011/2011-25.htm

Stathopoulos, K., & Voulgaris, G. (2015). The importance of shareholder activism: The case of say-on-pay. *Corporate Governance an International Review, 24*(3), 359–370. https://doi.org/10.1111/corg.12147

Stout, L. (2012). *The shareholder value myth: How Putting Shareholders First Harms Investors, Corporations, and the Public*. Berrett-Koehler Publishers.

Thomas, R. S., & Van der Elst, C. (2015). Say on pay around the world. *Washington University Law Review, 92*(3), 653–707.

The Guardian. (2012a, April 17). Citigroup shareholders reject CEO pay plan. *The Guardian*. https://www.theguardian.com/business/2012/apr/17/citigroup-shareholders-reject-ceo-pay

The Guardian. (2012b, May 3). Aviva rocked by shareholder rebellion over executive pay. *The Guardian*. https://www.theguardian.com/business/2012/may/03/aviva-shareholder-rebellion-executive-pay

The Wall Street Journal. (2012, April 17). Citigroup shareholders reject executive pay plan. *The Wall Street Journal*. https://www.wsj.com/articles/SB10001424052702304299304577349931225459386

Chapter 23
The Role of Corporate Governance on Investor's Confidence: A Comparative Study from the Perspective of the Banking Sector in India

Moumita Banerjee

Abstract In recent times, corporate governance has received considerable attention. It has developed as an important mechanism for improving a company's performance and reducing the risk of corporate collapses and accounting fraud. The banking sector acts as a backbone of modern business and plays a pivotal role in promoting stable economic growth in a country. The functioning of banks has ramifications for the sound operation of firms, capital allocation, and industrial expansion. Effective banking governance is more likely to allocate capital efficiently and stimulate productivity growth, thereby increasing the prosperity of an economy. The protection of data of the depositor and the risk of bank failures are quiet serious reasons that make the banking governance different from most of the non-financial firms. To understand how banks are structured, it is important to first understand their corporate governance. Part A of the study discusses the structural overview of corporate governance in the banking system and detailed computation of corporate governance scores of the top 10 commercial banks in India based on their market capitalization. Part B of the chapter intends to analyze the investment behavior by examining how the stock price fluctuations of these banks are affected by the corporate governance scores. The enhancement of investors' confidence can be better studied by analyzing fluctuations in stock prices. It unravels how the investors' confidence changes with respect to corporate governance in the banking sector and thereby explores how the shareholders' value maximizes their investment level in a sustainable way based on banking governance.

Keywords Corporate governance · Banking sector · Banking governance · Investors' confidence · Stock price

JEL classification C01 · C23 · G3 · G17 · G21

M. Banerjee (✉)
Department of Economics, South Calcutta Girls' College, University of Calcutta, Kolkata, West Bengal, India
e-mail: moumita@southcalcuttagirlscollege.com

V. Kandpal et al. (eds.), *Developments in Corporate Governance*,
Approaches to Global Sustainability, Markets, and Governance,
https://doi.org/10.1007/978-981-96-6366-8_23

23.1 Introduction

Effective corporate governance is contingent upon transparency in ethics and morality in the financial disclosure of a company. Governance is synonymous with discipline and is consensus-oriented, which follows the rule of law. The leadership, operational strategy, communication, and policies of a company depend on corporate governance to a great extent. As stated by the OECD, Corporate Governance is described as "the procedures and processes according to which an organization is directed and controlled. The corporate governance structure specifies the distribution of rights and responsibilities among the different participants in the organization, such as the board, managers, shareholders, and other stakeholders. It lays down the rules and procedures for decision-making." The World Bank furnishes a different perspective on corporate governance, which is more about elevating corporate impartiality, accountability, and transparency.

A well-organized banking system in a country is imperative for the economic growth of that country. As banks play a pivotal role in the cash-keeping of their depositors, they are morally liable for funds and finances that they disseminate inside an economy. Research indicates that the 1997 Asian crisis led to the failure of numerous banks and business enterprises in Thailand, Malaysia, South Korea, and Indonesia. These were primarily caused by inadequate corporate governance. This meltdown caused the majority of the world to adopt improved corporate governance and make it a priority. The Basel Committee Report 1999, Section 49 of corporate governance rules, laid the banks to disclose the paradigmatic practices of corporate governance in their financial performance and maintain transpicuousness in their balance sheets. In growing countries, banks are administered mainly by government bodies and laws contingent on hackneyed procedures.

Many studies have focused on examining the effect of corporate governance on stock prices, mostly in developed nations. Many studies also capture the same in the manufacturing sector in developing countries. This research delves into exploring how corporate governance influences stock prices, especially in India's banking sector. It needs to be cited that this is the first time this sort of investigation has been executed on panel data of banking in India.

23.2 Research Objective

- How does the functioning of different banks depend on corporate governance principles, and to what extent?
- Does the volatility of bank stock prices correlate with the corporate governance index?

23.3 Literature Review

Most organizations' management decisions and management structure determine their performance (Abrams, 1951). Company ownership and government bodies mainly monitor those decisions (Edirin & Ekwueme, 2015). The primary concerns of corporate governance are the position of the CEO, the size of the board, and its structure (Chong et al., 2009). A company's corporate governance procedures contribute to achieving a high degree of security for its stakeholders (Shleifer & Vishny, 1997). In the case of large firms, the shareholders do not have either the resources or time to monitor the actions of the managers as the shareholders are often dispersed. Hence, the monitoring instrument becomes the responsibility of the board of directors. The board may experience information asymmetry since it may not have as much awareness about what is happening within the company (Bebchuk, 2002). In order to extract private benefits or hide possible managerial errors, managers might have an interest in limiting the information flow to the board, which ultimately leads to information asymmetry between the directors of the board and the management of the firm. Management is able to control the information provided to the board and hide certain information from the board (Becht et al., 2007). For banks, the degree of information asymmetry is associated with the complexities and opaqueness of the banking business (Andres & Valleledo, 2008). These complexities include the trading of complex financial products, complex institutional structures, and the intriguing involvement of banks with each other (Adams et al., 2010). The complex financial products make it more difficult for customers to understand the products. For example, the bank's creation of credit default obligations (CDOs) became too complex to understand after the outbreak of the financial crisis of 2008 (Becht et al., 2011).

In banks, the flow of information is processed through the subsidiary boards before reaching the board of directors and management in the holding company. That is, the information is filtered through more actors before it reaches the top management of the bank. This structure makes the information flow and decision-making process in banks more complex, which calls for transparency and opaqueness in banking governance (Adams et al., 2010). Banks take risks as part of their business, and since the financial crisis, the role of risk management in banks has received more and more attention (Brogi & Lagasio, 2018). The regulatory bodies in banking have proposed the principles of good corporate governance that are long overdue, where several measures have been taken to improve supervisory control over risk management to guarantee effective risk management in the face of shifting conditions and stricter guidelines for directors' duties (Brogi, 2011). An empirical study of ownership concentration and profitability in banks demonstrates that a bank's ability to minimize risk is significantly improved by concentrated ownership (Shehzad et al., 2010). A study on Malaysian banks from 1997 to 1998 was conducted to examine the effect of ownership structure on efficiency levels. The authors used generalized Least Squares multivariate regression, and the result showed that it may indicate better efficiency in the performance of Malaysian banks (Adnan et al., 2011). Whenever a disparity exists between the ultimate controlling owner's cash flow and control rights, a larger gap correlates with diminished bank performance. We observe that where there is no disparity between the ultimate controlling owner's cash flow rights and control rights, a non-monotonic relationship exists between ownership concentration and the bank's performance (Azofra & Santamaría, 2011). Research indicates that corporate governance characteristics more effectively elucidate financial success than loan quality. We observe substantial evidence of a negative correlation between leverage and both financial performance and loan quality. CEO dualism is inversely correlated with financial performance. The magnitude of CEO incentive compensation correlates positively with financial performance while demonstrating a negative correlation with loan quality over the long term. A concave connection exists between financial success and both board size and average director age. We present little evidence of a correlation between anti-takeover mechanisms, the frequency of board meetings, and the affiliated character of committees and financial success (Grove et al., 2011). Particularly during the US financial crisis, a clear positive correlation between concentrated ownership and bank risk-taking was observed (Beltratti & Stulz, 2012).

The findings indicate a favorable correlation between these characteristics and corporate governance. This link significantly diminishes after accounting for national variations. The correlation between corporate governance and business performance is examined, as shown by net profit margin and return on equity (Bauer et al., 2004). The study compared the stock returns of companies with sound governance with those of companies with weak governance from 1997 to 2002. The research utilizes Deminor Corporate governance ratings, where they found that businesses with better shareholder rights are more successful, see faster sales growth, and require less capital investment than those with weaker shareholder rights (Drobetz et al., 2003; Gompers et al., 2001). Another study based on Regression and Correlation Analysis methods on 30 firms listed on the Nigeria Stock Exchange was conducted by Uwuigbe (2013) to examine the connection between corporate governance practices and share price. The result indicates that the corporate auditing committee has a positive effect on a firm's share prices, while firm ownership structure and share prices are negatively related. The effect of corporate governance, Return on Equity (ROE), enterprise value, Earnings per Share (EPS), and Dividends on the volatility of stock prices are examined for 30 Indian companies listed in BSE SENSEX for the FY 2017–2018. The study paid heed to improved corporate governance in order to achieve better stock performance (P.S. & Jain, 2020). Researcher tested the hypothesized association between corporate governance and liquidity in the stock exchange market on observations of 500 BSE-listed corporations during FY2013-FY2017 through a panel data regression model. The study provided evidence that corporate governance played an important role in the board of directors to improve financial disclosures, which would consequently promote the liquidity of the stock market (Sidhu & Kaur, 2019).

23.4 Data and Methodology

Purposive sampling is used, where the top 10 banks are chosen based on their market capitalization. The corporate governance index value is computed for each bank based on some parameters whose data is taken from secondary data sources like annual reports, corporate governance reports, and sustainability reports. The historical stock prices of the banks are observed from the Money Control website, and their yearly closing prices are considered.

Corporate Governance Index (CGI) values of banks are compared and studied to understand the banking governance structure. The comparison is discussed through graphs and tables. According to the literature, good governance indicates better stock performance in a company. Through empirical study in Stata software, it is analyzed whether the CGI index is a useful tool to predict stock prices.

23.5 Part A

23.5.1 Comparative Study on Corporate Governance

In this section, India's top 10 commercial banks, based on their market size, are chosen for the study period FY2019 to FY2024. The Corporate Governance Index (CGI) numbers of these banks are computed year-on-year, and a detailed comparative analysis is conducted (Table 23.1).

Table 23.1 The following table showcases the top 10 commercial banks in India, ranked by their market capitalization (as of June 2024)

Commercial banks	Type	Market Capitalization (Crore INR)
HDFC Bank	Private	1,251,536.12
ICICI Bank	Private	869,004.45
SBI	Public	765,330.10
Axis Bank	Private	396,977.39
Kotak Mahindra	Private	368,049.30
Bank of Baroda	Public	137,506.52
Punjab National Bank	Public	134,995.05
Indian Overseas Bank	Public	118,593.73
IndusInd Bank	Private	111,863.85
Union Bank of India	Public	102,946.81

Source Market Capitalization, Investing.com

23.5.2 Formulation of Corporate Governance Index

The CGI values are self-constructed for each of the selected banks over six financial years. The index utilizes information on five important corporate governance dimensions: accountability, Transparency, Fairness, Independence, and Ethics.

To construct the index value, some attributes under each of these five dimensions are taken within a specified governance mechanism, and each of these attributes is scored using binary coding. If an attribute under a specific dimension is present in the corporate governance disclosure report or annual report of a bank, then Yes = 1, otherwise No = 0. Then, a summation of these binary codes is taken to calculate the corporate governance score of that bank for a particular financial year (Table 23.2).

$$\text{CGI} = (\text{Corporate Governance score/Total attributes}) * 100$$

Accountability: The accountability principle states that the board is responsible for an organization's operations and the outcomes of its behavior. The board or company leadership must communicate important, relevant issues to shareholders and is responsible for evaluating the company's capability, potential, and performance. Conducting annual general meetings, the percentage of non-executive directors, and the percentage of non-executive directors in the Risk management committee are the important attributes to be concentrated on. According to the SEBI

Table 23.2 Components of corporate governance

Corporate governance dimensions	Attributes
Accountability	Annual General Meeting
	Composition of the audit committee's non-executive board members
	Composition of the Risk Management Committee's non-executive board members
Transparency	Annual Report
	Audit Committee
	Risk Management Committee
Fairness	Female on the main board
	The proportion of females on the main board
Independence	Non-executive board (in the main board)
	The proportion of non-executive boards (in the main board)
Ethics	Green technology
	Charity or donation
	CSR programs

Source Author's own work

Listing Agreement (clause 49), good corporate governance ensures 2/3rd of the non-executive board in the audit committee and 2/3rd of the non-executive board in the Risk management committee.

Transparency: The proper disclosure of the annual report, audit committee, and Risk Management committee reflects transparency in governance. Banking transparency helps reduce uncertainty and wild stock price fluctuations. During the global financial crisis of 2007–2008, risk management failure was the primary reason for the collapse of financial markets. It is imperative to incorporate particular risk management strategies and disclosures in the corporate governance framework.

Fairness: Better corporate governance is achievable when the presence of female directors and at least 30% of women on the main board of directors reflects fairness in governance. All selected banks have at least one female director on the board and score 1 in the study.

Independence: Independence is measured on the basis of two attributes- (i) the existence of a non-executive board in the main board and (ii) more than 2/3rd of a non-executive board in the main board.

Ethics: Ethics encompasses initiatives to assess responsibility for impact on society and the environment. Green banking uses innovative financing techniques to accelerate the deployment of clean energy technologies. Corporate Social Responsibility (CSR) programs include sustainable practices and engagement in philanthropic endeavors. As responsible corporate institutions, banking in India is actively contributing to social welfare (Table 23.3).

23.5.3 Comparative Analysis of CGI Values

- The Board of Directors manages an institution's operations. The non-executive director is an independent director who helps the company improve corporate credibility and governance standards. The non-executive director does not have any relation with the organization, which may affect the independence of his/her judgment. Low state ownership and more diverse and effective independent directors are essential for good banking governance.
- The strength of the board structure rests on the optimum composition of executive and non-executive directors, where non-executive directors make up at least 2/3 of the board. HDFC Bank scores less mainly due to the absence of this optimum combination of executive and non-executive directors. Moreover, the audit committee and the risk management committee do not fulfil the criteria for the required proportion of independent directors.
- The CGI value of ICICI Bank has improved over time. The composition of the non-executive board directors in the risk management committee has reached the criteria of its required proportion from FY 2020–2021.
- In the case of SBI, the non-executive directors in the audit and risk management committees became less than 2/3rds of the total directors in both committees in FY2021 and FY2022 compared to FY19 and FY2020. As a result, the CGI value

Table 23.3 Corporate Governance Index (CGI) for Indian commercial banks

Commercial banks	Year						
	2018–2019	2019–2020	2020–2021	2021–2022	2022–2023	2023–2024	Average CGI
HDFC Bank	69.23	69.23	69.23	69.23	69.23	69.23	69.23
ICICI	76.92	76.92	92.3	92.3	92.3	92.3	87.17
SBI	61.54	84.61	61.54	61.54	84.61	84.61	73.075
Axis Bank	92.31	92.31	92.31	92.31	92.31	92.31	92.31
Kotak Mahindra	69.23	69.23	84.61	92.31	92.31	92.31	83.33
Bank of Baroda	69.23	69.23	76.92	61.53	84.61	84.61	74.35
Punjab National Bank	69.23	69.23	76.92	76.92	84.61	84.61	76.92
Indian Overseas Bank	69.23	61.54	76.92	76.92	84.61	92.31	76.92
IndusInd Bank	69.23	76.92	84.61	92.31	92.31	92.31	84.61
Union Bank	69.23	69.23	76.92	76.92	76.92	76.92	74.36

Source Author's own computation based on the Annual reports

decreased to 61.54 in both FY2021 and FY2022 from 84.61 in FY2020. Later, the composition of the committees reached the optimum level, and the value strengthened to 84.61 again in FY2023 and FY2024.

- Axis Bank's CGI values remain stable each year, with an index value of 92.31, which is a commendable governance score. Axis Bank fulfills all the criteria for the attributes of governance principles except the presence of 30% female members as board directors.
- CGI values of Kotak Mahindra Bank have been improved with time. The bank has strengthened its commitment to Environmental, Social, and Governance (ESG) policy to promote long-term sustainable growth.
- Green banking in the Bank of Baroda has been strengthened since FY2021. The CGI value diminished and became the lowest in FY 2021–2022 due to the inadequate board composition of the audit and risk management committees.
- PNB and Indian Overseas Bank have improved their CGI value over time, which reflects the improved composition of their risk management committee.
- Union Bank attained medium CGI value. The risk management committee needs to be strengthened.
- Volatility in CGI values is observed in the case of SBI and Bank of Baroda.
- HDFC Bank maintains low stability, and Axis Bank maintains high stability in CGI values.

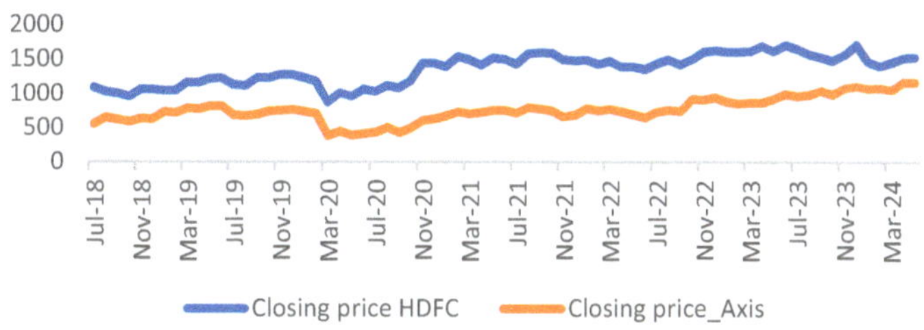

Fig. 23.1 Portfolio with high average CGI vs Portfolio with low average CGI. *Source* Author's own calculation based on data from moneycontrol.com

23.6 Part B

23.6.1 Empirical Findings and Analysis

The monthly stock closing prices of the bank with the lowest average CGI value and the bank with the highest average CGI value are studied to analyze how corporate governance affects stock price volatility.

- The Fig. 23.1 the stock price volatility of HDFC Bank and Axis Bank. HDFC Bank is associated with the lowest average CGI, and Axis Bank is associated with the highest average CGI among the selected banks. The graph depicts that a low average CGI value indicates more volatility in stock price than a portfolio with a high average CGI.
- The graph also indicates that HDFC has comparatively higher stock closing prices than Axis Bank. The stocks with low closing prices are highly risky as they are infrequently traded. Meanwhile, stocks with high closing prices are less risky due to their reputation and market capitalization. However, stocks that are priced lower attract potential investors because they have the opportunity to purchase more stocks in limited amounts.
- Again, low-priced stocks often attract investors due to their lower P/E ratios, which means investors have to pay less money to purchase them.
- Low-priced stocks are generally more volatile. However, the study depicts that the stock price of Axis Bank, which has had a very high and stable CGI index over the years, suggests less volatility. This opposite result is perhaps due to the bank's good corporate governance index value.

23.6.2 Testing for Granger Causality

Economic modeling for detecting Granger Causality in panel datasets was developed by Dumitrescu and Hurlin (2012). The underlying regression equation for the DH test is given by

$$y_{i,t} = \alpha_i + \sum_{k=1}^{K} Y_{ikYi,t-k} + \sum_{k=1}^{K} \beta_{ikxi,t-k} + \varepsilon_{i,t} \qquad i = 1,2,\ldots N \text{ and } t = 1, 2, \ldots T$$

which is used to test whether x causes y.

If past values of x are significant predictors of the current value of y even when the past values of y were included in the model, then the variable x has a causal influence on y. For individual I in period t, the two stationary variable observations are denoted by y_{it} & x_{it}. A balanced panel is required, and K is the lag order, and it is presumed that it is the constant for all cross-section units.

According to (Granger, 1969) testing for the significant impact of past values of X on Y is the process to determine the existence of a causal relationship between variables, while the DH test assumes that for some individuals, there can be causality but not necessarily for all individuals.

To run the Granger Causality test, the precondition is that the variable must be non-stationary at the level, i.e., the variable must have a unit root at the level. However, when it is converted into the first difference, the variable becomes stationary. xtcointtest implements a variety of cointegration tests like Kao, Pedroni, and Westerlund for large N large T-case.

To check the stationarity in yearly stock closing prices of selected banks in a panel data framework, Levin-Lin-Chu unit root test is first conducted with lag 1. The hypothesis is formulated as

H_0 : Panels contain unit roots.
H_1 : Panels are stationary.

The test result shows–

	Statistic	p-value
Unadjusted t	−29.7199	
Adjusted t*	−31.8543	0.0000

The test statistic shows that the p-value is significant; hence, we can reject the null hypothesis and state that the stock closing prices do not have a unit root at the level, i.e., the data is stationary at the level.

Now, another method of unit root test called the Hadri LM test is conducted where the formulated hypothesis is

H_0 : All panels are stationary.
H_1 : Some panels contain unit roots.

The test result shows–

	Statistic	p-value
z	2.2261	0.0130

Here, the p-value is significant as it lies below 5% level of significance. Hence, the null hypothesis is rejected, which means closing price data at the level containing the unit root. So, the data is checked at first difference. The result shows–

Hadri LM test for D. Closing price

	Statistic	p-value
z	−0.8398	0.7995

Here, the p-value of the z statistic is 79.95%, which means the null hypothesis is accepted, i.e., the closing price data at the first difference is stationary.

If the Hadri LM test is assumed as the benchmark, then it can be concluded that the stock closing price data is non-stationary at level but stationary at first difference.

Now, the DH test is run by the Stata command xtgcause for the panel set of 90 observations. Since xtgcause allows panel data for at least 8 periods, 3 more financial years have been included, and their respective CGI values were calculated.

The hypothesis for causality in DH test is –

H_0 : stock closing price does not Granger-cause CGI.
H_1 : Stock closing price does Granger-cause CGI for at least one panel.

Here, CGI is the independent variable, and closing price is the dependent variable. The hypothesis examines whether the corporate governance index of banks is useful for forecasting their stock price. The result indicates a significant p-value (0.000) at a 95% critical level in bidirectional causality. It dictates the importance of CGI on stock price and vice-versa.

23.7 Corporate Governance on Investors' Confidence

Investment decisions play a vital role and are most critical for all the stakeholders due to the involvement of large amount of money. The quality of corporate governance practices ensures reliability and improved quality of financial information and enhances the investor's confidence. In the real world, investors often make wrong investment decisions that are compelled by emotions rather than rationality. As long as investors become more optimistic about an organization, the management tends to invest more. As a result, corporate investment decisions vary with the level of confidence among the shareholders. Sound corporate practices are reflected through the corporate governance index computed based on governance principles. A higher index value of an organization indicates higher investor confidence, which is why the share price of the company soars. At the same time, weak governance plummets the value of the stock. That means, good management and administration of companies affect the confidence level of investors heavily.

The empirical findings of the chapter indicate a bidirectional causality between CGI values and stock closing prices of banks. It indicates that good banking governance leads to a higher stock valuation for these banks. On the other hand, higher stock prices boost the internal management of banks and enrich their corporate governance. Moreover, Part A of the study analyzed how weak banking governance is reflected through volatility in stock prices. However, the investors must consider the bank's financial ratios, such as profitability, return on equity (ROE), and non-performing assets (NPAs). However, they must study and be well informed about the corporate governance structure of banks before purchasing bank shares.

23.7.1 A Case Study on YES Bank Crisis

India's fastest-growing private bank, YES Bank, witnessed a severe financial crisis in 2019–2020, which triggered panic among the bank's stakeholders, investors, the general public, and deposit holders. The plight was the result of poor corporate governance, which consisted of a lack of transparency, bad lending practices, lack of due diligence, and gender discrimination on the board.

23.7.2 Background of the Crisis

YES Bank Pvt. Ltd., one of India's largest banks (current market capitalization of INR 564.37 billion), started its banking business in the year 2004. The trio, Mr. Rana Kapoor, Mr. Ashok Kapur, and Mr. Harkirat Singh, co-founded Rabo India Finance in partnership with Rabo Bank of Netherlands in the year 1977–78. In early 2003, following the differences with Rabo Bank, the trio led the way for the existence of YES Bank by selling off their stakes in Rabo India Finance. Mr. Singh quit his role in 2003 when Mr. Ashok Kapoor was appointed as a non-executive chairman. During the 26/11 terrorist attacks in Mumbai, Asok Kapur was killed when Rana Kapoor became the sole promoter and started acting as managing director and CEO of YES Bank. The bank earned various prestigious recognitions both at national and internal levels and achieved most of the growth during 2010–2015.

Mr. Kapoor turned out to be an aggressive lender to corporations, and the bank used to charge high interest rates compared to other public and private commercial banks. YES Bank was the first institution in the world to raise a loan through the A/B loan facility of the International Finance Corporation. The bank's book of accounts revealed a loan of Rs.55,633 crores, and the book of deposit reflected Rs. 74,192 crores in March 2014. During that time, loans increased enormously and reached Rs. 2.25 trillion as of September 30, 2019. A global financial service company, UBS, in

Table 23.4 Credit-Deposit ratio (CDR) of YES bank from FY16 to FY20 (in Rs. Crore)

Financial Year	2016	2017	2018	2019	2020
Total advances	117,087	132,262.68	203,533.86	2,41,500	1,71,4,33.90
Total deposits	132,375.8	142,873.86	200,738.15	227,610	105,364
CDR (%)	88.45	92.57	101.39	106.10	162.71

Source Author's own computation based on the Annual reports of YES Bank

2015, made an analysis and pointed out that YES bank was facing serious governance issues for which they were providing loans to corporate companies like DHFL, CCD, Essel group, Reliance group of industries, etc., and to some struggling sectors of India during past years like non-banking financial companies, real estate firms, and construction sector which could not repay the loan in a stipulated timeframe. The financial position of YES Bank deteriorated to such an extent that the share price of the bank stood at just Rs. 16 as of 6 March 2020, from Rs. 400 in 2018. Bank run situations resulted due to panic rather than true insolvency when the bank showed steady withdrawal by the depositors on the balance sheet (Table 23.4).

Credit-Deposit ratio is calculated by CDR = Total advances/Total deposits. The ratio conveys what proportion a bank lends each rupee it receives in deposits. The optimum credit-deposit ratio is 80% to 90%. CDR of 100% indicates a high liquidity crunch. The alarming situation caused from FY18, where CDR exceeds 100% and leads to a full-blown crisis by FY20.

The question was raised about the expropriation of financial resources by CEO Rana Kapoor, and RBI asked him to vacate his position by January 31, 2019. The independent auditors confirmed irregularities in the banking operations. The reputation of the bank was even corroded when one of the bank's independent directors resigned, alleging the compromising of corporate governance standards. In March 2020, RBI initiated a reconstruction plan for YES Bank and took over the management of the bank. Mr. Prasant Kumar became the new CEO, and two additional directors were appointed to the board of directors by RBI.

23.7.3 Data Analysis of the Case Study

The Corporate Governance Index (CGI) for YES Bank is computed using the formula given in Sect. 5.2. The following Table 23.5 shows that the value has sharply decreased during the financial years 2019 and 2020.

Accountability is an important parameter in corporate governance, which helps build trust with shareholders and investors. In 2018–2019, the board of YES Bank was not sufficiently accountable, and the audit committee and risk management committee lacked the required proportion of non-executive members on the board. This resulted in poor scrutiny of the actions of the executive management team. Moreover, the main board also lacked the required proportion of female directors in

Table 23.5 Corporate governance index for YES Bank from FY2016 to FY2020

Bank	Year					
	2015–2016	**2016–2017**	**2017–2018**	**2018–2019**	**2019–2020**	**Average CGI**
YES Bank	53.84	61.53	61.53	46.15	46.15	53.84

Source Author's own computation based on the annual reports

all 5 years. According to the Gender socialization theory, women are more caring toward ethical issues and more concerned with the welfare of the stakeholders. Gender discrimination in the corporate board of directors impacts the transparency and decision-making process of the board and positively affects board cohesion and conflict. The independent director and chairman of the bank, Mr. Uttam Prakash Agarwal, cited concerns about mismanagement of handling investor identities and binding agreements with the board. He resigned on 10 January 2020.

23.7.4 Regulatory Responses by RBI

- To take over the bank's management, RBI primarily used powers under Section 36ACA of the Banking Regulation Act, 1949.
- RBI imposed a moratorium on 5th March 2020 for a period of 30 days and restructured the board of directors to improve the level of transparency in governance.
- On October 23, 2017, RBI imposed a monetary penalty of Rs. 60 million on the bank for delayed reporting of an information security event involving YES bank ATMs and non-compliance with the guidelines issued by RBI on IRAC (Income Recognition Asset Classification) norms.
- Stricter regulatory measures have been implemented to monitor the bank's risk management and lending practices and prevent future governance failures.

23.8 Conclusion

The corporate governance framework encourages the efficient use of resources and requires accountability in the administration of those resources. For the smooth running and fair distribution of benefits among the stakeholders, institutions need a governing structure with a prescription comprising decrees and doctrine. The principles of corporate governance encircle accountability, fairness, transparency, independence, and social responsibility. The operations and profitability of a firm are called into question by poor corporate governance. The investors who turn a deaf ear to corporate governance are taking undue risk.

This study attempts to construe the importance of idiosyncrasies of corporate governance in the stock markets and thus enhance the decision-making of the investors. According to the two-way causal effect, banks with sound corporate governance and financial practices disclose information more effectively, which eventually improves investor returns. On the other hand, the improved stock performance will provide internal management incentives to expedite the governance structure's improvement.

References

Abdullah, S. (2004). Board composition, CEO duality and performance among Malaysian listed companies. *Corporate Governance: THe International Journal of Business in Society, 4*(4), 47–61.

Abrams, F. W. (1951). Management's responsibilities in a complex world. *Harvard Business Review*, 29–34.

Adams, R. H. (2010, March). The role of boards of directors in corporate governance: A conceptual framework and survey. *Journal of Economic Literature, 48*(1), 58–107.

Adnan, M. H. (2011). A panel data analysis on the relationship between corporate governance and bank efficiency. *Journal of Accounting, Finance and Economics, 1*(1), 1–15.

Andres, P. (2008). Corporate governance in banking: The role of the board of directors. *Journal of Banking and Finance, 32*(12), 2570–2580.

Azofra, V., & Santamaría, M. (2011). Ownership, control and pyramids in Spanish commercial banks. *Journal of Banking & Finance, 35*(6), 1464–1476.

Barry, T. A. (2011). Ownership structure and risk are publicly held and privately owned. *Journal of Banking & Finance, 35*(5), 1327–1340. https://doi.org/10.1016/j.jbankfin.2010.10.004

Bauer, R., Guenster, N., & Otten, R. (2004). Empirical evidence on corporate governance in Europe: The effect on stock returns, firm value and performance. *Journal of Asset Management, 5*, 91–104.

Bebchuk, L. A. (2002). Asymmetric Information and the Choice of Corporate Governance Arrangements. *Harvard Law and Economics Discussion Paper No. 398*. Available at SSRN: https://ssrn.com/abstract=327842 or http://dx.doi.org/10.2139/ssrn.327842

Becht, M., Bolton, P., & Roell, A. (2007). Corporate law and governance. *Columbia Law School Scholarship Archive*, 829–943.

Becht, M., Bolton, P., & Roell, A. (2011). Why bank governance is different. *Oxford Review of Economic Policy, 27*(3), 437–463.

Beltratti, A., & Stulz, R. M. (2012). The credit crisis around the globe: Why did some banks perform better? *Journal of Financial Economics, 105*(1), 1–17. http://www.sciencedirect.com/science/articlc/pii/S0304405X11002790

Bennis, W. (2000). Don't hire the wrong CEO. *Harvard Business Review, 78*(3), 170–176.

Berger, A. N. (2014). Executive board composition and bank risk taking. *Journal of Corporate Finance, 28*, 48–65. https://doi.org/10.1016/j.jcorpfin.2013.11.006

Boyd, B. (1994). Board control and CEO compensation. *Strategic Management Journal, 15*(5), 335–344.

Brick, I. P. (2006). Journal of corporate finance. *CEO compensation, director compensation and firm performance: Evidence of cronyism?, 12*(3), 403–423.

Brickley, J. C. (1997). Leadership structure: Separating the CEO and chairman of the board. *Journal of Corporate Finance, 3*(3), 189–220.

Brogi, M. (2011). Board, governance and firm performance: Are financial intermediaries different? *Corporate Ownership and Control, 8*(3), 60–68. https://doi.org/10.22495/cocv8i2sip6

Brogi, V., & Lagasio, V. (2018). Environmental, social, and governance and company profitability: Are financial intermediaries different? *Corporate Social Responsibility and Environmental Management*, 1–12.
Busta, I. S. (2014). Ownership concentration and market value of European banks. *Journal of Management & Governance, 18*(1). https://doi.org/10.1007/s10997-012-9223-8
C.W.J, G. (1969). Investigating causal relations by econometric models and cross-spectral methods. *Econometrica, 37*, 424–438.
Caprio, G. L. (2007). Governance and bank valuation. *Journal of Financial Intermediation, 16*(4), 584–617. https://doi.org/10.1016/j.jfi.2006.10.003
Chen, K. A. (2007). Audit committee, board characteristics, and auditor switch decisions by Anderson's clients. *Contemporary Accounting Research, 24*, 1085–1117.
Chen, Y. M. (2005). Audit committee composition and the use of an industry specialist audit firm. *Accounting and Finance, 45*, 217–239.
Chong, A. G. (2009). Corporate governance reform and firm value in Mexico: An empirical assessment. *Journal of Economic Policy Reform, 12*(3), 163–188.
Cohen, J. K. (2004). The corporate governance mosaic and financial reporting quality. *Journal of Accounting Literature, 23*, 87–152.
Donaldson, L. (1991). Stewardship theory or agency theory: CEO governance and shareholder returns. *Australian Journal of Management, 16*(1), 49–64.
Drobetz, W. S. (2003, January). Corporate governance and expected stock returns: Evidence from Germany. *SSRN Electronic Journal.*
Dumitrescu, E., & Hurlin, C. (2012). Testing for Granger non-causality in heterogeneous panels. *Economic Modelling, Elsevier, 29*(4), 1450–1460.
Edirin, J. (2015). Interest rate regime and the performance of the Nigerian capital market. *Studies and Scientific Researches, 1*(22), 43–54.
Elsayed, K. (2007). Does CEO duality really affect corporate performance? *Corporate Governance: An International Review, 15*(6), 1203–1214.
Gompers, P. I. (2001). Corporate governance and equity prices. *National bureau of economic research.* http://www.nber.org/papers/w8449
Granger, C. (1969). Investigating causal relations by econometric models and cross-spectral methods. *Econometrica, 37*(3), 424–438.
Grove, H. P. (2011). Corporate governance and performance in the wake of the financial crisis: Evidence from US commercial banks. *Corporate Governance: An International Review, 19*(5), 418–436.
Levine, R. (2004). The corporate governance of banks: A concise discussion of concepts and evidence. *World Bank Policy Research Working Paper No. 3404.* https://papers.ssrn.com/sol3/papers.cfm?abstract_id=625281
Livne, G. M. (2011). Bankers' compensation and fair value accounting. *Journal of Corporate Finance, 17*(4), 1096–1115. https://doi.org/10.1016/j.jcorpfin.2011.04.010
Luu, H. N. (2015). Corporate governance, regulation and bank risk-taking behaviour in developing Asian countries. *Journal of Governance and Regulation, 4*(1–2), 201–223.
Macey, J. &. (2003). The corporate governance of banks. *Economic Policy Review, 9*(1), 91–107.
Medis, A. (2019). Corporate governance: A review of theoretical and practical implications. *The International Journal of Business Management and Technology*, 63–68.
Nair-Reichert, U. W. (63). Causality tests for cross-country panels: A new look at FDI and economic growth in developing countries. *Oxford Bulletin of Economics and Statistics, 63*, 153–171.
P.S., R. & . (2020). Impact of corporate governance on stock performance-evidence from BSE sensex. *The Indian Journal of Industrial Relations, 55*(3), 543–558.
Raheja, C. (2005). Determinants of board size and composition: A theory of corporate boards. *Journal of Financial and Quantitative Analysis, 40*(2), 283–306.
Samontarary, D. (2010). Impact of corporate governance on stock prices of the nifty 50 broad index listed companies. *International Research Journal of Finance and Economics*(41).

Shehzad, C., & d. (2010). The impact of bank ownership concentration on impaired loans and capital adequacy. *Journal of Banking & Finance, 34*(2), 399–408. https://doi.org/10.1016/j.jbankin.2009.08.007
Shleifer, A. (1997). A survey of corporate governance. *Journal of Finance, 52*, 737–783.
Sidhu, M. (2019). Effect of corporate governance on stock market liquidity: Empirical evidence from Indian companies. *Decision, 46*(3), 197–218.
Stearns, L. (1993). Board composition and corporate financing: The impact of financial institution representation on borrowing. *Academy of Management Journal, 36*(3), 603–618.
Turley, S. (2004). The corporate governance effects of audit committees. *Journal of Management and Governance, 8*, 305–332.
Uwuigbe,. (2013). Corporate governance and share price: Evidence from listed firms in Nigeria. *African Research Review, 7*. https://doi.org/10.4314/afrrev.v7i2.8
Zahra, S. A., & Pearce, J. A. (1989). Boards of directors and corporate financial performance: A review and integrative model. *Journal of Management, 15*(2), 291–334.

Chapter 24
Developments of Corporate Governance Regulations

Rajiv Iyer, Vedprakash Maralapalle, and Poornima Mahesh

Abstract The chapter gives a detailed analysis of the events that defined the innovations in the regulations of corporate governance, particularly in the emergent markets. Speaking of key corporate governance regulations, it is high time to grasp the main objectives and principles, which include the initiatives to exercise transparency, accountability, sufficient protection of shareholders' rights, etc. Then, the chapter outlines the evolution of corporate governance regulations, sectioning the regulations based on their historical past and the factors that influenced them. Subsequently, the chapter expands on the prospects and issues of corporate governance regulations in emerging markets to embark on the development and implementation processes. The study examines how international standards such as the IFC's Corporate Governance Development Framework and the OECD Standards of Corporation Administration have influenced the reform agenda and provided a roadmap for implementing effective corporate governance frameworks. In the last section of the chapter, the authors discuss future prospects of corporate governance and reveal tendencies and potential threats that can influence the development of the corresponding regulations in emerging countries. This review looks at more critical ESG factors in today's business environment and the emergence of new technologies, as well as a greater emphasis on sustainability and corporate social responsibilities and how they affect the structures and regulations of companies. Thus, this chapter has shed substantial light on the changes in corporate governance regulations in emerging markets while making recommendations that can benefit the policymakers, the regulators, and the industry players as well.

R. Iyer (✉)
Department of CSE, Amity School of Engineering and Technology, Amity University Maharashtra, Mumbai, India
e-mail: rajivkjs@gmail.com

V. Maralapalle
Department of Civil Engineering, Amity School of Engineering and Technology, Amity University Maharashtra, Mumbai, India

P. Mahesh
University of Mumbai, Mumbai, India

V. Kandpal et al. (eds.), *Developments in Corporate Governance*,
Approaches to Global Sustainability, Markets, and Governance,
https://doi.org/10.1007/978-981-96-6366-8_24

Keywords Corporate governance · Emerging markets · Regulatory innovations · Economic evolution · Transparency · Shareholder rights

24.1 Introduction

Corporation governance is now virtually an important item on the global business agenda because companies around the world are faced with unprecedented tests. It is therefore significant to note that corporate governance is a crucial component of the management of corporations with the aim of creating sustainable business practices that uphold shareholders' and other stakeholders' interests as well as supporting the health of the economy (Aguilera & Florackis, 2019). It is especially noteworthy that within the last few years, authorities and scholars have paid much attention to the establishment of regulations in the sphere of company authority, especially in developing marketplaces. They have attained a high rate of economic liberalization and globalization and equally suffered a lot of vices in the areas of transparency, accountability, and rights of shareholders. Thus, numerous EMs have established new CG regulations or have substantially overhauled the existing rules (Bebchuk & Tallarita, 2020). The objective of this chapter is to present the main changes in the regulations concerning corporate governance and, in particular, within emerging markets. Next, the focus will be on the modern development of corporate governance frameworks due to the economic, social, and political shifts and the consequent shifts in accounting for business and organizational performance. We will also examine the effectiveness of the regulations regarding corporate governance as influenced by international organizations and standards for emerging markets and offer future trends of corporate governance regulations in a nutshell (Cuomo et al., 2021).

24.1.1 The Corporate Governance Regulations

Corporation governance regulations thus have an important function of assisting in the management and regulation of organizations. These regulations map out the manner in which firms are managed and operated, besides explaining the legal relations between firms and different parties in the market, such as shareholders, directors, and managers. Analyzing the list of rules and requirements of the corporate governance regulations, one can point out that they contribute to establishing effective standards for the corporations' behavior and actions, as well as to enhancing transparency and shareholder rights' protection (De Masi et al., 2022). In recent years, awareness of the significance of corporations' actions and the need for their regulations in the form of governance rules has increased notably, especially after colossal scandals and financial failures. The recent corporate frauds such as Enron and WorldCom and the global financial crisis, which started in 2008, exposed the frailty of business authority systems and the need for better corporate mechanisms to avoid frauds, poor management, and exorbitant risk-taking (Ferrarini & Zhu, 2018).

24.1.2 Emerging Markets: The Usual Suspects

The necessity of corporate governance regulations at the present stage is rather obvious. However, the problems, opportunities, and threats of the emerging markets differ from those of the developed ones. Common features of EM include higher economic growth rates, a quickly developing Foreign Direct Investment, and an environment that is changing rather quickly in terms of legislation (Fiandrino et al., 2019). Thus, one of the more pressing issues dominating the agenda of emerging markets is the issue of how to meet the conflicts of interests of different parties involved, such as foreign investors, domestic shareholders, and the government. The distribution of ownership is still high. How is it that in many EMs most of the shares are still owned by a few families or state-owned enterprises? This could lead to a spillover of ownership from one clearly defined area to the other, leading to conflicts between controlling shareholders and minority shareholders; enforcement of corporate governance regulations becomes complicated in such structures (García-Meca & Palacio, 2018).

24.1.3 International Organizations and Standards

Irrespective of the challenges highlighted above, emerging markets have noted growth in the implementation and development of company regulations in the last few years. On this and the following grounds, one can notice that the participation of international organizations and the adoption of international standards have become the leading drivers of this progress (Grewal et al., 2021). Certain types of international organizations, such as the Organization for Economic Co-operation and Development (OECD) and the International Finance Corporation (IFC), have been especially active in the policy-making process for good corporate governance practices in emerging markets. For example, many nations, including many OECD developing markets, have embraced the corporate governance guidelines that the Organization for Economic Cooperation and Development has developed. These principles outline the best practices for creating sound corporate governance structures and include elements such as shareholders' rights and responsibilities and, the duties of participants, and revelation, as well as responsibilities of the boards of directors (Grosman et al., 2019).

24.2 Principles and Goals of Corporate Governance Regulations

Due to this, many countries have formulated corporate governance regulations that aim to regulate the ways through which companies are managed so as to enhance the provision of accountability to shareholders as well as the protection of their interests. Such regulation offers the manner and conduct in which companies are managed and governed and specifies how various stakeholders are entitled to and required to behave and perform, including shareholders and directors as well as managers (Haque & Jones, 2020; McNulty & Nordberg, 2021).

In the area of corporate management and governance, the main goals of the law are to combat corporate fraud, protect the interests of participants such as shareholders, and promote economic growth. In this way, the very existence of governance regulations establishes high ethical benchmarks of corporate conduct (Merendino & Sarens, 2020).

24.2.1 Objectives of Corporate Governance Regulations

The key objectives of corporate governance regulations are to:

a. Promote transparency and accountability: To address the issue of a lack of data disclosure and poor-quality states, regulatory regulations require corporations to report financial performance, significant risks, and other key facts. (Nguyen et al., 2020).
b. Protect shareholder rights: There is evidence that supports claims that corporate governance regulations assist in protecting shareholders and providing them with information regarding the amounts and/or other ways the company is performing or making decisions (Puchcta-Martínez & Gallego-Álvarez, 2019).
c. Prevent corporate misconduct: In this respect, the regulations of corporate governance facilitate the setting of rules and procedures that can minimize fraud, mismanagement, and risky behavior (Schnatterly et al., 2021).
d. Promote long-term value creation: Many of the corporate governance regulations provide guidelines for corporations to practice the idea of long-term wealth generation as opposed to immediate earnings. This involves buying equipment, research and development, and training of employees as well as Acquiring new and better technologies for the business to operate (Schwartz-Ziv & Weisbach, 2021).

Figure 24.1 illustrates the key components of good corporate governance using a house-like structure.

According to the figures, good corporate governance comprises and has its foundation in these three practices. The situation means that when one wishes to manage a company's affairs in a correct, ethical, and efficient manner, one should ensure

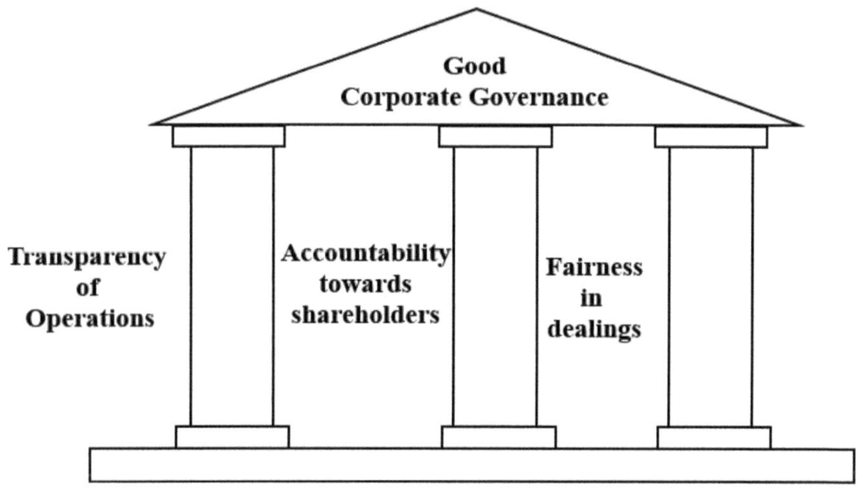

Fig. 24.1 Good governance pillars

transparency, accountability, and fairness to all the shareholders as well as other stakeholders. The structure implies that development requires equal levels of importance from the three pillars working together within structures of good governance. A compromised element within any part of the corporate governance structure makes it possible to affect the entire framework according to (Tarmuji et al., 2020).

24.2.2 Role of International Organizations and Standards

Global corporate governance legislation follows directions established by global organizations through their developed standards according to this research investigation. The OECD and IFC jointly developed worldwide corporate governance best practices through their principles and guidelines to promote good corporate governance practices internationally (Zattoni & Cuomo, 2020). The OECD Fundamentals of the Company Administration contains a standard system to develop robust corporate governance frameworks around the world. The framework establishes requirements regarding shareholder privileges as well as equality treatment of clients and includes provisions regarding stakeholder involvement in management, while defining disclosure protocols and accountability frameworks, and setting protocols for the chairman council (Habib, 2024). In the same way, the IFC's Corporate Governance Development Framework is a systematic approach toward measuring and enhancing CG practices in emerging markets. These IFRs and CAFs have contributed to increasing people's awareness about the significance of good CG and assisted in outlining measures in which international CG standards may be adopted globally.

24.2.3 Challenges and Future Outlook

It is vital to assess today's level of corporate governance regulations, which, even though they have matured in the last several years, still have a long way to go. The main issue identified is the issue of stakeholder management, which is whereby the organization has to manage the various demands of shareholders, employees, customers, and society (Ahmad & Mahmood, 2024). A further difficulty under this aspect is the imperative to constantly update corporate governance frameworks and adjust them to novel phenomena and trends, including ESG factors and AI and blockchain technologies.

In order to overcome these challenges, there is a need for policymakers and regulators to further strengthen and elaborate the rules governing the corporate governance of companies, either by the codification of laws and rules, standard setting, or benchmarking with international standards yet localizing them to fit the country's circumstances. This will indeed call for concerted efforts, especially from the governments, companies, and investors, among other players in the corporate world, to observe good governance and practice sustainable business (Li & Zaiats, 2018). On this basis, one might suggest that by investing in the realization of the objectives aimed at the improvement of the corporation's governance system, countries can secure a long-term perspective for themselves in the global economy. Effective corporate governance and certainly corporate social responsibility solutions assist in the attraction of foreign investment, development of economic stability, and welfare of society. It remains a crucial element of the thriving and dynamic business climate necessary for the survival and prosperity of companies in present and future administrations and legislation (Gilson, 2018).

24.3 Historical Context of Industrial Atmosphere Regulations

A number of economic, social, and political factors have caused a shift in corporate governance metrics during the last few decades. It is important to note that the origins of corporate governance regulations can be dated back to the early twentieth century; consequently, in response to the growing volume and complexity of commercial activity and the ensuing need to give readers and investors pertinent and correct information about the company, the first legal documents pertaining to the framework of corporate governance were published.

24.3.1 Early Developments

The modern theories of corporate governance originated in the early twentieth century as the conception of the first rules and principles in this field appeared; the primary purpose of those regulations was to prevent the concentration of power in the hands of individuals or companies and protect the shareholders and other stakeholders. The earliest corporate governance rules, which were developed in the 1920s and 1930s in response to the need to control business behavior, may be used to trace the history of corporate governance.

24.3.2 Post-War Period

More specifically, the subject of corporate governance regulations started to develop in the context of post-war reconstruction and the aftermath of the Second World War, mainly in the fifties and the sixties. The appearance of new technologies and the expansion of multinational corporations require more extensive regulation and control.

24.3.3 International Initiatives

It is worth pointing out that both OECD and IFC have been one of the most influential players in establishing the framework of corporate governance rules across the world. The OECD Principles of Corporate Governance, for instance, outline how to create strong corporate governance frameworks. Among the important subjects covered are the Rights of shareholders, nondiscrimination against shareholders, involvement of stakeholders in corporate governance, transparency and reporting, and the duty of the board of directors (Schnatterly et al., 2021).

24.3.4 Emerging Markets

The implementation of corporate governance in emerging markets has been characterized by different evidence peculiar to the environment. They have achieved impressive levels of economic development and globalization over the last decade; however, these markets have encountered multiple issues such as opacity, unsuitable corporate governance, and inadequate shareholders' rights protection. By providing guidance and support for the creation of appropriate regulations, the OECD has been actively promoting the raising of corporate governance standards, particularly in emerging nations.

24.3.5 *Future Outlook*

Therefore, with future insight, corporate governance regulations will continue to play a very significant role in placing emphasis on the business environment. In the future, there is a likelihood of a higher emphasis on corporate governance frameworks due to the globalization of the economy, issues of sustainability, and social responsibility. The following are some significant issues and trends that will probably influence how corporate governance laws are drafted in the future:

a. The higher relevance of Environmental, Social, and Governance (ESG) aspects for investment activities and business management.
b. The increasing awareness around the world regarding environmental sovereignty and the exigency of organizations to perform convergence issues like climate change, human rights, and corruption (Iyer et al., 2024).
c. The increasing application of innovative features, such as artificial intelligence and blockchain, as well as the necessity for new corporate governance rules, take into consideration the current state of the world (Merendino & Sarens, 2020).
d. The rising expectations by investors, consumers, and various stakeholders for disclosure and reporting of the events and actions of corporations (Schnatterly et al., 2021).
e. The requirement to pay attention to the outstanding issues influencing the EMs, including the high ownership centrality, the underdeveloped institutional environment, and the cultural and political limitations of the CG (Tarmuji et al., 2020).

In tackling these challenges, emergent markets will remain crucial in enhancing and strengthening the regulation of corporate governance, integrating best practices from across the globe into the emerging markets environment. This will have to be achieved through the collective cooperation of governments, firms, investors, and other groups, guided by the principles of good governance and sustainable business practices (Habib, 2024).

24.4 Corporate Governance Regulations in Emerging Markets

Several corporate governance reforms have been implemented across abundant emerging markets in recent years due to a pressing demand relating to the framework of foreign investment attraction, economic growth stimulation, and shareholders and other stakeholders' protection. The procedure for formulating and executing corporate governance policies in developing economies has not been easy as the markets encountered a number of difficulties mainly due to the concentrated ownership structure, a problem with the legal and institutional framework, and a lack of institutional capacity and knowledge (Ahmad & Mahmood, 2024).

Among the current problems, one of the crucial issues arising from the globalization of businesses in the world markets is the conflict of interests of a new type of player, which includes foreign investors, domestic shareholders, and the government. It is a common observation in most of the emerging economies where the dominance of the market is found within a few families or state-owned enterprises. The conflicts of interest that arise between the majority owners and the minority shareholders create agency issues and make it extremely challenging to put corporate governance laws into practice (McNulty & Nordberg, 2021). The other is the element of capacity and experience at the institutional level for implementing corporate governance. Most emergent countries do not possess robust legal and normative environments accompanied by a pool of skilled professionals and organizations required to enforce corporate governance codification proactively. This can result in unequal treatment of corporations and no one being accused of legal wrongdoing despite clearly being in the wrong (Merendino & Sarens, 2020).

However, much can still be done, and several other issues and scopes exist in the corporate governance regulation of emerging markets. A significant issue is the necessity to refine the existing corporate governance models to incorporate new trends and technologies, such as ESG factors influencing investors' decisions, AI, and blockchain.

24.4.1 Unique Challenges and Opportunities

Analyzing the issues and challenges in the field of corporate governance, one can assess that the specifics of the emerging markets are boldly revealed in this aspect. On the one hand, such markets are characterized by high growth rates and profitability for foreign investors. However, they also harbor some immense vulnerabilities in the areas of transparency, accountability, and especially the assertion of shareholders' rights (Nguyen et al., 2020). Foreign investors' demands, as well as domestic shareholders and government interests, inevitably make a balance a critical issue that emerging markets have to solve. One of the more noteworthy characteristics of many emerging markets is that they are highly concentrated, with major blocks of shares often held by a few families or state-owned enterprises. This can lead to a conflict of interest between the controlling shareholders and the minority shareholders. At the same time, it becomes a challenge to implement the regulation on corporate governance (Pucheta-Martínez & Gallego-Álvarez, 2019). The other problem is the lack of Institutional capability and knowledge of corporate governance. Most EMs have not developed comprehensive legal frameworks and professional undertakings qualified to enforce CG regulations and policies for organizational setups. This can result in unequal application of the rules, and also no one is being held responsible for corporate wrongdoings.

Another important on-venture benefit that can be connected with emerging markets is the theme of innovativeness and venturing. Despite this, the population of emerging markets mainly has high literacy, a higher degree of technological readiness, and is ready to expand or improve the usage of new models. Thus, emerging markets can develop a base that would make them leaders in the formation of new technologies and revolutionary business strategies (Schnatterly et al., 2021).

To effectively harness these opportunities and overcome the challenges that will come their way, emerging markets will need to strengthen and increase the stringency of the rules concerning the corporate governance regulation systems that are already in place as well as to develop new ones using key foreign experiences while taking into consideration certain specific conditions of the given countries. It will take a collective effort of politicians, business organizations, shareholders, and the public to encourage the/methods of good governance and sustainable management of business organizations (Habib, 2024). Particular emphasis should be placed on the concern as to how best to institute sound methods of shareholders' communication and safeguard mechanisms. In the case of emerging markets, specifically, minority shareholders should have a say in the decision-making process of the corporation as well as have the protection of their rights, and with this movement, more foreign investors should be attracted to invest in the markets (Ahmad & Mahmood, 2024).

24.4.2 Regulatory Approaches in Different Emerging Markets

The types of regulation that have emerged increasingly in different emerging markets have largely changed over the past couple of years due to the desire to cultivate FDI and encourage economic growth, as well as investor and shareholders' protection. Newly developed economies, especially those of the emerging markets, have had peculiar issues in developing corporate governance codes and enforcing them, issues like concentrated ownership structures, relatively and absolutely weak legal and institutional structures, and lack of institutional and professional capacity and experience. Another problem is that organizations' capacity and knowledge of corporate governance have to be developed. Some of the emerging markets have not developed sophisticated legal and regulatory bodies as well as experienced professionals and institutions to properly implement and enforce corporate governance regulations. This can result in the rules being applied arbitrarily and no one being held responsible for corporate wrongdoings (Gilson, 2018).

The structure implies that development requires equal levels of importance from the three pillars working together within structures of good governance. A compromised element within any part of the corporate governance structure makes it possible to affect the entire framework according to Tarmuji et al. (2020).

24.4.3 Role of International Organizations and Standards

Global corporate governance legislation follows directions established by global organizations through their developed standards according to this research investigation. The OECD and IFC jointly developed worldwide corporate governance best practices through their principles and guidelines to promote good corporate governance practices internationally (Zattoni & Cuomo, 2020). The OECD Fundamentals of the Company Administration contains a standard system to develop robust corporate governance frameworks around the world. The framework establishes requirements regarding shareholder privileges as well as equality treatment of clients and includes provisions regarding stakeholder involvement in management while defining disclosure protocols and accountability frameworks and setting protocols for the chairman council. (Habib, 2024). Among all the best practices that have recently become popular in the member countries of the emerging markets, the practice of creating independent boards of directors deserves special attention. Most countries have put in place measures in the form of legal provisions that seek to prescribe the minimum number or the proportion of independent directors that should be on the board for the purpose of bringing efficiency and independence to the board. Independent directors have several responsibilities in relation to monitoring the management, dealing with related party transactions, and safeguarding the interests of other minority shareholders (Habib, 2024).

The other improvement that has found its way into the management of corporations in emerging markets is the establishment of board committees such as the audit, nomination, and remuneration committees. These committees can also assist in strengthening the board in the sense that they are responsible for specific areas and do not have the board's inherent conflict of interest or lack of independence. For instance, audit committees can be a significant factor in the corporate unprotected fiscal report in addition to the societal oversight of constitutions and systems of internal control and risk administration (Schnatterly et al., 2021). Again, there has been development in the area of transparency and disclosure standards within emerging markets. Most nations have passed measures on the annual reporting obligation of ownership details, related party dealings, leadership remunerations, and other significant issues. This information is important in that it enables investors to determine probabilities and possibilities of a business making an ideal investment decision.

However, these achievements are accompanied by issues where corporate governance regulations for emerging markets should still be developed more extensively. One of these areas is the protection of minority shareholders' rights. Market share concentration in many EMs is still high since many companies are owned by a few families or state giants. Conflicts of interest arise between dominant owners and smaller shareholders as a result, and implementing corporate governance regulations will be extremely difficult (Tarmuji et al., 2020). In response to this vice, some of the emerging markets have adopted laws that seek to guarantee minority shareholder rights. For instance, it is common for some countries to impose regulations where controlling shareholders must seek the permission of other shareholders who are minorities for specific transactions such as related party transactions or any move to de-list the company. Different countries have adopted tag-along rights, which entitle minority shareholders to join the controlling shareholder in the sale of their shares (Zattoni & Cuomo, 2020).

24.5 Role of International Organizations and Standards

The use of gears to signify these components implies that they complement one another and operate in the system of transnational business management. Every 'gear' is a task that would aggregate the entire operation in setting and enforcing good governance policies at the international level (Fiandrino et al., 2019). Likewise, the IFC has formulated its own corporate governance methodology and tool kit that it applies when assisting different companies in emerging markets to enhance their corporate governance systems. The IFC's, for example, the Corporate Governance Development Framework, is a systemized model for evaluating and strengthening corporate governance structures and has been implemented by firms in over one hundred countries.

In addition to these international organizations, there has been an emphasis on the recommendation and implementation of global standards and benchmarks in corporate governance. International bodies such as the ICGN and GRI have purposefully prepared principles and standards that aim to uphold good governance and sustainability worldwide.

Herzen, international standards, and the principles that are described as the best practices have helped significantly in the formation of regulations on corporate governance in emerging markets. They too have served the useful purpose of giving a standard model to measure as well as enhance corporation governance standards and thus play a key role in thc awareness of good governance and have acted as a guide to the effective regulatory measures. Another way through which international organizations have assisted the emergence of corporate governance regulations in emerging markets is through technical cooperation services. The OECD, for example, instructs some of the emerging market countries on how to formulate and enact corporate governance codes and regulations. The IFC has also assisted firms and authorities in emergent economies by offering technical support and training on matters of corporate governance in order for them to develop the capacity that will enable them to implement and enforce the regulations on corporate governance.

However, there may still be issues with the organizational and legal measures related to the corporate governance laws being implemented in the EM nations today. In most countries, there is a lack of infrastructure and adequate skills required for implementing and subsequently enforcing such regulations; there often are significant cultural and political nuances that hinder corporate governance (Grewal et al., 2021).

To deal with these issues, it will therefore be crucial for such international organizations to further extend support and wisdom to such EM countries. This could encompass extended technical support and development, encouragement of countries to already adhere to international standards regarding corporate governance, and assistance in tackling problems and particularities that hinder the efficient implementation of corporate governance systems (Pucheta-Martínez & Gallego-Álvarez, 2019).

24.6 Outlook for Corporate Governance Regulations in Emerging Markets

Following the continuation of the growth of emerging markets and their integration into the global system, the future trend of corporate governance regulations in these markets is indeed one of the challenges as well as opportunities. On one hand, some contingencies are peculiar to emergent economies and have significantly slowed the efforts to establish good CG practices; these include high ownership concentration, feeble legal systems, and inadequate institutional capacity and knowledge. On the contrary, the rising role of these markets in the world, along with the greater attention to sustainability and social responsibility, can turn into a valuable chance for emerging markets to upgrade their legislators' enactments concerning corporate governance and introduce themselves as leaders on the mentioned subject (Merendino & Sarens, 2020).

It was found that the consideration of ESG issues is one of the most significant forces that may shape the future of corporate governance legislation in emerging nations. Such themes as climate change, human rights, and corruption remain very relevant, and more and more people expect companies to respond to these themes and act sustainably and responsibly. Best practice ESG for emerging markets is consequently one whereby the national corporate governance frameworks are receptive to ESG principles and hence can effectively court foreign investments and ensure sustainable corporate development. A further element that might have a substantial impact on the future development of corporate governance laws in emerging economies is the trends in the use of new technologies, such as blockchain and artificial intelligence. These technologies may affect the sphere of company management and increase the role of corporate governance and responsibility; thus, emerging markets will need to adjust their legislation accordingly. For instance, the application of the blockchain would allow for strengthening the transparency and the traceability of the supply chains, and artificial intelligence would help to bring additional value to risk management and the making of decisions (Gilson, 2018).

Regarding the aforementioned opportunities and challenges brought about by globalization, policymakers and regulators should continue to improve corporate governance laws and regulations by drawing on global experience while also taking into account the unique characteristics of national contexts. This will involve all the stakeholders, such as the government and policymakers, corporations and investment houses, and other agencies acting in unison to champion good governance and proper business practices (Hertig, 2018) (Judge & Talaulicar, 2017).

Corporate governance regulations in emerging markets face multiple upcoming trends while facing various significant challenges leading to their future development. ESG considerations have emerged as one of the leading developments because they now determine investment choices and corporate strategic directions. ESG considerations have become crucial for successful governance frameworks through their integration in businesses of emerging markets. Both environmental regulatory compliance and social issue resolution related to labor rights and community engagement and human rights form the core responsibility of this approach. The governance structure requires companies to maintain transparency together with accountability functions and proper board member selection and executive compensation levels and systematic risk management.

The implementation of ESG principles throughout operations allows businesses to boost stakeholder perception while lowering their risk exposure to environmental and social requirements and achieving sustainable growth. For instance:

1. Organizations need to combat climate change through their implementation of renewable energy systems and green supply chain operations to lower carbon footprints.
2. The practice of social responsibility involves maintaining fair employee working conditions in supply networks as well as supporting community development through charitable initiatives and employment generation.
3. A robust system of governance helps executive bodies of organizations monitor sustainability matters in order to maintain operational alignment with strategic enterprise decisions.

24.6.1 Future Trends

Emerging markets will experience substantial changes in corporate governance due to multiple future trends that will impact this evolution.

1. Secure technological tools based on blockchain enable both transparent transaction records that cannot be modified or altered. Risk management systems experience better analysis through artificial intelligence (AI) because it processes extensive data sets with greater efficiency than human capabilities.
2. International organizations will continue directing regulatory developments throughout different national borders because of both globalization trends and standardization efforts. The standards used by emerging markets should follow the best international practices, which require local context adaptation.
3. Organizations will intensify their stakeholder involvement beyond shareholder relations to include employees and suppliers as well as customers and their engagement in corporate decision-making. Through such an inclusive method,

different groups representing business stakeholders create trust regarding operations.
4. The evolving business environment will force regulatory bodies to develop fast adaptations for maintaining effective supervision while protecting innovation.

Emerging market entities have chances to expand their growth through developing governance practices which adhere to international regulatory standards.

The implementation of ESG factors extends business longevity by satisfying stakeholder requirements. Advanced technologies provide better compliance tracking capabilities to regulatory bodies while existing legislation must be properly adapted to honor their implementation. Emerging markets have a unique opportunity to position themselves as leaders in sustainable business practices by integrating ESG considerations into their corporate governance frameworks while leveraging technological advancements for enhanced transparency and efficiency in compliance monitoring systems (Hertig, 2018).

24.6.2 ESG Regulations and Digital Innovations

Business operations in emerging markets need to integrate Environmental Social Governance (ESG) considerations into their corporate governance structures due to rising importance. Business operations undergo substantial changes because ESG regulations demand companies to work on environmental matters, starting with climate change solutions, and social needs, regarding labor rights, along with community engagement, and governance requirements to enhance transparency and build accountability.

24.6.2.1 Impact of ESG Regulations

The application of ESG regulations directs business operations into sustainable directions. Core principles which companies implement within their operations allow stakeholders to trust the business more while lowering non-compliance risks and building lasting sustainable structures. For instance:

1. Organizations must handle environmental concerns through climate change action, which includes using sustainable practices to decrease carbon release, deploying renewable energy systems, and developing green value chains.
2. Social Responsibility involves active promotion of fair work conditions in supply chains combined with making positive contributions to local communities through money donations and creating employment opportunities.
3. A strong board oversight system for sustainability issues guarantees these matters receive strategic integration in corporate decision-making activities.

Role of Global Regulatory Bodies.

Essential to ESG governance policy development stand the global regulatory bodies whose main task is to establish these frameworks.

1. The Securities and Exchange Commission under the United States has prioritized ESG disclosure transparency by making it mandatory for public companies to reveal specific environmental threats.
2. The Organisation for Economic Co-operation and Development through its guidelines supports nations to create solid corporate governance systems with ESG principles integrated.
3. Organizations under IFRS (International Financial Reporting Standards) maintain the Sustainability Disclosure Standards Board to establish global reporting guidelines for sustainability information.

Several organizations create international standards which inspire national regulatory systems in emerging markets.

24.6.2.2 Digital Innovations

Blockchain technology operates as a digital innovation that ensures secure transaction records which are tamper-proof. Low human comparison rates transform because artificial intelligence works to evaluate extensive data collections at rates better than what humans can accomplish. These technologies enable better compliance oversight through the provided tools, while companies need to integrate them properly under current law.

24.7 Conclusion

The emerging markets experienced corporate governance reform shifts during recent years to achieve three main goals: attracting foreign investors while developing economies alongside shareholder and stakeholder protection. Newly industrialized countries face multiple obstacles in their efforts to establish solid corporate governance practices through the combination of large corporate holdings and inadequate legal structures and inexperienced organizational management. Several emerging markets are now showing better results in the adoption of international standards for implementing corporate governance regulations. Recent corporate governance changes incorporated more internal board committees while increasing independent director numbers and enhancing the level and transparency elements. International organizations along with regulation standards represent the most significant elements which influence this particular subject. The OECD and IFC as part of multinational organizations have provided emerging market corporate governance regulations with assistance through both models and strategies. They also promote the development of international best practices while enhancing institutional capacity and expertise development. In the field of emerging market corporate governance regulations, multiple prospective developments will either create threats or generate fresh business prospects.

Developing markets enjoy expanding potential to enhance their corporate governance through recent trends, which include environmental, social, governance (ESG) considerations, and technological innovations, and sustainability-based social responsibility programs. In order to harness these opportunities, emerging markets will have to see advancement in the corporate governance codes adopted by them by incorporating global standards in their own environment. It will need the cooperation of governments, corporations, and other stakeholders, which will support good governance and favorable business climates. With a focus on the efficient mechanisms of corporate governance, emerging markets have the capability to encourage FDI in their economy, help in the growth and stability of the economy, and, in turn, contribute positively to the welfare of society. There can be no doubt that it is an essential element of a dynamic competitive commercial climate and one in which there will remain much focus from those in political authority and the civil service.

References

Aguilera, R. V., & Florackis, C. (2019). Global corporate governance: The maturing of a field. *Journal of Management, 45*(7), 2685–2692. https://doi.org/10.1177/0149206318822718

Ahmad, Z., & Mahmood, Z. (2024). Corporate governance regulation in regulatory space of a developing country: Actors, strategies and outcomes. *Journal of Accounting in Emerging Economies, 14*(1), 99–124. https://doi.org/10.1108/JAEE-01-2023-0004

Bebchuk, L. A., & Tallarita, R. (2020). The illusory promise of stakeholder governance. *Cornell Law Review, 106*, 91–178.

Cuomo, F., Mallin, C., & Zattoni, A. (2021). Corporate governance codes: A review and research agenda. *Corporate Governance: An International Review, 29*(1), 89–117. https://doi.org/10.1111/corg.12356

De Masi, S., Słomka-Gołębiowska, A., & Paci, A. (2022). Women on boards and corporate social responsibility: A systematic literature review. *Journal of Management & Governance, 26*(1), 241–278. https://doi.org/10.1007/s10997-021-09558-4

Ferrarini, G., & Zhu, S. (2018). Board compensation and ownership structure: Empirical evidence for UK listed firms. *European Business Organization Law Review, 19*(3), 595–629. https://doi.org/10.1007/s40804-018-0122-9

Fiandrino, S., Devalle, A., & Cantino, V. (2019). Corporate governance and financial performance for engaging socially and environmentally responsible practices. *Social Responsibility Journal, 15*(2), 171–185. https://doi.org/10.1108/SRJ-04-2017-0078

García-Meca, E., & Palacio, C. J. (2018). Board composition and firm reputation: The role of business experts, support specialists and community influentials. *Business Research Quarterly, 21*(2), 111–123. https://doi.org/10.1016/j.brq.2017.12.001

Gilson, R. J. (2018). From corporate law to corporate governance. In J. N. Gordon & W. Ringe (Eds.), *The Oxford handbook of corporate law and governance* (pp. 334–362). Oxford University Press.

Grewal, J., Serafeim, G., & Yoon, A. (2021). Shareholder activism on sustainability issues. *Journal of Applied Corporate Finance, 33*(3), 66–77. https://doi.org/10.1111/jacf.12457

Grosman, A., Aguilera, R. V., & Wright, M. (2019). Lost in translation? Corporate governance, independent boards and blockholder appropriation. *Journal of World Business, 54*(4), 258–272. https://doi.org/10.1016/j.jwb.2019.03.004

Habib, A. M. (2024). Financial development and corporate governance efficiency: An overview of MENA. In *Handbook of banking and finance in the MENA Region* (pp. 147–169). https://doi.org/10.1007/978-3-030-90547-0_7

Haque, F., & Jones, M. J. (2020). European firms' corporate biodiversity disclosures and board gender diversity from 2002 to 2016. *The British Accounting Review, 52*(2), Article 100893. https://doi.org/10.1016/j.bar.2019.100893

Hertig, G. (2018). Governance by institutional investors in a stakeholder world. In J. N. Gordon & W. Ringe (Eds.), *The Oxford handbook of corporate law and governance* (pp. 821–838). Oxford University Press.

Iyer, R., Maralapalle, V., Mahesh, P., & Patil, D. (2024). Generative AI and LLM: Case study in e-commerce. In *Generative AI and LLMs: Natural language processing and generative adversarial networks* (pp. 253–272). De Gruyter. https://doi.org/10.1515/9783111425078-013

Judge, W. Q., & Talaulicar, T. (2017). Board involvement in the strategic decision making process: A comprehensive review. *Annals of Corporate Governance, 2*(2), 51–169. https://doi.org/10.1561/109.00000008

Li, T., & Zaiats, N. (2018). Information environment and earnings management of dual class firms around the world. *Journal of Banking & Finance, 92*, 249–268. https://doi.org/10.1016/j.jbankfin.2018.05.002

McNulty, T., & Nordberg, D. (2021). Ownership, activism and engagement: Institutional investors as active owners. *Corporate Governance: An International Review, 29*(2), 209–219. https://doi.org/10.1111/corg.12358

Merendino, A., & Sarens, G. (2020). Crisis? What crisis? Exploring the cognitive constraints on boards of directors in times of uncertainty. *Journal of Business Research, 118*, 415–430. https://doi.org/10.1016/j.jbusres.2020.07.042

Nguyen, T., Locke, S., & Reddy, K. (2020). Does board gender diversity influence firm performance? Empirical evidence from Vietnam. *Equality, Diversity and Inclusion: An International Journal, 39*(6), 737–755. https://doi.org/10.1108/EDI-04-2019-0120

Pucheta-Martínez, M. C., & Gallego-Álvarez, I. (2019). An international approach of the relationship between board attributes and the disclosure of corporate social responsibility issues. *Corporate Social Responsibility and Environmental Management, 26*(3), 612–627. https://doi.org/10.1002/csr.1717

Schnatterly, K., Clark, B. B., Howe, J., & DeVaughn, M. L. (2021). Regulatory oversight and fraud: A review of the empirical evidence. *Journal of Management, 47*(6), 1546–1577. https://doi.org/10.1177/01492063211013205

Schwartz-Ziv, M., & Weisbach, M. S. (2021). What do boards really do? Evidence from minutes of board meetings. *Journal of Financial Economics, 139*(2), 528–549. https://doi.org/10.1016/j.jfineco.2020.08.014

Tarmuji, I., Maelah, R., & Tarmuji, N. H. (2020). The impact of environmental, social and governance practices (ESG) on economic performance: Evidence from ESG score. *International Journal of Supply Chain Management, 9*(3), 1074–1084.

Zattoni, A., & Cuomo, F. (2020). Why adopt codes of good governance? A comparison of institutional and efficiency perspectives. *Corporate Governance: An International Review, 28*(3), 297–317. https://doi.org/10.1111/corg.12318

Part IV
ESG Reporting, Risk Management, and Sustainability

Chapter 25
ESG Reporting and Strategy for Sustainable Marketing

Abhishek K. Singh and Maryam Vaziri

Abstract Environmental, social, and management issues (ESG) have become the main focus of activity and the focus of investors' attention. The ESG reporting landscape is ambiguous: hundreds of reporting standards form a so-called alphabetical list of abbreviations without a single common framework for controlling disclosure by companies. ESG corporate data varies greatly and provides comparable and decision-making information to organisations, which need to properly take ESG considerations into account when making their decisions on sustainable growth. Environmental, social, and governance (ESG) is the basic conceptual approach that is usually applied to understand and evaluate an organisation's business and marketing strategies and performance on a range of ethical and sustainability issues. ESG certainly plays a very significant role in the long-term goals of the organisation and its market positioning. ESG also provides corrective measures to forecast the risks and opportunities in the real areas of marketing and business development. In the present global capital market, some stakeholders also identify the ESG factors to analyse the global trends of marketing and consumer behaviour, as certainly it would help the organisations visualizers to plan business strategies for smooth growth and to assess future risks. In global business scenarios, the mission of the ESG program is to ensure accountability and implement systems and processes to manage company impacts, such as corporate social responsibility. ESG initiatives mainly contribute to business sustainability for responsible business management. This chapter examines the dynamic landscape and forecasts significant changes in ESG reporting regulation in the coming years.

Keywords ESG reporting · Marketing strategy · Sustainability · Business development · Social growth

A. K. Singh (✉)
School of Liberal Studies and Media, University of Petroleum and Energy Studies, Dehradun, India
e-mail: asingh8319@gmail.com

M. Vaziri
DaneshPajoohan Pishro Higher Education Institution, Isfahan, Iran

V. Kandpal et al. (eds.), *Developments in Corporate Governance*,
Approaches to Global Sustainability, Markets, and Governance,
https://doi.org/10.1007/978-981-96-6366-8_25

25.1 Introduction

In recent years, environmental, social and management reporting (ESG) and sustainable marketing have become increasingly important key elements of corporate strategy. ESG reports and reporting contain information on environmental impacts, social audits, and governance based awareness practices. Sustainable marketing aims to market products and services in a socially and environmentally responsible manner. This chapter mainly examines the importance of ESG reporting, the role of sustainable marketing, and their interaction in enhancing corporate responsibility and consumer confidence (Lagasio & Cucari, 2019).

The roles of ESG reporting are the framework that allows companies to disclose information about their activities in three critical areas:

- Environmental responsibility,
- Social responsibility and
- Management practices.

Companies that actively participate in ESG reporting are better prepared to comply with these regulatory requirements and avoid possible legal and financial penalties. The role of sustainable marketing means the promotion of products and services that are not only economically, but also socially and environmentally responsible. The role of sustainable marketing can be summarised in the following sections:

1.1 Consumer demand: Demand for environmentally friendly and ethically produced products and services are growing. Sustainable marketing responds to this demand by emphasising the environmental and social benefits of the company's offering, thereby attracting consumers who care about the environment.

1.2 Brand differentiation: In a competitive market, sustainable marketing can distinguish a brand from its competitors. By focusing on sustainable development, companies can create a unique value proposition that resonates with consumers who promote ethical consumption (Khatib et al., 2022).

1.3 Long-term growth: Sustainable marketing is not limited to short-term benefits. The goal is to build long-term relationships with consumers based on trust and shared values. Companies using sustainable marketing methods are more likely to retain customer loyalty and achieve sustainable growth.

1.4 Innovation and efficiency: The implementation of the principles of sustainable development often promotes innovation, as companies want to develop new products and processes that minimise their impact on the environment. This can increase efficiency and save money, which can be highlighted in marketing campaigns to further enhance the brand's reputation. The interaction between ESG commitment and Sustainable Marketing, Corporate responsibility and customer loyalty are at the heart of ESG reporting and sustainable marketing. This transition is crucial for several reasons:

- Integrated Strategy: Companies that adapt their ESG reports to sustainable marketing create a coherent strategy that reinforces their commitment to sustainable development. This holistic approach ensures that sustainability is integrated into both business practices and communication with consumers.
- Add Credibility: ESG reports provide a factual basis for sound marketing statements. If companies can protect their marketing messages with reliable information from ESG, this will increase their faithfulness and credibility, as well as reduce the risk of misunderstanding.
- Stakeholder Collaboration: ESG reporting and sustainable marketing are stakeholder collaborations. ESG reports address the concerns of investors, regulators, and communities, while sustainable marketing is about consumers. Together, it will lead to the development of a comprehensive engagement strategy that meets the needs and expectations of all stakeholders.
- Continuous improvement: ESG reports often highlight areas where a company can improve its sustainability. This can contribute to innovation and improvements, which are then passed on through sustainable marketing. This creates a continuous cycle of improvement and communication that benefits both the company and its stakeholders.

The integration of ESG factors into marketing strategies and valuation models is significant, especially in the field of sustainable and responsible marketing planning, although challenges remain in bridging the gap between the depth of brand adoption and promotion. The present chapter mainly demonstrated the research gap associated with the implementation of the ESG approach, a holistic approach, from the highest levels to balance the marketing outlooks for sustainability and to recognise the importance of sustainability issues and take appropriate action (Bawa et al., 2024). This continuous path requires a deep understanding of ESG factors and their effectiveness for long-term value by investors and enterprises, which leads to a more strategic and detailed approach to corporate governance, strategy, and marketing decisions. The chapter also tries its best to point out the factors and essence of improving ESG reporting through technological innovation and mentions that accurate and thorough data collection is the basis for successful ESG reporting. The researchers have compared and used advanced analytics to evaluate the data, providing useful information and highlighting related models to make decisions, which can improve both strategic planning and ESG performance for sustainable and effective marketing. ESG surely leads to sustainable marketing, which primarily refers to environmental, based marketing strategies for organisational growth. ESG's strategies are very effective in creating a positive brand image and market positioning through connecting stakeholders with similar values, demonstrating the company's commitment to sustainable and conscientious business. This opens up the opportunity for companies to attract socially oriented customers, increase brand awareness and establish long-term relationships (Huang, 2019). Transparency and authenticity are at the heart of this extensive marketing structure. Companies must demonstrate their positive impact on our world by avoiding accusations of environmentally friendly laundry and creating a positive brand image.

25.2 Conceptual Framework

In 2004, the United Nations Global Compact and the Swiss Ministry of Foreign Affairs published a report which discussed ESG certainly leads to sustainable marketing, which primarily refers to environmental, based marketing strategies for organisational growth. ESG's strategies are very effective in creating a positive brand image and market positioning through connecting stakeholders with similar values, demonstrating the company's commitment to sustainable and conscientious business. The report also pointed out the ESG factors as environmental, social and governance factors that companies take into account when managing their activities, and investors take into account when investing, based on the risks, consequences and opportunities that arise from them, among other things (Healy et al., 1992):

- Environmental issues: possibly the actual changes happening in the physical and natural environment (e.g., pollution, impacts on biodiversity, carbon dioxide emissions, climate change, use of natural resources)
- Social issues: possible changes in the community and people (e.g., health and safety, supply chain);
- Governance: The governance structures that allow companies to follow governmental rules and accountability. (e.g., board structure and diversity, ethics, risk management), including the management's strategies, social policies, and procedures.

The ESG standard means the performance standard that defines the client's accountability in managing environmental and social risks and governance methods that define measures' to evaluate and improve customer satisfaction. ESG, in a real sense, covers a wide range of issues and maximises the possibilities of social progress, such as climate change and economic stability (Lagasio & Cucari, 2019). In fact, corporate associations are using ESG as an effective tool for managing social, environmental, and political risks, raising public awareness as part of social responsibilities, and looking for opportunities to develop proper strategies to manage possible risks.

In the present global scenario, ESG reporting has a deep influence on the working ideology of companies and their stakeholders. In a general sense, ESG reports are meant to describe the process by which community's measure, publish, and report on the impact of their activities on socio-political and environmental development and its progress. The term ESG is mainly associated with environmental, social, and governance-related issues, and ESG reports focus on indicators that facilitate the assessment of community enactment in these three areas. Environmental standards scrutinise the impact of community action on the environment and the mechanisms for mitigating these effects. Social criteria examine the impact of a company on social parameters, which may include social responsibilities, human rights and values, social unification, and the eradication of social illness (Nollet et al., 2016). Besides, the management will focus on developing effective mechanisms to support social development by minimising conflicts of interest and promoting social values such as accountability, transparency, moral values, and human interest. In the present multicultural environment of almost all major enterprises, ESG factors are considered one of the most important factors in marketing and financial decisions. Many companies now voluntarily submit reports on their ESG performance. As per the KPMG-Global Organisation Report, around 90% of the major multinational companies in the world provide an annual report on their ESG contributions and plans for sustainable social development. This would positively intensify the interest in ESG factors to their potential value, which can give a more intense focus on the commercial and social position of the company in the market and provide better recognition among consumers as social considerations are crucial for the intense marketing and future strategic planning of the companies.

The present chapter mainly wanted to focus on the significance of ESG reporting criteria's on sustainable marketing. The methods and models used in the chapter pointed out that the believed of the market and the companies are very similar, which will surely have an impact on the consumers. The chapter also added theories to identify the various factors to relate the influence of a different variable in ESG reporting to improve sustainable marketing approaches (Paolone et al., 2018). Basically, the theories reflect the evolving situations and the inflections of company's actions on ESG issues to impress and attract consumers for better brand positioning.

Sustainable marketing is key for success in any potential organisation, the KPMG addressed this concern efficiently and concluded that the due to the increasing pressure to disclose information about Environment, Social and governance (KPMG, 2024).

- KPMG sustainability survey in February 2024 It is pointed out that for the development of global organisations and for sustainable business growth, ESG reporting is taking a serious and very crucial shape in almost every growing multinational brand. As per the KPMG survey analysis, most of the organisations view this engagement as an essential performance both now and in the in the future. This shows the fact that organisations must invest in social facts for social acceptance and manage the top ways to integrate the sustainable goal with the overall organisation's objectives.

Considering all these facts and future visualisations, KPMG has analysed the future top priorities of companies, and surely ESG reporting is very crucial in this respect, as mentioned below: -

<u>Rapid Investment in sustainability is top priority</u>

Top areas of future investment are:

- Dedicated ESG Personnel- 43%
- ESG Specific Software – 40%
- Employee Training and Education- 38%

} 90% increase in ESG investment in Next 3 years

The present chapter essentially wanted to focus on the significance of ESG reporting criteria's on sustainable marketing, as the progress of organisations also very much depends on sustainability.

The methods and models used in the chapter pointed out that the ideology of the market and the companies are very similar, which will undoubtedly have an impact on the consumers. The chapter also added theories to identify the various factors to relate the influence of a different variable in ESG reporting to improve sustainable marketing approaches (Provasi & Harasheh, 2021). Basically, the theories reflect the evolving situations and the inflections of company's actions on ESG issues to impress and attract consumers for better brand positioning.

It is believed that the focus on ESG reports has now shifted from a simple risk management tool to recognition as an effective value creation mechanism. The goal of this chapter is very clear: to provide insights on sustainable marketing growth and ESG reporting for constructive business development, as well as its perspectives in the context of sustainable development. The development of environmental, social and management reporting (ESG) marks a dynamic transition from a pure approach to risk management to their recognition as a key element of value creation in organisations. This kind of shift is normally characterised by strong regulatory compliance and effective risk management to use the ESG initiative to increase productivity, innovation, and brand value. These kinds of plans are very significant for the marketing strategy of any growing organisation, as investors also have a demand for non-financial enterprises, which has a direct impact on the company's valuation and market positioning and this is core reason behind the growing demand of ESH factors for the company's financial performance and strategy (Shakil et al., 2020). There are many multinational companies (MNCs) that have a focused motive to shift from risk-based ESG strategies to value-based strategies and identify the role of marketing sections in taking appropriate initiatives to integrate sustainability into the business structure that can pave the way towards sustainable growth of organisations through sustainable practices. These can not only reduce the risk factors but also foster innovation and better consumers reach by strengthening improved stakeholder relationships. The integration of ESG factors into marketing strategies and assessment models is a substantial achievement in the field of sustainable and responsible marketing. Although there are many challenges in the market and present competitive global business, the brand awareness and depth of marking campaign implementation largely depend on the constructive management of ESG reporting and strategies.

In fact, there is various literature that has specifically pointed out that modern businesses are countering the growing demands of stakeholders and regulatory pressure by integrating aspects of ESG into their core marketing communication and strategies. Organisational planning on ESG reporting for sustainable development primarily benefits from new opportunities for innovation and expansion. This strategic approach contributes to the creation of long-term value and also leads to the development of sustainable business models, efficient operations and sustainable goods and services. ESG reporting will continue to be important for companies that want to achieve sustained success. Among the companies that have skilfully transformed their ESG reporting from a simple marketing management tool into an effective value creation mechanism are ACCENTURE, E.ON, and KMPG. Accenture's approach to ESG reporting has evolved to reflect sustainability as a critical factor of competitiveness and a driver of growth. They have implemented a "360-degree reporting approach" that integrates economic, environmental, social, and management indicators into the results of the 2021 fiscal year. These initiatives are based on global collaboration within the organisation and aim to provide clear, transparent, and comprehensive reporting on the sustainability and financial performance of the company. This is in line with Accenture's sustainability promise, which aims to integrate sustainability into all aspects of operations and customer service to create both business value and long-term impact. Actually, these examples very explicitly demonstrated the mechanism of ESG reporting as a holistic approach, from the highest level of management to every section of organisations. (Smith et al., 2007). This is also important as it leads to the realisation of sustainable development and the accurate identification of issues and marketing challenges. This continuous effort and strategies give business planners and investors a path a path to a deep understanding of ESG reporting for the long-term sustainable development of marketing strategic plans and effective approaches for quality corporate governance and investment decisions. In the conceptual development of ESG reporting for sustainable marketing growth, it is very important to understand the methods of ESG integration. In this context, the value-based approach (VBA) has been widely taken as an innovative approach for investment decisions and to innovate new associations for business models and competitive positions (Torchia & Calabro, 2016). This approach relates various ESG issues, their analysis, and their impact on changing market trends and values; it also qualifies their impact on corporate factors. However, this prospective path to full ESG integration faces obstacles such as limited data and difficulties in linking ESG issues with value factors.

25.3 The Marketing-Sustainability Relationship

The relationship between sustainable marketing and ESG can be divided into three phases. It all starts with the environment, the wellbeing of the social agencies and surroundings, and the issues of society such as pollution, depletion of natural resources, social awareness, and other social evils. This affected social development, the strategies of environmental growth, and issues of human interest (Usman & Amran, 2015).

This state of environmental marketing is focused on the development of products along with a message of environmental protection. It is the same as the idea of corporate social responsibility (CSR) with excellent socio-ecological characteristics in order to promote the environmentally friendly consumer and their responses; indeed, it has a deep competitive advantage. This was very successful and covered the major arenas of audiences and industries. This represented the social interest of consumers towards environmental issues and awareness, and it has a major role in sustainable marketing (Huang, 2019).

The last and third era of sustainable marketing encompasses the transformation of markets and marketing, significant progress that involves environmental, social, and governance (ESG), and its reporting to enhance society and its regular affairs. The corporate perspective of incorporating these external factors becomes the major deciding variable for sustainable marketing, which has reached the ultimate sustainable state. The market orientation and changing patterns of corporations identified the ethical issues and integrated the era of green marketing, awareness-based campaigns, and most importantly, social togetherness, focusing on relationship marketing to form a long-lasting value-based relationship with customers. The other side of such a relationship is also to promote contributions towards social justice and social growth. However, in practice, Environmental, Social & Governance (ESG) issues are so closely associated within the framework of the sustainable marketing agenda that it is giving a new shape to the strategic mark.

25.4 Theoretical Framework

The theoretical basis Traditional marketing plans are not enough to fully describe the various aspects of a company's business and customers' accessibility. For this reason and in accordance with the policy, large companies must have some plans for social responsibility and to solve sustainable developmental problems. Using social growth and its related issues, the companies must also place emphasis on sustainability reports for socio-political development (Wang et al., 2017).

This additional disclosure will reduce data asymmetry and marketing costs, as well as the negative effects of agency theory. In addition, as the theory of signals suggests, the sustainability report sends a positive signal to the market about the company's commitment to the principles of sustainable development, which can have a positive impact on the company's performance (Pulino et al., 2022).

- The theory of agency is based on two basic principles:

  i the association concerning the principal and the agent; and
  ii the differences between ownership and control are that the principal, represented by the owner of the company, transfers their administrative authority to an agent who works in the interest of the principal but also works on their own goals to support the interests of the principal.

On the other hand, administrative mechanisms can reduce agency costs and conflicts, especially when exploring a wide range of mechanisms. At this stage of the story, there is a conflict of interest between the principal and the agent, because they want to realise different interests. Therefore as mentioned in the Fig. 25.1 the ESG performance is directly decides the Marketing performance.

The principal has a long-term horizontal trend that contradicts the short-term horizontal trend of the agent. On one side, the agent is more focused on the opportunities it can get in the short term. On the other side, the principal uses the disclosure of marketing and social information to reduce agency costs for the separation of

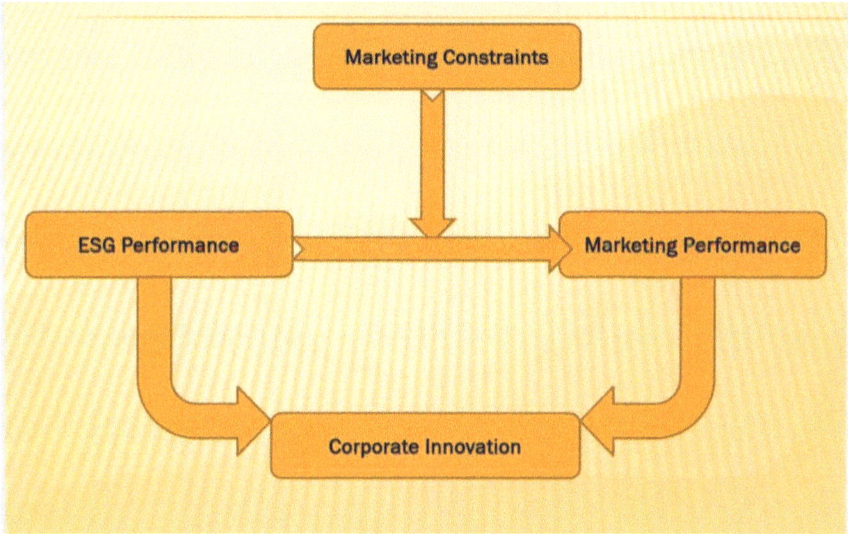

Fig. 25.1 Agency theory

ownership and control (information asymmetry). The higher the degree of disclosure of social information, i.e. ESG, the more satisfied the client will be with greater transparency.

- **Signalling theory:** Signal theory mainly involves the role of information in social transactions, and managers normally use impactful information to reduce the information asymmetry in society and provide a scenario of information-rich shareholders. In general, companies are making a huge investment in marketing tools to earn social resources in the form of loyal consumers and spread valuable information about environmental, social, and political governance and its value for sustainable society growth. Considering all these associated factors, signal theory is based on four elements: signal, alarm, receiver, and feedback (Pulino et al., 2022). The signal is represented by a rivulet of information, i.e., data, that is diffused as a signal in a very specific way by the internal management to the beneficiary, as signified by external stakeholders.

This signalling theory basically deals with the interaction based on information between the signals and the receivers. May authors also supported the idea of the information discloser for the long-term sustainability initiatives to enhance the marketing purpose for the maximum interactions of society and companies ideas, so that company can make a clear sign of commitment to society, the environment, and the stakeholders (Hart & Ahuja, 1996). In this way, the managers of the company resolve the confusion and maintain the information balance between the company and its internal and external stakeholders (Fig. 25.2).

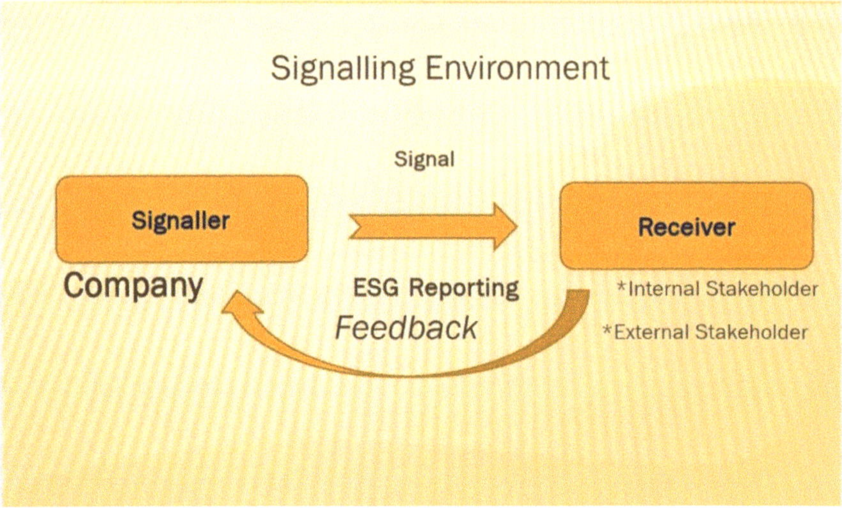

Fig. 25.2 Signal theory

Agency theory and signal theory suggest the basics of ESG as part of the corporate social responsibilities of companies to reach out to the most corners of the market and plan for marketing sustainability. The main effect is reflected in the indicators of effective marketing planning, that is, its turnover, and, consequently, in an increase in annual profit.

ESG reporting has become very crucial for corporate marketing to attract social consciousness, as ESG marketing provides an innovative course of interaction with social elements, and companies are helping to solve social issues to attract socially oriented customers. This surely permits corporate organisations to show their positive impact on the environment, society, and political affairs. This would help the organisations improve their reputation and strengthen the value of trust among stakeholders. ESG is very significant in the present scenario to attract investors, which highlights approaches for sustainable development organisations. Sustainable marketing is a key to the organisation's growth and progress. The strategies to communicate ESG initiatives positively demonstrate the organisation's vision of success. Ultimately, ESG marketing highlights the social values of the company, and in today's competitive business world, customers are becoming more demanding and socially conscious (Gjergji et al., 2021).

In today's scenario, people are very aware, and their requirements are more than the products and services; actually, they are more interested in the organisation's values and principles towards society and the wellbeing of social capital. They are more interested in knowing how the organisations are advocating the structure and ideology of a better world. By adopting ESG planning and strategies, the companies are actually making a stronger path for marketing with the objective of sustainable growth and development and making a significant difference in their identity among the other competitive companies. It also gives a lesson to future generations about the values of social issues and togetherness for a sustainable future. The planned use of the ESG approach shall give better marketing strategies, such as:

- Integrating planned ESG is certainly used as a better tool of marketing to achieve companies' goals and promote their sustainability practices and environmental projects. This may lead to the environmentally friendly development of society and products, but it is also effective in minimising waste and supporting environmental measures.
- Integrating the social advantages through marketing planning may provide ground to promote social diversity, respect for people at the workplace, and the conduct of socio-political campaigns for social awareness about social rights and well-being. The social impact of these activities definitely has an effective brand promotion and generates consumer interest to support the company's mission for development.
- Sustainable marketing is very important for the brand positioning of the company in the market. The interaction of company management is very important to reach the people, as it clearly shows the commitment and responsible behaviour of the brand and is important for public relations (Gerged, 2020). This certainly demands effective market communication with consumers, stakeholders, and companies social mechanisms against social awareness on some crucial issues like fraud and corruption. Brands are using these efforts to create brand image by showing their commitment to effective governance and moral behaviour.

25.5 Successful ESG Marketing Campaigns

ESG reporting is very successful in presenting brand names and strengthening the methodology of sustainable marketing. There are many epic examples of this context, but the Patagoia is a very famous example, showing their commitment to environmental friendliness and taking many measures to create social awareness on environmental issues. The "Don't Buy This Jacket" campaign in Patagonia has created awareness of overconsumption, encouraged shoppers to consider the environmental impact of their purchases, and motivated buyers to consider environmental issues before buying products (Lange & Busch, 2012).

The message was very clear in encouraging people to think about the effect of consumerism and its environmental concerns. There are deep meaning and issues related to ESG reporting and corporates, in this respect, the European marketing director, Jonathan Petty stated that "It has helped to establish a strong community of people who appreciate the brand's values and its products. Our customers expect very high quality and that's why they always come back to us".

In another case, ESG was intensively used by the brand and focused on various videos to demonstrate Roots Flower Farm's need to make effective use of its unique farm-to-table approach. Working closely with owners Michelle and Mike Elston, it was found that demonstrating the daily aspect of showing the commitment to the relationship between South-Central Pennsylvania and unique flower patterns (Cucari et al., 2018).

After a full strategy and planning, the brand has devoted itself to every detail of the day, from early morning to dinner, and created a fascinating video detailing the roots and reputation of the farm. The impact of ESG marketing was very good on reputation-building. The implementation of ESG marketing strategies can have a significant impact on a company's reputation and digital footprint.

As more and more consumers and stakeholders expect companies to be socially and environmentally responsible, brands certainly have to make some plans to fulfil consumers' expectations in this respect and to gain trust and social l loyalty. This example clearly shows that a positive brand reputation and market credibility not only attract people and convert them into customers but also have a deep impact on employee satisfaction and create a friendly environment of working with confidence (Cheng et al., 2013). The companies need to respond positively and consider these initiatives to respond to ESG issues, not just strengthen the external public but also the internal public, and of course, with these measures, the companies are able to deal with most of their legal, financial, and operational problems.

25.6 Challenges in Implementing ESG Reporting

The global business scenario is becoming increasingly complex due to the changing global trends, and it is actually providing a high stakeholder expectation and needs to afford a long-term vision for sustainable growth. In this context, ESG is a hallmark, which transcended the marketing concept at global level and given a significant momentum in the corporate world, representing a fundamental shift towards sustainable business practice. Significantly, meeting the global framework and standards necessitates significant challenges too. There are many challenges in implementing ESG reporting for sustainable marketing. It can also be remarkable here to point out that ESG reporting has some advantages but also bears challenges, such as:

- Ensuring regular ESG initiatives for a positive image and maintaining its reputation in the market is a challenge.
- Quantifying and amplifying the ESG performance can be Sometimes, it can be difficult for organisations to accurately determine and report on the number of ESG efforts, especially if they do not have reliable data collection and reporting mechanisms.
- There may be questions and anxiety from stakeholders about the motives of ESG planning and initiatives. Such efforts must be well justified and implemented for social good and long-term commitments.
- Companies operating in many countries may find it difficult to keep up with changing ESG standards, regulations, and best practices. Meeting complex global environmental and social norms can be challenging for multinational corporations.

- Companies should be wary of "greening", a practice in which companies embellish or distort their environmental impact, as this can have serious consequences and damage the reputation of the brand.

Environmental, Social, and Governance (ESG) reporting has a meaningful role in the strategic requirement for the effective regulation of companies. In the present era of the global marketing system, companies have realised that the value of ESG transparency is very vital in attracting investors, building trust, and minimising risk. It's really important to appreciate the different parameters to maintain ESG principles, and in a real sense, effective ESG can strengthen the company's reputation, ensure long-term profits, and stand out in the global competitive market with unique marketing approaches. In order to have an effective marketing strategy for sustainable global marketing growth, ESG surely plays a role in identifying the challenges and areas for improvement by setting meaningful sustainability goals and keeping track of them. The standard regulatory systems, such as GRI (EU Global Reporting Initiative), SASB (Council for Accounting Standards in the Field of Sustainable Development), and CSRD (Directive on Corporate Sustainability Reporting), are effectively able to provide standards that global companies use to compile reports on environmental, social issues, and governance (ESG) and communicate with stakeholders, investors, and global partners about the growth and its related crisis. Depending on the nature and criteria of changing global challenges, the standard regulatory framework, such as GRI, provides a comprehensive overview of the impact of sustainable development; SASB emphasises the business-specific economic drivers of ESG, while CSRD gives operational parameters for ESG to maintain a sustainable marketing approach locally and globally. These frameworks significantly enhance corporate transparency by dealing with the challenges and possible limitations. It also provides a constructive platform with detailed information on environmental, social, and governance (ESG) practices to show the clear vision of sustainability efforts for constructive decision-making and enhances the credibility of ESG disclosure to stabilise marketing sustainability and corporate global reputations. The ESG reporting field is rapidly evolving to ensure transparency and accountability. Considering the entire standard framework, corporates must navigate the holistic approach of ESG reporting to deal with the sustainable global marketing, crisis management, and fulfilment of stakeholder expectations. Ultimately, effective ESG reporting is not just about verifying compliance with requirements but a strategic necessity that can lead to long-term value creation.

25.7 Case Study on ESG and Marketing Strategy

25.7.1 Unilever Presentation

Unilever, a global consumer goods company, is an important example of how integrating environmental, social, and management principles (ESG) into a marketing strategy can bring significant commercial benefits. Unilever's environmental, social, and management principles (ESG) cover an extensive range of connections with society and its social agencies. This clearly provides the criteria for managing socio-political and environmental-based relationships with internal and external stakeholders and many other members of society every day. Unilever's assurance of an extraordinary level of entrepreneurial behaviour is a fundamental part of our business traditions (Alareeni & Hamdan, 2020). This is defined in business principles and easily incorporated into many business practices. The company is also excellently able to identify the actual progress towards social progress, which means they are doing their best to examine social practices, formulate policies, and demonstrate effectiveness. Unilever is very unique in operating in a competitive market and understands the need for their marketing approaches to maximise the number of customers and their social performance. Responsibility at the local level is the foundation of a company's activities and the cornerstone of its success. Collecting clear data on the social performance of a wide variety of companies creates a lot of problems when accounting for activities on the scale of Unilever. As a result, in 1999 and 2000, nine Unilever companies worldwide, representing more than a quarter of the entire company, tested a system for evaluating and monitoring the effectiveness of corporate social responsibility in performing their activities under ESG. It was also well understood that society's expectations are changing and that Unilever is increasingly faced with the need to play a more active social role, be open about our social impact, and strive for a sustainable future (Barnett & Salomon, 2012).

The result is this first international social review of Unilever's approach to social responsibility. The purpose of the study is threefold. Unilever explains what CSR they wanted to initiate. It traces current policies and practices regarding the environmental, social, and political impact of their activities and illustrates examples of local businesses (initial overview). In fact, this gives a starting point for determining the stages of future actions.

This social report is an appendix to the Environmental Report for 2000 and is based on the significant progress made over the past five years in managing global environmental impacts. The environmental indicators for 2000 also reflect Unilever's commitment to sustainable development, in particular the work the company is doing in the fields of sustainable agriculture, water management, and the use of renewable marine resources. The Sustainability Plan developed by Unilever, in 2020 demonstrates the company's commitment to the principles of sustainable development and its impact on brand image, consumer engagement, and overall company performance (Hart & Ahuja, 1996).

25.8 Conclusion

ESG reporting and sustainable marketing are integral to modern corporate strategy. ESG reporting enhances transparency, manages risks, attracts investors, and ensures regulatory compliance. Sustainable marketing meets consumer demand, differentiates brands, fosters long-term growth, and drives innovation. Together, they create a powerful synergy that promotes corporate responsibility, builds consumer trust, and supports sustainable development. As businesses continue to navigate the complexities of the modern world, the integration of ESG reporting and sustainable marketing will be essential in achieving long-term success and positive societal impact (Acar & Çalıyurt, 2020). The present chapter mainly wanted to focus on the significance of ESG reporting criteria's on sustainable marketing. The methods and models used in the chapter pointed out that the ideology of the market and the companies are very similar, which will certainly have an impact on the consumers. The chapter also added theories to identify the various factors to relate the influence of a different variable in ESG reporting to improve sustainable marketing approaches. Basically, the theories reflect the evolving situations and the inflections of company's actions on ESG issues to impress and attract consumers for better brand positioning (Lange et al., 2022).

Therefore, the present chapter also clearly states, first, the significance of ESG performance for sustainable marketing values, allowing a cross-country analysis that would highlight similarities and disparities. Second, there are various social and political measures of performance that might be investigated to gain a deeper understanding of the business benefits of ESG disclosure. Thirdly, organisations must emphasise the ESG disclosures and data available and their operational implementation, especially in the future, to better understand the long-term implications and new requirements. The chapter also used the case study of the famous brand Unilever as the classic case, which demonstrated the deep value of ESG reporting for effective and sustainable marketing to have a strong connection with the stakeholder (Alareeni & Hamdan, 2020). As evidenced by the discussion and the related variables, ESG reporting is a big industrial benchmark, as organisations must determine and visualise the future scopes to establish the right balance between business and social development by creating expectations that companies will adopt ESG reporting to inform and highlight potential opportunities that might affect the company's best interests in civil society.

To sum up, ESG reporting is rapidly becoming a key element of modern marketing approaches and serves as a guide for any organisation to deal with its various challenges. Sustainable marketing and ESG reporting, along with its business strategy, are important for a company's future and its positioning and have a positive impact on an equitable society. It is undoubtedly very imperative to consider ESG reporting as a management tool to maintain sustainable marketing for value creation, and undeniably, it is vital for the long-term sustainability and profitability of companies.

References

Acar, E., Çalıyurt, K. T., & Zengin-Karaibrahimoglu, Y. (2021). Does ownership type affect environmental disclosure? *International Journal of Climate Change Strategies and Management, 13*, 120–141.

Accenture. (2024). *ESG Reporting.* Accenture. https://www.accenture.com/us-en/case-studies/about/esg-reporting

Alareeni, B. A., & Hamdan, A. (2020). ESG impact on performance of US S&P 500-listed firms. *Corporate Governance: The International Journal of Business in Society, 20*, 1409–1428. Available online:

Barnett, M. L., & Salomon, R. M. (2012). Does it pay to be really good? Addressing the shape of the relationship between social and financial performance. *Strategic Management Journal, 33*, 1304–1320.

Bawa, P., Molero, E., & Knickle, K. (2024) Green Quadrant: *ESG and sustainability consulting.* https://www.ey.com/content/dam/ey-unified-site/eycom/en-gl/about-us/analyst-relations/documents/ey-gl-verdantix-green-quadrantesg-sustainability-consulting-03-2024

Carnini Pulino, S., Ciaburri, M., Magnanelli, B. S., & Nasta, L. (2022). Does ESG Disclosure influence firm performance? *Sustainability, 14*, 7595. https://doi.org/10.3390/su14137595

Cheng, B., Ioannou, I., & Serafeim. (2013). G. Corporate social responsibility and access to finance. *Strategic Management Journal, 35*, 1–23.

Cucari, N., Esposito de Falco, S., & Orlando, B. (2018). Diversity of board of directors and environmental social governance: Evidence from Italian listed companies. *Corporate Social Responsibility and Environmental Management, 25*, 250–266.

de Lange, D. E., Busch, T., & Delgado-Ceballos, J. (2012). Sustaining sustainability in organizations. *Journal of Business Ethics, 110*, 151–156.

E. ON SE. (2024) *Sustainability reporting.* E.ON. https://www.eon.com/en/about-us/sustainability/reporting.html

Gerard, B. (2019). ESG and socially responsible investment: A critical review. *Beta, 33*, 61–83.

Gerged, A. M. (2020). Factors affecting corporate environmental disclosure in emerging markets: The role of corporate governance structures. *Bus. Strategy Environ., 30*, 609–629.

Gjergji, R., Vena, L., Sciascia, S., & Cortesi, A. (2021). The effects of environmental, social and governance disclosure on the cost of capital in small and medium enterprises: The role of family business status. *Business Strategy and The Environment, 30*, 683–693.

Hart, S. L., & Ahuja, G. (1996). Does it pay to be green? An empirical examination of the relationship between emission reduction and firm performance. *Bus. Strategy Environ., 5*, 30–37.

Healy, P. M., Palepu, K. G., & Ruback, R. S. (1992). Does corporate performance improve after mergers? *Journal of Financial Economics, 31*, 135–175. https://shura.shu.ac.uk/27475/9/MostafizBoardGenderDiversity%28AM%29.pdf

Huang, D. Z. (2019). Environmental, social and governance (ESG) activity and firm performance: A review and consolidation. *Accounting and Finance, 61*, 335–360.

Khatib, S. F., Abdullah, D. F., Elamer, A., & Hazaea, S. A. (2022). The development of corporate governance literature in Malaysia: A systematic literature review and research agenda. *Corporate Governance: The International Journal of Business in Society, 22*, 1026–1053.

KPMG International. (2023). ESG: Moving from risk to value creation. *KPMG.* https://kpmg.com/xx/en/blogs/home/posts/2023/02/esg-moving-from-risk-tovalue-creation.html

Lagasio, V., & Cucari, N. (2019). Corporate governance and environmental social governance disclosure: A meta-analytical review. *Corporate Social Responsibility and Environmental Management, 26*, 701–711.

Nollet, J., Filis, G., & Mitrokostas, E. (2016). Corporate social responsibility and financial performance: A non-linear and disaggregated approach. *Economic Modelling, 52*, 400–407.

Paolone, F., Cucari, N., Wu, J., & Tiscini, R. (2022). How do ESG pillars impact firms' marketing performance? A configurational analysis in the pharmaceutical sector. *The Journal of Business and Industrial Marketing, 37*, 1594–1606.

Provasi, R., & Harasheh, M. (2021). Gender diversity and corporate performance: Emphasis on sustainability performance. *Corporate Social Responsibility and Environmental Management, 28*, 127–137.

Shakil, M. H., Tasnia, M., & Mostafiz, M. I. (2020). Board gender diversity and environmental, social and governance performance of US banks: Moderating role of environmental, social and corporate governance controversies. *International Journal of Bank Marketing, 23*, 125–133.

Smith, M., Yahya, K., & Marzuki Amiruddin, A. (2007). Environmental disclosure and performance reporting in Malaysia. *Asian Review of Accounting, 15*, 185–199.

Torchia, M., & Calabro, A. (2016). Board of directors and financial transparency and disclosure. Evidence from Italy. *Corporate Governance: The International Journal of Business in Society, 16*, 593–608.

Usman, A. B., & Amran, N. A. B. (2015). Corporate social responsibility practice and corporate financial performance: Evidence from Nigeria companies. *Social Responsibility Journal, 11*, 749–763.

Wang, Z., Hsieh, T. S., & Sarkis, J. (2017). CSR performance and the readability of CSR reports: Too good to be true? *Corporate Social Responsibility and Environmental Management, 25*, 66–79.

Chapter 26
A Study of Interrelationship of Environmental and Social Disclosure with Corporate Attributes

Preeti Sharma and Priyanka Panday

Abstract This research paper investigates the relationship between environmental and social disclosures and corporate attributes among top-listed Indian companies. The study aims to determine the impact of market performance, profitability, foreign institutional investors' (FIIs) stake, and leverage on the level of environmental and social disclosures. The findings indicate that higher profitability and market performance are positively associated with increased environmental and social disclosures, suggesting that more profitable firms are more inclined to engage in stakeholder-oriented practices. However, FIIs' stake and leverage show a negative and insignificant relationship with these disclosures. The study underscores the importance of environmental and social disclosures for building stakeholder trust and managing risks related to environmental and social factors. The results have implications for investors, regulators, and policymakers, emphasizing the significance of non-financial disclosures in assessing a firm's long-term sustainability and performance. Further research opportunities are suggested, including exploring combined environmental and social disclosure scores and conducting panel data analyses.

Keywords Environmental and social disclosures · Corporate attributes · Market performance · Profitability · Foreign Institutional Investors (FIIs)

26.1 Introduction

Over the last two decades, the need for non-corporate disclosures by corporations and their stakeholders has been imperative due to the high emphasis on environmental and social issues at the global level that took a central role (Gray, 2009; Ingram &

P. Sharma
DIT University, Dehradun, India

P. Panday (✉)
DBS Global University, Dehradun, India
e-mail: priyanka.pandey13@gmail.com

V. Kandpal et al. (eds.), *Developments in Corporate Governance*,
Approaches to Global Sustainability, Markets, and Governance,
https://doi.org/10.1007/978-981-96-6366-8_26

Frazier, 1980). Issues such as water scarcity, strikes and disputes among employees, and depletion of natural resources required urgent attention (Ioannou & Serafeim, 2010; Setiawan & Darmawan, 2011). More and more corporations voluntarily report non-financial reporting, encompassing environmental and social issues, and factors. Corporations do not do so for accountability towards the society and community; corporate reputation and firm growth are the most significant motivations (Lungu et al., 2011); Sharma et al., 2020). Furthermore, the linkages of these disclosures with several corporate attributes are becoming a crucial area of research for academicians and researchers. Social and environmental disclosures generally comprise activities about the environment and society. No organization can survive without community support; a poor brand image and negative sentiments related to the company can harm its overall performance. Most research is confined to fuel usage, clean energy, equivalent opportunity, and corporate governance. Environmental and disclosure might take place through alternative media. Simultaneously, most such disclosures focus on data from the company's annual reports. Other mediums, such as advertising, booklets, and school education, might have been utilized.

Verrecchia (1983) elucidates that the firms with greater performance on the ecological and community front have more scope to disclose selective and convincing evidence as per their need and requirements. There is an ongoing and strong debate about the relevance of these disclosures being relevant. While some scholars find empirical evidence to stigmatize it as a 'legal tool' (Cho & Patten, 2007; Gray et al., 1995), others support the disclosures as value-relevant. (Aerts et al., 2007; Clarkson et al., 2011; Cormier and Magnan 2014) argued that disclosures enhance a firm's brand value and convey financial benefits to the company, including increased share value. The rationale behind the present paper is to measure environmental and social disclosures and then examine the impact of corporate attributes such as market performance, profitability, FIIs' stake, and leverage on their disclosure level. The paper includes a literature review, research methodology, empirical findings, discussion, and conclusion.

26.2 Literature Review

Companies increasingly use social and environmental disclosures to communicate with the shareholders. The existing literature has a contradiction about its effectiveness in bringing benefits to the organization. As per Konar and Cohen (2001), company action in extenuating pollution and related environmental issues will pay off positively in the short run. Margolis and Walsh (2003) highlighted no relationship between disclosures and financial performance. Contrary to that, Cordeiro and Sarkis (1997) explained an inverse relationship between these disclosures and firm performance. He further argued the potential negative impact and threat to the firm's existence if the focus is short-term. Financial disclosures are being scandalized due to poor framework; greater demand for such disclosures is inviting for the broader research work in this area. This is primarily due to the increasing value of ESG factors and a major threat to companies with only profit-making motives. A study depicts the direct relationship between enhanced financial performance and ESG disclosures that build big brand value. (Brammer & Pavelin, 2006; Orlitzky et al., 2003).

Voluntary disclosures are making a strong place in firms' valuation process (Agostini-Costa et al., 2012). In the old context, the major decision of investments comes from financial performance. However, a major shift in this old mindset can be easily seen in the minds of people. Now evaluation is being done on financial performance vis-a-vis social and environmental performance. Multiple stakeholders, including government agencies, are now keeping a close eye on the ESG disclosures. ESG disclosers also act as a bridge the manage CSR deviation for maintaining the sustainability of the business. The major ESG disclosure includes environmental, social, and governance scores. Environmental indicators include pollution, gas emissions, unauthorized exploitation of water bodies, wastewater management, etc. These include some company costs that affect the firm's long-term sustainability and must evaluate the firm's performance. The social factor includes the societal contribution of the company. The major indicators of social performance are labor practices, facilities, human rights, etc. The governance indicator requirement includes the management structure, accountability towards shareholders, and various management committees; corporate characteristics are highly significant for a strong and effective corporate governance system.

Financial reports often do not capture the information, which is also essential for a healthy and successful environment that has faced trouble, such as Enron. Financial valuation may vary if it includes ESG risks as well. India receives foreign funds on a very large scale, and surprisingly, the area remains unexplored considering the investors' interests in the Indian financial market. The results can be relevant for the market regulators and policymakers as a rising number of scams and increasing demands for green projects make it a win–win situation for the government and the companies.

The ESG disclosures are derived from two major theories: The stakeholders and the Legitimate theory (Socio-Political aspect). Legitimacy theory explains the need for social disclosures for the society in which it operates. In a nutshell, firm disclosures are the outcome of the firm's value. The shareholder's theory wherein the company's stakeholders satisfy their interest by environment and social disclosure score.

The sizeable body of literature has emphasized the importance of non-financial disclosures for a firm for multiple purposes. Although significant differences are visible within the literature, authors have concluded that higher-return companies show a strong disposition towards ESG disclosures. The source of ESG disclosure is found to be in SRI (Socially Responsible Investing). ESG disclosures are also called the foundation of sustainability; Disclosure practices help organizations improve their credibility and effectiveness concerning improved shareholder trust (Kolk & Perego, 2010; Simnett et al., 2009).

Richardson and Welker (2001) highlighted two groups of investors: the first includes people who believe in ethical behavior and trust that an ethical practice promotes better financial performance. The other one believed in only financial performance but later started believing in non-financial disclosures (Agostini-Costa et al., 2012), improved CSR performance is a strong, attractive point for customers who want to spend their money on sustainable development and voluntary disclosure leads to higher sales and better productivity. Gunasekaran and Spalanzani (2012) highlighted the role of sustainable manufacturing and business activity in increasing global warming and carbon emission issues. Value enhancement and further assessment will certainly add to the financial performance of the company.

Due to environmental degradation in the last few years, the investors are more inclined to be in the line of sustainable development supporters, and the same is getting fulfilled in this due process. The focus is more on adopting cleaner technologies and investing in promoting the same.

There are tremendous studies that found a positive and significant relation between CSR practices and corporate attributes. In their study, McGuire, Sundgren et al. (1988) assessed the financial performance of firms and found that CSR is more closely related to past performance than subsequent performance. Waddock and Graves (1997) investigated the relationship between corporate social performance and financial performance (ROA, ROE, and ROS) by controlling variables such as size, risk, and industry. The researchers found a significant positive relationship was found. Another empirical study by Stanwick and Stanwick (1998) found that environmental disclosures, such as pollutant emissions, positively affected the firm's Corporate Social Performance (CSP). Tsoutsoura (2004) and Lepoutre and Heene (2006) have found similar positive and statistically significant results. The study found that small businesses faced more difficulties than their larger firms while engaging in social responsibility activities due to a lack of financial resources.

In the last many years, the public has overwhelmingly accepted companies with responsible social and environmental disclosures (Carroll & Shabana, 2010). In this regard, non-financial disclosures have become highly significant for companies. Consequently, companies have also made such disclosures an essential part of their regular disclosures. It has become a focal point of decision-making in this regard, specifically after the global financial crisis.

Mallin et al., (2013) recommended that companies should supply a structured and calculated direction, which would be vital for their long-term survival and new avenues. Guided by a long-term growth standpoint, a rising number of companies willingly began to announce non-financial information in their reports. It is highly recommended to measure and recognize the environmental impact on society and make the plan accordingly. In almost all market segments, the consideration of environmental issues has risen like never before in the last decades. (Labatt & White, 2002). All stakeholders have become increasingly emphasizing corporate activities related to the environment.

Furthermore, due to the highly integrated economy, companies' pressure is high to follow international standards and maintain a high reputation in disclosing non-financial disclosures (Hooghiemstra, 2000). Moreover, a non-financial disclosure provides better insight and control over the disclosure process (Adams & Zutshi, 2004). In theory, the information should be correct and follow all the quality standards. These standards might differ according to the sectors, so they can't be mentioned specifically (Aguilera et al., 2006).

Some studies found mixed results on relations between CSR practices and different corporate attributes. Aras et al. (2010) explored a similar relationship in developing countries. In Istanbul stock market for the period 2005–2007. The study found a strong relationship between the size of the company and the socially responsible firm's responsible behavior, but no relation was detected with profitability. Similarly, the CSR activities of selected Casino, Hotel, and airline companies on their financial performance. The study found mixed results in different selected sectors. The study found no relationship between CSR activities and firm performance in the Casino industry. In the Hotel Industry group, the study found a positive impact of positive CSR activities on firm value. On the other hand, positive CSR had a negative impact on the airline industry. The present study individually examined the linkage between corporate attributes and environmental and social disclosures by formulating two models. The objectives of the work are as follows:

1. To examine the linkage between environmental disclosures and corporate attributes (market performance, profitability, FIIs stake, and leverage).
2. To examine the linkage between the social disclosures and corporate attributes (market performance, profitability, FIIs stake, and leverage).

Based on objectives, the following hypotheses are formulated:

H_{01}: *There is no significant association between the environmental disclosure score and corporate attributes of sample companies.*

H_{02}: *There is no significant association between the social disclosure score and corporate attributes of sample companies.*

26.3 Research Methodology

This work has adopted corporate annual reports and standalone reports of firms. The sample consists of the top hundred listed companies. The final sample consists of 80 companies after excluding the financial and banking sectors due to their least environmental influence for 2015–2016; to achieve this objective, content analysis is adopted. A similar method was adopted by Milne and Adler (1999). However, an environmental and social disclosure index, including environmental and social factors, is constructed. Furthermore, a dichotomous procedure is adopted to find the environmental and social disclosure score. Multiple regression analysis is used to examine the linkage between environmental and social disclosures and selected corporate attributes (see Table 26.1). The following two equations are formulated:

$$EDS = \alpha_1 + \beta_1 MBV + \beta_2 ROA + \beta_3 FIIs\ \ stake + \beta_4 Leverage + \beta_5 LTS + \beta_5 IND + \varepsilon_i \tag{26.1}$$

$$SDS = \alpha_1 + \beta_1 MBV + \beta_2 ROA + \beta_3 FIIs\ stake + \beta_4 Leverage + \beta_5 LTS + \beta_5 IND + \varepsilon_i \tag{26.2}$$

Table 26.1 Equation description

Variables	Description
EDS	Overall Environmental disclosure score
SDS	Overall Social disclosure score
α_1	The intercept term
MBV	Market to book value of common stock
ROCE	Return on Capital Employed
ROA	Return on Assets
FIIs stake	FIIs stake
Leverage	Debt-Equity ratio
LTS	Log value of sales
Industry	Type of industry
Bn	Coefficient
ε_i	The error term

26.4 Empirical Analysis and Findings

The empirical analysis of the paper shows a significant relationship between environmental and social disclosures and corporate attributes, highlighting positive and negative associations.

26.4.1 Environmental Disclosure (ED) and Corporate Attributes

Table 26.2 highlights key correlations between environmental disclosure (EDS) and corporate attributes. The study found a positive relation between EDS and log total sales (LTS) which suggests that larger companies are more likely to disclose their environmental activities, possibly due to greater regulatory scrutiny and stakeholder expectations. However, the results show a negative relation between EDS and market performance measured by MBV which implies that firms with higher market valuations are less concerned about the level of environmental disclosures. Furthermore, leverage (LVG) and foreign institutional investment (FII), show a weaker relation with environmental disclosure.

Table 26.3 presents the results of multiple regression analysis for Eq. (26.1) which reveals several significant relationships. A moderate explanatory power is indicated by the model's ability to explain 34.5% of the variation in EDS ($R^2 = 0.345$). EDS is strongly positively and significantly impacted by the log value of total sales (LTS) ($B = 20.337$, $p < 0.01$), supporting the notion that larger companies disclose more information about their environmental practices because they are subject to more regulatory scrutiny and stakeholder pressure. Financially stable companies are more

Table 26.2 Correlation coefficients for Eq. (26.1)

	EDS	MBV	ROA	FII	LVG	LTS	IND
EDS	1						
MBV	−0.226*	1					
ROA	0.012	0.547**	1				
FII	−0.113	−0.052	−0.102	1			
LVG	−0.112	−0.222*	−0.251*	−0.086	1		
LTS	0.456**	−0.195	-0.303**	−0.052	0.020	1	
IND	−0.118	−0.156	−0.238*	−0.088	0.190	0.048	1

** Significant at the 0.01 level (2-tailed), * Significant at the 0.05 level (2-tailed)

inclined to invest in environment disclosures, according to profitability (ROA), which likewise exhibits a positive and substantial influence (B = 0.529, $p < 0.05$).

Conversely, market performance (MBV) significantly reduces EDS (B = −0.966, $p < 0.01$), suggesting that companies with greater market valuations might prioritize short-term financial objectives over environmental transparency. The absence of significant effects for leverage (LVG), foreign institutional investment (FII), and industry classification (IND) may be due to several factors, including varying industry-specific reporting practices, a lack of mandatory disclosure requirements, or a lack of investor activism related to ESG. There are no significant autocorrelation issues, according to the Durbin-Watson statistic (2.024), and multicollinearity is not a problem, according to collinearity diagnostics (VIF values less than 2). All things

Table 26.3 Multiple regression analysis for Eq. (26.1)

Equation		Unstandardized coefficients		T	Sig.	Collinearity statistics	
		B	Std. Error			Tolerance	VIF
26.1	(Constant)	−171.480	47.485	−3.611	0.001	–	–
	MBV	−0.966	0.334	−2.890	0.005	0.692	1.446
	ROA	0.529	0.239	2.212	0.030	0.616	1.623
	FII	−0.332	0.214	−1.548	0.126	0.966	1.035
	LVG	−2.548	2.172	−1.173	0.244	0.897	1.115
	LTS	20.337	4.144	4.908	0.000	0.903	1.108
	IND	−0.773	0.634	−1.218	0.227	0.915	1.093
	R = 0.587	$R^2 = 0.345$		Adjusted $R^2 = 0.291$			
	Std. Error of the Estimate = 20.27			Durbin-Watson = 2.024, F = 6.411 (0.000)			

Note Dependent Variable: EDS

considered, these results show how business size and profitability influence environmental disclosures, but they also imply that market performance could impede transparency initiatives.

26.4.2 Social Disclosure (SD) and Corporate Attributes

To examine the relationship between social disclosure score (SDS) and selected attributes of companies, the following regression Eq. (26.2) is formulated:

$$SDS = \alpha_1 + \beta_1 MBV + \beta_2 ROA + \beta_3 FIIs\ stake + \beta_4 Leverage + \beta_5 LTS + \beta_5 IND + \varepsilon_i \quad (26.2)$$

Table 26.4 depicts the positive and significant correlation between SDS and the total sales log value (LTS). A negative and significant correlation is observed between SDS and MBV. The results also show that the correlation between predictors is not very high (none of the values > 0.8, Brooks (2008). Hence, no co-linearity is found; the regression equation can be accepted.

Multiple regression analysis has been conducted to understand the relationship of the independent variables of Eq. (26.2) with SDS. Table 26.5 presents the results of multiple regression analysis for Eq. (26.2).

As per the results of the multiple regression analysis, the model has modest explanatory power, explaining 33.6% of the variation in SDS ($R^2 = 0.336$). The most significant factor, like environmental disclosure, is the log value of total sales (LTS) ($B = 16.291$, $p < 0.01$), indicating that larger companies are more likely to make social disclosures because of their greater public visibility, regulatory compliance, and stakeholder expectations. Moreover, profitability (ROA) has a positive and significant impact on SDS ($B = 0.506$, $p < 0.05$), suggesting that financially sound companies devote more funds to social reporting and corporate social responsibility (CSR).

Table 26.4 Correlation matrix for Eq. (26.2)

	SDS	MBV	ROA	FII	LVG	LTS	IND
SDS	1						
MBV	−0.244*	1					
ROA	0.022	0.547**	1				
FII	−0.044	−0.052	−0.102	1			
LVG	−0.048	−0.222*	−0.251*	−0.086	1		
LTS	0.454**	−0.195	−0.303**	−0.052	0.020	1	
IND	−0.118	−0.156	−0.238*	−0.088	0.190	0.048	1

** Significant at the 0.01 level (2-tailed), * Significant at the 0.05 level (2-tailed)

Table 26.5 Multiple regression analysis for Eq. (26.2)

Equation		Unstandardized Coefficients		t	Sig.	Collinearity statistics	
		B	Std. Error			Tolerance	VIF
26.2	(Constant)	−124.560	38.108	−3.269	0.002	–	–
	MBV	−0.830	0.268	−3.094	0.003	0.692	1.446
	ROA	0.506	0.192	2.636	0.010	0.616	1.623
	FII	−0.122	0.172	−0.710	0.480	0.966	1.035
	LVG	−0.683	1.743	−0.392	0.696	0.897	1.115
	LTS	16.291	3.326	4.899	0.000	0.903	1.108
	IND	−0.611	0.509	−1.199	0.234	0.915	1.093
	R = 0.580	$R^2 = 0.336$		Adjusted $R^2 = 0.281$			
	Std. Error of the Estimate = 16.26			Durbin-Watson = 1.938, F = 6.155 (0.000)			

Note Dependent Variable: SDS

However, market performance (MBV) has a negative and significant impact on SDS (B = −0.830, $p < 0.01$). Foreign institutional investment (FII) and leverage (LVG) show no significant effects on SDS. This can be the result of sector-specific differences in disclosure standards, a lack of social reporting requirements, or a lack of investor activity in India that is focused on social factors. There are no significant autocorrelation problems, according to the Durbin-Watson statistics (1.938), and multicollinearity concerns are not present, according to collinearity diagnostics (VIF values less than 2). These results imply that social disclosures are driven by corporate size and profitability, but market performance factors can serve as a disincentive.

26.5 Discussion and Practical Implication

The main objective of this paper was to examine the relationship between environmental and social disclosures and corporate attributes. The four primary attributes—market performance, profitability, foreign institutional investors' stake, and financial risk- were recognized as important determinants of environmental and social disclosures. The studies' results show that companies' profitability and market performance are positively associated with their level of environmental and social disclosures. The result of the study has been supported by various previous studies (Iqbal et al., 2014). It is concluded that firms' high profitability and performance lead to high environmental and social disclosures. Firms with high profitability are more capable and willing to invest in stakeholder engagement practices (Qiu et al., 2014). However, the result also detected a negative and insignificant relationship between foreign institutional investors (FIIs) and leverage with the level of environmental and social disclosures. This result is consistent with the study of Gulzar et al. (2019), which also found an insignificant relationship between foreign institutional investors and the CSR practices of Chinese companies. A number of factors contribute to the insignificant impact of FIIs on environment and social practices: FIIs tend to prioritize short-term financial returns over long-term CSR goals, which limits their influence on CSR initiatives; they lack direct control over corporate governance, especially when their stake is small; they may avoid companies with aggressive CSR strategies because they perceive them to be riskier; cultural and regulatory differences, as well as a diversified investment strategy, further dilute the impact of FIIs on CSR; and in some markets, CSR is more influenced by government regulations than investor pressure. Few studies explored the factors associated with the environmental and social practices of Indian companies; thus, the present study would help fill the research gaps.

The result of the study provides a further scope of improvement in the level of relevant environmental and social practices of Indian companies. The findings suggested that investors should focus more on the market performance and profitability of the firms in India because more profitable firms focus more on their environmental and social disclosure levels. Companies with good environmental and social performance develop trustworthiness among their stakeholders. Nowadays, it is widely accepted by institutional and retail investors that long-term performance can be improved by a company with lower environmental and social risks. The chances of manipulation, financial fraud, misrepresentation, etc., can be minimized with a good disclosure score. The demand for environmental disclosures has increased due to the rise in the adaptation of the factors of climate change and the regulatory framework concerning these factors. Similarly, media and other sources have also noticed emerging labor disputes, unrest, strikes, and working conditions. Thus, the disclosures related to environmental and social factors provide risk management to mitigate the risk.

The major implications of this research have been revealed in two ways. First, the study helps to understand the level of environmental and social disclosures of Indian companies and how the different corporate factors affect their environmental and social performance. Second, this study highlights the fact that profitability and performance significantly affect the level of environmental and social disclosures. The more profitable companies are more inclined to spend on environmental and social activities.

Further research can be done by using a combined environmental and social disclosure score as a dependent variable. A panel data study can yield better results.

26.6 Conclusion

The relationship of environmental and social disclosure (dependent variable) with market performance, profitability, FIIs stake, and leverage (independent variables) has been examined. Two equations (26.1 and 26.2) are formulated by taking measurements of market performance, profitability, FIIs stake, and leverage. The results found that Eq. (26.1) is statistically significant with a p-value < 0.01. The results of Eq. (26.2) found that both equations are statistically significant with a p-value < 0.01. Overall, the results reveal that MBV has a negative and significant impact on SDS for both the equations and profitability, both positively and significantly. The study found a positive and significant effect of companies' market performance and profitability on environmental and social disclosures. However, environmental and social disclosures have a negative and insignificant impact on FII's stake and leverage. This is in the same line with previous work done by Waddock and Graves (1997). The disclosure of value depends on legal policies, and this score depicts the firm's competitiveness and irregularity among stakeholders. Further research can be done by taking a combined environmental and social disclosure score as a dependent variable. A panel data study can opt for better results.

Conflict-of-Interest Statement The Authors have no conflict of interest to declare.

References

Adams, C., & Zutshi, A. (2004). Corporate social responsibility: Why a business should act responsibly and be accountable. *Australian Accounting Review, 14*(34), 31–39.

Aerts, W., Cormier, D., & Magnan, M. (2007). The association between web-based corporate performance disclosure and financial analyst behaviour under different governance regimes. *Corporate Governance: An International Review, 15*(6), 1301–1329.

Agostini-Costa, T. D. S., Wondraceck, D. C., Rocha, W. D. S., & Silva, D. B. D. (2012). Carotenoids profile and total polyphenols in fruits of Pereskia aculeata Miller. *Revista Brasileira De Fruticultura, 34*(1), 234–238.

Aguilera, R. V., Williams, C. A., Conley, J. M., & Rupp, D. E. (2006). Corporate governance and social responsibility: A comparative analysis of the UK and the US. *Corporate Governance: An International Review, 14*(3), 147–158.

Aras, G., Aybars, A., & Kutlu, O. (2010). Managing corporate performance investigating the relationship between corporate social responsibility and financial performance in emerging markets. *International Journal of Productivity and Performance Management, 59*, 229–254.

Brammer, S. J., & Pavelin, S. (2006). Corporate reputation and social performance: The importance of fit. *Journal of Management Studies, 43*(3), 435–455.

Brooks, C. (2008). *Introductory econometrics for finance*. Cambridge University Press.

Carroll, A. B., & Shabana, K. M. (2010). The business case for corporate social responsibility: A review of concepts, research and practice. *International journal of management reviews, 12*(1), 85–105.

Cho, C. H., & Patten, D. M. (2007). The role of environmental disclosures as tools of legitimacy: A research notes. *Accounting, Organizations and Society, 32*(7), 639–647.

Clarkson, P. M., Overell, M. B., & Chapple, L. (2011). Environmental reporting and its relation to corporate environmental performance. *Abacus, 47*(1), 27–60. https://doi.org/10.1111/abac.2011.47, https://doi.org/10.1111/j.1467-6281.2011.00330.x

Cordeiro, J. J., & Sarkis, J. (1997). Environmental proactivism and firm performance: Evidence from security analyst earnings forecasts. *Business strategy and the environment, 6*(2), 104–114.

Cormier, D., & Magnan, M. (2014). The impact of social responsibility disclosure and governance on financial analysts' information environment. *Corporate Governance*.

Gray, R., Kouhy, R., & Lavers, S. (1995). Corporate social and environmental reporting: A review of the literature and a longitudinal study of UK disclosure. *Accounting, Auditing & Accountability Journal, 8*(2), 47–77.

Gray, T. (2009). Investing for the environment? The limits of the UN Principles of Responsible Investment. Paper was presented at the Oikos Foundation/UN Principles of Responsible Investment Young Scholars Finance Academy, 2009.

Gunasekaran, A., & Spalanzani, A. (2012). Sustainability of manufacturing and services: Investigations for research and applications. *International journal of production economics, 140*(1), 35–47.

Hooghiemstra, R. (2000). Corporate communication and impression management—New perspectives why companies engage in corporate social reporting. *Journal of Business Ethics, 27*(1/2), 55–68.

Ingram, R. W., & Frazier, K. B. (1980). Environmental performance and corporate disclosure. *Journal of Accounting Research*, 614–622.

Ioannou, I., & Serafeim, G. (2010). *What drives corporate social performance?: International evidence from social, environmental and governance sources*. Harvard Business School.

Iqbal, N., Ahmad, N., Hamad, N., Bashir, S., Sattar, W. (2014). Corporate social responsibility and its possible impact on firm's financial performance in banking sector of Pakistan. *Arabian Journal of Business and Management Review (Oman Chapter), 3*(12), 150.

Kolk, A., & Perego, P. (2010). Determinants of the adoption of sustainability assurance statements: An international investigation. *Business Strategy and the Environment, 19*(3), 182–198.

Labatt, S., & White, R. R. (2002). *Environmental finance: A guide to environmental risk assessment and financial products*. Wiley.

Lepoutre, J., & Heene, A. (2006). Investigating the impact of firm size on small business social responsibility: A critical review. *Journal of Business Ethics, 67*(3), 257–273.

Lungu, C. I., Caraiani, C., Dascalu, C., Guse, R., & Sahelian, D. (2011, August 11). *Corporate social and environmental reporting: Another dimension for accounting information*. SSRN: https://ssrn.com/abstract=1447247 or https://doi.org/10.2139/ssrn.1447247

Mallin, C., Michelon, G., & Raggi, D. (2013). Monitoring intensity and stakeholders' orientation: How does governance affect social and environmental disclosure? *Journal of Business Ethics, 114*(1), 29–43.

Margolis, J. D., Walsh, J. P. (2003). Misery loves companies: Rethinking social initiatives by business. *Administrative Science Quarterly, 48*(2), 268–305. https://doi.org/10.2307/3556659

McGuire, J., & A. Sundgren, et al. (1988). Corporate social responsibility and firm financial performance. *Academy of Management Journal*, 854–872. http://dx.doi.org/10.2307/256342

Milne, M., & Adler, R. (1999). Exploring the reliability of social and environmental disclosures content analysis. *Accounting, Auditing and Accountability Journal, 12*(2), 237–256.

Orlitzky, M., Schmidt, F. L., & Rynes, S. L. (2003). Corporate social and financial performance: A meta-analysis. *Organization Studies, 24*(3), 403–441.

Qiu, Y., Shaukat, A., & Tharyan, R. (2016). Environmental and social disclosures: Link with corporate financial performance. *The British Accounting Review, 48*(1), 102–116.

Richardson, A. J., & Welker, M. (2001). Social disclosure, financial disclosure and the cost of equity capital. *Accounting, Organizations and Society, 26*(7–8), 597–616.

Sharma, P., Panday, P., & Dangwal, R. C. (2020). Determinants of environmental, social and corporate governance (ESG) disclosure: A study of Indian companies. *International Journal of Disclosure and Governance, 17*(4), 208–217.

Simnett, R., Vanstraelen, A., & Chua, W. F. (2009). Assurance on sustainability reports: An international comparison. *The Accounting Review, 84*(3), 937–967.

Stanwick, P. A., & Stanwick, S. D. (1998). The relationship between corporate social performance, organization size, financial performance, and environmental performance: An empirical examination. *Journal of Business Ethics., 17*(2), 195–204.

Tsoutsoura, M. (2004). *Corporate social responsibility and financial performance.* Hass School of Business. University of California, Berkeley. https://escholarship.org/uc/item/111799p2

Verrecchia, R. (1983). Discretionary disclosure. *Journal of Accounting and Economics, 5*, 179–194.

Waddock, S. A., & Graves, S. B. (1997). The Corporate social performance—Financial performance link. *Strategic Management Journal, 18*(4), 303–319.

Chapter 27
Investigating the Interplay Between Corporate Governance and Corporate Sustainability in India

Raghuveer Katragadda, Rubee Singh, and Vinay Kandpal

Abstract Corporate governance is essential to a company's operational support. As sustainable development gains prominence, corporate governance extends beyond mere firm management to encompass environmental, social, and financial obligations. In light of the increasing prevalence of the sustainability concept, corporate governance is now emphasizing the promotion of corporate sustainability adoption. A unique and uncharted domain exists in the relationship between corporate governance and corporate sustainability for Indian enterprises. This chapter addresses this gap by elucidating the connection between corporate governance and corporate sustainability, providing a novel perspective. This chapter enhances information on corporate governance and corporate sustainability in India, as less research has examined the relationship between these variables in the context of Indian firms.

Keywords Corporate governance · Corporate sustainability · Organizational sustainability · Panel data analysis · Environmental resilience

27.1 Introduction

Global interconnectivity and rapid dynamism have been buzzing since the globalized phenomena started, and similarly, there has been mounting pressure for businesses all over the world to operate environmentally and socially responsibly (Aboud & Yang, 2022; Pillai et al., 2023). In the era of sustainable development, stakeholders call

R. Katragadda
School of Management, V. R. Siddhartha Engineering College, Kanuru, India

R. Singh (✉)
Institute of Business Management, GLA University, Mathura, India
e-mail: dr.rubeerajput@gmail.com

Department of Management Studies, Kumaun University, Nainital, India

V. Kandpal
Department of Management Studies, Graphic Era (Deemed to Be University), Dehradun, India

V. Kandpal et al. (eds.), *Developments in Corporate Governance*,
Approaches to Global Sustainability, Markets, and Governance,
https://doi.org/10.1007/978-981-96-6366-8_27

for companies to become more aware of their corporate responsibilities, including addressing human rights and global warming issues (Tjahjadi et al., 2021). Over the last few years, researchers described the detrimental effects intensively acknowledging the environmental abuse by organizations and demanding increased accountability from corporations (Ren et al., 2022; Yu et al., 2023). Henceforth, with this increasing demand for accountability, organizations that claim to be stakeholder-sensitive and shareholder-responsible have surely felt the pressure to spend more on CSR initiatives (Aras & Crowther, 2008; Singh et al., 2021). Many attempts were made to deliver the meaning of sustainability, as well as to explain its importance for the effectiveness of societies.

Sustainability is a principle that we are all obligated to uphold. It pertains to decisions that encompass current standings and yield future benefits or mitigate potential risks (Scoones, 2016). Therefore, CSR branches into threefold activities, i.e., Social, Environmental, and economic sustainability, where sustainability is the means of responsible stewardship of resources of nature and ensuring they are available for every generation in the future. It includes measures to control harmful elements, for instance, carbon, effective wastes, and other chemicals, and letters controlling the loss of ecosystem services (Pezzey, 2017). Social sustainability emphasizes the improvement of interpersonal relationships within society to promote social equity. Economic sustainability encompasses the establishment of a sustainable framework for ongoing, healthy, and long-term economic growth and development, therefore minimizing adverse effects on other aspects of the local environment, community, and culture (Rai et al., 2021).

Corporate Sustainability (CS) can be noticed as the regimes employed by the business entities when they are only implementing systems for which they are agents, which serve either to destabilize or consolidate the systems in question (Meuer et al., 2019). The purpose is to meet the Direct and Indirect stakeholders' appreciation and at the same time while not putting the organization in a compromising position so as to meet more demands of the further stakeholders (Engert et al., 2015). Such processes can be regularly assessed with the backing of the "triple bottom line" which lays prominence on the socio-environmental and economic results. However, such an approach being implemented by the corporations is not very clear and in unison (Vaid et al., 2016). While many researchers exclude the financial aspect when discussing sustainability (Gleißner et al., 2022), there are actually four significant dimensions to sustainability: The domain covering the business and society concerns includes environmental, social, and organizational culture; financial is the major concerns (Aras & Crowther, 2008). The environmental dimension is crucial, focusing on the strategic use of ecological assets due to their limited availability and high value. Companies operate to ensure waste reduction and maximum resource utilization, often by focusing on the usage of renewable energy and decreasing emissions (Lloret, 2016; Nicolăescu et al., 2015).

The concept of social sustainability for organizations involves promoting well-being on a public scale, not just meeting the needs of their employees. Externally, it considers social development, growth, and justice. Internally, it contributes to learning, growth, safety, security, and community development. When it comes to organizational culture, corporate sustainability includes values, leaders' and employees' attitudes and mental state, leadership, and social factors. These elements impact the management and motivation of integrating corporate sustainability (Assoratgoon & Kantabutra, 2023). Therefore, sustainability implementation in corporations must eliminate conflicts between these dimensions and ensure a balanced impact on each dimension, considering both the present and the future.

This strategic decision-making necessitates corporations to develop Corporate Governance (C.G.) codes, which are a set of rules and practices that guide a business entity in ensuring the integration of sustainability into the company's operations and decision-making processes. As a system, corporate governance evolves and develops a relationship between the company's shareholders, board, management of the firm, and other stakeholders; it is crucial to meet the needs of the organization, exercise decision-making and policymaking, and work toward the best interest of stakeholders (Mubeen et al., 2023). As the current business world is causing climate issues like global warming, the potential positive influence of improved C.G. on society and sustainability is immense. Though across the world, corporate governance is becoming more assertive in Asian countries like India, the presence of more dominance of family-controlled companies, lack of adequate regulations, absence of transparency, high corruption, more dependence on loans and grants, and slight pressure from NGOs or other international communities results in weakening the status of corporate governance (Mahmood et al., 2018; Majeed et al., 2015; Masud et al., 2017, 2018; Shirodkar et al., 2016; Subramaniam et al., 2017). However, with the growing preference of stakeholders to be part of corporations wherein more focus is on sustainability, concerted efforts are being made to improve corporate governance status. These efforts include the introduction of stricter regulations, increased transparency, and the promotion of ethical practices (Aras & Crowther, 2008). Good corporate governance could help corporations meet the welfare of society even with limited resources; thus, nowadays, corporate governance is prevalent in India.

Corporate governance is closely associated with environmental development. Emphasizing business sustainability may enhance the resolution of environmental issues and the management of hazards. Neglecting environmental issues may tarnish the organization's reputation and perhaps subject the corporation to regulatory and financial liabilities. Current research predominantly examines the relationship between corporate governance and sustainability, although there is a paucity of studies addressing the condition of Indian enterprises. Simultaneously, a research deficit exists in analyzing the influence of corporate governance indicators on the sustainability of Indian corporations. Consequently, the present study seeks to overcome this gap by empirically assessing the relationship and assisting India in tackling its prevailing corporate governance challenges by determining their importance. Investigating the correlation between C.G. and C.S. in India is the main focal point of this study; hence, it is deemed necessary to understand the ethical significance of aligning corporate governance practices with sustainability goals. This ensures that companies take into account different socio-economic and environmental factors in which businesses operate. Although the practice of integrated sustainability reporting is new in India, the government encourages its adoption (Sun et al., 2022). Understanding this relationship is not only important from the perspective of understanding regulatory frameworks, but as well it enables to induce investors' confidence. Developing such strategies based on the study's findings is not only about improving the risk management capacity of the business entity but also necessary for the long-term sustainability of organizations and their impact on society. On the other hand, this study will as well enable to through light on the long-term viability of the connectivity between C.G. and C.S. in India. Further, the study connection can improve firms' effectiveness, competitive advantage, and societal impact, thereby contributing to the overall welfare of society.

27.2 Literature Review

27.2.1 A Fusion of Socio-environmental and Financial Performance

Good corporate sustainability performance means that the company operates in a manner that helps create value for stakeholders through sound governance, as well as conserving societal well-being and ecological integrity in its environment. For the assessment of corporate sustainability performance, there is a need to have an evaluation of both qualitative and quantitative indicators that assess corporate governance, social responsibility, and environmental stewardship (Barbosa et al., 2023; Katragadda, 2013). However, as the environmental dimension is an external resource for an organization therefore most companies are concerned with society and the organization's resources (Aras & Crowther, 2008). However environmental resources influence sustainability. The lack of focus on minimization of pollution, reduction of waste, promoting the usage of renewable resources, and development of sustainable techniques results not only in impacting society but also affecting the future of the company. Therefore, by focusing on environmental resources, the company could support the company in operating in the future effectively. Also, the focus on environmental resources by reducing waste and the investment in renewables techniques raise initial investment costs but, over time, could reduce the cost for the company, thus generating more revenue (Dewi & Widyawati, 2023; Wu et al., 2020). This reveals the presence of an association between financial and environmental performance. Apart from this, good environmental performance also contributes to benefitting societies by raising well-being. The improvement in well-being results in raising motivation thus, finally, the productivity of an individual, which again will benefit the organization (Aras & Crowther, 2008). On the other hand, a higher level of financial performance has been traced back to sustainable companies, which also demonstrated better environmental and social practices as per the research carried out by Martínez-Ferrero and Frías-Aceituno (2015). However, the study also revealed a worrisome trend for non-sustainable companies, indicating a potential negative bidirectional relationship between social, environmental, and financial performance. This could be due to the companies' failure to allocate adequate economic resources toward the improvement of sustainable behavior, resulting in potential penalties. Therefore, every business entity needs to understand that their performance is not a one-dimensional aspect; instead, the financial performance of the entities has a direct link with that of socio-economic and environmental factors.

27.2.2 Corporate Governance and Sustainability Relationship

Corporate governance within a company includes the rules and procedures that support decision-making. It involves specific activities and rules followed to achieve corporate objectives and measure the result. Hence, an initiating role in sustainability practices in a business organization, C.G. plays a vital role. The results achieved through this combination of C.G. and sustainability practices are significant, providing reassurance about their effectiveness. As a result of the changes mentioned above, organizations have been forced to modify their approach, which in turn has led to the following outcomes. These are regulator evolution, enhancing the organizational sustainability goals, improved quality of products, enhanced customer satisfaction, and, at the same time, reduced costs of production (Naciti et al., 2022). This shows that good governance contributes to the improvement of environmental and social performance because social and illegitimate activities that are unwanted are discouraged in order to protect the image of the Company (Munir et al., 2019).

Earlier studies have pointed out that factors such as the board size of the firm, CEO duality or board composition have an impact on sustainability performance. On the other hand, few studies have pointed out that an independent director's role in a company has a positive effect on the company's sustainability performance (Arora & Dharwadkar, 2011). Other measures that enhance the best practices of corporate governance are also useful in the improvement of the economy, environment, and society and are also advantageous to the corporation's image and performance. As postulated by the stakeholder theory, it is clear that corporate governance has the critical role of safeguarding stakeholders through incorporating various interest groups within the company's policies and actions (Galbreath, 2016). Additionally, a number of scholars have pointed out that corporate governance is essential in improving the organization's reputation and the assurance of its efficiency. Therefore, organizations can apply the principles of corporate governance, such as accountability, fairness, responsibility, and transparency, for the achievement of corporate objectives and strategies (Aras & Crowther, 2008).

27.2.3 *Need for Corporate Governance and Corporate Sustainability in India*

Protecting the shareholders and their investments has historically been the focal point of corporate governance. Nevertheless, in recent years, ideas about C.G. have been expanded to encompass oversight of the organizations' socio-environmental impact (Naciti et al., 2022). Therefore, one has to understand that the idea of strong C.G. is not something that should be relevant only for developed countries but also for developing economies. As noted by Bae et al. (2018), it is just as important in emerging economies, such as India, where governance is poor, stakeholders' involvement is low as well as corruption is high, which negatively affects the country's economic performance. It also symbolizes a weak institutional environment, lack of corruption control, and absence of the rule of law. Corporate governance for firms is low due to political interference and a need for more knowledgeable human resources. However, the study also found that, similar to other countries, the studies for exploring environmental reporting in India have increased, but most of these studies' results are inconclusive. Sabharwal and Bansal (2015) in their study explored the reporting for companies and identified that for some companies, CSR expenditure has increased, and also sectors like software, IT, energy, and mining are reporting their CSR expenditure, but many other sectors like banking are not much focusing on CSR expenditure. Jha and Mehra (2015), in their research work, discussed that understanding corporate governance is essential for making organizations successful and competitive. Another study by Mishra (2023) identified the need for immediate action to drive a shift toward a sustainable and responsible approach. All the existing studies identified the presence of linkage, but the results are not conclusive enough to define the influence on sustainability and promoting corporate responsibility and transparency. In response to stakeholders demanding the inclusion of sustainability practices in operations, it is essential to comprehend how corporate governance contributes to the corporate sustainability of Indian companies.

27.3 Empirical Review

An extensive research study on 88 listed firms covering a period of 2009–2016 in Bangladesh, India, and Pakistan by Bae et al. (2018), examined that sustainability and annual reports submitted to the "Global Reporting Initiative" database supported the hypothesis that corporate governance and sustainability are closely related, particularly to the disclosure of operating practices and codes of business. The authors also identified that there are some C.G. factors, including institutional shareholding, directors' shareholding, shareholding patterns by foreigners or foreign entities, size of the board, and board independence, that have a significant impact on the level of corporate sustainability as well as the disclosure of sustainability report. This therefore means that communication has the capacity to enhance the market by curing informational asymmetry.

However, in another study, Munir et al. (2019) employed structural equation modeling to analyze the performance of 425 firms that are listed in the Australia Securities Exchange. The study has shown that there is a strong affirmative affiliation between the two variables namely C.G. and C.S. performance. It also identified the size of the board, especially the number of directors on the audit committee and the total number of directors on the board, as other modifiable variables that can be used in enhancing sustainability in the organization. Furthermore, the study found that the proportion of non-executive directors on the remuneration committee and the frequency of board meetings were factors that affected CSP in a moderated manner. Some of the factors are environmental management, socio-tenet commitment, socio-tenet capability, community development, eco-efficiency, and health and safety management systems. Crifo et al. (2019), a group of researchers, conducted research on 120 firms that are listed on the French stock exchange. They are also able to prove the linkage between C.G. and C.S. in the company, information that can be useful for our audience. The authors also derived the same relationship using internal directors as one of the factors, thus enhancing the robustness of the study.

In a recent research work by Tjahjadi et al. (2021) examined the influence of the C.G. on the C.S. performance of business entities listed in the Indonesia Stock Exchange. The research collected secondary data from the business entities filing reports submitted to the statutory bodies, audited financial statements, and sustainability reports of 117 non-financial firms from the years 2013 to 2017. To establish the connectivity of the educational background of the board of commissioners on economic sustainability, the authors employed the multiple regression methods. They observed a significant relationship between the academic achievements of the board of commissioners and economic viability. This can be explained by factors like profitability (Return on Asset), sales growth, firm size, firm age, and leverage. However, the study also established that the composition in terms of education of the board of commissioners influences the performance of socio-environmental sustainability negatively.

In addition, the research has found that the education level of the CEO and environmental and social sustainability performance have no relationship, while economic sustainability is negatively influenced by the education level of the CEO. Lastly, the study also discovered that the size of the top management team has no effect on social sustainability performance but has a negative impact on economic and environmental sustainability performance.

For instance, in a recent study, Lu (2021) examined 456 top US public companies and collected unbalanced panel data from 2007 to 2011. These researchers correlated the relationship between C.G. and C.S. by applying a multiple regression model, where the implications of this study are noteworthy and point to the fact that finance can be a critical driver of guidance and supervision of management and strategic sustainable investment by the companies. In detail, characteristics of board corporate governance measures like the independent board size, CEO duality, and women directors have a clear effect on the corporate sustainability of the company, thus underlining the significance of these factors in corporate decisions.

Furthermore, in a study by Drempetic et al. (2020), a sample of 6000 companies from the ASSET4 database of Thomson Reuters indicates a positive association between governance and sustainability of corporations. The study revealed that ESG reporting by a company is an indicator of the company's legitimacy. Further, the study finds that firm size enhances this relationship. The authors propose that larger firms focus on corporate sustainability and report it more frequently, as they have more data available, which is required by the measurement systems of the ESG rating agencies. Also, Gerwing et al. (2022) suggest that corporate governance enhances mandatory sustainability reporting quality. The authors also found that the relationship is enhanced by engagement in CSR activities based on compliance with the standards and codes of directives. Kocmanová et al. (2016) suggested through data from 211 corporations with Environmental Management Systems through structural equation modeling that corporate governance contributes to value creation and sustainability. The authors were also suggestive that socio-environmental and corporate governance indicators impact the environmental efficiency element of corporate sustainability positively.

Furthermore, in a rigorous study, Kumar et al. (2021) undertook a comprehensive analysis of companies' sustainability disclosure practices to understand the impact of C.G. To achieve this, the researchers carefully chose 57 firms listed on the National Stock Exchange of India (NSE) from the mining and energy industries and applied a panel analysis test. Based on the authors' analysis, some important corporate governance variables, including size, sustainability reporting, and market capitalization, positively affect the level of corporate sustainability.

27.3.1 Conceptual Framework

The above framework shows the complex relationship between the independent and dependent variables in the proposed study. The independent variables, which include the company's ownership structure, firm size, and firm efficiency, therefore, influence the dependent variable, which is corporate sustainability. This dynamic relationship points to the richness and interdependence of the subject at hand and should, therefore, generate the interest of the target readership, which includes academic researchers, business practitioners, and students interested in the fields of corporate governance and sustainability (Fig. 27.1).

27.3.1.1 Research Gap

Numerous studies identified that it is possible to identify the connection between corporate governance and sustainability based on previous works (Crifo et al., 2019; Lu, 2021; Tjahjadi et al., 2021). Thus, there is a research gap that needs to be filled

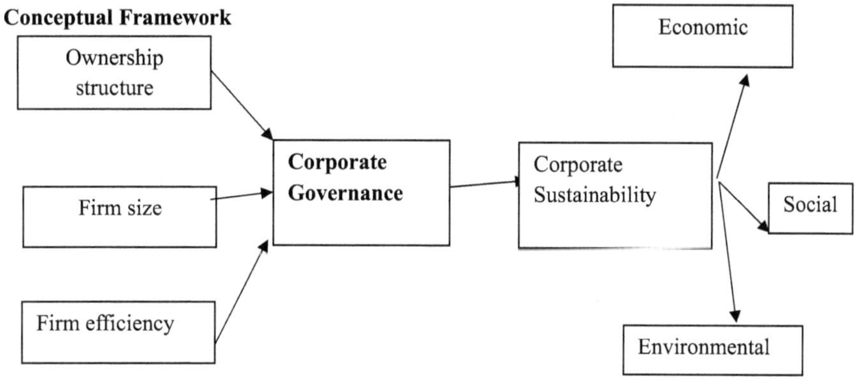

Fig. 27.1 Conceptual framework

concerning Indian companies. Most of the earlier research has investigated the link between corporate governance and the sustainability of corporations in the Indian context by considering big business houses. Nevertheless, there is a need to establish how firm size affects the sustainability management practices in the country. Also, from the literature review, it can be seen that India implements both voluntary and mandatory regulations in corporate sustainability (Mahmood et al., 2018; Majeed et al., 2015; Masud et al., 2017, 2018; Shirodkar et al., 2016; Subramaniam et al., 2017). Hence, there is a need to examine the part played by corporate governance in compliance with different regulations and the incorporation of sustainable development in environmental, economic, and social dimensions. Thus, the purpose of the present study is to fill these gaps in the literature and investigate the relationship between C.G. and C.S. in India.

27.4 Methodology

Numerous research studies have examined the relationship between C.G. and C.S. However, some studies still need to demonstrate this connection, often due to poor governance within companies. This issue is particularly prevalent among Indian companies, where poor corporate governance remains a persistent problem. We argue about the approach by stating the choice of poor governance companies is not an optimal methodology for assessing linkage. Alternatively, it seems reasonable to assume that corporate governance plays a significant role in the sustainability of Indian companies. For fulfilling the purpose of the study, the simple evaluation does not seem fitting; thus, herein, a panel regression model was developed using STATA. With the collection of data from the top 10 NIFTY 50 firms based on market capitalization from 2017 to 2024, the relationship between C.G. and C.S. was examined by using the fixed-effect panel regression model. A study by Bai et al.

(2023), a highly respected and influential researcher in the field, confirmed that well-established firms tend to focus more on CSR practices, leading to improved firm performances financially in both the short and long term. Therefore, selecting firms based on market capitalization appeared suitable. We have chosen to use a rigorous quantitative analysis to validate the connection between the variables and confirm our proposed approach. Our study will focus on statistical analysis of C.S. aspects in India.

To investigate the sustainability extent and the corporate governance the data was taken from the annual reports of the companies. Herein, 4 indicators, i.e., promoters' shareholding, Sales, profit after tax, and ROCE, were taken as the indicators of corporate governance as they help in measuring not just the performance of the company but also its efficiency and the status of ownership. The selection of the parameters was based on Kumar et al. (2021) study. Further, to represent the corporate sustainability status, the CSR expenditure of the company was taken as the measure. Based on the selected variables, the requirement was to build the linkage between corporate governance and corporate sustainability to understand their association. However, herein, initially, descriptive analysis and trend analysis was done for a brief overview of data. The descriptive analysis helped in summarizing the entire data while trend analysis defined the movement of CSR expenses for each of the companies. Following this, the regression model was developed to test the linkage between different C.G. indicators and sustainability. The proposed model for the study could be stated as

$$\boldsymbol{CS_{it}} = \boldsymbol{\beta_0} + \boldsymbol{\beta_1 PS_{it}} + \boldsymbol{\beta_2 FS_{it}} + \boldsymbol{\beta_3 P_{it}} + \boldsymbol{\beta_0 E_{it}} + \boldsymbol{\varepsilon_{it}}$$

Wherein

CS: Corporate sustainability
PS: Promoter shareholding
ES: Firm size
P: Profitability of firm
E: Efficiency of firm
Using the proposed model, the below-stated hypothesis was tested by a researcher at a 5% level of significance
$\mathbf{H_0}$: Good corporate governance does not have a significant impact on corporate sustainability
$\mathbf{H_A}$: Good corporate governance significantly impacts corporate sustainability

The collected panel data for 10 companies for 8 years was analyzed using Stata as the tool of analysis. The researcher initially focused on assumptions check, i.e., normality, multicollinearity. As the data was panel data and the assumption of normality was not fulfilled, the focus of the researcher shifted toward the identification of the panel regression model which seemed to be suitable for assessing the linkage. For this, the Hausman test was done to define whether the random-effect or fixed-effect model was suitable.

Based on the results, the optimal model was selected and testing of the results was done at the selected significance level of 5% to make the deduction about the linkage between corporate governance and corporate sustainability for Indian firms. As all assumptions were tested before building the model therefore the results derived from the study were efficient. Also, instead of randomly selecting the model, the focus of the researcher was on following the complete statistical procedure for the identification of the suitable regression model; thus, herein, the reliability and validity of the results were derived.

27.5 Findings of Study

With the focus of the study on assessing the linkage, the researcher began the procedure by having a brief overview of the data. For this, the selected data for all the companies was examined using descriptive statistics. Descriptive statistics is defined as the means of evaluating the data to provide information in a more organized, sensible, and summarized format (Privitera, 2023).

Table 27.1 identified that the mean value of the promoter shareholding is 0.49, i.e., 49%, with the minimum value being 13% and the maximum being 73%. The sales of the selected firms have a mean of Rs 134057.40 crores showing more capacity of the firms to generate revenue. Further, the profit after tax for these firms was Rs 18040.80 crores showing higher profitability of the selected large capitalization firms. The minimum value being negative reveals that some selected firms have also witnessed a loss of Rs 36088 crore. The ROCE reveals the 27% efficiency of the company in utilizing its capital to generate profits. The maximum value herein reveals that even some firm has an ROCE of 120% showing higher efficiency of the company. Lastly, the mean CSR expense is Rs 350.18 crore which is about 2% of the average profit after tax. Even with a maximum CSR expense of Rs 1271 crore, the percentage is 2% of the average profit after tax, revealing that despite being a large market capitalization company, the contribution to CSR is minimal, i.e., 2% of the profit after tax. The descriptive statistics identified the average contribution of CSR but for understanding the contribution of each company, the line graph for the CSR expenses is plotted (Fig. 27.2).

Table 27.1 Descriptive statistics

Variable	Mean	Std. Dev	Min	Max
Promoter shareholding	0.49	0.19	0.13	0.73
Sales	134,057.40	112,235.10	31,890.00	534,534.00
Profit after tax	18,040.80	16,982.24	−36,088.00	61,077.00
ROCE	0.27	0.26	−0.03	1.20
CSR expense	350.18	310.51	0.65	1271.00

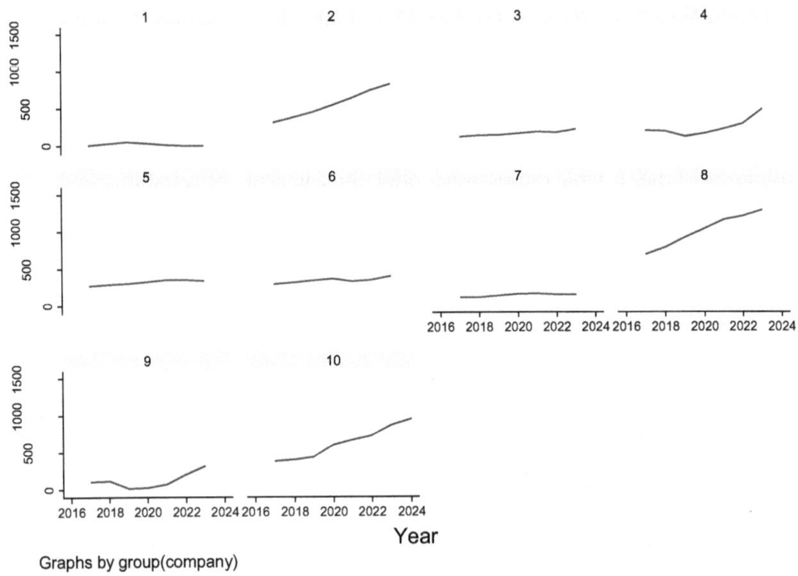

Fig. 27.2 Trend analysis

The line graph of the CSR expenses reveals that for most of the companies, there has been consistency in the CSR expenses only some companies like company 2 (HDFC bank), company 8 (Reliance Industries), and company 10 (TCS) witnessed over the years growth in CSR expenses. Contrary to these were the companies, i.e., company 4 (ICICI bank) and company 9 (SBI), wherein an initial decrease in CSR expense was seen, but from 2020, there has been a rise in their contribution toward CSR. This rise in focus toward CSR expense shows that the firms are gaining more recognition of the relevance of fulfilling their corporate sustainability.

The descriptive and trend analysis provided a brief overview of the data but for having the assessment of the impact, the assumption tests need to be done. Burton (2021) stated that for fitting in an OLS regression model to evaluate the linkage between the variables, there is a need to fulfill some basic assumptions of the model, i.e., linearity, no autocorrelation, normal data, no heteroskedasticity, and no multicollinearity. In case the data fails to meet the assumptions of the OLS regression, the model cannot be used for evaluating the impact; thus, herein, firstly, the normality of the data is tested using the Shapiro-Wilk test.

Table 27.2 shows that the significance value for all the variables except profit after tax is less than 0.00. As the significance level of the study is 0.05 thus, herein for the Shapiro-Wilk test, the null hypothesis of having normal distribution is rejected. This reveals that for most of the variables, the data is not normally distributed, resulting in the OLS regression not being suitable for the analysis. Further, the multicollinearity among the variables is also checked by assessing the VIF value with the inclusion of the CSR expense as the dependent variable. In contrast, all other variables, i.e., promoter shareholding, sales, profit after tax, and ROCE, are independent variables (Table 27.3).

The multicollinearity test reveals that the VIF value for all the variables is less than 10. As the value is less than 10 and even more than 0.2, thus this shows that there is no multicollinearity in the dataset (Kim, 2019). Even the examination of the heteroskedasticity using the Breusch Pagan test showed that the significance value is 0.00, which is less than 0.05, showing that the null hypothesis of having constant variance is rejected. This reveals that heteroskedasticity exists in the dataset. As the assumptions of the OLS regression are not fulfilled the OLS regression is not suitable for assessing the relationship between C.G. and C.S. Thus the panel regression model needs to be built to evaluate the impact. The initial requirement for building the panel regression model is to define which model needs to be developed. With the panel data, there is the possibility of having unobserved individual heterogeneity related to the regressors. To overcome this, Hausman has proposed a test that helps in detecting the existence of an effect in the data. The Hausman test is the main test used to understand whether the model has a fixed or random effect. The test focuses on comparing the individual-specific effects model within the estimator and the random-effect estimator. Observing the statistical difference between these two estimators, the nature of the model is determined using the Hausman test (Table 27.4).

The Hausman test revealed a probability value of 0.30, higher than the required significance level of 0.05. This implies that the null hypothesis of the Hausman test, which suggests a fixed-effect model, is not rejected. Consequently, the fixed-effect model is deemed more suitable than the random-effect model for understanding the impact of corporate governance on corporate sustainability. With this in mind, we have constructed a panel regression model with a fixed effect, with CSR expense as

Table 27.2 Normality test

Variable	W	V	z	Prob > z
Promoter shareholding	0.91	5.37	3.63	0.00
Sales	0.77	15.80	6.05	0.00
Profit after tax	0.97	2.02	1.54	0.06
ROCE	0.78	13.97	5.75	0.00
CSR expense	0.87	8.17	4.57	0.00

Table 27.3 Multicollinearity test

Variable	VIF	1/VIF
Sales	2.19	0.46
Profit after tax	2.01	0.50
ROCE	1.54	0.65
Promoter shareholding	1.29	0.78

Table 27.4 Hausman test

Ch-square	Prob
2.43	0.30

the dependent variable. The model incorporates promoter shareholding, sales, profit after tax, and ROCE as the independent variables, each playing a crucial role in our analysis.

Table 27.5 shows that the value of adjusted R square in the fixed-effect regression model is 0.565, which shows that about 56.5% of the variation in CSR expenses is due to the selected corporate governance indicators. Also, the model's F-value is 14.600, which is greater than 1. Thus, the analysis confirms that the chosen independent variables are precise enough to measure the connection between corporate governance and corporate sustainability. Upon scrutinizing the significance value of each independent variable, it was found that sales and after-tax profit have a significance value of less than 0.05. This suggests that these variables have an impact on the firm's CSR expenses. However, it was observed that promoter shareholding and ROCE do not have any significant impact, as the p-value for these variables is very high. In the case of ROCE, the significance value is also very high, so the fixed-effect regression model was rerun after removing ROCE from the model. The modified model regression results are shown in Table 27.6.

Table 27.5 Fixed effect regression model

CSR expense	Coefficient	Standard Error	T-stat	*P*-value	Adjusted R square	F-value
Promoter shareholding	792.113	591.191	1.340	0.187	0.565	14.600
Sales	0.002	0.000	4.520	0.000		
Profit after tax	0.004	0.002	2.490	0.016		
ROCE	−18.911	91.316	−0.210	0.837		
Constant	−314.507	291.958	−1.080	0.287		

Table 27.6 Modified fixed-effect regression model

CSR expense	Coefficient	Standard Error	T-stat	P-value	Adjusted R square	F-value
Promoter shareholding	752.461	553.477	1.360	0.181	0.564	19.870
Sales	0.002	0.000	4.570	0.000		
Profit after tax	0.004	0.002	2.510	0.016		
Constant	−299.005	279.248	−1.070	0.290		

The modified model results show that the adjusted R square has reduced very slightly, i.e., from 56.5% to 56.4% still showing that the corporate governance variables denote more than 50% of the variation in the CSR expense. Herein, the F-value has increased, showing more precision in the model derived by eliminating ROCE. However, the significance value for only sales and profit after tax is still lower than 0.05. Thus, the null hypothesis that good C.G. has no impact on C.S. is rejected. This indicates that, for the selected Indian firms, good corporate governance has contributed to influencing CSR expenditures. Herein, even the coefficient level reveals that with a 1% rise in sales, the CSR expense rises by 0.002%, while with a rise in profit after tax by 1%, the CSR expense rises by 0.004%. This shows that the contribution of Indian firms with large market capitalization is mainly dependent on their sales and profit after tax. The existence of more profitability or higher firm size could result in raising their efforts toward corporate sustainability.

27.6 Discussion

In today's society, sustainable development is widely acknowledged as an important area that is receiving increasing attention from researchers. Given the prevalence of numerous business scandals, corporate governance emphasizes encouraging corporate sustainability practices to restore investors' trust in companies (Achim et al., 2022). Studies (Aguilera et al., 2007; Jamali et al., 2008) suggested that the presence of corporate governance not only helps in maintaining the balance between social and economic goals but also contributes toward fulfilling the community and individual needs. The focus of the companies on corporate governance could help support the companies that have an interest in fulfilling their social responsibility and engaging in corporate sustainability practices. The following findings have been identified in previous studies (Naciti et al., 2021; Zubeltzu et al., 2018): There is an interconnectedness between corporate governance and sustainability, but the complex nature of this association has hindered empirical evaluation in most studies. However, some studies that managed to assess this relationship have found a positive correlation between the variables, such as a study by Suharyono and Laili Romanza Hs. (2023) in Indonesia concluded that audit committees, environmental scales, or boards of directors do not influence the sustainability performance of the company. However, factors such as management ownership and institutional ownership had an impact on sustainability. Another study by Ong and G. Djajadikerta (2017) explored the linkage between the Australian resource industry and identified that the presence of broad diversity contributes to better corporate governance, leading to a rise in the extent of sustainability disclosure.

Several studies, including Jouha (2015), Munir et al. (2019), Abdulrahim et al. (2020), Tjahjadi et al. (2021), Benson and Ganda (2022) and Bahdur et al. (2023), have examined the relationship between C.G. and C.S. focusing on the performance of companies in Indonesia, Australia, China, South Africa, and Oman. The research has shown a significant emphasis on examining the relationship between C.G. and

C.S. in India. However, the specific studies chosen have yet to delve into this correlation. This study aims to fill this gap. After evaluating the top 10 firms based on market capitalization, it was determined that corporate governance has a significant impact on corporate sustainability. The analysis revealed that financial metrics such as sales and post-tax profit greatly influence the companies' CSR expenditure. These findings align with previous studies indicating that corporate governance positively affects corporate sustainability. Moreover, Katragadda and Sreeram (2018), and Fadilah et al. (2022) emphasized the positive impact of firm size on sustainability reporting, suggesting that larger firms tend to prioritize corporate sustainability. Thus, the study confirms the essential role of corporate governance and indicates that as profitability and firm size increase, Indian firms are more inclined to prioritize corporate sustainability.

27.7 Implications

The findings of the study have both practical and theoretical implications. In a theoretical context, the study was able to establish the direct linkage between C.G. and C.S. in the Indian context, thus enhancing the information in the field of corporate sustainability. Also, as the new directions of the research were identified, the scope for future work was stated, which is valuable for future researchers. The practical implications of the study are in terms of providing more information to policymakers about the means of raising the sustainable development status of the country. This deduction would help policymakers design the strategy in a manner that benefits corporate governance. Policymakers could use corporate governance as leverage to encourage companies to make more contributions to sustainable development. These continuous efforts toward corporate governance would result in more options for human development and also improve environmental performance. The study also provides investors with insight into the factors that help raise the corporate sustainability of companies. The contribution of a company toward corporate sustainability confirms to the investor that the company has good profitability, resulting in raising investors' confidence in the company's performance. The rise in confidence would motivate investors to invest more and even improve the company's performance. Apart from investors, this information is also made available to other stakeholders of the company, like the public, employees, or society, which in turn provides a clearer picture of the position of the company in the market and the efforts of the company toward sustainability. Therefore, the study findings are not just valuable theoretically but also have meaningful contributions to improving company performance and motivating policymakers to divert efforts toward sustainable development.

27.8 Conclusion

The researcher aimed to explore the relationship between corporate governance and corporate sustainability. The study focused on 10 NIFTY 50 Indian firms and found that strong corporate governance is associated with corporate sustainability. Additionally, the research determined that strong C.G. could enhance a company's impact on C.S. The study's findings underscore the positive influence of firm size and company profitability on corporate sustainability. This suggests that companies with larger sizes and higher profits are more likely to allocate funds to support corporate sustainability efforts. The study, while providing valuable information about the relationship between corporate governance and corporate sustainability, has some limitations. It made novel contributions by exploring the connection between the variables in the context of India, but its scope was narrow, focusing only on the top 10 firms of NIFTY 50. This restricts the generalization of the findings to all Indian firms. However, these limitations present exciting opportunities for future research. A more diverse range of Indian firms, including small and medium-sized enterprises (SMEs), could be studied to better understand the role of corporate governance. Furthermore, the inclusion of other internal and external factors as control variables could lead to a more comprehensive assessment of the role of corporate governance in corporate sustainability in India.

References

Abdulrahim, M., Sukoharsono, E., Saraswati, E., & Subekti, I. (2020). The impact of corporate governance on sustainability performance with strategic orientation as intermediate variable. *International Journal of Economics Business and Management Research, 4*(7), 89–114.

Aboud, A., & Yang, X. (2022). Corporate governance and corporate social responsibility: New evidence from China. *International Journal of Accounting & Information Management, 30*(2), 211–229. https://doi.org/10.1108/IJAIM-09-2021-0195

Achim, M. V., Văidean, V. L., Sabau, A. I., & Safta, I. L. (2022). The impact of the quality of corporate governance on sustainable development: An analysis based on development level. *Economic Research-Ekonomska Istraživanja, 36*(1), 930–959. https://doi.org/10.1080/1331677X.2022.2080745

Aguilera, R. V., Rupp, D. E., Williams, C. A., & Ganapathi, J. (2007). Putting the S back in corporate social responsibility: A multilevel theory of social change in organizations. *The Academy of Management Review, 32*(3), 836–863. https://doi.org/10.5465/amr.2007.25275678

Aras, G., & Crowther, D. (2008). Governance and sustainability: An investigation into the relationship between corporate governance and corporate sustainability. *Management Decision, 46*(3), 433–448. https://doi.org/10.1108/00251740810863870

Arora, P., & Dharwadkar, R. (2011). Corporate governance and corporate social responsibility (CSR): The moderating roles of attainment discrepancy and organization slack. *Corporate Governance: An International Review*. https://doi.org/10.1111/j.1467-8683.2010.00843.x

Assoratgoon, W., & Kantabutra, S. (2023). Toward a sustainability organizational culture model. *Journal of Cleaner Production, 400*. https://doi.org/10.1016/j.jclepro.2023.136666

Bae, S. M., Masud, M. A. K., & Kim, J. D. (2018). A cross-country investigation of corporate governance and corporate sustainability disclosure: A signaling theory perspective. *Sustainability, 10*(8), 2611. https://doi.org/10.3390/su10082611

Bahdur, Y. M. M. Al, Azmi, I. B. A. G., Mohamaf, M. T., & Makhalef, A. F. M. B. (2023). The impact of corporate governance on business sustainability in industrial companies: A field study. *International Journal of Academic Research in Business and Social Sciences, 13*(12), 1–10. https://doi.org/10.6007/IJARBSS/v13-i12/12290

Barbosa, A., Crispim, M. C., Silva, L., Morioka, S., & Souza, V. (2023). Integration of environmental, social, and governance (ESG) criteria: Their impacts on corporate sustainability performance. *Humanities & Social Sciences Communications, 10*(1). https://doi.org/10.1057/s41599-023-01919-0

Bai, K., Ullah, F., Arif, M., Erfanian, S., & Urooge, S. (2023). Stakeholder-centered corporate governance and corporate sustainable development: Evidence from CSR practices in the top companies by market capitalization at Shanghai Stock Exchange of China. *Sustainability, 15*(4), 1–25. https://doi.org/10.3390/su15042990

Benson, A., & Ganda, F. (2022). The impact of corporate governance and corporate sustainability on the performance of South African listed companies. *International Journal of Research in Business and Social Science, 11*(5), 155–175. https://doi.org/10.20525/ijrbs.v11i5.1838

Burton, A. L. (2021). OLS (Linear) regression. *The Encyclopedia of Research Methods in Criminology and Criminal Justice*, 509–514. https://doi.org/10.1002/9781119111931.ch104

Crifo, P., Escrig-Olmedo, E., & Mottis, N. (2019). Corporate governance as a key driver of corporate sustainability in France: The role of board members and investor relations. *Journal of Business Ethics, 159*(4). https://doi.org/10.1007/s10551-018-3866-6

Dewi, A., & Widyawati, L. (2023). The relationship between sustainability performance and financial performance with external assurance as the moderating variable in Indonesian listed companies. *IOP Conference Series: Earth and Environmental Science, 1199*(1), Article 012025. https://doi.org/10.1088/1755-1315/1199/1/012025

Drempetic, S., Klein, C., & Zwergel, B. (2020). The influence of firm size on the ESG score: Corporate sustainability ratings under review. *Journal of Business Ethics, 167*(2). https://doi.org/10.1007/s10551-019-04164-1

Engert, S., Rauter, R., & Baumgartner, R. (2015). Exploring the integration of corporate sustainability into strategic management: A literature review. *Journal of Cleaner Production, 112*. https://doi.org/10.1016/j.jclepro.2015.08.031

Fadilah, F., Uzliawati, L., & Mulyasari, W. (2022). The effect of firm size and firm age on sustainability reporting and the impact on earnings management. *Journal Riset Akuntansi Terpadu, 15*, 84. https://doi.org/10.35448/jrat.v15i1.14510

Galbreath, J. (2016). Is board gender diversity linked to financial performance? The mediating mechanism of CSR. *Business & Society, 57*(5), 863–889. https://doi.org/10.1177/0007650316647967

Gerwing, T., Kajüter, P., & Wirth, M. (2022). The role of sustainable corporate governance in mandatory sustainability reporting quality. *Journal of Business Economics, 92*(3). https://doi.org/10.1007/s11573-022-01092-x

Gleißner, W., Günther, T., & Walkshäusl, C. (2022). Financial sustainability: Measurement and empirical evidence. *Journal of Business Economics, 92*(3), 467–516. https://doi.org/10.1007/s11573-022-01081-0

Jamali, D., Safieddine, A., & Rabbath, M. (2008). Corporate governance and corporate social responsibility synergies and interrelationships. *Corporate Governance: An International Review, 16*, 443–459. https://doi.org/10.1111/j.1467-8683.2008.00702.x

Jha, V., & Mehra, V. (2015). Corporate governance issues, practices, and concerns in the Indian context: A conceptual study. *ICTACT Journal on Management Studies, 1*, 93–102. https://doi.org/10.21917/ijms.2015.0014

Jouha, F. (2015). Effect of corporate governance on corporate financial and market performance with sustainability reporting as intervening variable. *South East Asia Journal of Contemporary Business, Economics and Law, 6*(1), 1–6.

Katragadda, R. (2013). Corporate governance and firm performance: A theoretical review. *International Journal of Advance Research in Computer Science and Management Studies, 1*(7), 368–371.

Katragadda, R., & Sreeram, A. (2018). Shareholding patterns and its impact on firm performance: A contemporary study of Indian NIFTY 50 companies. *Asian Journal of Managerial Science, 7*(1), 31–41. https://doi.org/10.51983/ajms-2018.7.1.1291

Kocmanová, A., Dočekalová, M. P., Škapa, S., & Šiřáňová, L. (2016). Measuring corporate sustainability and environmental, social, and corporate governance value added. *Sustainability, 8*(9), 945. https://doi.org/10.3390/su8090945

Kim, J. H. (2019). Multicollinearity and misleading statistical results. *Korean Journal of Anesthesiology, 72*(6), 558–569. https://doi.org/10.4097/kja.19087

Kumar, K., Kumari, R., & Kumar, R. (2021). The state of corporate sustainability reporting in India: Evidence from environmentally sensitive industries. *Business and Society Review, 126*(4), 513–538. https://doi.org/10.1111/basr.12247

Lloret, A. (2016). Modeling corporate sustainability strategy. *Journal of Business Research, 69*(2), 418–425. https://doi.org/10.1016/j.jbusres.2015.06.047

Lu, L. (2021). The moderating effect of corporate governance on the relationship between corporate sustainability performance and corporate financial performance. *International Journal of Disclosure and Governance, 18*(3), 1–14. https://doi.org/10.1057/s41310-020-00099-6

Mahmood, Z., Kouser, R., Ali, W., Ahmad, Z., & Salman, T. (2018). Does corporate governance affect sustainability disclosure? A Mixed Methods Study. *Sustainability, 10*(1), 207. https://doi.org/10.3390/su10010207

Majeed, S., Aziz, T., & Saleem, S. (2015). The effect of corporate governance elements on corporate social responsibility (CSR) disclosure: Empirical evidence from listed companies at KSE Pakistan. *International Journal of Financial Studies, 3*(4), 530–556. https://doi.org/10.3390/ijfs3040530

Martínez-Ferrero, J., & Frias-Aceituno, J. V. (2015). Relationship between sustainable development and financial performance: International empirical research. *Business Strategy and the Environment, 24*(1), 20–39. https://doi.org/10.1002/bse.1803

Masud, M. A. K., Bae, S. M., & Kim, J. D. (2017). Analysis of environmental accounting and reporting practices of listed banking companies in Bangladesh. *Sustainability, 9*(10), 1717. https://doi.org/10.3390/su9101717

Masud, M. A. K., Nurunnabi, M., & Bae, S. (2018). The effects of corporate governance on environmental sustainability reporting: Empirical evidence from South Asian countries. *Asian Journal of Sustainability and Social Responsibility, 3*(1), 1–26. https://doi.org/10.1186/s41180-018-0019-x

Meuer, J., Kölbel, J., & Hoffmann, V. (2019). On the nature of corporate sustainability. *Organization & Environment, 33*(3). https://doi.org/10.1177/1086026619850180

Mishra, M. K. (2023). The emerging role of corporate governance on environmental sustainability. *SSRN, Rochester*. https://doi.org/10.2139/ssrn.4656662

Mubeen, S., Mahar, K., & Gunday, I. (2023). An investigation into the relationship between corporate governance and corporate sustainability: Evidence from Indian IT sector. *European Chemical Bulletin, 12*(S6), 40–45. https://doi.org/10.31838/ecb/2023.12.si6.004

Munir, A., Khan, F., Usman-Taunsvi, M., & Khuram, S. (2019). Relationship between corporate governance, corporate sustainability, and financial performance. *Pakistan Journal of Commerce and Social Sciences, 13*(4), 915–933.

Naciti, V., Cesaroni, F., & Pulejo, L. (2022). Corporate governance and sustainability: A review of the existing literature. *Journal of Management and Governance*, 1–20.

Naciti, V., Cesaroni, F., & Pulejo, L. (2021). Corporate governance and sustainability: A review of the existing literature. *Journal of Management and Governance, 26*(3), 1–20. https://doi.org/10.1007/s10997-020-09554-6

Nicolăescu, E., Alpopi, C., & Zaharia, C. (2015). Measuring corporate sustainability performance. *Sustainability, 7*(1), 851–865. https://doi.org/10.3390/su7010851

Pezzey, J. (2017). Sustainability: An interdisciplinary guide. *The Economics of Sustainability, 1*(4), 321–362. https://doi.org/10.3197/096327192776680034

Privitera, G. J. (2023). Introduction to statistics. In *Statistics for the Behavioral Sciences*. ISBN: 978-1-5443-6281-6.

Pillai, K. A. C. S. R., Krishnakumar, A. G. R., & Chand, S. G. (2023). Evaluating the convergence of CSR and corporate governance towards sustainability: A case study of Oil and Natural Gas Corporation, India. *European Chemical Bulletin, 12*(10), 7832–7845.

Rai, S. S., Rai, S., & Singh, N. K. (2021). Organizational resilience and social-economic sustainability: COVID-19 perspective. *Environment, Development and Sustainability, 23*(8). https://doi.org/10.1007/s10668-020-01154-6

Ren, X., Li, Y., Shahbaz, M., Dong, K., & Zudi Lu. (2022). Climate risk and corporate environmental performance: Empirical evidence from China. *Journal of Environmental Management, 30*, 467–477.

Singh, R., Dsilva, J., & Kumar, R. S. (2021, July). Modelling the CSR initiatives on firm performance: A context of emerging economies. *Empirical Economics Letters, 20*(3).

Sabharwal, S., & Bansal, A. (2015). Corporate governance and sustainability reporting: Responsibility towards stakeholders and society. A study of top 10 Indian companies. *Business Analyst, 35*(2).

Scoones, I. (2016). The politics of sustainability and development. *Annual Review of Environment and Resources, 41*. https://doi.org/10.1146/annurev-environ-110615-090039

Shirodkar, V., Beddewela, E., & Richter, U. H. (2016). Firm-level determinants of political CSR in emerging economies: Evidence from India. *Journal of Business Ethics, 148*(3), 673–688. https://doi.org/10.1007/s10551-016-3022-0

Subramaniam, N., Kansal, M., & Babu, S. (2017). Governance of mandated corporate social responsibility (CSR): Evidence from Indian government-owned firms. *Journal of Business Ethics, 143*(3), 543–563. https://doi.org/10.1007/s10551-015-2804-0

Suharyono, & Hs, R. (2023). Analysis of the effect of corporate governance on corporate sustainability performance. *ICAESS*. https://doi.org/10.4108/eai.5-10-2022.2325905

Sun, Y., Davey, H., Arunachalam, M., & Cao, Y. (2022). Towards a theoretical framework for the innovation in sustainability reporting: An integrated reporting perspective. *Frontiers in Environmental Science, 10*. https://doi.org/10.3389/fenvs.2022.935899

Tjahjadi, B., Soewarno, N., & Mustikaningtiyas, F. (2021). Good corporate governance and corporate sustainability performance in Indonesia: A triple bottom line approach. *Heliyon, 7*(3), Article e06453. https://doi.org/10.1016/j.heliyon.2021.e06453

Vaid, J., Vaid, D. K., & Kesharwani, S. (2016). Corporate governance and sustainability tourism marketing: An analysis and review of recent literature. *Business Analyst, 40*(1), 59–85.

Wu, W., Ullah, R., & Shah, S. (2020). Linking corporate environmental performance to financial performance of Pakistani firms: The roles of technological capability and public awareness. *Sustainability, 12*(4), 1446. https://doi.org/10.3390/su12041446

Yu, H., Bansal, P., & Arjaliès, D. (2023). International business is contributing to environmental crises. *Journal of International Business Studies, 54*(6), 1151–1169. https://doi.org/10.1057/s41267-022-00590-y

Zubeltzu, E., Andicoechea, L., & Etxeberria, I. (2018). Corporate social responsibility and corporate governance and corporate financial performance: Bridging concepts for a more ethical business model. *Business Strategy and Development, 1*(3), 214–222. https://doi.org/10.1002/bsd2.29

Chapter 28
ESG Reporting and Investment Choices: A Comparative Study of ESG Indices in India and Other Safe-Haven Assets

Magesh Kumar, G. Delina, and Srinivasan Pazhamalai

Abstract The landscape of Environmental, Social, and Governance (ESG) reporting and investment choices in India remains largely unexplored. This study investigates the relationship between ESG reporting and investment choices from a fast-emerging market such as India by conducting a comparative analysis of ESG indices in India, including the Nifty 100 Enhanced ESG Index., Nifty 50, and other safe-haven assets, such as gold, silver, and 10-year Treasury bond yield. From the results, it has been found that the average returns of the Nifty50 and Nifty100 ESG index are higher than the other assets, such as gold, silver, and 10-year treasury bonds. The results of the Generalized Autoregressive Conditional Heteroscedasticity (GARCH) model confirm the presence of volatility clustering in all study samples. Exponential GARCH (E-GARCH) indicates a presence of leverage effect only for the equity indices Nifty 50 and Nifty100 ESG Enhanced index. The outcome of the Threshold GARCH (T-GARCH) signifies that bad news has a significant effect on volatility on Nifty 50, Nifty 100 ESG Enhanced, Silver, and Treasury bonds compared with good news except for gold. Thus, the findings of the study infer that investors could generate better returns by investing in companies that practice ESG. It also provides more avenues for the companies to raise capital, which strongly reports ESG practices. The outcome of the results can help investors diversify their portfolios and rebalance their investment strategies.

Keywords ESG reporting · Sustainable investment · Safe-Haven assets · Environmental · social · Governance (ESG) · Emerging markets

M. Kumar (✉)
School of Business and Management, Christ University, Bengaluru, India
e-mail: mageshkumar.ramalingam@christuniversity.in

G. Delina
Department of Business Administration, Kalasalingam Academy of Research and Education (Deemed to Be University), Srivilliputhur, India
e-mail: g.delina@klu.ac.in

S. Pazhamalai
CMS Business School, Jain (Deemed to Be University), Bengaluru, India
e-mail: Dr.srinivasan_pazhamalai@cms.ac.in

V. Kandpal et al. (eds.), *Developments in Corporate Governance*,
Approaches to Global Sustainability, Markets, and Governance,
https://doi.org/10.1007/978-981-96-6366-8_28

28.1 Introduction

ESG reporting is a significant practice followed by business organizations to disclose their performance and impact in key non-financial areas. ESG reporting is important to understand the impact that the activities of business organizations have on society and the environment as a whole. It is a vital measure to know and understand how the ESG initiatives of the business organization compare with the industry standards on the same. Raghavan (2022) identified the growing prominence of ESG reporting among companies, investors, and all its stakeholders. ESG reporting can increase the credibility of business firms through the public disclosure of their operational impact on the environment, society, and governance (Darnall et al., 2022). ESG reporting is an outcome-oriented approach to the initiatives taken by businesses toward environmental, social, and governance matters, unlike action-oriented approaches like corporate governance and corporate social responsibility.

ESG has been linked to sustainable business practices, and its contribution to the field of sustainability cannot be ignored. It has scaled to the level of global standards in financial markets, paving the way for a radical shift toward sustainability (Chopra et al., 2024). ESG reporting is, hence, known to be an evolutionary concept of "sustainability reporting" (Diwan & Sreeraman, 2024). ESG is considered to be an investment strategy termed a "sustainable investment" by researchers owing to its capabilities of linking returns with benefits integrating the environmental, social, and governance concerns (Jain et al., 2019; Heenetigala et al., 2017). On these lines, ESG indices are also termed sustainability indices. Therefore, using ESG indices to assess the performance of companies on environmental, social, and governance factors and making investment choices interlinked with ESG reporting aids organizations both with good financial returns and achievement of non-financial objectives.

As ESG has evolved into a "corporate sustainability concept," ESG reporting has become vital for communicating a company's performance and its future directions toward sustainability practices in line with ethically wired economic, social, and governance responsibilities (Asokan, 2023). Previous research (Bosi et al., 2022; Mohammad & Wasiuzzaman, 2021; Oprean-Stan et al., 2020; Pulino et al., 2022) also suggests the positive effects of ESG reporting on improving company performance. Research indicates that Asian firms adopt ESG reporting practices not only as a mode of expressing their accountability to their stakeholders but also as an ideal practice to be established as firms meeting public expectations and, thereby, achieve a legitimate status in the society substantiating the desirability and utility of their products and services (Rahman & Alsayegh, 2021).

Recent research by Kumar and Kakran (2023) recognized improvements in the country's ESG scores. Sharma et al. (2020) observed an increase in ESG reporting in India and emphasized the need for an increase in activities concerned with ESG reporting among Indian companies. Focusing on ESG factors can benefit corporations in two ways: it can draw in capital from investors and improve their financial performance, which will increase the possibility of socially conscious investing expanding in India (Jasuja et al., 2021). The research further advocated that investor decision-making can be aided by performance evaluation and comparison of ESG indicators, leading to the development of sustainable objectives.

Diwan and Sreeraman (2024) have asserted the need for empirical studies on ESG parameters. Further, owing to the benefits of ESG stocks and their safe-haven quality, ESG reporting can be employed as a strategy to attract potential investors and can be seen as a tool to influence investors' investment choices. The central theme of this chapter is built on this notion, drawing evidence from the Indian context.

28.2 Review of Literature

ESG reporting involves a company disclosing data that is concerned with environmental, social, and governance criteria. As ESG reporting relies on a self-reporting mechanism, information disclosure by business organizations can vary across the globe (Bosi et al., 2022); research studies have highlighted the need for reliable and transparent reporting of ESG metrics (Arvidsson & Dumay, 2022; Friedman et al., 2021; Kaplan & Ramanna, 2021). In the Indian context, Bhattacharya (2024) identified the variation between firms in ESG reporting and reported the rapid expansion of ESG practices in India. Abhishek et al. (2024) observed significant improvements in the quality of ESG reporting practices among select companies over time.

Conversely, there is limited research in India on ESG funds (Dutta et al., 2023) and the area of sustainable investing, particularly in ESG Indices. Tripathi and Bhandari (2014) observed that sustainability indices surpassed the broad market index Nifty 50 when compared with other indices such as Greenex and Carbonex based on risk-adjusted return. Jain et al. (2019), based on their findings, reported no differences in terms of financial returns among ESG indices when compared with the emerging and developed economies. This demonstrates that ESG investing is a feasible alternative to the conventional broad market indices. Sudha (2015) deployed the T-GARCH model to analyze the conditional volatility of the Nifty ESG Index.

Tripathi and Kaur (2020) found that India outperformed in times of crisis because of its better risk-adjusted return in the domain of sustainable investments. Broadstock et al. (2022) investigated the performance of ESG indices during the pandemic period. The outcome of the event study found major changes in the returns and volatility during the regular and pandemic periods. The study concluded that the portfolios with higher investing in ESG outperformed the portfolios with less ESG investing during the crisis. ESG disclosures help the organization gain a competitive advantage and provide sustainable and ethical solutions (Porter et al., 2019).

The advantages of adopting ESG for business organizations are overarching. Research evidenced that ESG performance lowered stock return volatility during the COVID-19 epidemic in India (Beloskar & Rao, 2023). Likewise, ESG equities have also shown safe-haven qualities during the pandemic outbreak (Rubbaniy et al., 2022). The study by Kaiser and Welters (2019) affirms the safe-haven nature of ESG investing. ESG reporting, an outcome of ESG screening, has been linked to long-term risk reduction effects (Verheyden et al., 2016).

Ellili (2022) has ascertained a positive relationship between ESG disclosure and investment efficiency. Sood et al. (2023) affirmed ESG factors as influencers of investment decisions made by investors. Research by Upadhyaya et al. (2023) focused on the impact of ESG on sustainable investments in India and found that investors preferred investing in companies that complied with ESG standards. Research evidence from India (Mulchandani et al., 2022) also suggests that ESG reporting creates long-term value for investors, thereby building their confidence.

The merits of Socially Responsible Investing (SRI) through ESG have attracted more investors to use ESG-based stocks to mitigate the risks of asset investments, especially during crisis times (Barson et al., 2024). The research also provided substantive evidence about the safe-haven nature of ESG stocks. A study in the Arab region also pronounced the safe-haven quality of ESG for investment in comparison with other normal indexes, particularly during crisis times (Mousa et al., 2021). Numerous studies carried out in developed countries signify a strong positive relationship between ESG reporting and organizational performance. This is because more ESG disclosure can lead to a lower risk as it makes the information transparent to all the stakeholders (Alareeni & Hamdan, 2020; Lee et al., 2016; Nguyen et al., 2022). Contrary to the developed countries, emerging countries exhibit mixed results in the case of ESG reporting and organizational performance.

Developing markets are desirable for foreign investors because of their strong economic growth development (Bahadori et al., 2021; Jyoti & Khanna, 2021; Naffa & Fain, 2022). Zhou et al. (2022) identified that ESG performance had a significant impact on market capitalization among the listed firms in China. The securities market regulator is the driving force for mandatory disclosures and ESG reporting, which has enhanced the importance of sustainable investing. Thus, ESG investing will help investors in managing their portfolio risk and diversification and fulfill the social goals of sustainability (Jasuja et al., 2021). Therefore, the study investigates the relationship between Environmental, Social, and Governance (ESG) reporting and investment choices from a fast-emerging market such as India by comparing its ESG indices with traditional safe-haven assets.

28.3 Methodology

Nifty100 Enhanced ESG is considered to be the benchmark Index for companies with an ESG risk score. This index aims to measure the performance of the top 100 companies in Nifty based on their ESG risk score. Companies with ESG scores of less than 60 are not included in the index. Further, the companies involved in the manufacturing of alcohol, tobacco, and weapons are exempted from the index.

Nifty 50 is considered a barometer of the Indian capital market, and the daily closing price of Nifty50, 10g of gold and silver, is extracted from the official website of the National Stock Exchange (NSE). Also, the daily frequency of 10-year Treasury

Table 28.1 Top constituents of Nifty 100 Enhanced ESG

Company's name	Weightage (%)
HDFC Bank Ltd	6.77
Infosys Ltd	6.35
ICICI Bank Ltd	4.29
Axis Bank Ltd	4.09
Bharti Airtel Ltd	2.88
HCL Technologies Ltd	2.75
Tata Consultancy Services Ltd	2.56
Tata Motors Ltd	2.38
Tech Mahindra Ltd	2.32
Wipro Ltd	2.11

Source NSE Website

bond yield is obtained from the portal of Investing.com. The time frame of the study is diligently chosen based on the post-Russian-Ukraine war crisis (i.e., 1 Mar 2022–1 Mar 2024), as there was a significant impact on the global market during this period. Thus, the study consists of 497 observations based on the selected time frame. Table 28.1 represents the companies from the Nifty 100 Enhanced ESG Index.

The returns of the selected variable have been computed by using the formula:

$$R_t = \mathrm{Ln}\ (P_t/P_{t-1})$$

Ln is the log value. P_0 is the price for the current period, and P_1 indicates the previous period.

28.3.1 ARCH and GARCH

The Auto-Regressive Conditional Heteroscedasticity (ARCH) was found by Engle (1982) to identify the volatility in the time series. Before the ARCH model, the volatility was found based on the assumption of constant variance. However, the variance was not constant all the time. This led to the development of the ARCH model, which was found to help handle errors and get a better estimation of the model. However, the error in the conditional variance is based on the lagged error square term. Thus, a higher order in the ARCH model may lead to a computational problem. However, this problem was overcome by the Generalized Autoregressive Conditional Heteroscedasticity (GARCH) Model.

Engle & Bollerslev (1986) found the GARCH (1, 1) in 1986. The conditional variance of the model is based on the past period of all the variables. Hence, the GARCH model is more robust and overcomes all the errors of the ARCH model. The present study uses the widely accepted method of GARCH (1, 1), and it is represented in the following equation:

$$h_t - \omega_1 + \sum_{i=1}^{p} \alpha_1 \varepsilon_{t-i}^2 + \sum_{j=1}^{q} \beta_1 h_{t-j}$$

In the above equation, $\alpha_1\ \varepsilon^2{}_{t-i}$ represents the ARCH term, and $\beta_1 h_{t-j}$ represents the GARCH term.

28.3.2 The GARCH-In-Mean (GARCH-M) Model

The return of the financial securities may be affected due to its volatility. Hence, the GARCH-M model was introduced to incorporate the conditional variance in the mean equation of the model. Thus, it can be stated as below:

$$\text{Mean Equation:}\quad R_t = a_0 + a_1 R_{t-1} + b_1 \varepsilon_{t-1} + \lambda\sigma^2 t + \varepsilon_t \tag{28.1}$$

$$\text{Variance Equation:}\quad h_t = \omega_1 + \sum_{i=1}^{p} \alpha_1 \varepsilon_{t-i}^2 + \sum_{j=1}^{q} \beta_1 h_{t-j} \tag{28.2}$$

In the above equation, a_1 and b_1 indicate the Auto-Regressive (AR) and Moving Average (MA) terms of the model.

The term λ, as shown in the mean equation, is known as risk premium. A positive λ indicates the return series is related positively to the volatility of the series. It means an increase in conditional variance, which is a proxy of risk, can lead to a rise in the mean return series.

28.3.3 The Exponential GARCH (E-GARCH) Model

One of the significant disadvantages of the GARCH model is that it cannot react to asymmetric information of the returns series. Thus, asymmetric models such as Exponential GARCH (E-GARCH) and Threshold GARCH (T-GARCH) were established to capture the asymmetric changes in the market. Therefore, E-GARCH (1, 1) was used to study the asymmetric volatility of the model.

The primary disadvantage of the symmetric GARCH model is that the conditional variance cannot respond to asymmetrical changes in the stock returns. As a result, a variety of asymmetric models have been developed to address the issue, including E-GARCH and T-GARCH, which are used to capture asymmetric behavior. Thus, the asymmetric volatility and return of the stock market indices are studied using the E-GARCH (1, 1) and T-GARCH (1, 1) models. Thus, the E-GARCH (1, 1) model is represented below:

$$\text{Mean Equation:} \quad R_t = a_0 + a_1 R_{t-1} + b_1 \varepsilon_{t-1} + \lambda \sigma^2 t + \varepsilon_t \tag{28.3}$$

$$\text{Variance Equation:} \quad \ln(\sigma_t^2) = \omega_1 + \beta_1 \ln(\sigma_{t-1}^2) + \alpha_1 \left| \frac{\varepsilon_{t-1}}{\sigma_{t-1}} \right| + \gamma_1 \frac{\varepsilon_{t-1}}{\sigma_{t-1}} \tag{28.4}$$

In the E-GARCH (1, 1) model, the conditional variance includes a log value as it does not require a non-negative parameter restriction. The left-hand side of the model indicates the log conditional variance, and the coefficient "γ_1" is known as asymmetric or leverage term. Thus, it is the extension of the traditional GARCH model, and the γ_1 term captures the effect of asymmetric changes. Therefore, $\gamma < 0$, there exists a leverage effect, and if $\gamma \neq 0$, it confirms a symmetric effect.

28.3.4 *Threshold GARCH (T-GARCH) Model*

The equation for the T-GARCH (1, 1) model is specified below. The asymmetric effect changes are based on Zaokian (1994) model.

$$\text{Mean Equation:} \quad R_t = a_0 + a_1 R_{t-1} + b_1 \varepsilon_{t-1} + \lambda \sigma^2 t + \varepsilon_t \tag{28.3}$$

$$\text{Variance Equation:} \quad \sigma_t^2 = \omega + \alpha_1 \varepsilon_t^2 - 1 + \gamma_1 d_{t-1} \varepsilon_{t-1}^2 + \beta_1 \sigma_{t-1}^2 \tag{28.8}$$

From the above model, the leverage coefficient can be estimated by the γ_1 term. Hence, good news and bad news can have a significant and differential impact on the variable. If $\varepsilon_{t-1} > 0$, there is a presence of good news, and vice versa if bad news $\varepsilon_{t-1} < 0$.

28.4 Empirical Results

Figure 28.1 illustrates the pattern of the time series data of the selected variables. From Fig. 28.1, the index value of the Nifty100 Enhanced ESG and the benchmark index of Nifty 50 move in tandem. However, the prices of gold, silver, and treasury bonds are less similar during the sample period.

The descriptive statistics of the selected variable return series are displayed in Table 28.2. It has been found that the average returns of the Nifty50 and Nifty100 ESG index (around 0.06%) are higher than the other assets, such as gold, silver, and 10-year treasury bonds. Gold, which is considered to be a very safe-haven asset, has an average return of 0.04%, which is much less than silver (0.009%) and treasury bonds (0.007%).

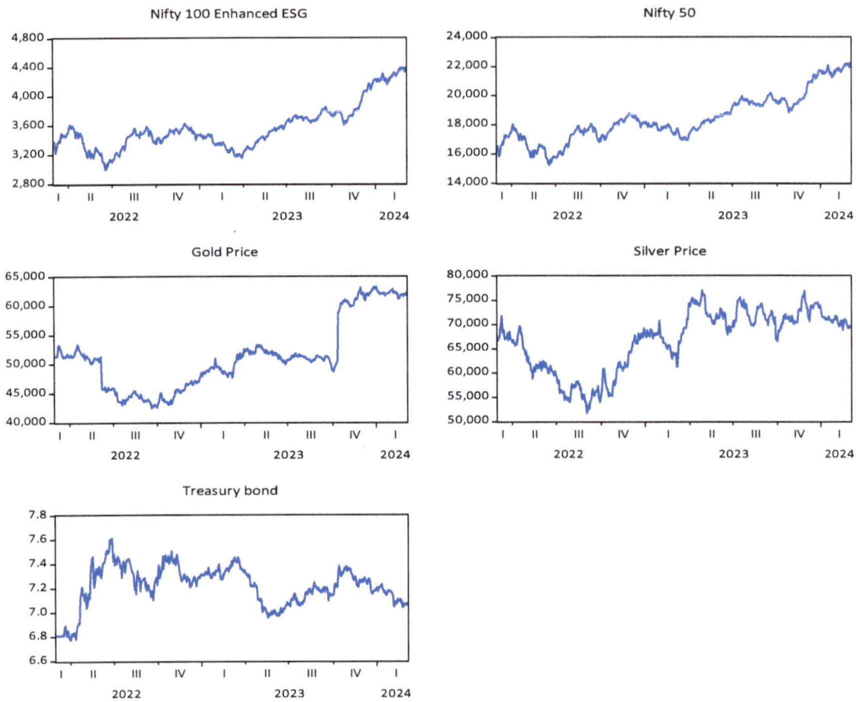

Fig. 28.1 Trend pattern of the samples

Table 28.2 Descriptive statistics

	Nifty 50	Nifty100 Enhanced ESG	Gold	Silver	Treasury bond
Mean (%)	0.0597	0.0530	0.0420	8.95E-03	7.14E-03
Median (%)	0.0954	0.0920	0.0000	−5.98E-03	0.0270
Maximum (%)	2.8482	2.533	15.4173	6.4853	3.5735
Minimum	−0.026891	−0.031264	−0.109188	−0.065441	−0.022214
Std. Dev	0.008177	0.008284	0.011051	0.014076	0.005171
Skewness	−0.137495	−0.327200	3.565734	−0.083297	0.704387
Kurtosis	3.874860	3.830821	96.51747	5.731878	10.35435
Jarque–Bera	17.41571	23.16229	181425.2	155.1243	1161.139
Probability	0.000165	0.000009	0.000000	0.000000	0.000000

Source Computed by author

Table 28.3 Unit root test

Variables	ADF value	*p* value
Nifty50	−20.967	0.000***
Nifty100 ESG Enhanced Index	−20.667	0.000***
Gold	−21.700	0.000***
Silver	−22.575	0.000***
Treasury bond	−23.007	0.000***

Note *** Indicates significance at 1% level. *Source* Computed by author

This indicates that treasury bonds are a very safe asset with less risk and less returns as their standard deviation is 0.005%. Among the maximum % of return, the equity indices provide a higher return but with more risk as their standard deviations are higher than gold, silver, and 10-year treasury bonds. Further, both equity indices suggest a negative skew, indicating there is a higher probability of lower returns as the distribution shows a fatter tail on the left side. At the same time, gold and treasury bonds are positively skewed, suggesting a greater possibility of higher returns, while silver has a stronger incline toward lower returns due to its negative skewness.

28.4.1 Unit Root Test

The unit root test is applied to measure whether the data are stationary or non-stationary. The data in the time series must be stationary or made stationary before applying any model. Thus, the Augmented Dickey-Fuller (ADF) test is used to check the unit root of the time series. The results of the ADF test are provided in Table 28.3. The probability (*p*) value is significant at 1%. This indicates that the null hypothesis of having a unit root is rejected, and the return series of the variable are stationary at level.

28.4.2 Volatility Clustering

Based on the volatility clustering from Fig. 28.2, it suggests that all the variables, such as the Nifty 50, Nifty100 ESG Enhanced index, gold, silver, and 10-year treasury bonds, show a strong ARCH (Auto-Regressive Conditional Heteroscedasticity) effect, which is a prerequisite for using the GARCH models.

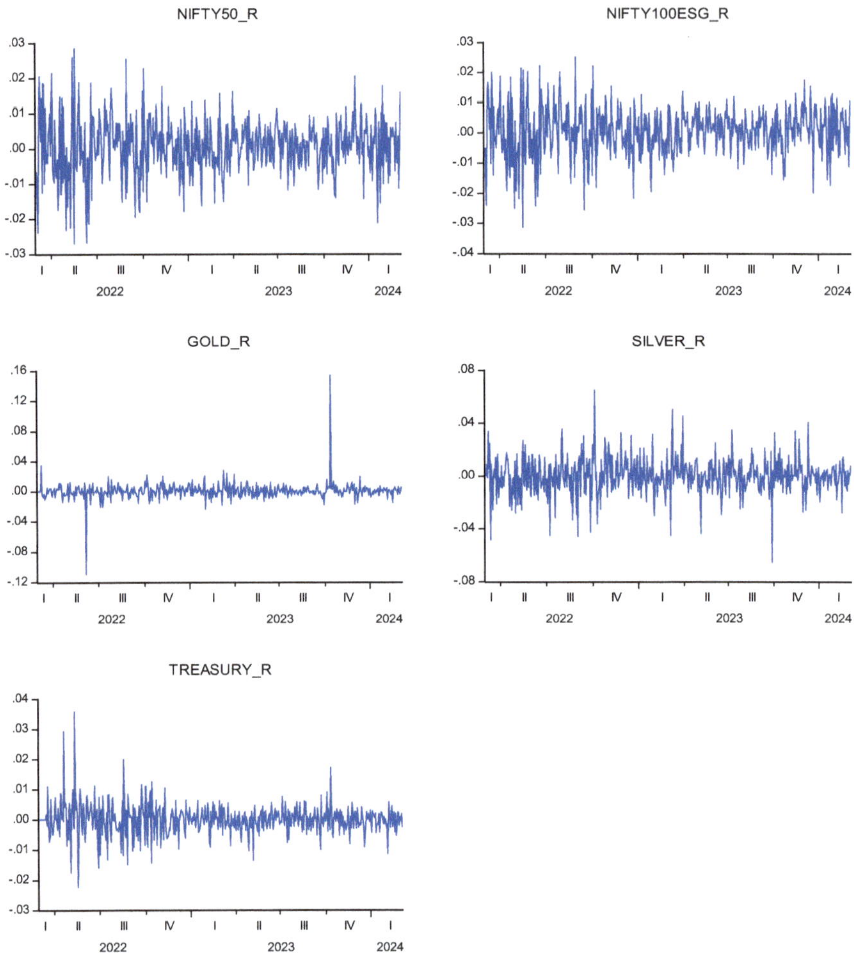

Fig. 28.2 ARCH effect of the selected variables

28.4.3 Heteroscedasticity Test

Before applying the GARCH models, it is necessary to examine the presence of heteroscedasticity in the data residuals. For this, the Lagrange Multiplier (LM) test is deployed (Table 28.4).

Table 28.5 displays the result of the GARCH (1, 1) model. The mean returns of the Gold, Silver, and Treasury bonds are statistically insignificant (*p*-value > 0.05). Similarly, the one-day lag returns are significant (*p*-value < 0.05) for the Nifty100 ESG Enhanced index only, suggesting that yesterday's returns do not influence today return for Nifty 50, Gold, Silver, and Treasury bond. Further, the coefficient of the GARCH parameter is highly significant at a 1% level for all variables except Gold.

Table 28.4 Heteroscedasticity Test

Variables	LM statistics	*p* Value
Nifty50	22.4085	0.0000***
Nifty100 ESG Enhanced *Index*	22.6470	0.0000***
Gold	10.0279	0.0110**
Silver	12.6527	0.0408**
Treasury bond	15.9897	0.0206**

Note *** and ** Indicates significance at 1% and 5% respectively
Source Computed by author

The sum of the coefficient for the ARCH (α) and GARCH (β) estimates are very close to one, revealing that significant changes in the returns are usually followed by more considerable changes and vice versa. This confirms the presence of volatility clustering in all the study samples. Also, the GARCH coefficient, which is estimated for all the variables, is higher than the ARCH in the variance model, indicating it has a long memory and is sensitive to any new information in the market. Thus, the Nifty50, Nifty100 ESG Enhanced index, Gold, Silver, and Treasury bonds showed volatility persistence during the Russian-Ukraine war crisis period. Further, the model fit is measured by ARCH-LM test statistics to check that there is no autocorrelation among the residuals. The samples Nifty50, Nifty100 ESG Enhanced index, Gold, Silver, and Treasury bonds have a significance value ($p > 0.05$) greater than 5%, indicating the estimated model is fit and free from autocorrelation.

Table 28.6 explains the estimated results of GARCH in the Mean model. The return of the financial securities may be affected due to its volatility. Hence, the GARCH-M model was introduced to incorporate the conditional variance in the mean equation of the model. The estimate of conditional variance in the mean equation for silver is negative and statistically insignificant. It indicates there is no risk-return trade of relation for asset silver. However, for the equity variables and treasury bonds, the conditional variance, which is a proxy of risk, is positive and statistically significant at 1%. This means that an increase in the conditional variance of the Nifty 50 and Nifty 100 ESG Enhanced index can lead to a rise in the mean return series. Further, yesterday's returns do not influence the current returns except the Nifty100 ESG Enhanced index, as it is statistically significant at 5%. On the other side, the coefficient of the GARCH-M model, which includes α (ARCH effect) and β (GARCH effect) of the samples, is significant at 1%. Also, the coefficients of α and β are closer to one, revealing that more significant changes and vice versa usually follow significant changes in the returns. This confirms the presence of volatility clustering in all the study samples.

Table 28.7 presents the estimated results of the E-GARCH model. In the E-GARCH (1, 1) model, the conditional variance includes a log value as it does not require a non-negative parameter restriction. The sum of α and β coefficients for all variables is greater than one except silver. This indicates that the conditional variance of the variables is explosive and statistically significant at 1%. Based on the results of E-GARCH (1, 1), there is a presence of leverage effect only for the equity

Table 28.5 GARCH (1, 1) Model

Coefficients	Nifty50	Nifty ESG Enhanced Index	Gold	Silver	Treasury bond
Mean Equation					
M	0.000641 (0.0482)	0.000612 (0.0681)	0.000345 (0.6171)	−4.45E-05 (0.9463)	−5.44E-05 (0.8048)
R_{t-1}	0.068740 (0.1074)	0.091308 (0.0377)	0.016919 (0.8671)	−0.011411 (0.8294)	−0.058694 (0.2771)
Variance Equation					
ω	6.27E-07	5.45E-07 (0.0119)	5.87E-05 (0.6030)	1.74E-05 (0.1813)	9.33E-07 (0.0000)
α	0.020536 (0.1124)	0.018231 (0.1379)	−0.005982 (0.0000)	0.020327 (0.1556)	0.087549 (0.0002)
β	0.964733 (0.0000)	0.968261 (0.0000)	0.585518 (0.4652)	0.888860 (0.0000)	0.880460 (0.0000)
Model diagnostics					
ARCH-LM test statistics	4.265336	2.316011	0.005820	0.338388	0.138664
Prob. Chi-square (1)	0.1389	0.1280	0.9392	0.5608	0.7096

Source Computed by author

indices as the coefficient of the leverage effect is negative and statistically significant at 1% for the Nifty 50 and Nifty100 ESG Enhanced index. However, for the other assets, such as gold, silver, and treasury bonds, there is no leverage effect. Further, the ARCH-LM test does not indicate an autocorrelation, suggesting that the model is well-specified. Also, the ARCH-LM test does not indicate the presence of a significant ARCH effect in the residual series, which shows that the conditional variance equation of the GARCH-type models of the study is well-specified.

Table 28.8 presents the results of the T-GARCH Model. From the T-GARCH model, the leverage coefficient can be estimated by the γ_1' term. The term leverage helps us understand how good news and bad news can have a significant and differential impact on the variable. The leverage term (γ1) is positive for Nifty 50 (0.129248), Nifty100 ESG Enhanced Index (0.219411), Silver (0.038923), and Treasury bond (0.895434) and statistically significant at 1%, which suggests there are asymmetries in the news except for Gold. This indicates that bad news has a significant effect on the volatility of Nifty 50, Nifty100 ESG Enhanced, Silver, and Treasury bonds compared with good news. Also, the ARCH-LM test confirms that there is no ARCH effect in the estimated model of T-GARCH.

Table 28.6 GARCH-M Model

Coefficients	Nifty50	Nifty100 ESG Enhanced Index	Gold	Silver	Treasury bond
Mean Equation					
M	0.011172 (0.0045)	0.010376 (0.0062)	0.000272 (0.6499)	−0.020061 (0.0807)	0.000656 (0.0000)
R_{t-1}	0.066545 (0.1173)	0.090258 (0.0394)	0.020387 (0.7351)	−0.010501 (0.8437)	−0.050503 (0.1930)
λ	0.001057 (0.0076)	0.000980 (0.0100)	2.649589 (0.7498)	−0.002345 (0.0826)	0.000656 (0.0000)
Variance Equation					
ω	5.87E-07 (0.0722)	5.13E-07 (0.0793)	3.74E-07 (0.0000)	2.91E-05 (0.3735)	6.81E-08 (0.0248)
α	0.023102 (0.0859)	0.023626 (0.0915)	−0.007520 (0.0000)	0.016310 (0.3354)	−0.014328 (0.0000)
Β	0.963711 (0.0000)	0.964612 (0.0000)	1.003969 (0.0000)	0.832322 (0.0000)	1.006543 (0.0000)
Model diagnostics					
ARCH-LM test statistics	2.421309	1.895062	0.096993	0.287631	1.242609
Prob. Chi-square (1)	0.1355	0.1686	0.7555	0.5917	0.1717

Source Computed by author

28.5 Discussions

The present study aims to compare the performances of ESG investing with other safe-haven assets (Gold, Silver, and Treasury bonds) by deploying the GARCH (1, 1) and the extensions of the GARCH model. The results reveal unique findings in the risk-return trade-off of these asset classes. From the results of data analysis, it has been found that the average returns of the Nifty50 and Nifty100 ESG index (around 0.06%) are higher than the other assets such as gold, silver, and 10-year treasury bonds. Among the maximum % of return, the equity indices provide higher returns but with more risk, as their standard deviations are higher than gold, silver, and 10-year treasury bonds. Similar results are also consistent with the other studies by Vasal (2009) and Śliwiński and Łobza (2017), who suggest slightly better risk-adjusted returns of ESG investing. Further, both equity indices suggest a negative skew, indicating there is a higher probability of lower returns as the distribution shows a fatter tail on the left side. This indicates that ESG investing generates higher returns when compared with safe-haven assets such as gold, silver, and 10-year treasury bonds.

Due to Sustainable investing being in its early stage in India, Investors lack trust in generating returns, particularly in ESG reporting companies. Hence, there exists

Table 28.7 E-GARCH Model

Coefficients	Nifty50	Nifty100 ESG Enhanced index	Gold	Silver	Treasury bond
Mean Equation					
M	0.000535 (0.1070)	0.000580 (0.0856)	0.000450 (0.2998)	5.77E-05 (0.9277)	−6.14E-05 (0.7907)
$R_t - 1$	0.089695 (0.0427)	0.125622 (0.0072)	−0.019990 (0.7388)	−0.046581 (0.3109)	−0.062607 (0.2340)
Variance Equation					
Ω	−0.386529 (0.0105)	−0.598239 (0.0056)	−2.003433 (0.0000)	−11.56949 (0.0000)	−1.061282 (0.0000)
A	0.101715 (0.0179)	0.121939 (0.0198)	0.379519 (0.0000)	0.014622 (0.8192)	0.205587 (0.0001)
B	0.968900 (0.0000)	0.948753 (0.0000)	0.936720 (0.0000)	0.947890 (0.0039)	0.914121 (0.0000)
Γ	−0.053766 (0.0375)	−0.077205 (0.0050)	0.559635 (0.0000)	0.335653 (0.0000)	0.055271 (0.0231)
Model diagnostics					
ARCH-LM test statistics	3.601746	0.849772	0.098158	0.613816	0.112668
Prob. Chi-square (1)	0.0577	0.3566	0.7541	0.4334	0.7371

Source Computed by author

a need to analyze the volatility and conditional variances based on ESG companies. Thus, E-GARCH and T-GARCH were used to measure the asymmetric behavior of the assets. The result of the GARCH (1, 1) model confirms the presence of volatility clustering in all the study samples. This finding is also in line with the existing studies by Sudha (2015) and Tripathi and Kaur (2020), who concluded that volatility clustering exists in socially responsible investing (SRI) indices. E-GARCH (1, 1) indicates a presence of leverage effect only for the equity indices Nifty 50 and Nifty100 ESG Enhanced index. However, for the other assets, such as Gold, Silver, and Treasury bonds, there is the absence of a leverage effect. T-GARCH (1, 1) leverage term (γ1) is positive and significant for Nifty 50, Nifty100 ESG Enhanced index, Silver, and Treasury bond indicating there are asymmetries in the news. This shows that bad news has a significant effect on the volatility of Nifty 50, Nifty100 ESG Enhanced, Silver, and Treasury bonds compared with good news.

Table 28.8 T-GARCH Model

Coefficients	Nifty50	Nifty100 ESG Enhanced Index	Gold	Silver	Treasury bond
Mean Equation					
M	0.000565 (0.0882)	0.000445 (0.1874)	−0.000468 (0.1731)	−1.21E-05 (0.9856)	3.39E-05 (0.8783)
R_{t-1}	0.108548 (0.0156)	0.161003 (0.0006)	−0.261032 (0.0000)	−0.008193 (0.8768)	−0.073257 (0.1587)
Variance Equation					
ω	3.27E-06 (0.0066)	6.04E-06 (0.0001)	4.04E-05 (0.0000)	6.14E-06 (0.0855)	8.69E-07 (0.0000)
α	−0.001470 (0.9613)	−0.025660 (0.4568)	4.809255 (0.0000)	−0.009953 (0.2261)	0.112997 (0.0000)
β	0.877225 (0.0000)	0.813867 (0.0000)	0.024910 (0.2174)	0.957919 (0.0000)	−0.076726 (0.0003)
γ_1	0.129248 (0.0032)	0.219411 (0.0006)	−4.751795 (0.0000)	0.038923 (0.0416)	0.895434 (0.0000)
Model diagnostics					
ARCH-LM test statistics	2.313813	0.213853	0.197087	0.296714	0.018053
Prob. Chi-square (1)	0.1282	0.6438	0.6571	0.5859	0.8931

Source Computed by author

28.6 Managerial Implications

The outcome of the study is of greater significance, particularly for fund managers and investors who are seeking to diversify by incorporating ESG-based investments into their portfolios. The findings imply that the ESG Index and benchmark Index generate greater average returns as compared to the traditional asset categories such as gold, silver, and treasury bonds. However, higher returns come with additional risk, as evidenced by the ESG indices, which display higher standard deviation and negative skewness. Even though the results emphasize that the investments in ESG can generate lucrative returns it is exposed to higher volatility and downside risk, particularly during the period of adverse conditions in the market. Hence, fund managers must evaluate all these aspects more carefully when recommending ESG investments by balancing the risk-return trade-off with their clients.

Also, the result of the GARCH, E-GARCH, and T-GARCH models confirms the presence of volatility clustering and as well as leverage effects in the ESG indices. This advocates that the markets are more sensitive to negative news than compared with positive news. Hence, this finding is highly relevant for managers who want to devise risk management strategies. Fund managers should formulate appropriate hedging strategies and risk-adjusted return approaches to reduce the influence of negative shocks on investments in ESG. Moreover, given the absence of leverage effect in the safe-haven assets, incorporating ESG with traditional assets such as gold, silver, and treasury bonds can provide optimal diversification, which could reduce the portfolio risk. Hence, understanding the nuances of the asymmetric behavior of various asset classes can provide deep insights into portfolio construction and the achievement of a maximum return at lower risk. Thus, the findings of the study highlight the need for financial literacy and awareness of the advantages and risks of ESG investments in India.

Further recommendations based on the findings of the study include the following. Quantify the metrics of ESG by integrating them with international standard practices. Thus, the reporting of ESG should consist of the disclosure details such as climate pact, social influence, and good governance actions. Also, by improving the capacity of the Securities and Exchange Board of India (SEBI), the ESG disclosure can be met. This could be achieved by enhancing the monitoring practices, conducting regular audits, and charging penalties for non-compliance actions. Another important managerial inference is necessary to critically evaluate India's ESG of SEBI's Business Responsibility and Sustainability Report (BRSR) guidelines with the global standards such as the European Union (EU) Sustainable Finance and Disclosure Regulation (SFDR) and Securities Exchange Commission (SEC) ESG requirements. The SEBI's BRSR focuses on "principle-based" reporting, whereas SFDR makes mandatory the "principal adverse impact (PAI)" indicators. Thus, the BRSR should mirror the screening parameters that are required for the EU taxonomy. Also, the BRSR lacks the standard templates for reporting the ESG disclosures. Thus, it warrants the need to align with the "European Single Electronic Format (ESEF)" for ease of use, understanding, and comparability. Finally, the SFDR mandates the requirement of reporting the sustainability risk and how it could affect the financial performance. Such similar actions are also required for BRSR to link the ESG performance with the financial metrics.

28.7 Conclusion

Nifty100 Enhanced ESG is considered to be the benchmark Index for companies with an ESG risk score. The objective of the paper is to compare the risk-return trade of the ESG sustainability indices with other safe-haven assets such as gold, silver, treasury bonds, and benchmark indices. Based on the result of the GARCH (1, 1) model, the mean returns of gold, silver, and treasury bonds are statistically insignificant. The sum of the coefficient for the ARCH (α) and GARCH (β) estimates are very close to one, confirming the presence of volatility clustering in all the study samples. GARCH-M model results indicate that conditional variance in the mean equation for silver is negative and statistically insignificant, indicating there is no risk-return trade-off relation for the asset silver. However, for the equity variables and treasury bonds, the conditional variance is positive and statistically significant. Results of E-GARCH (1, 1) indicate there is a presence of leverage effect only for the equity indices as the coefficient of the leverage effect is negative and statistically significant at 1% for the Nifty 50 and Nifty100 ESG Enhanced index. However, for the other assets, such as Gold, Silver, and Treasury bonds, there is the absence of a leverage effect. From the T-GARCH model, the leverage effect is positive for all variables except gold. This indicates that bad news has a significant effect on the volatility of Nifty 50, Nifty100 ESG Enhanced, Silver, and Treasury bonds compared with good news.

Thus, the sustainability index (ESG) outperforms other assets such as gold, silver, treasury bonds, and the benchmark index in terms of return and risk. Thus, the findings of the study infer that investors could generate better returns by investing in companies that practice ESG. It also provides more avenues for companies to raise capital who strongly adhere to and report ESG practices. The outcome of the results can help investors diversify their portfolios and rebalance their investment strategies. Thus, diversifying across various asset classes, including investing in ESG, can help investors to manage their portfolio risk and maximize returns.

28.8 Scope for Further Research

While the present study has provided deep insights into ESG investments and traditional assets, there lies a greater scope for further research. One important area for further expansion is the effect of ESG factors on long-term investment among various geographical markets. As ESG investment is a nascent stage in India, future studies could focus on regulatory changes, transparency in ESG reporting, corporate governance, and sustainability at the firm levels for further research. The present study has only used the conventional broad market index, Nifty 50, which is considered to be the barometer of the Indian capital market. However, other broad market, strategy, and thematic indices could also be considered.

Fewer studies have revealed that the expected returns based on ESG will vary based on diversified sectors and industries. So, the sectoral-based ESG could be explored in the future. India is a fast-emerging country, but there are emerging countries that include ESG as an important parameter in investing. Thus, a cross-country analysis of emerging and developed economies of ESG could be a future scope of study as part of sustainable investment. Lastly, future research could explore more advanced models apart from the GARCH family models to measure the non-linear patterns, regime shifts, and higher-order moments, which could further explain the behavior of ESG investments in a volatile equity market.

References

Abdul Rahman, R., & Alsayegh, M. F. (2021). Determinants of corporate environment, social and governance (ESG) reporting among Asian firms. *Journal of Risk and Financial Management, 14*(4), 167.

Abhishek, N., Suraj, N., & Divyashree, M. S. (2024). Quality of ESG reporting and disclosure practices of Indian companies. In *ESG and Ecosystem Services for Sustainability* (pp. 197–234). IGI Global.

Alareeni, B. A., & Hamdan, A. (2020). ESG impact on the performance of US S&P 500-listed firms. *Corporate Governance: THe International Journal of Business in Society, 20*(7), 1409–1428.

Arvidsson, S., & Dumay, J. (2022). Corporate ESG reporting quantity, quality and performance: Where to now for environmental policy and practice? *Business Strategy and the Environment, 31*(3), 1091–1110.

Asokan, A. (2023). An overview of ESG reporting in India: Practices and challenges. *Sustainable Boardrooms: Democratising Governance and Technology for Society and Economy*, 19–39.

Barson, Z., Ofori, K. S., Junior, P. O., Boakye, K. G., & Ampong, G. O. A. (2024). Time-varying connectedness between ESG Stocks and BRVM traditional stocks. *Journal of Emerging Market Finance*, 09726527241233920.

Bahadori, N., Kaymak, T., & Seraj, M. (2021). Environmental, social, and governance factors in emerging markets: The impact on firm performance. *Business Strategy & Development, 4*(4), 411–422.

Beloskar, V. D., & Rao, S. N. (2023). Did ESG save the day? Evidence from India during the COVID-19 crisis. *Asia-Pacific Financial Markets, 30*(1), 73–107.

Bhattacharya, P. (2024). ESG reporting practices in India: A study on selected listed companies. In *Contemporary issues in Accounting, Finance and Management* (pp. 72–76). Rohini Nandan.

Bosi, M. K., Lajuni, N., Wellfren, A. C., & Lim, T. S. (2022). Sustainability reporting through environmental, social, and governance: A bibliometric review. *Sustainability, 14*(19), 12071.

Broadstock, D. C., Chatziantoniou, I., & Gabauer, D. (2022). Minimum connectedness portfolios and the market for green bonds: Advocating socially responsible investment (SRI) activity. *Applications in energy finance: The energy sector, economic activity, financial markets, and the environment* (pp. 217–253). Springer International Publishing.

Chopra, S. S., Senadheera, S. S., Dissanayake, P. D., Withana, P. A., Chib, R., Rhee, J. H., & Ok, Y. S. (2024). Navigating the challenges of Environmental, Social, and Governance (ESG) Reporting: The path to broader sustainable development. *Sustainability, 16*(2), 606.

Darnall, N., Ji, H., Iwata, K., & Arimura, T. H. (2022). Do ESG reporting guidelines and verifications enhance firms' information disclosure? *Corporate Social Responsibility and Environmental Management, 29*(5), 1214–1230.

Diwan, H., & Amarayil Sreeraman, B. (2024). From financial reporting to ESG reporting: A bibliometric analysis of the evolution in corporate sustainability disclosures. *Environment, Development and Sustainability, 26*(6), 13769–13805.

Dutta, S., Sharma, D., & Singh, S. (2023). Performance review of ESG investing in India. *Rabindra Bharati University Journal of Economics, 27*(10), 121–131.

Ellili, N. O. D. (2022). Impact of ESG disclosure and financial reporting quality on investment efficiency. *Corporate Governance: THe International Journal of Business in Society, 22*(5), 1094–1111.

Engle, R. F. (1982). Autoregressive conditional heteroscedasticity with estimates of the variance of United Kingdom inflation. *Econometrica: Journal of the econometric society*, 987–1007.

Engle, R. F., & Bollerslev, T. (1986). Modelling the persistence of conditional variances. *Econometric reviews, 5*(1), 1–50.

Friedman, H. L., Heinle, M. S., & Luneva, I. (2021). A theoretical framework for ESG reporting to investors. *Available at SSRN 3932689.*

Heenetigala, K., Armstrong, A., Lokuwaduge, C. D., & Ediriweera, A. (2017). Environmental social and governance reporting: A case study of the Australian mining sector. In *CSR, Sustainability, and Leadership* (pp. 172–198). Routledge.
Jain, M., Sharma, G. D., & Srivastava, M. (2019). Can sustainable investment yield better financial returns: A comparative study of ESG indices and MSCI indices? *Risks, 7*(1), 15.
Jasuja, D., Prosad, J. M., & Nautiyal, N. (2021). Comparative performance analysis of sustainability-themed indices in India: Markov regime switching approach. *FIIB Business Review*, 23197145211052805.
Jyoti, G., & Khanna, A. (2021). Does sustainability performance impact financial performance? Evidence from Indian service sector firms. *Sustainable Development, 29*(6), 1086–1095.
Kaiser, L., & Welters, J. (2019). Risk-mitigating effect of ESG on momentum portfolios. *The Journal of Risk Finance, 20*(5), 542–555.
Kaplan, R. S., & Ramanna, K. (2021). How to fix ESG reporting. *Harvard Business School Accounting & Management Unit Working Paper* (22-005).
Kumar, A., & Kakran, S. (2023). ESG reporting landscape in India: Contrasting approaches and institutional frameworks. *Kakran S., & Kumar A. (2023). ESG reporting landscape in India: Contrasting approaches and institutional frameworks. Global Summit—2023 unlocking sustainability: G20 presidency paves the way for an ESG-driven new world order*, 95–102.
Lee, K. H., Cin, B. C., & Lee, E. Y. (2016). Environmental responsibility and firm performance: The application of an environmental, social and governance model. *Business Strategy and the Environment, 25*(1), 40–53.
Mohammad, W. M. W., & Wasiuzzaman, S. (2021). Environmental, Social and Governance (ESG) disclosure, competitive advantage and performance of firms in Malaysia. *Cleaner Environmental Systems, 2*, Article 100015.
Mousa, M., Saleem, A., & Sági, J. (2021). Are ESG shares a safe haven during COVID-19? Evidence from the Arab Region. *Sustainability, 14*(1), 208.
Mulchandani, K., Mulchandani, K., Iyer, G., & Lonare, A. (2022). Do equity investors care about environment, social and governance (ESG) disclosure performance? Evidence from India. *Global Business Review, 23*(6), 1336–1352.
Naffa, H., & Fain, M. (2022). A factor approach to the performance of ESG leaders and laggards. *Finance Research Letters, 44*, Article 102073.
Nguyen, D. T., Hoang, T. G., & Tran, H. G. (2022). Help or hurt? The impact of ESG on firm performance in S&P 500 non-financial firms. *Australasian Accounting, Business and Finance Journal, 16*(2), 91–102.
Oprean-Stan, C., Oncioiu, I., Iuga, I. C., & Stan, S. (2020). Impact of sustainability reporting and inadequate management of ESG factors on corporate performance and sustainable growth. *Sustainability, 12*(20), 8536.
Porter, M., Serafeim, G., & Kramer, M. (2019, October 16). Where ESG fails. *Institutional Investor*. https://www.institutionalinvestor.com/article/b1hm5ghqtxj9s7/Where-ESG-Fails. Accessed 13th December 2024.
Pulino, S. C., Ciaburri, M., Magnanelli, B. S., & Nasta, L. (2022). Does ESG disclosure influence firm performance? *Sustainability, 14*(13), 7595.
Raghavan, K. (2022). ESG reporting impact on accounting, finance. *Journal of Global Awareness, 3*(1), 9.
Rubbaniy, G., Khalid, A. A., Rizwan, M. F., & Ali, S. (2022). Are ESG stocks safe-haven during COVID-19? *Studies in Economics and Finance, 39*(2), 239–255.
Sharma, P., Panday, P., & Dangwal, R. C. (2020). Determinants of environmental, social and corporate governance (ESG) disclosure: A study of Indian companies. *International Journal of Disclosure and Governance, 17*(4), 208–217.
Śliwiński, P., & Łobza, M. (2017). Financial performance of socially responsible indices. *International Journal of Management and Economics, 53*(1), 25–46.
Sood, K., Pathak, P., Jain, J., & Gupta, S. (2023). How does an investor prioritize ESG factors in India? An assessment based on fuzzy AHP. *Managerial Finance, 49*(1), 66–87.

Sudha, S. (2015). Risk-return and volatility analysis of sustainability index in India. *Environment, Development and Sustainability, 17*, 1329–1342.

Tripathi, V., & Bhandari, V. (2014). Socially responsible investing-an emerging concept in investment management. *FIIB Business Review, 3*(4), 16–30.

Tripathi, V., & Kaur, A. (2020). Socially responsible investing: Performance evaluation of BRICS nations. *Journal of Advances in Management Research, 17*(4), 525–547.

Upadhyaya, C., Khan, S., Sandanshive, V., & Awasthi, M. G. (2023). A study of impact of behavioural factors on ESG investment. *The Journal of Contemporary Issues in Business and Government, 29*(1), 546–553.

Vasal, V. K. (2009). Corporate social responsibility & shareholder returns-Evidence from the Indian Capital Market. *Indian Journal of Industrial Relations*, 376–385.

Verheyden, T., Eccles, R. G., & Feiner, A. (2016). ESG for all? The impact of ESG screening on return, risk, and diversification. *Journal of Applied Corporate Finance, 28*(2), 47–55.

Zhou, G., Liu, L., & Luo, S. (2022). Sustainable development, ESG performance and company market value: Mediating effect of financial performance. *Business Strategy and the Environment, 31*(7), 3371–3387.

Chapter 29
Climate Change, Risk Management, and ESG: An Indian Perspective

Sunitha Abhay Jain and Neetha Kurian

Abstract Mr. Kofi Annan, the Secretary General of the United Nations Organization remarked that 'the world is reaching the tipping point beyond which climate change may become irreversible. If this happens, we risk denying present and future generations the right to a healthy and sustainable planet that the whole of humanity stands to lose' (RBI, Reserve Bank of India Publication, 2022). Climate change has negatively impacted humanity. It has become a focal point of discussion amongst researchers, academicians, and policymakers alike. It has become imperative for companies to assess and mitigate climate risks as a part of the sustainability journey. Many investors and lenders are integrating ESG aspects, including climate risk, into their investment decision-making processes. The Securities Exchange Board of India (SEBI) introduced Business Responsibility and Sustainability Reporting (BRSR), which is a comprehensive framework that seeks disclosures regarding the ESG performance of the companies (SEBI, 2021). However, one of the glaring issues is that most of the disclosures are voluntary in nature, and there are no penalties prescribed for non-disclosure by listed companies in India. In this chapter, we delve into risks posed by climate change on businesses and understand the significance of including the aspect of climate change in corporate planning and strategies. Another important aspect that has been discussed in the chapter is how ESG can be used to address the risks, issues, and challenges posed as a result of climate change. Further, a comparative approach has been adopted to understand the ESG policy framework in India and some prominent jurisdictions like the US, the UK, China, and Japan. We are of the view that framing policies related to mandatory ESG reporting for listed companies and compliance with ESG norms will be the deciding factor for the existence, future readiness, and sustainability of Indian businesses.

Keywords ESG · Climate change · ESG disclosures · Sustainability · Risk management

S. A. Jain (✉) · N. Kurian
School of Law, CHRIST University, Bangalore, India
e-mail: sunitha.abhay@christuniversity.in

N. Kurian
e-mail: neetha.kurian@christuniversity.in

V. Kandpal et al. (eds.), *Developments in Corporate Governance*,
Approaches to Global Sustainability, Markets, and Governance,
https://doi.org/10.1007/978-981-96-6366-8_29

29.1 Introduction

In today's corporate world, climate change and managing the risks associated with it have occupied a prominent place in policies, investment decisions, and strategies adopted by companies. Climate change profoundly impacts human life, health, the environment, and corporations. Due to climate change, the world is witnessing extreme weather conditions, storms, floods, rising sea levels, and other environmental disasters. Physical risks associated with climate change can destroy assets, disrupt supply networks, and reduce productivity. Companies are taking note of these ill-effects of climate change and are working towards net zero emissions. From the corporate perspective, it is imperative for corporate entities to factor in the losses caused as a result of climate change and also develop strategies to deal with the various risks involved therein. The ESG mandates disclosures related to the environment, society, and governance by companies in order to assess their business and sustainability practices and also to study how they impact the environment and society. This in turn helps them to improve their sustainability performance, gain a competitive edge and work towards higher financial growth. Risks that are covered under the environmental risks include carbon footprints, greenhouse gas emissions, water disposal, usage of water, and deforestation, amongst others. Social risks include wage equality, human rights violations, diversity, equity, and inclusion. Governance risks include ESG disclosures, board structure and diversity, corruption and fraud, amongst others.

Recently, climate change has emerged as a key parameter to measure corporate business performance. Further, ESG has garnered global attention in the corporate world and in the investment landscape. As corporates are coping with the risks arising due to climate change, governments across the world have started to adopt ESG performance indicators to assess corporate business practices. Many countries have included climate change as one of the key parameters in measuring the ESG performance of the companies. This has helped companies to strategize and plan risk management to handle the risks and, in turn, make profits. There are many studies which suggest that there is an inter-relationship between ESG performance and climate change (Li et al., 2024).

Corporates are important players in a country's economic growth and development and have realized the importance of integrating ESG while making investment decisions. The ESG landscape has reshaped due to the society's response to climate change. It involves considering a company's climate risk exposure, carbon footprint, and sustainability practices to identify investment opportunities aligned with long-term climate objectives. In 2004, the concept of ESG was proposed for the first time by the United Nations Global Compact (United Nations, 2004). ESG is a comprehensive framework that takes into consideration the non-financial factors in relation to a company's business practices and also its impact on the Environment, Society and Governance. This enables better and effective corporate governance and helps in greener development and ensures a more responsible corporate image branding.

In order to deal with the negative effects of climate change and to ensure better corporate accountability, many regulatory bodies, governmental authorities, and investors are evincing interest in ESG compliance. While taking investment decisions, many investors are factoring in the ESG performance of companies as it helps them navigate the non-financial risks. Additionally, many banks also rely on the ESG performance of companies while taking lending decisions. A corporation's ESG performance and sustainability investment can weaken due to the uncertainties caused by climate change. A corporation's ESG performance is a strategic response to the external environment, and it is imperative for companies to comply with the ESG disclosures and better their performance through sustainable business practices. Further, it might also be influenced by social, political, and economic factors, which play a major role in shaping the regulatory landscape.

29.2 Review of Literature and Hypothesis Development

A thorough literature review has been carried out to identify gaps and areas for further investigation, and also to provide an overview, analysis, and synthesis of relevant scholarly sources.

Until the mid-1990, the duties of the board of directors of a company generally revolved around enhancing shareholder rights which were reflected in their governance as emphasized by Clarkson (Clarkson, 1995). During 2000 the duties of the board of directors underwent a paradigm shift resulting in protecting the interest of all stakeholders of the company. This extended to the society and casted a duty to protect the environment and practice social governance as highlighted in the research of Chenxi Wang (2024). Madden holds that ESG criteria have high relevance today as it is a deciding factor in analysing the value of companies and their business investment decisions (2022). According to Sandberg, companies, when designing their ESG framework, must ensure that there is a balance between sustainable goals and their corporate goals (Sandberg et al., 2022). A company's ESG policy can also bring a competitive advantage to it over its competitors (Zhang et al., 2021). According to Harasheh and Provasi, ESG integration is possible only by consciously integrating environment and societal goals in one's business decision-making process (Harasheh & Provasi, 2023). Puchniak and Varottil emphasize the importance of ESG and hold that climate change will turn out to be the issue of this century and argue that while shareholder activism would have enabled institutional shareholders to prevent the board of directors who resist climate change solutions from continuing, the traditional 'acting in concert' rules act as a hindrance (Puchniak & Varottil, 2024). Kabra highlights that corporate social responsibility reforms alone will not help Indian companies in achieving sustainability goals by arguing that the top twenty-five Fortune India five hundred companies with their impressive corporate social responsibility commitments were ranked at high risk with respect to their ESG commitments, as they did not comply with several other obligations under Indian laws (2024). Through an empirical study conducted in 272 listed companies on the

National Stock Exchange from the period 2015 to 2021 regarding ESG disclosure and financial performance, Malik and Kashiramka concluded that companies with better ESG policy frameworks outperform their competitors and have better value and enhance investor confidence (2024).

29.3 Significance of the Study

This study will be helpful to all the stakeholders, such as companies, government, policymakers, legal professionals, environmentalists, and researchers at the global level, as it gives them an understanding of the importance of ESG compliance and how they can use it to mitigate the ill-effects of climate change. It will also help key managerial personnel in prioritizing resource allocation to ESG-related activities, thereby enhancing companies' market value and better investment opportunities.

29.4 Concept of ESG

The term ESG is an acronym for Environmental, Social, and Governance framework used for evaluation of a company's sustainability performance and its impact on the world around it. In order to assess the robustness of corporate governance mechanisms, countries across the world have come up with a performance evaluation criterion to manage their environmental and social impact. It examines how companies deal with matters such as resource conservation, waste management, and Net Zero Carbon Emission. Broadly speaking, in addition to corporate governance, it takes into account the company's response to climate change, its internal policies and code of conduct, how the company treats its employees, the nature of training given to staff, and most significantly, its corporate culture and working environment. ESG, therefore, includes multiple non-financial aspects that add value to a firm. While the environmental factors include how a company deals with its pollution and other climate risk activities, its social aspect will include its engagement with the community around which it operates. This will involve its relationships with employees, dealers, suppliers, and society. The term 'governance' refers to an organization's internal structure, which includes its leadership style, management, and governance culture. A strong ESG reputation helps an organization to navigate the challenges and maintain stakeholder support during difficult times. Aligning with ethical and sustainable values through the implementation of ESG, improves a company's reputation and attracts ethically conscious investors, employees, and customers.

The importance of ESG has been emphasized globally through many empirical researches and studies. The current global trend has seen a paradigm shift from corporate social responsibility to business sustainability. This also becomes relevant, considering the pledge of the United Nations Organization which developed seventeen sustainable goals, which urges member nations to pledge for a sustainable future by 2030 (United, 2015).

Environmental, social, and governance investment practices gained traction during the nineteenth century, based on the notion that success of a business organization must be predicted by assessing not only the financial but also the non-financial aspects. One of the most prominent advocates of this approach was John Elkington (Megan Morgan, 2022) who identified that there are many non-financial factors that should be included in assessing the long-term prospects of a business. Elkington coined the term 'triple bottom line' and included factors such as the environment, social, and financial aspects while calculating the success of a business organisation (Jeurissen, 2000). Elkington's proposition has been modernized as ESG, with the addition of the 'governance' factor. Governance includes the manner in which a business organizations are administered and the way in which the above factors can affect the bottom-line of a business. For instance, it is important to note that the manner in which a company treats its employees can impact the morale of the employee and in turn the loyalty, which in turn might affect its ability to retain talent.

There is a general view internationally that in order to make an accurate financial assessment of a company, it is important to access ESG-related information. Many investors across the world are factoring in the ESG aspects while making investment decisions. Governments across the world have taken steps to make disclosures relating to ESG mandatory and also try to implement these regulations. For instance, international organizations such as the United Nations' Principles for Responsible Investment (Principles of Responsible Investment, 2006) have been actively promoting and supporting legal developments in this field. In order to build a sustainable business economy, ESG investing has gained prominence in a fast-growing economy like India.

29.5 ESG Framework in India

The government of India has taken proactive steps towards environmental protection which is evident from the policies and programmes of the government such as the National Environment Policy (Forest, 2006), National Conservation strategy and policy statement on environment and development (Ministry of Environment and Forests, 1992), policy statement for abatement of pollution (Ministry of Environment and Forests G. o., 1992), Eco-labeling Scheme (Government of India M. o., Ecolabelling Scheme, 1992) amongst others. In the world economy, India occupies a very prominent place, and it becomes important for the government to take proactive steps to tackle the issues relating to climate change. Corporate social responsibility was gaining traction, and corporations started evolving strategies to protect the environment and also work for the betterment of society. In line with corporate social responsibility, the ESG principles evolved and started gaining importance in both the national and international arenas as countries became mindful of implementing sustainable business practices.

Though ESG is gaining importance in the corporate sector, there is no single piece of legislation laying down the compliance requirement, and it's regulated through various laws such as the Companies Act, 2013, Securities and Exchange Board of India (Listing Obligations and Disclosure Requirement) Regulations, 2015; environmental legislations, labour laws and various guidelines as laid down by the regulators. In 2007, Mr. Manmohan Desai, the then PM of India proposed the ten-point charter (Taxmann, 2024). This was proposed to deal with various instances of violations by companies in relation to social, economic, and environment. Earlier attempts in implementing ESG reporting commenced with the Ministry of Corporate Affairs directions on CSR guidelines, which made CSR reporting voluntary in 2009 (Affairs M. o., 2009). The Voluntary Corporate Social Responsibility guidelines which were issued at a later point of time highlighted the importance of ethical business practices, welfare of workers, respect for human rights and the environment, inclusive development and protection of stakeholder's interest. In 2011, the Ministry of corporate affairs issued the National Voluntary guidelines on Social, Environment, and Governance which aimed at integrating ESG by corporates. The Companies Act 2013, provided for Corporate Social Responsibility which further strengthened the ESG principles (Affairs M. o., 2021). In 2011, the Ministry of Corporate Affairs released the National Voluntary Guidelines on Social, Environmental, and Economic responsibilities of business (Government of India M. o., 2011) which emphasized the need for responsible business conduct. These guidelines further underscored the need for business entities to conduct and govern themselves ethically and also ensure transparency and accountability in their business transactions and governance. The guidelines also stressed upon the need and importance of protection of environment. In 2015, the National Voluntary guidelines were revised in alignment with the Sustainable Development Goals and the UN guiding principles. These new principles are called the National guidelines on responsible business conduct (NGRBC).

The Business Responsibility Reporting (BRR), Corporate Social Responsibility (CSR), and the latest Business Responsibility and Sustainability Report (BRSR) are some of the initiatives taken to regulate Environment, Social and Governance reporting in India. SEBI introduced the Business Responsibility and Sustainability reporting which mandated sustainability disclosures in alignment with NGRBC. BRSR draws inspiration from the Global Reporting initiative is one of the most widely accepted reporting frameworks. It was bought into force by amending the Regulation 34 (2) of SEBI (Listing Obligation and Disclosure Requirements) Regulation 2015. Under the BRSR, listed entities are required to make disclosures on various parameters such as economic performance, environmental stewardship, governance, product responsibility, stakeholder's interest, etc. The amended framework of BRSR aimed at integrating companies' financial performance and ESG performance (SEBI, Business and Sustainability Reporting by Listed companies,| Circular No.: SEBI/HO/CFD/CMD-2/P/CIR/2021/562, 2021). These disclosures are helpful as it helps them to make better investment decisions. BRSR has replaced the existing BRR and under it the top 1000 listed companies have to mandatorily file the disclosures and other smaller listed companies and private firms are encouraged to voluntarily disclosures.

The Reserve Bank of India also mandates disclosures by companies to deal with financial risks which are posed by the climate change. RBI has aligned these standards on the lines of the global standards laid down by the Task Force on Climate Related Financial disclosures. These standards were developed by the Financial Stability Board and provide guidance on the different types of information that a company is required to disclose. These disclosures revolve around the impact of climate change and the financial position of the companies.

In order to enhance competitiveness and to occupy the leadership position in the global economy it has become very important for Indian business entities to embrace ESG. It is further necessary for innovation and collaboration and to ensure sustainable growth amongst businesses, policy makers and investors. One more critical step for Indian businesses is that while developing a robust ESG framework, it must be aligned to international standards.

29.6 International Practices and Guidelines on ESG

The ESG regulatory regime is fragmented and there are many guidelines laid down by different regulatory authorities. As there is no uniform framework to govern company disclosures and vary greatly, investors argue that they find it difficult to include ESG in their investment choices (UNDP, 2024). Globally, there are various ESG reporting frameworks in place such as the Global Reporting Initiative (GRI); Sustainability Accounting Standards Board (SASB); United Nations Sustainable Development Goals (UN-SDGs); Task force on climate related financial disclosures (TCFD). ISSB was formed in November 2021, by the IFRS (International Financial Reporting Standards) foundation in response to the strong market demands for setting up quality standards in sustainability disclosures that would benefit investors. While Global guidelines have been fragmented until 2022, only when ISSB introduced some uniform global guidelines for organizations to follow, ESG reporting was brought to the forefront. ISSB guidelines included the—IFRS S1, first draft in 2022 and the final version in June 2023 (IFRS, FRS S1 General Requirements for Disclosure of Sustainability-related Financial Information, 2023) which urged corporations for sustainability related disclosures. The focus of IFRS S1 was mainly on corporations revealing sustainability related risks and opportunities in businesses that could be useful for investors to make informed choices in providing resources to these entities (IFRS, FRS Sustainability Standards Navigator, 2023). The main areas of IFRS S1 included disclosure on organization's governance strategies-namely the choices, internal controls, procedures and its performance to mitigate sustainability related risks and opportunities. ISSB also introduced the IFRS S2, first draft in 2022 and final issue in June 2023, relating to Climate related disclosures which took into account the recommendations of the TCFD (Task force on Climate Related Disclosures). The main thrust of IFRS S2 was to urge corporations to disclose climate related physical risks, transition risks, and opportunities that are available to their businesses that would help interested investors to make informed decisions (IFRS, IFRS 2, Climate Related Disclosures, 2023).

29.6.1 ESG Framework Followed by Different Countries

The ESG framework of different jurisdictions are discussed under Sect. 6.1.1.

29.6.1.1 The United Kingdom

The U.K.'s Companies Act, 2006 the annual strategic report mandates disclosures about environment, diversity at workplace, human rights, and corporate culture, amongst others. Under the UK Corporate Governance Code, 2018 listed companies are required to make disclosures on various aspects such as leadership, remuneration, shareholder's interests, and accountability. Disclosures are also mandated under the Task force on Climate related financial disclosures which aims at transparency in reporting the risk related to climate. The UK companies with operations in EU are also required to comply with the EU Non-Financial reporting directive (NFRD) which requires the companies to disclose non-financial and diversity related information. Global Reporting Initiative (GRI) is widely accepted in UK for sustainability reporting which covers aspects relating to social, economic, and environmental related matter. Companies which are listed in London Stock exchange are mandatorily required to publish ESG report in accordance with the ESG reporting guidelines, 2020 which includes the greenhouse gas reporting, energy use, gender pay and modern slavery.

29.6.1.2 European Union

The Non-Financial Reporting Directive (NFRD) makes it mandatory for public interest entities with more than 500 employees to disclose non-financial and diversity information. The NFRD was replaced by the Corporate Sustainability Reporting Directive (CSRD) and was adopted in 2021. It contains more detailed reporting requirements and mandates the use of EU sustainability reporting standards developed by the European Financial reporting Advisory Group (EFRAG). In December 2022, the official journal of the European Union published the CSRD which came into force in January 2023 that mandated ESG reporting requirements for EU companies and to non-EU companies with significant activity in the EU region. CSRD is one of the major initiatives by the EU which also expands its non-financial reporting by making reporting on environmental, human rights and social impacts mandatory (Union, Corporate Sustainability Reporting, 2023). The guideline is applicable to all large EU companies regardless of their stock exchange listing and even to non-EU companies with 'significant activity' in EU wherein, 'significant activity' will include generating a turnover of 150 million EUR in EU, with at least one subsidiary or branch in EU (Union, Corporate Sustainability Reporting, n.d.).

Another initiative by the European Council was the corporate sustainability due diligence directive (CSDDD), which mandated companies identify potential adverse impacts on the environment and human rights due to their business operations. The directive also imposed liability for violation and urged companies to change their business models in compliance with the Paris Agreement (Nations, Paris Agreement, n.d.). Suggestions included a phased approach to bring changes and revision of civil liability provisions.

Another development in the EU was the EU Taxonomy regulation, which came into force in July 2020 and serves as a transparency tool for financial institutions and companies to disclose their taxonomy matters for public awareness, including sharing the sustainable economic activities of their investments or lending in their business. EU taxonomy is a system that defines criteria for economic activities that aim at net-zero emissions by 2050 and other environmental protection goals (Commission, n.d.). It urges companies to promote environmentally sustainable economic activities based on six environmental objectives which include aspects such as mitigation of climate change, sustainable use, and protection of water bodies and marine resources, transition to circular economy, prevention and control of environmental pollution, protection and restoration of biodiversity, and ecosystems.

The United States of America

In the United States of America, the ESG disclosures are influenced by the Securities and Exchange Commission (SEC) Regulations, 2021. These rules focus on standardized climate related disclosures and it mandates the public limited companies to report on matters such as ESG-related risks, transparency in relation to the appointment of directors and human capital management amongst others. Sustainability Accounting Standards Board (SASB) also provides for industry specific ESG standards and reporting. Some of the US companies also voluntarily publish a CSR or sustainability report which covers a wide array of ESG topics and are aligned with the global standards of GRI, SASB, or TCFD. In addition to this some states and local bodies of the US have their own reporting requirements. For example, California provides for disclosures relating to greenhouse gas emissions and sustainability practices.

China

The Environmental Information Disclosure Act, 2015, provides for disclosure by Chinese Corporations regarding their environmental performance and contribution to the society. Companies which are listed on the Shanghai Stock Exchange are required to make environmental disclosures. In 2021, the ESG disclosure framework was revised by the Chinese Securities Regulatory Commission in order to achieve transparency and accountability in reporting. The companies in their annual reports were required to disclose information on various parameters such as the level of pollution, carbon emissions, waste generation, etc.

Japan

In Japan, the Financial Services Agency takes care of promoting the ESG disclosures. In 2021, the Corporate Governance code was revised to enhance the ESG-related disclosures especially related to climate change. The Tokyo Stock Exchange also requires the listed companies to disclose the various initiatives that are taken by the company and also the sustainability practices that are in place. The Ministry of Environment of Japan has also issued guidelines in relation to environmental reporting and also mandates companies to disclose information regarding the environmental impact such as the greenhouse gas emissions, resource usage and the various pollution control measures in place. Japan has strongly been in favor of The Task Force on climate related financial disclosure (TCFD) guidelines. Other initiatives include mandatory reporting on global warming, compulsory reporting on energy use by large energy consuming companies, and also the counter measure adopted by specified companies to deal with the same. In order to bring in gender diversity in governance there is a requirement of reporting on the number of female participations in management and career advancement opportunities. Other provisions like Government Pension Investment Fund encourages ESG investing and requirement for asset managers to integrate the ESG factors into the investment processes, encourages companies to make disclosures.

29.7 Comparative perspective on Indian & International Initiatives

In addition to country-specific rules and regulations, there are various global standards and initiatives in place such as the Global Reporting Initiative (GRI), Sustainability Accounting Standards Board (SASB), Task Force on Climate Relating Financial disclosures (TCFD), Carbon Disclosure Project (CPD), Integrated Reporting (IR) framework, UN Sustainable Development Goals (SDGs), Principles for Responsible Investment (PRI), International Sustainability Board (ISSB), EU regulations relating to Non-Financial Reporting Directive (NFRD), Corporate Sustainability reporting directive (CSRD), EU Taxonomy, ISO 26000 amongst others to regulate ESG

In light of the above discussion, India's ESG reporting requirement is mandatory for the top 1000 listed companies according to the BRSR framework. In the US, SEC climate disclosure (2024) mandates ESG disclosure for public firms. In the

UK, TCFD-aligned disclosures are mandatory. The ESG reporting requirements are mandatory for high-emission industries in China but are to a limited extent. In Japan, TCFD-aligned disclosures are mandatory.

With regard to environmental-related reporting in India, the BRSR mandates climate reporting but is not mandatory for all. In the US, SEC requires Scope 1 and 2 emissions reporting as mandatorily but Scope 3 is voluntary in nature. In China, with regard to environmental parameters, there is strong environmental focus as carbon markets are growing. In Japan, mandatory TCF reporting for listed firms.

In terms of social parameters, CSR is mandatory for large firms. In the US, diversity and inclusion reporting is mandatory. In the UK, diversity and supply chain accountability are required. China is lagging behind in terms of reporting with regard to social parameters. There is weak enforcement of labour and human rights law. In Japan, steps are taken to improve the workforce and gender diversity.

Further with regard to governance factors, SEBI mandates board independence. However, family-oriented businesses dominate in India. In US, SEC requires board diversity and shareholder protections. In UK, the corporate governance rules are strict and board independence is required. In China is state-owned enterprises are dominating and there is less transparency. In Japan, there is improvement in Board independence, but gender diversity is low.

29.8 Issues and Challenges

India has made significant progress in ESG, but lags behind developed countries such US, UK, Japan, and China. The implementation of ESG is fraught with various challenges. One of the major challenges is with regard to absence of standardization. As there are various ESG reporting framework, it causes confusion resulting in inconsistency. Further it also increases the burden on the companies to meet up with those standards.

Another major problem faced globally in ESG implementation, is the inaccuracy of the ESG data collected. In many countries like India, certain disclosures such as information on environmental impact assessment, organization's contribution to the nation, employment-friendly policies that would benefit employees work-life balance, anti-competitive practices of companies, etc., are not mandatory. Another issue is the lack of a mechanism to evaluate the accuracy and reliability of data disclosed by the corporates. Many a times whether the data is original or fabricated cannot be determined. It is important to ensure that company boards use only high-quality data. To ensure this, companies highly depend on internal audits, external audits, and reliable third-party advisors. Checking the data before sharing gives boards more confidence and assurance. The rise of ESG disclosure has also resulted in giving prominence to greenwashing (Nations, Climate Action, n.d.) and ESG rating providers (ERPs). ERPs are independent unregulated agencies that assess and rate organizations which in turn help investors to identify organizational commitment towards sustainability. Recently, SEBI has introduced a framework for the regulation of ERPs in India (SEBI, SEBI Master Circular for ESG Rating Providers, 2023).

Though ESG reporting has now been mandated in India, there is still a significant knowledge gap between ESG information and supply. Several variables contribute to this disparity, like—differing ESG reporting standards and frameworks, non-mandatory reporting regimes, and high data collection and reporting expenses. These factors may prevent efforts to provide investors with higher-quality facts on which to base their decisions. Companies may, fortunately, collaborate with professionals to design and implement ESG-balanced plans that improve the overall performance.

29.9 Conclusion and Suggestions

ESG is a recent phenomenon that has captured global interest. There is a promising opportunity for India in leveraging this phenomenon by taking into account the issues and challenges faced by the advanced countries like the European Union and the US who are the forerunners of it. Most of the reputed companies in India have elements of ESG in their existing risk frameworks, but the existing gaps and opportunities for improved management must be identified. They need to integrate ESG into their risk management framework across all risk types, and define an organization's risk appetite (Patalano, 2020). Organizations should take steps to understand how ESG might affect existing products and services or necessitate the launch of new ones. For comparing ESG maturity to peers, a risk readiness assessment can be used. This can then be used to create a road map for an integrated sustainability operating model. By taking an inclusive approach that includes stakeholder perspectives could be valuable for organizations. Companies can create long-term value for its business by identifying the key issues and challenges such as technological disruptions, impact of climatic changes and difficulties in tracking and collecting ESG data. This would help companies to develop new strategies in dealing with these disruptions and to be more competitive. Prioritizing on resolving ESG issues will help them in identifying their role for the societies they associate with. Though there is no global standard approach for assessing ESG risks, the US Sustainability Accounting Standard Board's (SASB) provisional sustainability accounting standards can be a useful reference for India. SASB provides provisional standards for seventy-nine industries across eleven sectors. The SASB Materiality Map (SASB, 2021) identifies and compares the possible sustainability issues across different industries and sectors. Another tool, the SASB Materiality Finder could also be a useful tool for corporates (SASB Materiality Map finder, 2021). Designing a similar tool in India would help in bringing a more standardized approach to measuring ESG risks in Indian companies. There is a need to frame uniform global guidelines for ESG reporting, to ensure sustainable business practices across organizations.

India also needs to accelerate its net-zero timelines and phase out coal faster. It can also incentivize companies to adopt renewables and hydrogen energy. Strong enforcement of gender diversity laws and worker protections must improve and be aligned with global labour standards. The whistle-blower and anti-corruption laws need to be strengthened. Mandatory ESG supply chain audits need to be introduced.

References

Affairs, M. o. (2009, December). *Corporate social responsibility voluntary guidelines 2009.* Retrieved July 18, 2024, from https://www.mca.gov.in/Ministry/latestnews/CSR_Voluntary_Guidelines_24dec2009.pdf

Affairs, M. o. (2021, January 22). *Corporate social responsibility Amendment Rules, 2021.* Retrieved July 8, 2024, from https://www.mca.gov.in/Ministry/pdf/CSRAmendmentRules_23012021.pdf

Boffo, R., & Patalano, R. (2020). *ESG investing: Practices, progress and challenges.* OECD Paris, https://www.oecd.org/finance/ESG-Investing-Practices-Progress-Challenges.pdf. Accessed 13 July 2024.

Clarkson, M. E. (1995). A stakeholder framework for analysing and evaluating corporate social performance. *Academy of Management Review, 20*(92), 117. https://doi.org/10.5465/amr.1995.9503271994

Commission, E. (n.d.). *Sustainable finance.* Retrieved July 2, 2024, from https://finance.ec.europa.eu/sustainable-finance_en

Compact, G. (2004). *Who cares wins-connecting financial markets to a changing world.* Swiss Federal Department of Foreign Affair and United Nations.

EU taxonomy for sustainable activities, https://finance.ec.europa.eu/sustainable-finance/tools-and-standards/eu-taxonomy-sustainable-activities_en. Accessed 2 July 2024.

European Union, Corporate Sustainability Reporting, January 2023, https://finance.ec.europa.eu/capital-markets-union-and-financial-markets/company-reporting-and-auditing/company-reporting/corporate-sustainability-reporting_en. Accessed 12 July 2024.

Forest, M. o. (2006, May 18). *National environmental policy.* Retrieved July 17, 2024, from india.gov.in: https://ibkp.dbtindia.gov.in/DBT_Content_Test/CMS/Guidelines/201904111035 2141_National%20Environment%20Policy,%202006.pdf

Government of India, M. o. (1992). *Ecolabelling scheme.* Retrieved July 18, 2024, from https://iimm.org/ecolabeling-scheme/

Government of India, M. o. (1992, February 26). *Policy statement for abatement of pollution.* Retrieved July 14, 2024, from https://www.iitr.ac.in/wfw/web_ua_water_for_welfare/environment/Pollution_Policy_Statement_1992.pdf

Government of India, M. o. (2011, July 12). *National voluntary guidelines on social, environmental & economic responsibilities of business, 2011.* Retrieved July 14, 2024, from https://www.mca.gov.in/Ministry/latestnews/National_Voluntary_Guidelines_2011_12jul2011.pdf

Government of India, Ministry of Corporate Affairs, National Voluntary guidelines on social, environmental, & economic responsibilities of business, July, 2011. https://www.mca.gov.in/Ministry/latestnews/National_Voluntary_Guidelines_2011_12jul2011.pdf. Accessed July 14, 2024.

Government of India, Ministry of Environment and Forests, Eco labeling Scheme, 1991. https://iimm.org/ecolabeling-scheme/. Accessed 18 July 2024.

Government of India, Ministry of Environment and Forests, Policy Statement for Abatement of Pollution, 26th February 1992. https://www.mca.gov.in/Ministry/latestnews/CSR_Voluntary_Guidelines_24dec2009.pdf. Accessed 18 July 2024.

Green washing is a practice by a company to mislead the public to believe that it is doing more to protect the environment than it is doing actually. United Nations, *Green washing -The Deceptive Tactics Behind Environmental Claims.* https://www.un.org/en/climatechange/science/climate-issues/greenwashing. Accessed 17 July 2024.

Harasheh, M., & Provasi, R. (2023). A need for assurance: do internal control systems integrate environmental, social, and governance factors? *Corporate Social Responsibility and Environmental Management, 30*, 384–401. https://doi.org/10.1002/csr.2361

IFRS, General requirements for Disclosure of Sustainability related Financial Information, June, 2023. https://www.ifrs.org/issued-standards/ifrs-sustainability-standards-navigator/ifrs-s1-general-requirements/. Accessed 16 July 2024.

IFRS, General Requirements for Sustainability Standards, June, 2023. https://www.ifrs.org/issued-standards/ifrs-sustainability-standards-navigator/ifrs-s1-general-requirements/. Accessed 17 July 2024.

IFRS, IFRS 2, Climate Related disclosures, June 2023. https://www.ifrs.org/issued-standards/ifrs-sustainability-standards-navigator/ifrs-s2-climate-related-disclosures/. Accessed 14 July 2024.

IFRS. (2023a, June). *FRS S1 general requirements for disclosure of sustainability-related financial information.* Retrieved July 16, 2024, from https://www.ifrs.org/issued-standards/ifrs-sustainability-standards-navigator/ifrs-s1-general-requirements/

IFRS. (2023b, June). *FRS sustainability standards navigator.* Retrieved July 17, 2024, from https://www.ifrs.org/issued-standards/ifrs-sustainability-standards-navigator/

IFRS. (2023c, June). *IFRS 2, climate-related disclosures.* Retrieved July 14, 2024, from https://www.ifrs.org/issued-standards/ifrs-sustainability-standards-navigator/ifrs-s2-climate-related-disclosures/

India, M. o. (1992, June). *National conservation strategy and policy statement on environment and development, 1992.* Retrieved July 14, 2024, from ICED: http://indiaenvironmentportal.org.in/files/introduction-csps.pdf

India, R. B. (2022, July 27). *Publications.* Retrieved July 17, 2024, from Reserve Bank of India: https://www.rbi.org.in/Scripts/Publications.aspx

Jeurissen, R., & Elkington J. (2000). *Cannibals with forks: The triple bottom line of 21st century business. Journal of Business Ethics, 23*, 229–231. https://doi.org/10.1023/A:1006129603

Kabra, A. (2024). Locating the future of ESG in India's present sustainability framework. *Asian Journal of Business Ethics, 13*, 137–173. https://doi.org/10.1007/s13520-024-00191-y

KPMG Board Leadership Centre Survey. (2017). *ESG, strategy, and the long view*, https://assets.kpmg/content/dam/kpmg/lu/pdf/lu-en-esg-strategy-framework-for-board-oversight.pdf. Accessed 19 July 2024.

Li, T. L. (2024). The impact of climate change on corporate ESG performance: The role of resource misallocation in enterprises. *Journal of Cleaner Production.* Retrieved from https://www.sciencedirect.com/science/article/abs/pii/S0959652624007108?via%3Dihub

Li, Tang, Lang, & Wang. (2024, March). The impact of climate change on corporate ESG performance: The role of resource misallocation in enterprises. *Journal of Cleaner Production* (445). https://doi.org/10.1016/jclepro.2024.141263

Madden, B. J. (2022). Bet on innovation, not environmental, social and governance metrics, to lead the net zero transition. *Systems Research and Behavioral Science,* 417–428. https://doi.org/10.1002/sres.2915

Megan Morgan, L. H. (2022, December 22). *Wrap up on ESG reporting.* Retrieved July 18, 2024, from Bevrigde and Diamond, Publications: https://www.bdlaw.com/publications/2022-wrap-up-on-esg-reporting/

Ministry of Corporate Affairs, *Corporate Social Responsibility, Voluntary Guidelines 2009*, India Corporate Week.

Ministry of Environment and Forests, National Conservation Strategy and Policy Statement, June 1992. http://indiaenvironmentportal.org.in/files/introduction-csps.pdf. Accessed 14 July 2024.

Morgan, M., Hopkins, L., Gage, D., & Gruver, K. (2022). *Wrap up on ESG reporting, EU adopts new ESG reporting requirements and other global developments,* December, 2022. Beveridge & Diamond: Publications. https://www.bdlaw.com/publications/2022-wrap-up-on-esg-reporting/. Accessed 18 July 2024.

National Environment Policy, Government of India, Ministry of Environment and Forests, May 2006. https://ibkp.dbtindia.gov.in/DBT_Content_Test/CMS/Guidelines/20190411103521431_National%20Environment%20Policy,%202006.pdf. Accessed 17 July 2024.
Nations, U. (n.d.). *Climate Action*. Retrieved July 17, 2024, from Greenwashing the deceptive tactics behind environmental claims: https://www.un.org/en/climatechange/science/climate-issues/greenwashing
Nations, U. (n.d.). *Paris Agreement*. Retrieved July 13, 2024, from https://unfccc.int/process-and-meetings/the-paris-agreement
Malik, N., & Kashiramka, S. (2024, April 5). Impact of ESG disclosure on firm performance and cost of debt: Empirical evidence from India. *Journal of Cleaner Production, Science Direct, 448*, 141582.
Paris Agreement is a legally binding International Treaty on Climate Change adopted by 196 nations at the UN Climate Change Conference in Paris on Dec 12, 2015. Its overarching goal is to hold the increase in the global average temperature to well below 2° C above pre industrial levels and pursue efforts to limit the temperature increase to 1.5° C above pre-industrial levels. https://unfcc.int/process-and-meeting/the-paris-agreement. Accessed 13 July 2024.
Patalano, B. R. (2020). *ESG investing: Practices, progress and challenges, OECD, Paris*. Retrieved July 13, 2024, from http://www.oecd.org/finance/ESG-Investing-Practices-Progress-and-Challenges.pdf
Principles of Responsible Investment. (2006). Retrieved July 16, 2024, from PRI: https://www.unpri.org/
Puchniak, D. W., & Varottil, U. (2024, March 24), Climate-related shareholder activism as corporate democracy: A CALL TO REFORM "acting in concert" rules. Forthcoming in the Journal of Corporation Law (Vol. 50), NUS Law Working Paper No. 2023/23, European Corporate Governance Institute—Law Working Paper No. 773/2024.
Reserve Bank of India, Discussion Paper on Climate Risk, July 27, 2022. https://www.rbi.org.in/Scripts/PublicationsView.aspx?id=21071. Accessed 17 July 2024.
Sandberg, H., Alnoor, A., & Tiberius, V. (2022). Environmental, social, and governance ratings and financial performance: Evidence from the European food industry. *Business Strategy and the Environment,* 2471–2489. https://doi.org/10.1002/bse.3259
SASB. (2021, October). *SASB Materiality Map*. Retrieved July 19, 2024, from https://sasb.ifrs.org/standards/materiality-map/
SASB Materiality Map finder. (2021, October). Retrieved July 18, 2024, from https://sasb.ifrs.org/standards/materiality-map/
SEBI. (2023, July 12). *SEBI master circular for ESG rating providers*. https://www.sebi.gov.in/legal/master-circulars/jul-2023/master-circular-for-esg-rating-providers-erps-_73856.html. Accessed 10 July 2024.
SEBI, Business Responsibility, and Sustainability Reporting format, 2021. https://www.sebi.gov.in/sebi_data/commondocs/may2021/Business%20responsibility%20and%20sustainability%20reporting%20by%25listed%20entitiesAnnexure2_p.PDF. Accessed 16 July 2024.
SEBI, Circular on Business Responsibility and Sustainability Reporting by Listed Entities, May 10, 2021, Circular No.: SEBI/HO/CFD/CMD-2/P/CIR/2021/56. https://www.sebi.gov.in/legal/circulars/may-2021/business-responsibility-and-sustainability-reporting-by-listed-entities_50096.html. Accessed 15 July 2024.
SEBI. (2021a, May 10). *Business and sustainability reporting by listed companies,|Circular No.: SEBI/HO/CFD/CMD-2/P/CIR/2021/562*. Retrieved July 15, 2024, from https://www.sebi.gov.in/legal/circulars/may-2021/business-responsibility-and-sustainability-reporting-by-listed-entities_50096.html
SEBI. (2021b, May). *Business and sustainability reporting format*. Retrieved July 16, 2024, from SEBI: https://www.sebi.gov.in/sebi_data/commondocs/may-2021/Business%20responsibility%20and%20sustainability%20reporting%20by%20listed%20entitiesAnnexure1_p.PDF

SEBI. (2023, July 23). *SEBI master circular for ESG rating providers*. Retrieved July 10, 2024, from https://www.sebi.gov.in/legal/master-circulars/jul-2023/master-circular-for-esg-rating-providers-erps-_73856.html

Taxmann, Guide to ESG, Features, Principles and Reporting requirements, 6 March, 2024. www.taxmann.com. Accessed 19 July 2024.

Taxmann. (2024, March 6). *Guide to ESG in India, features, principles and reporting requirements*. Retrieved July 19, 2024, from Taxmann: https://www.taxmann.com/

The Companies (Corporate Social Responsibility) Rules, 2014, further amended in 2018 and 2021 bringing more stringent reporting compliance for companies. https://www.mca.gov.in/Ministry/pdf/CSRAmendmentRules_23012021.pdf. Accessed July 8 2024.

The United Nations Global Compact. (2004). *Who Cares Win, Connecting Financial markets to a changing world, 2004*. https://www.unepfi.org/fileadmin/events/2004/stocks/who_cares_wins_global_compact_2004.pdf. Accessed 18 July 2024.

UNDP, Theme 10, rethinking the governance of ESG, 2024. https://www.undp.org/future-development/signals-spotlight/rethinking-governance-esg. Accessed 15 July 2024.

UNDP. (2024). *Global, future of development*. Retrieved July 15, 2024, from Theme 10, Rethinking of governance of ESG: https://www.undp.org/future-development/signals-spotlight/rethinking-governance-esg

Union, E. (2023, January 5). *Corporate sustainability reporting*. Retrieved July 12, 2024, from https://finance.ec.europa.eu/capital-markets-union-and-financial-markets/company-reporting-and-auditing/company-reporting/corporate-sustainability-reporting_en

Union, E. (n.d.). *Corporate sustainability reporting*. Retrieved July 11, 2024, from https://finance.ec.europa.eu/capital-markets-union-and-financial-markets/company-reporting-and-auditing/company-reporting/corporate-sustainability-reporting_en

United Nations Principles for Responsible Investment; Under this, six principles have been developed by investors for responsible investing by incorporating ESG principles in investment practices. https://www.unpri.org/about-us/what-are-the-principles-for-responsible-investment. Accessed 16 July 2024.

United Nations, Sustainable Development Goals, 2015. https://sdgs.un.org/goals. Accessed 19 July 2024.

United, N. (2015). *17 SDG*. Retrieved July 19, 2024, from United Nations Department of Economic and Social Affairs: https://sdgs.un.org/goals

Wang, Y., Ghadimi, M., Wang, Q., Hou, L., Zeraatkar, D., Iqbal, A., Ho, C., Yao, L., Hu, M., Ye, Z., Couban, R., Armijo-Olivo, S., Bassler, D., Briel, M., Gluud, L. L., Glasziou, P., Jackson, R., Keitz, S. A., Letelier, L. M., … Guyatt, G. H. (2022). Instruments assessing risk of bias of randomized trials frequently included items that are not addressing risk of bias issues. *Journal of Clinical Epidemiology, 152*(218), 225. https://doi.org/10.1016/j.jclinepi.2022.10.018

Zhang, Y., Ruan, H., Tang, G., & Tong, L. (2021). Power of sustainable development: Does environmental management system certification affect a firm's access to finance? *Business Strategy and the Environment, 17*. https://doi.org/10.1002/bse.2839

Chapter 30
A Bibliometric Analysis of the Transition of Corporate Governance and Corporate Social Responsibility Toward ESG and Sustainability

Sunaina Kanojia, Nishant Sharma, and Neelam Jhawar

Abstract This study conducts a bibliometric analysis and detailed examination of the corporate governance and corporate social responsibility (CG-CSR) literature, focusing on significant trends, the intellectual structure of CG-CSR research, and the development of scientific knowledge from 2003 to 2022. Examining 1287 publications, we identify contemporary dynamics, core themes, and emerging trends. Our research highlights influential works, geographical contributions, prevalent keywords, and knowledge foundations. We developed thematic clusters based on bibliographic coupling and used co-citation analysis to outline the intellectual structure of CG-CSR research, forming the basis for future research directions using a representative sample from the Scopus database. Post-2007, there is a notable increase in publications focusing on corporate governance characteristics as determinants of CSR disclosure. Our findings reveal distinct thematic clusters in the field: "Institutional and Political Perspectives of Corporate Social Responsibility," "Impact of Corporate Social Responsibility on Firm Performance," "Quality of Corporate Governance and Corporate Social Responsibility Disclosures and Reporting," and "Board Diversity and Corporate Social Performance." We propose five prospective research trajectories, including the influence of board diversity on sustainable development goals (SDGs), theorizing triple bottom line (TBL), climate change, tensegrity of ESG, TBL, and SDGs, and disclosures, among others. This study directs scholarly attention to areas warranting further exploration in the evolving intersection of corporate governance and corporate social responsibility.

S. Kanojia (✉) · N. Jhawar
Department of Commerce, Faculty of Commerce and Business, Delhi School of Economics, University of Delhi, Delhi, India
e-mail: skanojia@commerce.du.ac.in

N. Jhawar
e-mail: neelamjhawar90@commerce.du.ac.in

N. Sharma
Jesus and Mary College, University of Delhi, Delhi, India
e-mail: nishantsharma@commerce.du.ac.in

V. Kandpal et al. (eds.), *Developments in Corporate Governance*,
Approaches to Global Sustainability, Markets, and Governance,
https://doi.org/10.1007/978-981-96-6366-8_30

Keywords Corporate Governance (CG) · Corporate Social Responsibility (CSR) · Review · Bibliometric analysis · Scopus · VOSviewer

30.1 Introduction

Businesses are expected to address social and environmental challenges (Elkington, 2006). In response to this, companies worldwide have begun implementing and disclosing corporate social responsibility (CSR) practices either voluntarily or mandatorily, to engage stakeholders and contribute to their well-being (Jhawar, 2020; Naciti et al., 2021; Zaman et al., 2022). The traditional outlook of corporate governance (CG), rooted in agency theory, emphasizes on maximizing shareholders' wealth and sets rules that govern the relationship among shareholders, board, and managers (Berle & Means, 1932). However, the contemporary literature sees this view as narrow and calls for CG to balance the needs of both the shareholders and the non-financial stakeholders (Jain & Jamali, 2016). Agency theory does not address the integration of social responsibility into corporate goals (Walls et al., 2012). On the contrary, stakeholder theory recognizes the broader impact of a company on different stakeholders and society (Bolourian et al., 2021). The integration of agency theory with stakeholder concepts surname the relationship between CG practices and CSR as a mechanism to decrease information gaps and maintain manager transparency and prevent selfish decisions. Effective corporate governance practices provide both support and enhancement for firm CSR initiatives toward all stakeholders according to Tran et al. (2021). The expanded CG framework now embraces the rights as well as responsibilities of every organization stakeholder according to Aoki (2000) while shaping financial together with non-financial results (Aguilera et al., 2015; Zaman et al., 2022). Researching academic perspectives about the CG-CSR link and its historical developments becomes essential because of urgent social and environmental needs.

Numerous scientific investigations focusing on diverse corporate governance systems which shape corporate social performance have produced conflicting results. Scholarly research produces conflicting results regarding board independence as an important governance element (Dwekat et al., 2022). Research results about independent board members demonstrate either a beneficial effect on CSR (Harjoto & Jo, 2011; Jizi, 2017) or an adverse relationship through lack of CSR knowledge or financial priority over CSR (Majeed et al., 2015; Sundarasen et al., 2016; Yasser et al., 2017). Other corporate governance mechanisms present inconclusive evidence about their impact according to Dwekat et al. (2022) regarding CEO duality and according to Velte (2022) and Baker et al. (2020) for board diversity and institutional ownership respectively.

The inconsistent findings from this research highlight the requirement to organize previous literature findings into major categories. Variables that have received excessive research attention specifically point to the requirement of studying additional influential factors. The study adopts a bibliometric analysis approach to evaluate and sum up the CG-CSR literature and generate an integrated overview. This research addresses four fundamental inquiries which include RP1: The Evaluation of CG-CSR Literature for Major Patterns Together with Research Themes and Publication Dates While Demonstrating Scholarly Affiliations. The research investigates how scientific knowledge grows through author identification and collaborative analysis of the present situation. This section evaluates vital research articles within CG-CSR research by determining which papers made the greatest impact on field development. The analysis of RP4 examines how core elements in CG-CSR research establish their relationship network through conceptual interconnections. The focus should be on the most recent publications which represent new trends in CG-CSR research. CG-CSR research findings will be summarized to develop proposals for future scholarly directions in this field.

Our study differs from previous reviews in several ways. First, earlier reviews focused either on CG (Aguilera et al., 2019) or CSR constructs (Fernandez-Gago et al., 2020). Second, existing studies typically review a specific dimension of either CG or CSR when analyzing their relationship such as the board of directors (Afeltra et al., 2022; Velte & Stawinoga, 2020) or CG as one of the determinants of CSR (Ali et al., 2022), or CG with one CSR dimension like environmental sustainability (Enciso-Alfaro & García-Sánchez, 2022). Third, the existing reviews on CG-CSR (Zaman et al., 2022) are systematic literature reviews limited to content analysis of a smaller sample of studies.

Table 30.1 compares earlier reviews with our study, highlighting several contributions to the existing literature. First, to the best of our knowledge, our study is the first to apply bibliometric analysis to CG-CSR research, covering 491 journals and 1287 research articles over 20 years. Previous reviews were limited in scope or focused on specific dimensions of CG and/or CSR. A bibliometric approach is beneficial when the scope of review is broad and data is too large for manual review (Donthu et al., 2021). Second, our study provides depth by applying methodological rigor to delineate the evolution of CG-CSR. We employ distinct bibliometric techniques, such as authorship analysis, keyword analysis, citation analysis, bibliographic coupling, and co-citation analysis, to map and evaluate the literature. This approach answers pertinent research questions about the most influential work, helping researchers assess studies more objectively. Third, we examine the past, present, and future of CG-CSR research by analyzing the seminal and current themes, outlining the intellectual structure of CG-CSR research, and presenting a comprehensive research agenda for future progress.

Our study not only aims to map the evolution of CG-CSR research but also provides insights that resonate with a global audience. Given the worldwide emphasis on sustainability and responsible business practices, this study holds broad relevance for scholars, practitioners, and policymakers. By analyzing a diverse sample of studies from various regions, it reflects global trends in corporate governance and CSR, making the findings applicable across both developed and developing economies. As businesses everywhere face growing pressure to adopt sustainable practices, this research offers a comprehensive view of the shift toward ESG and sustainability. Moreover, by examining the development of scientific knowledge

Table 30.1 Comparison of previous review studies on CG-CSR and our study

Basis of comparison	Period	Focus	Methodology	Sample studies
Afeltra et al. (2022)	No time limit	Relationship between the board of directors and corporate social reporting	Systematic Literature Network Analysis (SLNA)	1406
Ali et al. (2022)	1983–2021	Determinants of CSR Disclosure in Developing Economies	Systematic review	71
Enciso-Alfaro and García-Sánchez (2022)	1991–2022	Corporate governance and environmental sustainability	Bibliometric analysis	233
Velte and Stawinoga (2020)	2010–2020	Impact of CSR-related board expertise on CSR outputs	Systematic review, content analysis	48 quantitative peer-reviewed empirical
Zaman et al. (2022)	1989–2016	CG-CSR relationship	Systematic review, content analysis	218
This study	2003–2022	CG-CSR relationship	Bibliometric analysis	1287

Source Authors' compilation

and collaboration across different regions, the study highlights opportunities for international partnerships to address shared challenges in corporate governance and sustainability. This global perspective contributes to a deeper understanding of the evolving landscape, supporting the alignment of corporate strategies with international expectations, and promoting sustainable business practices worldwide.

30.2 Methodology

Utilizing a bibliometric approach akin to previous studies (Baker et al., 2020; Enciso-Alfaro & García-Sánchez, 2022), this study explores the intellectual structure of CG-CSR research, identifies knowledge gaps, and proposes avenues for further investigation. This study covers articles published from 2003 to 2022, a period marked by significant discourse on CG and CSR. Leveraging the Scopus database for its comprehensive coverage of international multidisciplinary research, we established a search protocol (see Table 30.2) that initially yielded 2113 documents.

Applying rigorous inclusion criteria by restricting the time frame (2003–2022), document type (articles), source (journals), language (English), and publication stage (final) resulted in 1438 documents. Areas not relevant to the study, with fewer than

Table 30.2 Search protocol and data retrieval strategy

	Database scopus	Search String	Search yield
Step 1	Initial Search	("Corporate Social Responsibility*" OR CSR OR "Corporate Responsibility/*" OR "Corporate Accountability" OR "Corporate Philanthropy" OR "Social Responsibility*") AND ("Corporate Governance")	2113
Step 2	First Stage Filters	Inclusions-Year: 2003–2022, Document Type: Articles, Source Type: Journals, Language: English, Stage: Final publications	1438
Step 3	Second Stage Filters	Excluded Categories: Nursing, Chemical Engineering, Agriculture and Biological Sciences, Material Science, Earth and Planetary Sciences, Physics and Astronomy, Pharmacology, Toxicology and pharmaceuticals, Chemistry, Biochemistry, Genetics and Molecular Biology, Multidisciplinary, Medicine and Computer science	1287

Source Authors' compilation

three publications during the study period (e.g., nursing, chemical engineering), were excluded, resulting in 1287 academic publications from 419 journals. Visualization and analysis were conducted using VOSviewer to map bibliographic linkages (Tunger & Eulerich, 2018; Yu et al., 2020), ensuring clarity and interpretability of the results (Van Eck & Waltman, 2010).

30.3 Bibliometric Analyses

The succeeding paragraphs provide detailed insights to answer RP1 (about publication trends) and RP2 (about productive and influential authors).

30.3.1 Publication Trend by Year

The publication growth accelerated from 2007 onwards, with significant contributions came during 2018–2021. Figure 30.1 exhibits a peak in publications in 2020 and 2021, as the COVID-19 pandemic lockdowns gave scholars more time to focus on their research, leading to a surge in studies (Ioannidis et al., 2021).

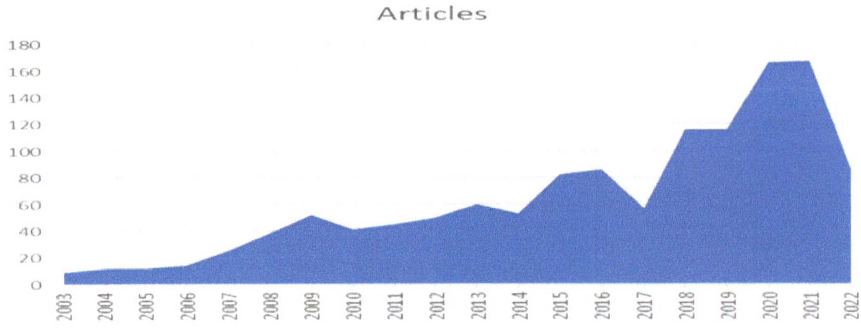

Fig. 30.1 CG-CSR research publication trend. *Source* MS-Excel generated by author

30.3.2 Journals with the Highest Publications and Citations

Figure 30.2(a) shows the network map of the 51 top-cited journals. Sustainability (Switzerland) journal has the largest node with 80 publications, followed by the Journal of Business Ethics and CSR and Environmental Management with 74 and 46 publications, respectively, as shown in Figure II(b).

To further address RP1 on publication trends, we considered journals publishing at least five articles and having at least ten citations. This limitation resulted in 51 journals, of which the top 15 journals are shown in Table 30.3.

30.3.3 Publication Activity by Country

Figure 30.3 shows that countries like the USA, the UK, China, Spain, and Australia dominate the field. Table 30.4 lists the top 15 countries in publications, with the USA leading, followed by the UK. The dominance is due to major CG scandals in the USA and the UK such as Pacific Gas and Electric Company, Enron, Arthur Anderson, WorldCom, and Equitable Life Assurance Society, MG Rover Group. These events highlighted governance issues and irresponsible behavior, prompting significant regulatory changes and increased corporate social responsibility in these countries. The Sarbanes–Oxley Act of 2002 in the USA also brought global attention to corporate governance and corporate responsibility.

The significant contribution from China, Spain, and Australia became noticeable post-subprime crisis when CG and CSR gained prominence. The Netherlands leads in citations per publication, surpassing the most productive countries like the USA, the UK, Spain, and China as well as those with high average citations like Canada and Germany. Country-level co-authorship analysis, shown in Fig. 30.3, highlights the global author associations, with larger nodes for leading countries like the USA, the UK, China, Spain, and Australia, indicating strong research collaboration links.

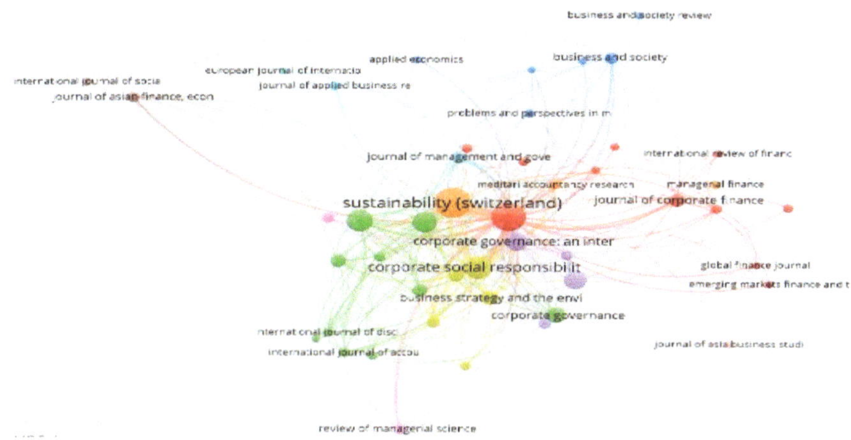

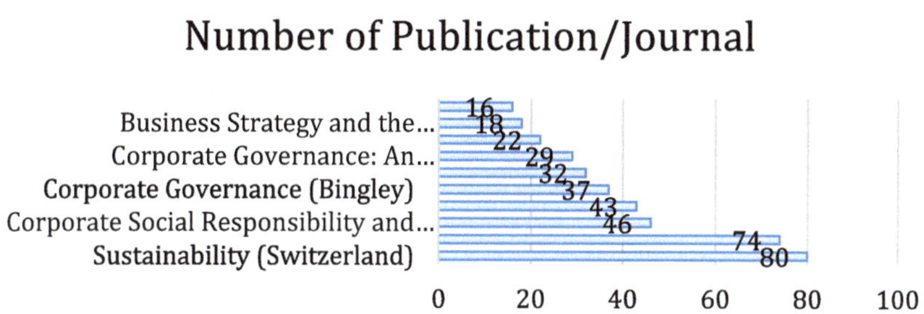

Fig. 30.2 **(a)** Journal citation network map. *Source* VOSviewer output. **(b)** Journals with maximum publications. *Source* MS-Excel output

Indian scholars, with 45 publications and 7.38 citations per article, have less global visibility and need required collaborations.

30.3.4 Publication Activity by Author and Organization

Authorship analysis classifies authors with high research output. Following Raza et al. (2021) and Tunger and Eulerich (2018), we identified the top authors dominating by publication count. Among 2652 authors, Sial M.S., Velte P., and García-Sánchez I.-M. lead in research output. The strongest link exists between Sial M.S. and Cherian J. followed by García-Sánchez I.-M. and Martínez-Ferrero J (see Table 30.5).

Overall, weak associations between authors indicate a lack of collaborative research. Tables 30.5 and 30.6 show that highly cited authors like Jo (1419), Harjoto (1412), Renneboog (1034), Jackson (1002), and Barnea and Rubin (each with 813 citations) do not publish many papers, suggesting that research output does not

Table 30.3 Most cited journals

Rank	Journal	ABDC ranking	Documents	Citations
1	Journal of Business Ethics	A*	74	8712
2	Corporate Governance: An International Review	A	29	2848
3	Corporate Social Responsibility and Environmental Management	C	46	2225
4	Journal of Banking and Finance	A*	8	1736
5	Sustainability (Switzerland)	NR	80	1191
6	Journal of Corporate Finance	A*	16	1153
7	Business Strategy and the Environment	A	18	1139
8	Social Responsibility Journal	B	43	1060
9	Corporate Governance (Bingley)	C	37	793
10	Journal of Cleaner Production	A	16	736
11	Corporate Governance	C	22	610
12	Management Decision	B	14	524
13	Business and Society	A	12	488
14	International Journal of Law and Management	C	16	472
15	Journal of Management and Governance	C	12	359

Source Authors' Elaboration of data from VOSviewer. ABDC ranking provided by the Australian Business Deans Council (Ranking A* - C, with A* as the top-rated followed by A, B, and C. NR: Not Ranked (listed) in ABDC

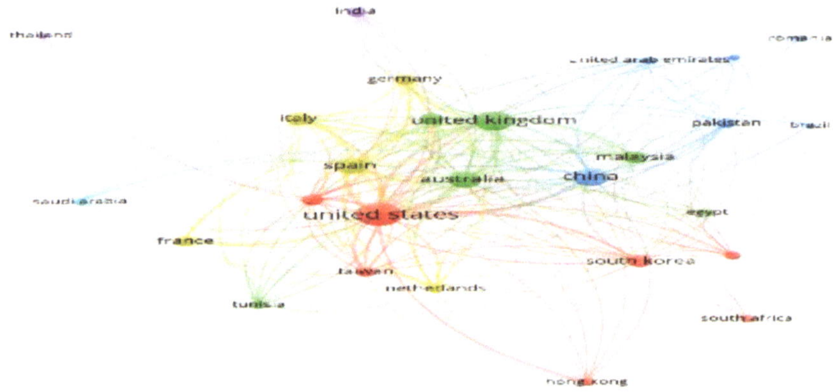

Fig. 30.3 Country co-authorship network. *Source* VOSviewer

Table 30.4 Top 15 countries with the highest publications

Rank	Countries	Documents	Citations	Citations per publication
1	United States of America (USA)	203	10,572	52.08
2	United Kingdom (UK)	146	6716	46
3	China	117	2314	19.78
4	Spain	111	4173	37.60
5	Australia	99	4583	46.29
6	Italy	72	1793	24.90
7	South Korea	52	910	17.5
8	Malaysia	52	1272	24.46
9	Indonesia	52	262	5.03
10	Canada	51	3259	63.90
11	France	50	1388	27.76
12	Germany	45	2139	47.53
13	India	45	332	7.38
14	Taiwan	37	1078	29.12
15	Pakistan	34	521	15.32

Source Authors' elaboration of data from VOSviewer

Table 30.5 Top authors based on the number of publications

Rank	Authors	No. of publications	Citations	Total link strength
1	Sial M.S	10	131	5
2	Velte P	9	328	0
3	García-Sánchez I.-M	8	333	5
4	Martínez-Ferrero J	7	208	5
5	Pucheta-Martínez M.C	7	127	3
6	Gallego-Álvarez I	6	78	3
7	Hussainey K	6	137	0
8	Jarboui A	6	45	0
9	Jiraporn P	6	66	0
10	Cherian J	5	71	5
11	Choi D	5	24	0
12	Gangi F	5	106	0
13	Gennari F	5	83	0
14	Giannarakis G	5	307	0
15	Lee J.-H	5	28	0

Source Authors' elaboration of data from VOSviewer

Table 30.6 Most influential authors

Rank	Authors	No. of documents	Citations	Total link strength
1	Jo H.	4	1419	391
2	Harjoto M.A.	3	1412	387
3	Renneboog L.	3	1034	86
4	Jackson G.	2	1002	82
5	Barnea A.	1	813	196
6	Rubin A.	1	813	196
7	Zhang C/	2	790	68
8	Li S/	3	721	74
9	Ter Horst J.	1	694	45
10	Garcia-Sanchez I.M.	3	687	156
11	Matten D.	3	634	55
12	Muttakin M.B.	3	628	180
13	Brammer S.	2	562	37
14	Kolk A.	2	554	75
15	Khan A.	2	548	160

Source Authors' elaboration of data from VOSviewer

necessarily equate to impact. The top five cited authors have 1–4 papers with 800–1450 citations, whereas the most productive authors have 1–10 papers with 100–350 citations. Thus, the most prolific authors are not necessarily the most cited.

Table 30.7 shows that the University of Salamanca (Spain), the University of Sfax (Tunisia), and Universiti Teknologi Mara (Malaysia) lead in research output with 24, 20, and 15 articles, respectively, followed by Pennsylvania State University (USA) with 13 articles. Despite most authors being from Anglo-American and European institutions, the topic has global interest.

Table 30.8 shows the most cited academic organizations in CG-CSR research. Although the top three most productive and most influential organizations differ, Santa Clara University, York University, Deakin University, University of Amsterdam, Lebanese American University and University of Groningen are both productive and influential. Only six out of the 20 top organizations appear in both lists, highlighting the distinction between high research output and strong influence.

30.3.5 Citation Network Analysis

It describes scholarly work when one author cites another and identifies the most influential work. To answer RP3 (most influential articles on CG-CSR), 489 publications have been cited more than 15 times. The most cited papers are by Barnea and Rubin (2010), Renneboog et al. (2008), and Jo and Harjoto (2011). Barnea and

Table 30.7 Research output by universities

Rank	Organizations	Country	Number of articles	Citations	Average citations
1	University of Salamanca	Spain	24	1230	51
2	University of Sfax	Tunisia	20	150	8
3	Universiti Teknologi Mara	Malaysia	15	353	24
4	Pennsylvania State University	USA	13	684	53
5	Universiti Utara Malaysia	Indonesia	12	84	7
6	Lebanese American University	Lebanon	11	568	52
7	University Of Zaragoza	Spain	11	173	16
8	University of Nigeria	Nigeria	10	23	2
9	University Of Portsmouth	UK	10	196	20
10	York University	Canada	9	1096	122
11	University of Groningen	Netherlands	8	567	71
12	Universitas Airlangga	Indonesia	8	68	9
13	Hanyang University	South Korea	8	122	15
14	American University of Beirut	Lebanon	7	130	19
15	La Trobe University	Australia	7	268	38

Source Authors' elaboration of data from VOSviewer

Rubin (2010) noted that CSR conflict can be viewed in two ways: over-investment in CSR can reduce firm value (negative) or advance a social agenda (positive). This affirms that CSR policy can create conflict between shareholders, requiring better alignment of corporate and social goals.

Further, Renneboog et al. (2008) reviewed the academic literature on socially responsible investments (SRI), highlighting various issues and challenges. While CSR is linked to higher shareholder value, the direction of causality is unclear. Moreover, the third most cited paper by Jo and Harjoto (2011) shows that CSR engagement positively influences firm value, as measured by industry-adjusted Tobin's q. CSR activities addressing internal social enhancement, such as employee diversity and product quality, enhance firm value more than external social enhancement, like community relations and environmental concerns.

Table 30.8 Most cited academic institutions

Rank	Organizations	Country	Number of articles	Citations	Average citations
1	Santa Clara University	USA	6	2002	334
2	Pepperdine University	USA	4	1420	355
3	Claremont Mckenna College	USA	4	1399	350
4	The University of Warwick	UK	3	1258	419
5	York University	Canada	9	1096	122
6	Tilburg University	Netherlands	3	1034	345
7	Flinders University	Australia	5	971	194
8	University Of Manchester	UK	5	901	180
9	Universidad Carlos iii De Madrid	Spain	5	864	173
10	Simon Fraser University	Canada	3	842	281
11	Universidad De Grenada	Spain	3	719	240
12	Deakin University	Australia	7	673	96
13	University of Amsterdam	Netherlands	7	583	83
14	Lebanese American University	Lebanon	11	568	52
15	University of Groningen	Netherlands	8	567	71

Source Authors' elaboration of data from VOSviewer

Hence, the top three cited papers provide mixed evidence. They either indicate no causality between SRI and shareholders' value or a positive impact of CSR engagement on firm value, though limited to internal social enhancement activities (focused on employees). Table 30.9 presents the top fifteen papers with the maximum citations. Years 2008–2014 are prominent, and articles published immediately after the subprime crisis received the maximum citations.

30.3.6 Co-citation Analysis

To address RP4 (the intellectual structure of current research on CG-CSR), we use co-citation analysis, exhibiting the relationship between the core ideas of a specific field of research (Small, 1973) and dividing the articles into several clusters to conclude

Table 30.9 Most cited journal articles

Rank	Authors	Article	Source	Citation
1	Barnea and Rubin (2010)	Corporate Social Responsibility as A Conflict Between Shareholders	Journal of Business Ethics	813
2	Renneboog et al. (2008)	Socially Responsible Investments: Institutional Aspects, Performance, And Investor Behavior	Journal of Banking and Finance	694
3	Jo and Harjoto (2011)	Corporate Governance and Firm Value: The Impact of Corporate Social Responsibility	Journal of Business Ethics	597
4	Brammer et al. (2012)	Corporate Social Responsibility and Institutional Theory: New Perspectives on Private Governance	Socio-Economic Review	535
5	Khan et al. (2013)	Corporate Governance and Corporate Social Responsibility Disclosures: Evidence from An Emerging Economy	Journal of Business Ethics	514
6	Jackson and Apostolakou (2010)	Corporate Social Responsibility in Western Europe: An Institutional Mirror or Substitute?	Journal of Business Ethics	467
7	Kim et al. (2014)	Corporate Social Responsibility and Stock Price Crash Risk	Journal of Banking and Finance	436
8	Jo and Harjoto (2012)	The Causal Effect of Corporate Governance on Corporate Social Responsibility	Journal of Business Ethics	417
9	Prior et al. (2008)	Are Socially Responsible Managers Really Ethical? Exploring The Relationship Between Earnings Management and Corporate Social Responsibility	Corporate Governance: An International Review	408
10	Kolk (2008)	Sustainability, Accountability and Corporate Governance: Exploring Multinationals' Reporting Practices	Business Strategy and the Environment	399
11	Harjoto and Jo (2011)	Corporate Governance and CSR Nexus	Journal of Business Ethics	398
12	Krüger (2015)	Corporate Goodness and Shareholder Wealth	Journal of Financial Economics	368
13	Jizi et al. (2014)	Corporate Governance and Corporate Social Responsibility Disclosure: Evidence from The US Banking Sector	Journal of Business Ethics	344

(continued)

Table 30.9 (continued)

Rank	Authors	Article	Source	Citation
14	Li and Zhang (2010)	Corporate Social Responsibility, Ownership Structure, And Political Interference: Evidence from China	Journal of Business Ethics	328
15	Frias-Aceituno et al. (2013)	The Role of The Board in The Dissemination of Integrated Corporate Social Reporting	Corporate Social Responsibility and Environmental Management	326

Source Authors' elaboration of data from VOSviewer

about the intellectual structure of a research field (Rossetto et al., 2018; Small, 1980). A co-citation exists if two documents are jointly cited by one or more articles (Smith, 1981). Compared to bibliographic coupling, it reviews scientific research structures rooted in citations and co-citations, exploring seminal themes and knowledge foundations (Donthu et al., 2021). A reference-based co-citation analysis and a source-based co-citation (Yu et al., 2020) have been performed to develop item-relatedness. Parts (a) and (b) of Fig. 30.4 show the reference-based and source-based co-citation analysis results, respectively.

Figure 30.4(a) shows three thematic clusters (red, green, and blue), based on 56 references with at least 20 citations. For each of the 56 cited references, the total strength of the co-citation links with other cited references has been calculated. Part (b) uses the same 20-citation threshold, with 412 sources meeting it. To identify a common theme within each cluster, we adopted the common practice of studying the top ten documents in each cluster (Baker et al., 2020; Fahimnia et al., 2015; Xu et al., 2018).

The prominent theme of the red cluster is "Corporate Governance Characteristics as Determinants of Corporate Social Responsibility Disclosure." Board characteristics are the most analyzed aspect of CSR disclosures (Bear et al., 2010; Eng & Mak, 2003; Haniffa & Cooke, 2002; Hillman & Dalziel, 2003; Michelon & Parbonetti, 2012). Haniffa and Cooke (2005), the most cited paper, highlighted that disclosures are influenced by country-specific culture and serve as a legitimation strategy to appeal to ethical investors. The green cluster's theme is "Influence of Corporate Governance on Corporate Social Responsibility." It posits a relationship between CG and CSR (Arora & Dhadwadkar, 2011; Jamali et al., 2008; Jo & Harjoto, 2011, 2012; Johnson & Greening, 1999). The most influential document is Jensen and Meckling's "Theory of the Firm: managerial behaviour, agency costs, and ownership structure" (1976), suggesting agency theory frequently frames CG-CSR studies. This aligns with the literature's focus on CSR-CG in the context of agency problems, given agency theory's dominance in corporate governance studies. However, the narrow principal-agent perspective in CG-CSR research can be limiting and biased. Embracing broader theoretical foundations like stewardship could advance the field more effectively.

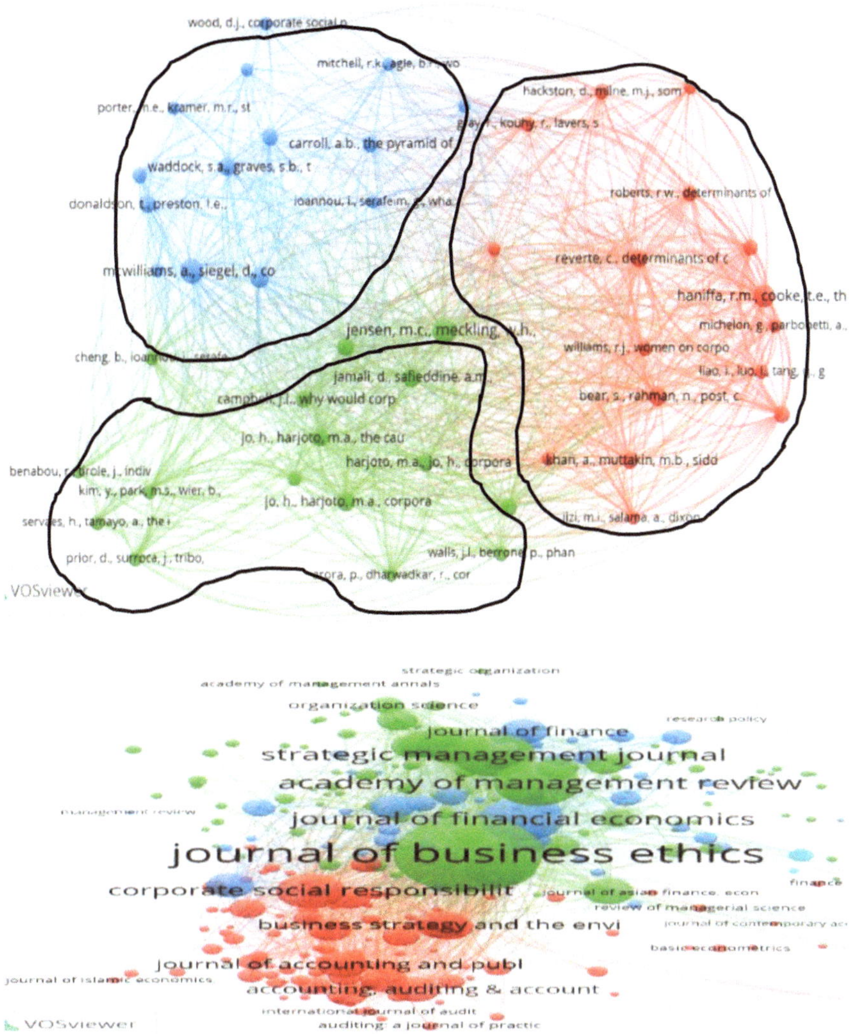

Fig. 30.4 **(a)** Reference-based co-citation network map. *Source* VOSviewer and authors' elaboration of data from VOSviewer. **(b)** Reference-based co-citation network map. *Source* VOSviewer

The blue cluster focuses on the relationship between Corporate Social Responsibility (CSR) and financial performance, highlighted by McWilliams and Siegel's (2001) "Corporate Social Responsibility: A Theory of the Firm Perspective." Research findings vary, showing positive (Orlitzky et al., 2003; Tsoutsoura, 2004; Waddock & Graves, 1997), negative (Wright & Ferris, 1997), and neutral (Mcguire et al., 1988; McWilliams & Siegel, 2000; Teoh et al., 1999) relationships. Despite assuming a link between CSR and financial performance, future research should aim to clarify when this relationship is positive, negative, or neutral.

Part (b) of Fig. 30.4 highlights the journals Journal of Business Ethics, Strategic Management Journal, Academy of Management Review, Journal of Financial Economics, and Corporate Governance: An International Review with the strongest links, underscoring their contribution to foundational themes.

30.3.7 Bibliographic Coupling

To gather insights for RP5 (emerging and currently the most popular theme among scholars), bibliographic coupling has been used. It groups articles based on similarities between citing documents referencing the same set of cited documents (Maseda et al., 2022). It can identify thematic clusters based on shared references (Zupic & Čater, 2015). This approach enhances visibility for current and niche publications (Donthu et al., 2021). Figure 30.5(a) and (b) depict results from document and source-based bibliographic coupling analysis. Part (a) shows four thematic clusters from 1287 documents, with 408 meeting a citation threshold of at least 20. Part (b) applies the same threshold for source-based clusters in Fig. 30.5.

The largest (green) cluster focuses on "Institutional and Political Perspectives of CSR," emphasizing how various institutional and political contexts shape CSR policies' design and impact on organizations. For instance, Jackson and Apostolakou (2010) observed that firms from liberal markets score more on more dimensions of CSR in comparison to their coordinated market counterparts. Kang and Moon (2012) and Daniel et al. (2012) also observed differences along the same lines. The theme of the second cluster (blue) focuses on the "Impact of CSR on Firm Performance." This cluster is based on the CSR and Firm (financial) Performance nexus (Ammann et al., 2011; Buchanan et al., 2018; Kruger, 2015; Lee, 2016; Yuan et al., 2016). The third (red) cluster focuses on "Quality of CG & CSR Disclosures." It asserts that firm CG characteristics are linked to CSR disclosures. Research generally shows a positive association (Adnan et al., 2018; Giannarakis, 2014; Lu & Wang, 2021; Mallin et al., 2013; Matuszak et al., 2019), though some variables influence CSR disclosure more than others (Habbash, 2016; Omair Alotaibi & Hussainey, 2016; Orazalin, 2019; Rao & Tilt, 2016; Said et al., 2009). The fourth (yellow) cluster focuses on "Board Diversity and Corporate Social Performance" (Amorelli & García-Sánchez, 2020; Al-Shaer & Zaman, 2016; Atif et al., 2021; Beji et al., 2021; Khatib et al., 2021; Kyaw et al., 2017; Sial et al., 2018), specifically analyzing the impact of gender diversity on firm performance (Byron & Post, 2016; Furlotti et al., 2019; Isidro & Sobral, 2015; Velte, 2016; Zaichkowsky, 2014).

Part (b) indicates that journals such as Journal of Business Ethics, Sustainability, Corporate Social Responsibility and Environmental Management, Corporate Governance: An International Review, Social Responsibility, Review of Management Science, Management Research Review, and Journal of Corporate Finance are significant contributors with strong links, underscoring their importance in publishing recent and niche documents in the field.

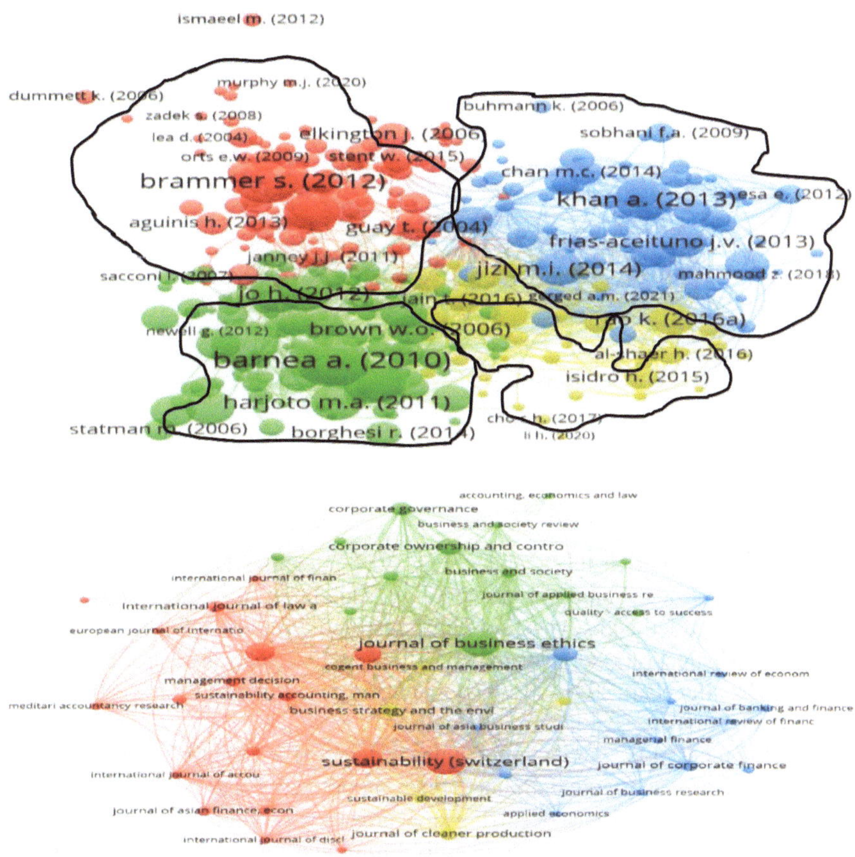

Fig. 30.5 **(a)** Document-based bibliographic coupling map. . *Source* VOSviewer *Source* VOSviewer and authors' elaboration of data from VOSviewer. **(b)** Source-based bibliographic coupling map

30.3.8 Keyword Analysis

Keyword analysis identifies central research areas (Liao et al., 2018), supporting RP4 and RP5 by ranking keywords from most to least frequent. Figure 30.6 and Table 30.10 illustrate keyword distribution, with larger nodes indicating higher frequency and connecting lines showing co-occurrences. The top 20 keywords highlight "corporate governance" as predominant, followed by "corporate social responsibility," sustainability, board of directors, sustainable development, stakeholders, and financial performance. Less frequent keywords include environmental disclosure, climate change, Islamic banks, triple bottom line, COVID-19, and ESG performance. This analysis underscores board attributes, sustainability, ESG, and climate change as future research priorities in CG-CSR studies. Figure 30.6 Keyword co-occurrence

Fig. 30.6 Keyword co-occurrence network map. *Source* VOSviewer

network map. To enrich the analysis, Fig. 30.7 integrates keyword analysis, co-citation analysis, and bibliographic coupling (Donthu et al., 2021), aligning with the themes identified and suggesting future research directions. To enrich the analysis, Fig. 30.7 integrates keyword analysis, co-citation analysis, and bibliographic coupling (Donthu et al., 2021), aligning with themes identified and suggesting future research directions.

The alignment of keywords with themes is depicted in Fig. 30.7: most occurring keywords correspond to foundational themes, moderately occurring keywords relate to emerging themes, and least occurring keywords suggest future research directions. For instance, prominent keywords like disclosure and financial performance align with foundational themes such as "corporate governance characteristics influencing CSR disclosure" and "the relationship between CSR and financial performance," respectively. Similarly, themes like board diversity and corporate social performance and the impact of CSR on firm performance feature keywords like firm performance and gender diversity. Current research primarily focuses on gender in board diversity, neglecting dimensions like race, ethnicity, nationality, and education, underscoring the need for future studies (Baker et al., 2020).

Through the least occurring keywords, we postulate that the future research directions shall be based on board diversity, climate change and environmental disclosure, Islamic banks, triple bottom line, SDGs, COVID-19, and ESG performance. The future research directions or themes can be formulated as the influence of board diversity on attaining sustainable development goals (SDGs), and theorizing TBL, climate change, and disclosures, among others. Thus, Fig. 30.7 integrates the review of past (through co-citation), present and emerging (through bibliographic coupling), and future research directions (by combining the results of co-citation and bibliographic coupling with keyword analysis).

Table 30.10 Top 20 keywords of CG-CSR research

Rank	Keyword	Occurrences	Total link strength
1	Corporate Governance	745	1649
2	Corporate Social Responsibility	602	1241
3	CSR	100	261
4	Social Responsibility	83	206
5	Sustainability	75	224
6	Organizational Culture	56	106
7	Board of Directors	47	134
8	Sustainable Development	46	132
9	*Stakeholder*	42	137
10	*Financial Performance*	42	116
11	Business Ethics	39	106
12	Stakeholder Theory	37	118
13	Agency Theory	31	86
14	Disclosure	29	89
15	Corporate Responsibility	29	87
16	Firm Value	29	77
17	Ethics	28	71
18	Corporate Social Performance	24	63
19	Stakeholder Engagement	22	78
20	Gender Diversity	22	60

Source Authors' elaboration of data from VOSviewer

Irrefutably CG-CSR research has evolved significantly, moving from CG as a pillar of CSR to CSR as a dimension of CG, and now toward addressing ESG concerns (Jamali et al., 2008).

The transition has significantly reshaped corporate governance in developed nations, which have swiftly adapted to these changes. However, Corporate Social Responsibility (CSR) remains in its early stages in developing and emerging countries. Tables 30.4 to 30.8 highlight that research concentration and publication citations are higher in developed nations, indicating their agility in responding to evolving business landscapes and fostering corporate sustainability. This trend is reinforced by emerging regulatory and social governance issues in these countries. Conversely, countries like China and India, which began privatization and corporatization of state-owned enterprises later, are now focusing on activities that promote social welfare and balanced regional growth. Consequently, integrating Corporate Governance and CSR in developing nation businesses is still in its infancy, necessitating a shift from shareholder to stakeholder value.

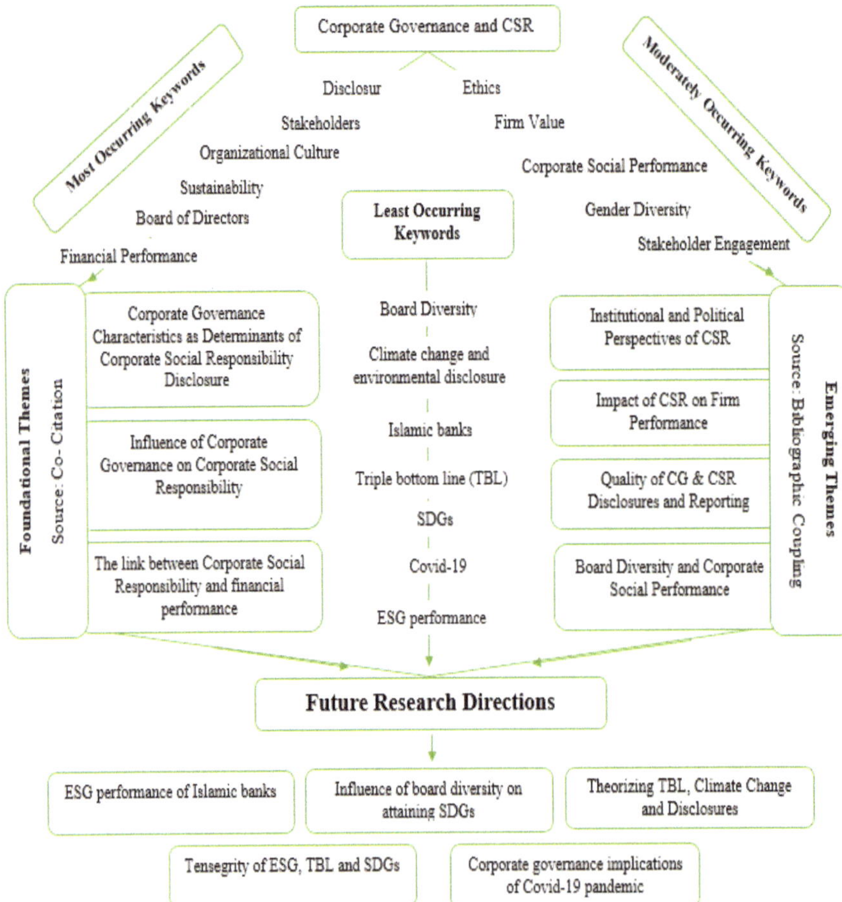

Fig. 30.7 Link between Keywords, Co-citation, and Bibliographic coupling. *Source* Authors' elaboration

30.4 Discussion and Future Research Direction

This section synthesizes the findings of our study and provides future research directions (RP6).

First, regarding RP1, the annual publication trend in CG-CSR research has shown significant growth since 2007, particularly accelerating after 2018, indicating heightened interest. The USA leads in research output, followed by the UK and China. Top journals include Sustainability (Switzerland), Journal of Business Ethics, and Corporate Social Responsibility and Environmental Management. Additionally, the most influential journals by citations are Journal of Business Ethics, Corporate Governance: An International Review, and Corporate Social Responsibility and Environmental Management. Second, the authorship analysis (RP2) identified Jo, Harjoto,

Renneboog, and Jackson as the most cited authors, while Sial, Velte, and García-Sánchez have the highest research output. Co-authorship analysis showed weak collaboration among authors and a lack of strong links between Asian countries (except China) and Anglo-American and European academics. Country-level co-authorship supports Liao et al.'s (2018) view that geographic proximity doesn't ensure collaboration. The most influential institutions are Santa Clara University, Pepperdine University, and Claremont McKenna College, while the most productive are the University of Salamanca, University of Sfax, and Universiti Teknologi Mara. Third, the three most influential articles (RP3) were on CSR conflict, socially responsible investments, and CSR engagement and firm value. Fourth, co-citation analysis (RP4) revealed three key themes in CG-CSR literature: corporate governance characteristics as determinants of CSR disclosure, the influence of corporate governance on CSR, and the link between CSR and financial performance. Finally, we found emerging and the currently most popular themes (RP5) in four major clusters using bibliographic coupling. The largest cluster addresses the institutional and political perspectives of CSR, followed by the nexus between CSR and firm performance, the quality of CG and CSR disclosures and reporting, and board diversity and corporate social performance. Additionally, the findings of keyword analysis logically support the seminal and emerging themes.

The findings of this study illuminate several critical factors that impede the future development of the CG-CSR field, thereby offering important implications for researchers, practitioners, and policymakers. By understanding these barriers, stakeholders can better navigate the complexities of corporate governance and social responsibility, guiding future research efforts and informing effective practices that address the evolving challenges within this domain.

1. Research in less-explored countries: The scarcity of research in developing countries, as highlighted in Sect. 4.4, hinders progress in CG-CSR studies. Existing literature is predominantly focused on Anglo-American and European contexts, offering limited insights from Asian and African nations. To advance the field, researchers should prioritize studies in developing regions to better understand their unique social, environmental, and governance challenges related to CSR. In doing so, particular attention should be given to country-specific cultural characteristics and board compositions, examining their influence on firms' overall financial performance. Furthermore, researchers must distinguish between research output and research influence, as this differentiation is crucial for assessing the relevance of both existing and emerging literature in the field. A deeper understanding of diverse regional contexts will enrich CG-CSR scholarship, fostering a more comprehensive and globally inclusive discourse. The quality and timing of disclosures vary in common law countries depending on their state of development. Emerging economies typically suffer from slow legal proceedings, feeble compliance, discreet disclosures, and political interference in the choice of CSR outlays. For practitioners, such research can provide valuable insights into how companies in these regions can effectively implement CSR strategies that are culturally and contextually relevant. Recognizing that

CSR encompasses multiple dimensions, firms in different markets may prioritize distinct aspects of CSR depending on the political and social environment. It could help broaden the scope of CSR implementation by helping firms identify and focus on the dimensions most suited to their business models and environmental contexts. Policymakers can also benefit from such studies, as they highlight the limitations of applying uniform CSR policies across diverse organizational and cultural landscapes. The direct adoption of policies from other regions may not always yield the desired outcomes due to contextual differences that impact their implementation and effectiveness. Instead, policymakers may design frameworks that support sustainable business practices tailored to local conditions, ultimately contributing to broader economic and social development goals.

2. Weak collaboration among authors: The findings of this study underscore the necessity of cross-country research collaborations to deepen our understanding of CG and CSR. Such collaborative networks can facilitate comparative studies, which can provide more comprehensive insights into CG-CSR practices across diverse institutional contexts. It not only enables the exchange of knowledge but also promotes the development of shared methodologies and best practices. Additionally, fostering international partnerships can lead to innovative solutions and strategies that leverage diverse experiences in addressing CG-CSR challenges. Researchers should also investigate the underlying reasons for weak collaborations, as understanding these barriers is crucial for strengthening global research networks. Policymakers can play a pivotal role in encouraging such collaborations through targeted funding and incentives, recognizing that robust research partnerships contribute to more effective governance frameworks informed by a broad spectrum of perspectives and experiences. Furthermore, practitioners, with the support of policymakers, can work toward fostering collaborative governance frameworks, particularly among companies operating within the same geographic regions. Given that neighboring countries, particularly in Europe and Asia, often share similar governance systems, philanthropic traditions, and cultural values, regional collaboration can enhance the effectiveness of CG-CSR initiatives. Additionally, policymakers can strive to develop common governance parameters and regulatory frameworks to facilitate ease of doing business and streamline governance practices. This approach is particularly relevant in regions where CG and CSR norms are still evolving, as harmonized governance standards can enhance regulatory coherence and improve corporate accountability on a broader scale.
3. Theory-based research**:** While a substantial body of literature exists, no single theoretical framework comprehensively explains the CG-CSR relationship. Researchers have primarily examined this relationship through the lenses of agency and stakeholder theories. The realm of even agency theory differs with the legal and social environment of different economies. However, this focus highlights the need to explore and integrate additional theoretical perspectives, such as stewardship theory, political and engaged shareholder theory, and new institutional theory. Incorporating these frameworks can provide a more nuanced

understanding of CG-CSR dynamics in diverse institutional and cultural contexts, thereby enriching academic discourse. Developing novel theoretical frameworks will be essential for understanding these relationships across varying governance structures and economic environments (Filatotchev et al., 2022).

In order to further enhance the academic understanding of the CG-CSR relationship, our research suggests the following five directions for future research.

1. Influence of board diversity on attaining SDGs: Future research should delve deeper into the influence of board diversity in driving CSR initiatives toward achieving the Sustainable Development Goals (SDGs). While existing literature primarily examines gender diversity, there is a pressing need to explore additional dimensions such as race, ethnicity, nationality, and educational background (Baker et al., 2020; Lin et al., 2020). Future studies could employ comparative analyses across diverse geographical contexts to understand how diverse perspectives at the board level can shape strategic decision-making in relation to ESG investments, thereby enriching the discourse on board effectiveness in promoting sustainable practices.
2. Theorizing TBL, Climate Change, and Disclosures: There is a significant gap in the theoretical exploration of the relationships between the Triple Bottom Line (TBL), climate change, and corporate disclosures. Future research should explore the inter-relationship, collect definitive evidence on whether the relationships are positive, negative, or neutral, and identify the specific conditions under which the interaction occurs. Scholars can provide valuable insights that contribute to both academic theory and practical managerial applications.
3. The research industry should explore how ESG, TBL, and SDGs combine through building tensegrity structures to advance sustainability goals. Regarding real-world application, researchers study both case samples alongside time-based data to observe organizations working with these frameworks and identify resulting partnerships between the integrated systems. The research would generate a complete understanding of sustainability strategy development processes enabled by ESG-TBL-SDG alignment framework.
4. Through their distinct ethical and social principles Islamic banks provide researchers with special opportunities to investigate their ESG performance metrics (Hashim et al., 2015). Analyzing how ESG factors have been integrated into operational frameworks of these institutions helps researchers study their sustainability practices together with the performance of their governance systems. Analyzing ESG performance between Islamic banking institutions and conventional banking organizations will reveal how Islamic finance supports global sustainability programs.
5. Corporate governance strategies must undergo adaptive transformations during crisis situations because the COVID-19 pandemic demonstrated the importance of such changes (Jebran & Chen, 2023). Researchers need to examine which governance systems enhanced organizational endurance and operational achievement when confronting such exceptional distressing situations. Analysis

of governance practices used by firms throughout different sectors and geographical regions during the pandemic will lead researchers to discover effective crisis management approaches useful for future business governance recovery planning.

Business sustainability requires entrepreneurs to cultivate CG-CSR within their organizations since both have become essential elements in startups that prioritize enduring success. Organizations must make CG-CSR their priority despite facing difficulties. The investigation seeks to establish research directions which support business sustainability by implementing proper governance structures. Despite challenges, prioritizing corporate governance and social responsibility is crucial. Consequently, this study aims to identify future research agendas that can contribute toward achieving business sustainability through the establishment of appropriate governance structures.

Finally, while the study provides a comprehensive assessment of existing literature by highlighting the intellectual structure of CG and CSR with five future directions, a few limitations may be noted. Firstly, the study pertains to using the Scopus database only, though exhaustive, and may not include some relevant publications indexed in other databases. Secondly, non-English publications could not be considered, and the time was limited to the last two decades (2003–2022). Nevertheless, these biases do not impact the validity of the results. Lastly, the analysis has been done solely on VOSviewer, which, despite being excellent software, suffers from its limitations. Future researchers may explore other bibliometric software such as CiteSpace, and Biblioshiny, among others.

References

Adnan, S. M., Hay, D., & van Staden, C. J. (2018). The influence of culture and corporate governance on corporate social responsibility disclosure: A cross-country analysis. *Journal of Cleaner Production, 198*, 820–832. https://doi.org/10.1016/j.jclepro.2018.07.057

Afeltra, G., Alerasoul, A., & Usman, B. (2022). Board of directors and corporate social reporting: A systematic literature network analysis. *Accounting in Europe, 19*(1), 48–77. https://doi.org/10.1080/17449480.2021.1979609

Aguilera, R. V., Desender, K., Bednar, M. K., & Lee, J. H. (2015). Connecting the dots: Bringing external corporate governance into the corporate governance puzzle. *Academy of Management Annals, 9*(1), 483–573. https://doi.org/10.5465/19416520.2015.1024503

Aguilera, R. V., Marano, V., & Haxhi, I. (2019). International corporate governance: A review and opportunities for future research. *Journal of International Business Studies, 50*(4), 457–498. https://doi.org/10.1057/s41267-019-00232-w

Ali, W., Wilson, J., & Husnain, M. (2022). Determinants/motivations of corporate social responsibility disclosure in developing economies: A survey of the extant literature. *Sustainability, 14*(6), 3474. https://doi.org/10.3390/su14063474

Al-Shaer, H., & Zaman, M. (2016). Board gender diversity and sustainability reporting quality. *Journal of Contemporary Accounting & Economics, 12*(3), 210–222. https://doi.org/10.1016/j.jcae.2016.09.001

Ammann, M., Oesch, D., & Schmid, M. M. (2011). Corporate governance and firm value: International evidence. *Journal of Empirical Finance, 18*(1), 36–55. https://doi.org/10.1016/j.jempfin.2010.10.003

Amorelli, M. F., & García-Sánchez, I. M. (2020). Critical mass of female directors, human capital, and stakeholder engagement by corporate social reporting. *Corporate Social Responsibility and Environmental Management, 27*(1), 204–221. https://doi.org/10.1002/csr.1793

Aoki, M. (2000). Information, corporate governance and institutional diversity: Competitiveness in Japan, the USA, and the transitional economies. *Oxford University Press.* https://doi.org/10.2307/3096479

Arora, P., & Dharwadkar, R. (2011). Corporate governance and corporate social responsibility (CSR): The moderating roles of attainment discrepancy and organization slack. *Corporate Governance: An International Review, 19*(2), 136–152. https://doi.org/10.1111/j.1467-8683.2010.00843.x

Atif, M., Hossain, M., Alam, M. S., & Goergen, M. (2021). Does board gender diversity affect renewable energy consumption? *Journal of Corporate Finance, 66*, Article 101665. https://doi.org/10.1016/j.jcorpfin.2020.101665

Baker, H. K., Pandey, N., Kumar, S., & Haldar, A. (2020). A bibliometric analysis of board diversity: Current status, development, and future research directions. *Journal of Business Research, 108*, 232–246. https://doi.org/10.1016/j.jbusres.2019.11.025

Barnea, A., & Rubin, A. (2010). CSR as a conflict between shareholders. *Journal of Business Ethics, 97*(1), 71–86. https://doi.org/10.1007/s10551-010-0496-z

Bear, S., Rahman, N., & Post, C. (2010). The impact of board diversity and gender composition on CSR and firm reputation. *Journal of Business Ethics, 97*(2), 207–221. https://doi.org/10.1007/s10551-010-0505-2

Beji, R., Yousfi, O., Loukil, N., & Omri, A. (2021). Board diversity and corporate social responsibility: Empirical evidence from France. *Journal of Business Ethics, 173*(1), 133–155. https://doi.org/10.1007/s10551-020-04522-4

Berle, A., & Means, G. (1932). *The modern corporation and private property.* Macmillan.

Bolourian, S., Angus, A., & Alinaghian, L. (2021). The impact of corporate governance on corporate social responsibility at the board level: A critical assessment. *Journal of Cleaner Production, 291*, Article 125752. https://doi.org/10.1016/j.jclepro.2020.125752

Brammer, S., Jackson, G., & Matten, D. (2012). Corporate social responsibility and institutional theory: New perspectives on private governance. *Socio-Economic Review, 10*(1), 3–28. https://doi.org/10.1093/ser/mwr030

Buchanan, B., Cao, C. X., & Chen, C. (2018). Corporate social responsibility, firm value, and influential institutional ownership. *Journal of Corporate Finance, 52*, 73–95. https://doi.org/10.1016/j.jcorpfin.2018.07.004

Byron, K., & Post, C. (2016). Women on boards of directors and corporate social performance: A meta-analysis. *Corporate Governance: An International Review, 24*(4), 428–442. https://doi.org/10.1111/corg.12165

Daniel, S. J., Cieslewicz, J. K., & Pourjalali, H. (2012). The impact of national economic culture and country-level institutional environment on corporate governance practices. *Management International Review, 52*(3), 365–394. https://doi.org/10.1111/j.1468-2370.2012.00300.x

Donthu, N., Kumar, S., Mukherjee, D., Pandey, N., & Lim, W. M. (2021). How to conduct a bibliometric analysis: An overview and guidelines. *Journal of Business Research, 133*, 285–296. https://doi.org/10.1016/j.jbusres.2021.04.070

Dwekat, A., Meqbel, R., Seguí-Mas, E., & Tormo-Carbó, G. (2022). The role of the audit committee in enhancing the credibility of CSR disclosure: Evidence from STOXX Europe 600 members. *Business Ethics, the Environment & Responsibility.* https://doi.org/10.1111/beer.12439

Elkington, J. (2006). Governance for sustainability. *Corporate Governance, 14*, 522–529.

Enciso-Alfaro, S. Y., & García-Sánchez, I. M. (2022). Corporate governance and environmental sustainability: Addressing the dual theme from a bibliometric approach. *CSR and Environmental Management.* https://doi.org/10.1002/csr.2403

Eng, L. L., & Mak, Y. T. (2003). Corporate governance and voluntary disclosure. *Journal of Accounting and Public Policy, 22*(4), 325–345. https://doi.org/10.1016/S0278-4254(03)00037-1
Fahimnia, B., Sarkis, J., & Davarzani, H. (2015). Green supply chain management: A review and bibliometric analysis. *International Journal of Production Economics, 162*, 101–114. https://doi.org/10.1016/j.ijpe.2015.01.003
Fernandez-Gago, R., Cabeza-García, L., & Nieto, M. (2016). Corporate social responsibility, board of directors, and stock price: Evidence from Spain. *Corporate Governance: THe International Journal of Business in Society, 16*(3), 505–520. https://doi.org/10.1108/CG-10-2015-0126
Frias-Aceituno, J. V., Rodriguez-Ariza, L., & Garcia-Sanchez, I. M. (2013). The role of the board in the dissemination of integrated corporate social reporting. *Corporate Social Responsibility and Environmental Management, 20*(4), 219–233. https://doi.org/10.1002/csr.1294
Furlotti, K., Mazza, T., Tibiletti, V., & Triani, S. (2019). Women in top positions on boards of directors: Gender policies disclosed in Italian sustainability reporting. *Corporate Social Responsibility and Environmental Management, 26*(1), 57–70. https://doi.org/10.1002/csr.1657
Giannarakis, G. (2014). The determinants influencing the extent of CSR disclosure. *International Journal of Law and Management*. https://doi.org/10.1108/IJLMA-05-2013-0021
Habbash, M. (2016). Corporate governance and corporate social responsibility disclosure: Evidence from Saudi Arabia. *Social Responsibility Journal*. https://doi.org/10.1108/SRJ-07-2015-0088
Haniffa, R. M., & Cooke, T. E. (2002). Culture, corporate governance and disclosure in Malaysian corporations. *Abacus, 38*(3), 317–349. https://doi.org/10.1111/1467-6281.00112
Haniffa, R. M., & Cooke, T. E. (2005). The impact of culture and governance on corporate social reporting. *Journal of Accounting and Public Policy, 24*(5), 391–430. https://doi.org/10.1016/j.jaccpubpol.2005.06.001
Harjoto, M. A., & Jo, H. (2011). Corporate governance and CSR nexus. *Journal of Business Ethics, 100*, 45–67. https://doi.org/10.1007/s10551-011-0772-6
Hashim, F., Mahadi, N. D., & Amran, A. (2015). Corporate governance and sustainability practices in Islamic financial institutions: The role of country of origin. *Procedia Economics and Finance, 31*, 36–43. https://doi.org/10.1016/S2212-5671(15)01129-6
Hillman, A. J., & Dalziel, T. (2003). Boards of directors and firm performance: Integrating agency and resource dependence perspectives. *Academy of Management Review, 28*(3), 383–396. https://doi.org/10.5465/amr.2003.10196729
Ioannidis, J. P., Salholz-Hillel, M., Boyack, K. W., & Baas, J. (2021). The rapid, massive growth of COVID-19 authors in the scientific literature. *Royal Society Open Science, 8*(9), Article 210389. https://doi.org/10.1098/rsos.210389
Isidro, H., & Sobral, M. (2015). The effects of women on corporate boards on firm value, financial performance, and ethical and social compliance. *Journal of Business Ethics, 132*(1), 1–19. https://doi.org/10.1007/s10551-014-2302-9
Jackson, G., & Apostolakou, A. (2010). Corporate social responsibility in Western Europe: An institutional mirror or substitute? *Journal of Business Ethics, 94*(3), 371–394. https://doi.org/10.1007/s10551-009-0269-8
Jain, T., & Jamali, D. (2016). Looking inside the black box: The effect of corporate governance on corporate social responsibility. *Corporate Governance: An International Review, 24*, 253–273. https://doi.org/10.1111/corg.12154
Jamali, D., Safieddine, A. M., & Rabbath, M. (2008). Corporate governance and corporate social responsibility synergies and interrelationships. *Corporate Governance: An International Review, 16*(5), 443–459. https://doi.org/10.1111/j.1467-8683.2008.00702.x
Jebran, K., & Chen, S. (2023). Can we learn lessons from the past? COVID-19 crisis and corporate governance responses. *International Journal of Finance & Economics, 28*(1), 421–429. https://doi.org/10.1002/ijfe.2428
Jensen, M. C., & Meckling, W. H. (1976). Theory of the firm: Managerial behavior, agency costs and ownership structure. *Journal of Financial Economics, 3*(4), 305–360. https://doi.org/10.1016/0304-405X(76)90026-X

Jhawar, N. (2020). CSR in India after The Companies Act 2013: An empirical study on its relationship with corporate governance and financial performance. *Corporate Governance Insight, 2*(1), 6–40. https://doi.org/10.58426/cgi.v2.i1.2020.6-40
Jizi, M. I., Salama, A., Dixon, R., & Stratling, R. (2014). Corporate governance and corporate social responsibility disclosure: Evidence from the US banking sector. *Journal of Business Ethics, 125*(4), 601–615. https://doi.org/10.1007/s10551-013-1929-2
Jizi, M. (2017). The influence of board composition on sustainable development disclosure. *Business Strategy and the Environment, 26*, 640–655. https://doi.org/10.1002/bse.1943
Jo, H., & Harjoto, M. A. (2011). Corporate governance and firm value: The impact of corporate social responsibility. *Journal of Business Ethics, 103*(3), 351–383. https://doi.org/10.1007/s10551-011-0869-y
Jo, H., & Harjoto, M. A. (2012). The causal effect of corporate governance on corporate social responsibility. *Journal of Business Ethics, 106*(1), 53–72. https://doi.org/10.1007/s10551-011-1052-1
Johnson, R. A., & Greening, D. W. (1999). The effects of corporate governance and institutional ownership types on corporate social performance. *Academy of Management Journal, 42*(5), 564–576. https://doi.org/10.5465/256977
Kang, N., & Moon, J. (2012). Institutional complementarity between corporate governance and CSR: A comparative institutional analysis of three capitalisms. *Socio-Economic Review, 10*(1), 85–108. https://doi.org/10.1093/ser/mwr025
Khan, A., Muttakin, M. B., & Siddiqui, J. (2013). Corporate governance and corporate social responsibility disclosures: Evidence from an emerging economy. *Journal of Business Ethics, 114*, 207–223. https://doi.org/10.1007/s10551-012-1336-0
Khatib, S. F., Abdullah, D. F., Elamer, A. A., & Abueid, R. (2021). Nudging toward diversity in the boardroom: A systematic literature review of board diversity of financial institutions. *Business Strategy and the Environment, 30*(2), 985–1002. https://doi.org/10.1002/bse.2665
Kim, Y., Li, H., & Li, S. (2014). CSR and stock price crash risk. *Journal of Banking & Finance, 43*, 1–13. https://doi.org/10.1016/j.jbankfin.2014.02.013
Kolk, A. (2008). Sustainability, accountability and corporate governance: Exploring multinationals' reporting practices. *Business Strategy and the Environment, 17*(1), 1–15. https://doi.org/10.1002/bse.511
Krüger, P. (2015). Corporate goodness and shareholder wealth. *Journal of Financial Economics, 115*(2), 304–329. https://doi.org/10.1016/j.jfineco.2014.09.008
Kyaw, K., Olugbode, M., & Petracci, B. (2017). Can board gender diversity promote corporate social performance? *Corporate Governance: The International Journal of Business in Society*. https://doi.org/10.1108/cg-09-2016-0183
Lee, M. T. (2016). CSR and stock price crash risk: Evidence from an Asian emerging market. *Managerial Finance*. https://doi.org/10.1108/mf-10-2015-0278
Li, W., & Zhang, R. (2010). Corporate social responsibility, ownership structure, and political interference: Evidence from China. *Journal of Business Ethics, 96*, 631–645. https://doi.org/10.1007/s10551-010-0488-z
Liao, H., Tang, M., Luo, L., Li, C., Chiclana, F., & Zeng, X. J. (2018). A bibliometric analysis and visualization of medical big data research. *Sustainability, 10*(1), 166. https://doi.org/10.3390/su10010166
Lin, K. J., Lu, X., Zhang, J., & Zheng, Y. (2020). State-owned enterprises in China: A review of 40 years of research and practice. *China Journal of Accounting Research, 13*(1), 31–55. https://doi.org/10.1016/j.cjar.2019.12.001
Lu, J., & Wang, J. (2021). Corporate governance, law, culture, environmental performance, and CSR disclosure: A global perspective. *Journal of International Financial Markets, Institutions and Money, 70*, Article 101264. https://doi.org/10.1016/j.intfin.2020.101264
Majeed, S., Aziz, T., & Saleem, S. (2015). The effect of corporate governance elements on CSR disclosure: Empirical evidence from listed companies at KSE Pakistan. *International Journal of Financial Studies, 3*(4), 530–556. https://doi.org/10.3390/ijfs3040530

Mallin, C., Michelon, G., & Raggi, D. (2013). Monitoring intensity and stakeholders' orientation: How does governance affect social and environmental disclosure? *Journal of Business Ethics, 114*(1), 29–43. https://doi.org/10.1007/s10551-012-1324-4

Maseda, A., Iturralde, T., Cooper, S., & Aparicio, G. (2022). Mapping women's involvement in family firms: A review based on bibliographic coupling analysis. *International Journal of Management Reviews, 24*(2), 279–305. https://doi.org/10.1111/ijmr.12278

Matuszak, Ł, Różańska, E., & Macuda, M. (2019). The impact of corporate governance characteristics on banks' corporate social responsibility disclosure: Evidence from Poland. *Journal of Accounting in Emerging Economies*. https://doi.org/10.1108/jaee-04-2017-0040

McGuire, J. B., Sundgren, A., & Schneeweis, T. (1988). Corporate social responsibility and firm financial performance. *Academy of Management Journal, 31*(4), 854–872. https://doi.org/10.5465/256342

McWilliams, A., & Siegel, D. (2000). Corporate social responsibility and financial performance: Correlation or misspecification? *Strategic Management Journal, 21*(5), 603–609. https://doi.org/10.1002/(sici)1097-0266(200005)21:5%3C603::aid-smj101%3E3.0.co;2-3

McWilliams, A., & Siegel, D. (2001). Corporate social responsibility: A theory of the firm perspective. *Academy of Management Review, 26*(1), 117–127. https://doi.org/10.5465/amr.2001.4011987

Michelon, G., & Parbonetti, A. (2012). The effect of corporate governance on sustainability disclosure. *Journal of Management & Governance, 16*(3), 477–509. https://doi.org/10.1007/s10997-010-9160-3

Naciti, V., Cesaroni, F., & Pulejo, L. (2021). Corporate governance and sustainability: A review of the existing literature. *Journal of Management and Governance*, 1–20. https://doi.org/10.1007/s10997-020-09554-6

Omair Alotaibi, K., & Hussainey, K. (2016). Determinants of CSR disclosure quantity and quality: Evidence from non-financial listed firms in Saudi Arabia. *International Journal of Disclosure and Governance, 13*(4), 364–393. https://doi.org/10.1057/jdg.2016.2

Orazalin, N. (2019). Corporate governance and corporate social responsibility (CSR) disclosure in an emerging economy: Evidence from commercial banks of Kazakhstan. *Corporate Governance: The International Journal of Business in Society*. https://doi.org/10.1108/cg-09-2018-0290

Orlitzky, M., Schmidt, F. L., & Rynes, S. L. (2003). Corporate social and financial performance: A meta-analysis. *Organization Studies, 24*(3), 403–441. https://doi.org/10.1177/0170840603024003910

Prior, D., Surroca, J., & Tribó, J. A. (2008). Are socially responsible managers really ethical? Exploring the relationship between earnings management and corporate social responsibility. *Corporate Governance: An International Review, 16*(3), 160–177. https://doi.org/10.1111/j.1467-8683.2008.00678.x

Rao, K., & Tilt, C. (2016). Board diversity and CSR reporting: An Australian study. *Meditari Accountancy Research*. https://doi.org/10.1108/medar-08-2015-0052

Raza, H., Gillani, S. M. A. H., Kashif, M. T., & Khan, N. (2021). The impact of corporate social responsibility on share price: A systematic review and bibliometric analysis. *Studies of Applied Economics, 39*(4). https://doi.org/10.25115/eea.v39i4.4311

Renneboog, L., Ter Horst, J., & Zhang, C. (2008). Socially responsible investments: Institutional aspects, performance, and investor behavior. *Journal of Banking & Finance, 32*(9), 1723–1742. https://doi.org/10.1016/j.jbankfin.2007.12.039

Rossetto, D. E., Bernardes, R. C., Borini, F. M., & Gattaz, C. C. (2018). Structure and evolution of innovation research in the last 60 years: Review and future trends in the field of business through the citations and co-citations analysis. *Scientometrics, 115*(3), 1329–1363. https://doi.org/10.1007/s11192-018-2709-7

Said, R., Zainuddin, Y. H., & Haron, H. (2009). The relationship between corporate social responsibility disclosure and corporate governance characteristics in Malaysian public listed companies. *Social Responsibility Journal*. https://doi.org/10.1108/17471110910964496

Sial, M. S., Zheng, C., Cherian, J., Gulzar, M. A., Thu, P. A., Khan, T., & Khuong, N. V. (2018). Does CSR mediate the relation between boardroom gender diversity and firm performance of Chinese listed companies? *Sustainability, 10*(10), 3591. https://doi.org/10.3390/su10103591

Small, H. (1973). Co-citation in the scientific literature: A new measure of the relationship between two documents. *Journal of the American Society for Information Science, 24*(4), 265–269. https://doi.org/10.1002/asi.4630240406

Small, H. (1980). Co-citation context analysis and the structure of paradigms. *Journal of Documentation, 36*(3), 183–196. https://doi.org/10.1108/eb026695

Smith, L. C. (1981). Citation analysis. *Library Trends, 30*(1), 83–106. https://doi.org/10.46895/lis.18.171

Sundarasen, S., Je-Yen, T., & Rajangam, N. (2016). Board composition and CSR in an emerging market. *Corporate Governance: THe International Journal of Business in Society, 16*(1), 35–53. https://doi.org/10.1108/CG-05-2015-0059

Teoh, S. H., Welch, I., & Wazzan, C. P. (1999). The effect of socially activist investment policies on the financial markets: Evidence from the South African boycott. *The Journal of Business, 72*(1), 35–89. https://doi.org/10.1086/209602

Tsoutsoura, M. (2004). *CSR and financial performance*. Retrieved from eScholarship: http://escholarship.org/uc/item/111799p2

Tunger, D., & Eulerich, M. (2018). Bibliometric analysis of corporate governance research in German-speaking countries: Applying bibliometrics to business research using a custom-made database. *Scientometrics, 117*(3), 2041–2059. https://doi.org/10.1007/s11192-018-2919-z

Van Eck, N., & Waltman, L. (2010). Software survey: VOSviewer, a computer program for bibliometric mapping. *Scientometrics, 84*(2), 523–538. https://doi.org/10.1007/s11192-009-0146-3

Velte, P. (2016). Women on management board and ESG performance. *Journal of Global Responsibility, 7*(1), 98–109. https://doi.org/10.1108/jgr-01-2016-0001

Velte, P. (2022). Which institutional investors drive corporate sustainability? A systematic literature review. *Business Strategy and the Environment*, 1–30. https://doi.org/10.1002/bse.3117

Velte, P., & Stawinoga, M. (2020). Do chief sustainability officers and CSR committees influence CSR-related outcomes? A structured literature review based on empirical-quantitative research findings. *Journal of Management Control, 31*(4), 333–377. https://doi.org/10.1007/s00187-020-00308-x

Waddock, S. A., & Graves, S. B. (1997). The corporate social performance–financial performance link. *Strategic Management Journal, 18*(4), 303–319. https://doi.org/10.1002/(sici)1097-0266(199704)18:4%3C303::aid-smj869%3E3.0.co;2-g

Wright, P., & Ferris, S. P. (1997). Agency conflict and corporate strategy: The effect of divestment on corporate value. *Strategic Management Journal, 18*(1), 77–83. https://doi.org/10.1002/(sici)1097-0266(199701)18:1%3C77::aid-smj810%3E3.0.co;2-r

Xu, X., Gong, Y., Jia, F., Brown, S., & Xu, Y. (2018). Supply chain finance: A systematic literature review and bibliometric analysis. *International Journal of Production Economics, 204*, 160–173. https://doi.org/10.1016/j.ijpe.2018.08.003

Yasser, Q. R., Al Mamun, A., & Ahmed, I. (2017). Corporate social responsibility and gender diversity: Insights from Asia Pacific. *Corporate Social Responsibility and Environmental Management, 24*(3), 210–221. https://doi.org/10.1002/csr.1400

Yu, Y., Li, Y., Zhang, Z., Gu, Z., Zhong, H., Zha, Q., Yang, L., Zhu, C., & Chen, E. (2020). A bibliometric analysis using VOSviewer of publications on COVID-19. *Annals of Translational Medicine, 8*(13). https://doi.org/10.21037/atm-20-4235

Yuan, R., Sun, J., & Cao, F. (2016). Directors' and officers' liability insurance and stock price crash risk. *Journal of Corporate Finance, 37*, 173–192. https://doi.org/10.1016/j.jcorpfin.2015.12.015

Zaichkowsky, J. L. (2014). Women in the board room: One can make a difference. *International Journal of Business Governance and Ethics, 9*(1), 91–113. https://doi.org/10.1504/ijbge.2014.062774

Zaman, R., Jain, T., Samara, G., & Jamali, D. (2022). Corporate governance meets corporate social responsibility: Mapping the interface. *Business & Society, 61*(3), 690–752. https://doi.org/10.1177/0007650320973415

Zupic, I., & Čater, T. (2015). Bibliometric methods in management and organization. *Organizational Research Methods, 18*(3), 429–472. https://doi.org/10.1177/1094428114562629

Index

V. Kandpal et al. (eds.), *Developments in Corporate Governance*, Approaches to Global Sustainability, Markets, and Governance, https://doi.org/10.1007/978-981-96-6366-8